Neutral and Indifference Portfolio Pricing, Hedging and Investing

Srdjan Stojanovic

Neutral and Indifference Portfolio Pricing, Hedging and Investing

With Applications in Equity and FX

Springer

Dr. Srdjan Stojanovic
Department of Mathematics
University of Cincinnati
PO Box 210025
45221-0025 Cincinnati Ohio
USA
srdjan@math.uc.edu

ISBN 978-1-4899-9781-4 ISBN 978-0-387-71418-9 (eBook)
DOI 10.1007/978-0-387-71418-9
Springer New York Dordrecht Heidelberg London

Mathematics Subject Classification (2010): 91G20, 91G30, 91G10, 91G80

Softcover re-print of the Hardcover 1st edition 2012

Printed on acid-free paper

Springer is part of Springer Science+Business Media (www.springer.com)

This book is dedicated to my children,
Juju and Max

Preface

> "Some will love you, son, and some will hate you," an anchovy tells his child. "It's always been that way with anchovies."
> Leo Cullum, Longtime New Yorker Cartoonist, NYTimes.com, October 25, 2010.

> MAY 1, 2010, 3:48 P.M. ET DOW JONES NEWSWIRES
> OMAHA, Neb.

> ... Berkshire Vice Chairman Charlie Munger ... took a swipe at the academic theories and quantitative formulas the ratings firms used to make their judgements. "I've yet to hear a single apology from business academia for its huge contribution to our present difficulties," he said, to rousing applause from the audience.
> (2010 Berkshire Hathaway Annual Shareholders Meeting)

This book was not written as an apology, but as a call—and a blueprint—for a more comprehensive *and* more feasible kind of financial mathematics.

We present herein an original, analytic approach to financial mathematics of incomplete markets that is very comprehensive, yet very amenable to symbolic and numerical implementations. Furthermore, this approach is fully consistent with (see Remark 5.2.2), yet completely independent of, the current prevailing abstract-probabilistic "equivalent martingale measure" approach. We rely solely on stochastic control, i.e., portfolio optimization, and differential equations. This is made possible by the discovery of what we call the "fundamental matrix of derivatives pricing and hedging" (see Theorem 4.6.1).

To that end, our objective was not mathematical "rigor," but mathematical correctness and financial usefulness. The extent our objective has been achieved, only time will tell.

The modern theory of pricing of financial contracts begun by the seminal works of Black, Scholes, and Merton in the early 1970s: if a financial contract is written on a perfectly tradable underlying asset (such as stock), if the price of the underlying asset obeys a simple (log-normal) stochastic differential equation with a known constant volatility, if the interest rate is constant, and if, through trading of the underlying asset, i.e., through hedging, the owner of the contract eliminates all risk, then the price of such a contract is uniquely, and relatively easily, calculated. The main idea is to use the possibility of elimination of risk. If it is possible to eliminate risk, then one does so, and the price is set so that arbitrage is prevented, or more explicitly, so that a portfolio consisting of the considered contract, the variable (hedging) quantity of the underlying asset, and of the cash account (used for financing the hedging transactions) behaves, i.e., appreciates, due to the market forces, in the same fashion as the cash account. In turn, this principle determines the prices of the considered contract.

On the other hand, what if, due to the breakdown of some of the simplifying assumptions above, the complete elimination of risk of holding a financial contract is not possible? The whole pricing argument of Black and Scholes is no longer applicable, or at least, if applied, it does not lead to a complete answer as to what the price of the contract should be, since it does not provide a method for determining the market price of risk, that is, the risk premium. Indeed, if perfect hedging is not possible in pricing financial contracts, then one has to be able to price the remaining risk.

It is only in the last decade or two that such problems, problems of pricing risk, have been formulated, and it is only recently that definitive answers in their full generality, yet applicable to all (diffusive, Markovian) financial models have been found.

For example,

(1) What should prices of assets, i.e., prices of financial contracts, be so that it is optimal to have neither a long nor a short position in such assets (see Theorems 4.7.2 and 4.9)?
(2) What should the price of a portfolio of assets be, or more precisely, how much cheaper (more expensive) should the assets be, so that having a certain long (short) position in it is by and large equivalent to not having it at all, from the point of view of investing in the rest of the market (see Theorems 4.8.1 and 4.10.1)?
(3) What should the price of a portfolio of assets be, or more precisely, how much cheaper (more expensive) should the asset be, so that having a certain long (short) position in it is in fact an optimal investing position (see Theorems 4.7.1 and 4.9.1; see also Remark 4.8.3)?
(4) How can one manage the risk of holding a position in a portfolio of assets, i.e., how can one hedge, when perfect (Black–Scholes) hedging is not possible (see Theorem 5.2.1)?
(5) How can one hedge if one is willing to allow for a bit more exposure to risk for the chance of greater profit (see Theorem 5.3.1)?

All of these questions are answered in the full generality of Markovian diffusive financial engineering models, and then they are applied in various situations, with emphasis on problems in equity portfolio valuation, foreign exchange rates, and foreign exchange derivatives, answering, for example, questions like these:

(6) Why and by how much do equity markets fall when investors become risk-averse (see Example 6.4.3)? What is the effect of the dividend policy (see

Section 6.4.2), and what is the effect of stock dilution/buyback on the equity value (see Section 6.9)?

(7) Which foreign exchange markets fall when investors ignore risks, and why, and by how much (see Remark 7.3.3)?

One of the key ingredients in answering most of the above questions in the framework of incomplete markets is the adopted type and intensity of the investors's risk-aversion. Risk aversion is qualified and quantified by means of the choice of a utility of wealth function. Most of the results presented in this book will first be derived for an arbitrary utility of wealth, and then we shall specialize to the more manageable ones.

The power of the general results presented in this work can be best understood and utilized with the help of some symbolic computer platform, such as *Mathematica* (http://www.wolfram.com). Indeed, we present many examples, and they were *all* derived using *Mathematica*, and many of them would not be feasible otherwise.

Acknowledgments

It took a long time to write this book. As a consequence, while it was being developed it was tested in front of (and written for) several kinds of audiences.

The early versions were presented, with published lecture notes (see [37] and [43]), on a number of occasions in front of financial professionals in London and New York, at two-day intensive courses organized by GARP (Global Association of Risk Professionals), for which I would like to express my gratitude to Mr. Andreas Simou, director at GARP.

Shorter versions, covering parts of the subject matter, were presented on numerous occasions at courses organized by RISK (Incisive Media), held in London and New York.

Then there were Seminars in Financial Mathematics—two-semester courses for graduate students at the University of Cincinnati—that ran a couple of times over several years while this book project was underway, for which I would like to thank Professor Timothy Hodges, the head of the department at that time.

Among my students at the University of Cincinnati, I would like in particular to acknowledge and thank Zhuang Kang, who recently defended his doctoral dissertation, and also Cheng Liu, who recently defended his master of science thesis.

Much progress on this book was made during my sabbatical leave from the University of Cincinnati during the academic year 2009/10, spent partially at Kanazawa University, in Japan, for which I would like in particular to thank Professor Seiro Omata.

This book took its final form while I was presenting semester-long courses on the subject matter at the Center for Financial Engineering, School of Mathematics, Suzhou University, in China, during the academic years 2009/10, and 2010/11, and also, concurrently, during a five-week lecture series at Tongji University, in

Shanghai, during the autumn semester of 2010. For these opportunities I would like in particular to thank Professor Lishang Jiang, director of the Center for Financial Engineering in Suzhou, and Professor Guojing Wang, deputy director, and also to Professor Jin Liang, of Tongji University. Among my many excellent students at Suzhou University I would like to acknowledge my PhD student Yaling Cui, my new students Jingxian Du and Wen Chen, as well as Xi Liu, Meng Zhang, Cuicui Bi, Yanan Zhang, and the rest of the 2010 class.

I would like to thank very much Professor P. K. Sen of the Department of Accounting, Business School, University of Cincinnati, for some recent very insightful discussions in regard to the results presented in this book, and in particular in regard to the results of Chap. 6 on equity valuation. Indeed, those discussions have now expanded into what, as it seems, will be the next phase of this research: a comprehensive empirical verification and implementation of the developed theory.

I would like to thank very much Ms. Ann Kostant, mathematics editor at Springer, and, more recently, Ms. Achi Dosanjh and Ms. Donna Chernyk, editors at Springer, for having the patience to consider my manuscript even after many long, yet worthwhile, I hope, delays.

It is my family to whom I owe the deepest gratitude, and most of all to my wonderful children, Lepa Juju Stojanovic and Max Dushan Stojanovic. I dedicate this work to them.

Cincinnati and Suzhou Srdjan D. Stojanovic

Contents

Chapter 1
Background Material

SDEs, PDEs, and Their Interplay

1.1 Mathematical Framework

In optimal portfolio theory, in derivative pricing and hedging theory, in equity valuation theory, in foreign exchange and foreign exchange derivatives theory, one has to model various underlying dynamic quantities – the underlying dynamics: prices of stocks, interest rates, dividends, various cash flows, and economic indicators.

Modeling dynamic quantities is the central theme of classical applied mathematics. Very often, it is done by means of (systems of) ordinary differential equations (ODEs). In some areas, such as mechanics and astronomy, the high level of precision of physical models and the high level of precision of measured data make ODEs a sufficiently precise modeling device so that the study of errors is of secondary importance, or may even be ignored.

In many other application areas, even physical ones such as meteorology, and in particular the nonphysical ones such as economics and finance, the mathematical models and the measured data are both insufficiently precise. Indeed, for such phenomena, reality is too complicated to even attempt to model them using deterministic mathematics. The remaining error, the remaining indeterminacy, is quite significant in size, and therefore has to be considered statistically. The mathematical vehicle for such a generalization of ODEs is stochastic differential equations (SDEs), which one can think of as randomly perturbed ODEs.

1.2 White Noise and Brownian Motion

To generate dynamic randomness in the most natural and most productive fashion, one uses white noise, or equivalently (standard) Brownian motion.

White noise is a *distribution*-valued stochastic process, where intuitively, the distribution can be thought of as the derivative of a nondifferentiable function:

S. Stojanovic, *Neutral and Indifference Portfolio Pricing, Hedging and Investing: With Applications in Equity and FX*, DOI 10.1007/978-0-387-71418-9_1,

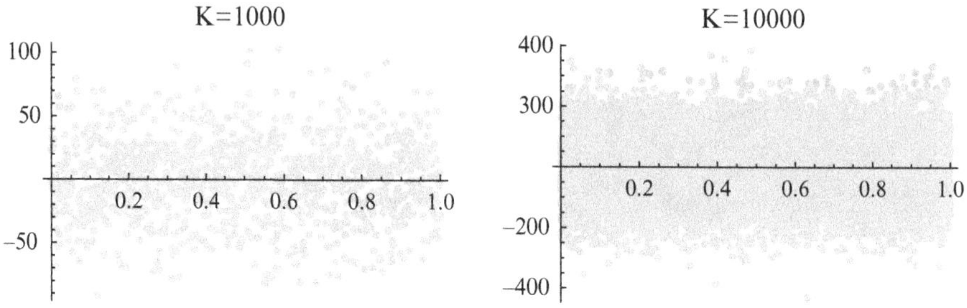

Fig. 1.1 (Approximate) White Noise

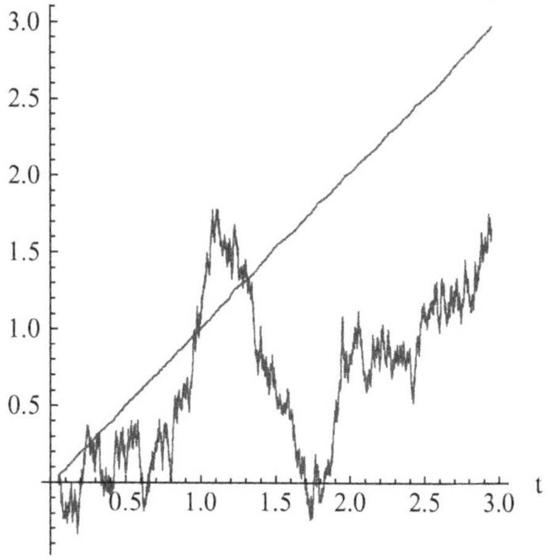

Fig. 1.2 (Approximate) Brownian Motion and its quadratic variation process

$W(t) = \mathrm{d}B(t)/\mathrm{d}t$. The (approximate) properties of the process W are that for any t, $W(t) \sim N\left(0, 1\Big/\sqrt{\mathrm{d}t}\right)$, i.e., $\mathrm{d}B(t) \sim N\left(0, \sqrt{\mathrm{d}t}\right)$; for any $t \neq s$, $W(t)$ and $W(s)$, i.e., $\mathrm{d}B(t)$ and $\mathrm{d}B(s)$, are independent. In Fig. 1.1 we generate the approximate trajectories of the white noise. The approximation refers to the fact that $\mathrm{d}t = 1/K$ is a fixed small number, instead of being set equal to zero; the white noise is obtained by passing to the limit $\mathrm{d}t \to 0$, i.e., by taking the limit $K \to \infty$.

Thus the trajectory of a white noise (as $\mathrm{d}t \to 0$) is not a function (for an application of white noise, see (7.4.1)). Nevertheless, by integrating the white noise's trajectory, one gets a trajectory of the standard Brownian motion $B(t)$, which is a Hölder continuous yet nondifferentiable function (see Fig. 1.2).

So in particular, $E(\mathrm{d}B(t))^2 = \mathrm{d}t$. A stronger property holds, which can be written as

$$(\mathrm{d}B(t))^2 = \mathrm{d}t \tag{1.2.1}$$

i.e., as $\int_0^t (\mathrm{d}B(\tau))^2 = \int_0^t \mathrm{d}t = t$. Indeed, in Fig. 1.2, an approximate trajectory of $B(t)$ and the corresponding approximate trajectory of $\int_0^t (\mathrm{d}B(\tau))^2$ are plotted. We observe that if $B(\tau)$ is a differentiable function, then $\int_0^t (\mathrm{d}B(\tau))^2 = \int_0^t (B'(\tau))^2 (\mathrm{d}t)^2 = 0$, and the difference between the two explains the need for an extension of the classical calculus and the introduction of the Itô stochastic calculus.

1.3 Multidimensional Brownian Motion

An m-dimensional process whose components are *independent* Brownian motions will be referred to as a multidimensional Brownian motion, and it will always be denoted by $B(t)$:

$$B(t) = \{B_1(t), \ldots, B_m(t)\}. \tag{1.3.1}$$

The component processes have the property that

$$\mathrm{d}B_i(t)\mathrm{d}B_j(t) = \begin{cases} 0, & i \neq j, \\ \mathrm{d}t, & i = j. \end{cases} \tag{1.3.2}$$

Often, we shall also use an m-dimensional process whose components are *correlated* (scalar) Brownian motions. Such a process will always be denoted by $\mathbb{B}(t)$:

$$\mathbb{B}(t) = \{\mathbb{B}_1(t), \ldots, \mathbb{B}_m(t)\}. \tag{1.3.3}$$

Its component processes have the property that

$$\mathrm{d}\mathbb{B}_i(t)\mathrm{d}\mathbb{B}_j(t) = \rho_{i,j}\mathrm{d}t, \tag{1.3.4}$$

where $\rho_{i,j} = \rho_{j,i}$ is the (instantaneous) correlation between Brownian motions $\mathbb{B}_i(t)$ and $\mathbb{B}_j(t)$. Given the (instantaneous) correlations $\rho_{i,j}$, process $\mathbb{B}(t)$ can be built from $B(t)$. Indeed, for example, if $m = 3$, and we are given the correlation matrix

$$P_3 = \begin{pmatrix} 1 & \rho_{2,1} & \rho_{3,1} \\ \rho_{2,1} & 1 & \rho_{3,2} \\ \rho_{3,1} & \rho_{3,2} & 1 \end{pmatrix}, \tag{1.3.5}$$

we can define (see the appendix)

$$\mathrm{d}\mathbb{B}(t) = \mathbb{S}_3 \cdot \mathrm{d}B(t) = \begin{pmatrix} 1 & 0 & 0 \\ \rho_{2,1} & \sqrt{1-\rho_{2,1}^2} & 0 \\ \rho_{3,1} & \dfrac{\rho_{3,2} - \rho_{2,1}\rho_{3,1}}{\sqrt{1-\rho_{2,1}^2}} & \sqrt{\dfrac{1 + 2\rho_{2,1}\rho_{3,1}\rho_{3,2} - \left(\rho_{2,1}^2 + \rho_{3,1}^2 + \rho_{3,2}^2\right)}{1-\rho_{2,1}^2}} \end{pmatrix} \cdot \mathrm{d}B(t). \tag{1.3.6}$$

If the correlation matrix P_m is constant, then by means of integration, (1.3.6) is equivalent to $\mathbb{B}(t) = \mathbb{S}_m \cdot B(t)$, but this is not true in general, whence the term *instantaneous* correlations.

1.4 Stochastic Differential Equations

Suppose $b(t,y) = \{b_1(t,y),\dots,b_n(t,y)\}$ is an n-vector-valued function of $\{t,y\} \in (-\infty,\infty) \times (-\infty,\infty)^n$, and suppose

$$c(t,y) = \{\{c_{1,1}(t,y),\dots,c_{1,m}(t,y)\},\dots,\{c_{n,1}(t,y),\dots,c_{n,m}(t,y)\}\}$$
$$= \begin{pmatrix} c_{1,1}(t,y) & \dots & c_{1,m}(t,y) \\ \dots & \dots & \dots \\ c_{n,1}(t,y) & \dots & c_{n,m}(t,y) \end{pmatrix} \tag{1.4.1}$$

is an $n \times m$ matrix (-valued function of $\{t,y\} \in (-\infty,\infty) \times (-\infty,\infty)^n$). Matrix c will be called the *diffusion matrix*.

Let $B(t) = \{B_1(t),\dots,B_m(t)\}$ be an m-dimensional Brownian motion. An equation of the form

$$\mathrm{d}y(t) = b(t,y(t))\mathrm{d}t + c(t,y(t)) \cdot \mathrm{d}B(t) \tag{1.4.2}$$

for $t > s \in (-\infty,\infty)$, with an initial condition $y(s) = x \in (-\infty,\infty)^n$, is called an Itô SDE (in n dimensions). For an in-depth account on SDEs we refer to [10]. Under certain conditions on the vector valued function b, and matrix valued function c, there exists a unique solution (a stochastic process) of (1.4.2), denoted by $\{y_{s,x}(t)\}_{s \le t < T}$.

We remark that the dot $\cdot$ in (1.4.2) and throughout denotes the usual (dot) product of vectors, matrices, or tensors. For example, matrix–matrix (dot) product is given by

$$\begin{pmatrix} a_{1,1} & a_{1,2} \\ a_{2,1} & a_{2,2} \end{pmatrix} \cdot \begin{pmatrix} b_{1,1} & b_{1,2} \\ b_{2,1} & b_{2,2} \end{pmatrix} = \begin{pmatrix} a_{1,1}b_{1,1} + a_{1,2}b_{2,1} & a_{1,1}b_{1,2} + a_{1,2}b_{2,2} \\ a_{2,1}b_{1,1} + a_{2,2}b_{2,1} & a_{2,1}b_{1,2} + a_{2,2}b_{2,2} \end{pmatrix}. \tag{1.4.3}$$

We shall also use apposition of quantities to indicate the *componentwise* multiplication of various compatible arrays. For example,

$$\begin{pmatrix} a_{1,1} & a_{1,2} \\ a_{2,1} & a_{2,2} \end{pmatrix} \begin{pmatrix} b_{1,1} & b_{1,2} \\ b_{2,1} & b_{2,2} \end{pmatrix} = \begin{pmatrix} a_{1,1}b_{1,1} & a_{1,2}b_{1,2} \\ a_{2,1}b_{2,1} & a_{2,2}b_{2,2} \end{pmatrix}. \tag{1.4.4}$$

1.5 Itô Chain Rule

Let $\{y_{s,x}(t)\}_{s\leq t}$ be the solution of the SDE (1.4.2)–(1.4.2), and let $\varphi(t,y)$ be a sufficiently differentiable function. Throughout, we shall use the following notation: $\varphi_t(t,y):=\partial\varphi(t,y)/\partial t:=\varphi^{(1,0)}(t,y)$; $\nabla_y\varphi(t,y)=\{\partial\varphi(t,y)/\partial y_1,\ldots,\partial\varphi(t,y)/\partial y_n\}$ is the gradient of $\varphi(t,y)$ with respect to $y=\{y_1,\ldots,y_n\}$; moreover, $\nabla_y\nabla_y\varphi(t,y)$ is the second gradient, i.e., the Hessian matrix. For example, if $n=2$,

$$\nabla_y\varphi(t,y) = \left\{\varphi^{(0,1,0)}(t,y_1,y_2),\varphi^{(0,0,1)}(t,y_1,y_2)\right\},$$

$$\nabla_y\nabla_y\varphi(t,y) = \begin{pmatrix} \varphi^{(0,2,0)}(t,y_1,y_2) & \varphi^{(0,1,1)}(t,y_1,y_2) \\ \varphi^{(0,1,1)}(t,y_1,y_2) & \varphi^{(0,0,2)}(t,y_1,y_2) \end{pmatrix}.$$

Since $\{y_{s,x}(t)\}_{s\leq t}$ is the solution of the SDE (1.4.2)–(1.4.2), and, therefore, its trajectories are not differentiable functions of time t, the infinitesimal Taylor expansion for $\mathrm{d}\varphi(t,y_{s,x}(t))$ has to include second-order terms in the $\mathrm{d}y_{s,x}(t)$ directions. More precisely,

$$\begin{aligned}
\mathrm{d}\varphi(t,y_{s,x}(t)) &= \varphi_t(t,y_{s,x}(t))\,\mathrm{d}t + \nabla_y\varphi(t,y_{s,x}(t))\cdot\mathrm{d}y_{s,x}(t) \\
&\quad + \frac{1}{2}\mathrm{d}y_{s,x}(t)\cdot\nabla_y\nabla_y\varphi(t,y_{s,x}(t))\cdot\mathrm{d}y_{s,x}(t) \\
&= \varphi_t(t,y_{s,x}(t))\,\mathrm{d}t + \nabla_y\varphi(t,y_{s,x}(t))\cdot\mathrm{d}y_{s,x}(t) \\
&\quad + \frac{1}{2}\mathrm{d}B(t)\cdot c(t,y_{s,x}(t))^T\cdot\nabla_y\nabla_y\varphi(t,y_{s,x}(t))\cdot c(t,y_{s,x}(t))\cdot\mathrm{d}B(t) \\
&= \varphi_t(t,y_{s,x}(t))\,\mathrm{d}t + \nabla_y\varphi(t,y_{s,x}(t))\cdot(b(t,y_{s,x}(t))\,\mathrm{d}t \\
&\quad + c(t,y_{s,x}(t))\cdot\mathrm{d}B(t)) \\
&\quad + \frac{1}{2}\mathrm{Tr}\left[c(t,y_{s,x}(t))^T\cdot\nabla_y\nabla_y\varphi(t,y_{s,x}(t))\cdot c(t,y_{s,x}(t))\right]\mathrm{d}t.
\end{aligned}$$

Furthermore,

$$\begin{aligned}
\mathrm{d}\varphi(t,y_{s,x}(t)) &= \Big(\varphi_t(t,y_{s,x}(t)) + \nabla_y\varphi(t,y_{s,x}(t))\cdot b(t,y_{s,x}(t)) \\
&\quad + \frac{1}{2}\mathrm{Tr}\left[c(t,y_{s,x}(t))^T\cdot\nabla_y\nabla_y\varphi(t,y_{s,x}(t))\cdot c(t,y_{s,x}(t))\right]\Big)\mathrm{d}t \\
&\quad + \nabla_y\varphi(t,y_{s,x}(t))\cdot c(t,y_{s,x}(t))\cdot\mathrm{d}B(t) \\
&= (\mathcal{L}\varphi)(t,y_{s,x}(t))\,\mathrm{d}t + \nabla_y\varphi(t,y_{s,x}(t))\cdot c(t,y_{s,x}(t))\cdot\mathrm{d}B(t), \qquad (1.5.1)
\end{aligned}$$

where Tr denotes the trace of a matrix, and where the differential operator $\mathcal{L}$ is of fundamental importance and is defined by

$$(\mathcal{L}\varphi)(t,y) := \varphi_t(t,y) + \frac{1}{2}\mathrm{Tr}\left[c(t,y)^T\cdot\nabla_y\nabla_y\varphi(t,y)\cdot c(t,y)\right] + \nabla_y\varphi(t,y)\cdot b(t,y). \qquad (1.5.2)$$

We observe that the principal part of the operator $\mathcal{L}$ can be computed alternatively as follows:

$$
\begin{aligned}
(\mathcal{E}\varphi)(t,y) &:= \frac{1}{2}\mathrm{Tr}\left[c(t,y)^T \cdot \nabla_y\nabla_y\varphi(t,y)\cdot c(t,y)\right] \\
&= \frac{1}{2}\mathrm{Tr}\left[c(t,y)\cdot c(t,y)^T \cdot \nabla_y\nabla_y\varphi(t,y)\right] \\
&= \frac{1}{2}\mathrm{Tr}\left[\nabla_y\nabla_y\varphi(t,y)\cdot c(t,y)\cdot c(t,y)^T\right] \\
&= \frac{1}{2}\sum_{i,j=1}^{n} a_{i,j}(t,y)\frac{\partial^2\varphi(t,y)}{\partial y_i\,\partial y_j},
\end{aligned}
\tag{1.5.3}
$$

where $a_{i,j} = \left(c\cdot c^T\right)_{i,j}$, the (i,j)th entry of the matrix $c\cdot c^T$.

Remark 1.5.1. The operator $\mathcal{L}$ is said to be *backward parabolic* if the operator $\mathcal{E}$ is *elliptic*. The operator $\mathcal{E}$ is said to be elliptic if there exists $\alpha > 0$, such that $\sum_{i,j=1}^{n} a_{i,j}\xi_i\xi_j = \xi\cdot c\cdot c^T\cdot\xi \geq \alpha|\xi|^2$, for all $\xi\in\mathbb{R}^n$. In another words, $\mathcal{E}$ is elliptic if $c\cdot c^T$ is a positive definite matrix. Of course, no matter what c is, $c\cdot c^T$ is nonnegative, i.e., $\xi\cdot c\cdot c^T\cdot\xi \geq 0$, yet it may often be singular, in which case $\mathcal{E}$ is not elliptic, and $\mathcal{L}$ is not parabolic. For the rigorous treatment of such differential operators the reader is referred to the classical Hörmander theory of hypoelliptic operators, or to the modern theory of subelliptic operators. □

Remark 1.5.2. If $B(t)$ is not a Brownian motion, but instead a differentiable function of t, then (1.5.1) becomes

$$
\begin{aligned}
\mathrm{d}\varphi\left(t,y_{s,x}(t)\right) = {}& \left(\varphi_t\left(t,y_{s,x}(t)\right)+\nabla_y\varphi\left(t,y_{s,x}(t)\right)\cdot b\left(t,y_{s,x}(t)\right)\right)\mathrm{d}t+\nabla_y\varphi\left(t,y_{s,x}(t)\right) \\
&\cdot c\left(t,y_{s,x}(t)\right)\cdot \mathrm{d}B(t).
\end{aligned}
\tag{1.5.4}
$$

Example 1.5.1. We compute $\mathcal{L}\varphi\left(t,y_1,y_2\right)$ if $c = \begin{pmatrix}1&0\\0&1\end{pmatrix}$ and $b = \{0,0\}$. We have

$$
\begin{aligned}
\nabla_y\varphi\left(t,y_1,y_2\right) &= \left\{\varphi^{(0,1,0)}\left(t,y_1,y_2\right),\varphi^{(0,0,1)}\left(t,y_1,y_2\right)\right\} \\
\nabla_y\nabla_y\varphi\left(t,y_1,y_2\right)\cdot c\cdot c^T &= \nabla_y\nabla_y\varphi\left(t,y_1,y_2\right) \\
&= \begin{pmatrix}\varphi^{(0,2,0)}\left(t,y_1,y_2\right) & \varphi^{(0,1,1)}\left(t,y_1,y_2\right)\\ \varphi^{(0,1,1)}\left(t,y_1,y_2\right) & \varphi^{(0,0,2)}\left(t,y_1,y_2\right)\end{pmatrix} \\
\mathrm{Tr}\left(\nabla_y\nabla_y\varphi\left(t,y_1,y_2\right)\cdot c\cdot c^T\right) &= \varphi^{(0,0,2)}\left(t,y_1,y_2\right)+\varphi^{(0,2,0)}\left(t,y_1,y_2\right) \\
\mathcal{L}\varphi\left(t,y_1,y_2\right) &= \frac{\partial\varphi\left(t,y_1,y_2\right)}{\partial t}+\frac{1}{2}\mathrm{Tr}\left[\nabla_y\nabla_y\varphi\left(t,y_1,y_2\right)\cdot c\cdot c^T\right] \\
&\quad+\nabla_y\varphi\left(t,y_1,y_2\right)\cdot b
\end{aligned}
$$

$$= \varphi^{(1,0,0)}(t,y_1,y_2) + \frac{1}{2}\Big(\varphi^{(0,0,2)}(t,y_1,y_2) + \varphi^{(0,2,0)}(t,y_1,y_2)\Big).$$

1.6 Relationship Between Itô SDEs and Second-Order Linear PDEs

Consider an Itô SDE system

$$\mathrm{d}y(t) = b(t,y(t))\mathrm{d}t + c(t,y(t))\cdot \mathrm{d}B(t) \tag{1.6.1}$$

in n dimensions for $t > s \in (-\infty,\infty)$, together with an initial condition $y(s) = x \in (-\infty,\infty)^n$, where the vector-valued function b and matrix-valued function c are such to guarantee the existence of a unique solution. Denote the solution by $y_{s,x}(t)$. Equation (1.5.1) is sufficient to establish a relationship between Itô SDEs and second-order linear PDEs with no zero-order terms. In general, we need one more step: let $\varphi(t,x)$ be a smooth function; then by means of the Itô product rule and Itô chain rule (observe that since the function $t| \to \int_s^t r(\tau, y_{s,x}(\tau))\,\mathrm{d}\tau$ is differentiable, the Itô product rule below, which can be considered a special case of the Itô chain rule, coincides with the classical product rule),

$$\begin{aligned}
d\left(\varphi(t,y_{s,x}(t))\,\mathrm{e}^{-\int_s^t r(\tau,y_{s,x}(\tau))\,\mathrm{d}\tau}\right) &= \mathrm{e}^{-\int_s^t r(\tau,y_{s,x}(\tau))\,\mathrm{d}\tau}\Big(\varphi_t(t,y_{s,x}(t))\,\mathrm{d}t \\
&\quad + \nabla_y\varphi(t,y_{s,x}(t))\cdot \mathrm{d}y_{s,x}(t) \\
&\quad + \frac{1}{2}\mathrm{d}y_{s,x}(t)\cdot\nabla_y\nabla_y\varphi(t,y_{s,x}(t))\cdot\mathrm{d}y_{s,x}(t)\Big) \\
&\quad - r(t,y_{s,x}(t))\,\varphi(t,y_{s,x}(t))\,\mathrm{e}^{-\int_s^t r(\tau,y_{s,x}(\tau))\,\mathrm{d}\tau}\mathrm{d}t \\
&= \mathrm{e}^{-\int_s^t r(\tau,y_{s,x}(\tau))\,\mathrm{d}\tau}\Big(\varphi_t(t,y_{s,x}(t))\,\mathrm{d}t + \nabla_y\varphi(t,y_{s,x}(t)) \\
&\quad \cdot\left(b(t,y_{s,x}(t))\,\mathrm{d}t + c(t,y_{s,x}(t))\cdot\mathrm{d}B(t)\right) \\
&\quad + \frac{1}{2}\mathrm{Tr}\big[c(t,y_{s,x}(t))\cdot c(t,y_{s,x}(t))^T \\
&\quad \cdot\nabla_y\nabla_y\varphi(t,y_{s,x}(t))\big]\,\mathrm{d}t \\
&\quad - r(t,y_{s,x}(t))\,\varphi(t,y_{s,x}(t))\,\mathrm{d}t\Big).
\end{aligned} \tag{1.6.2}$$

Integrating (1.6.2) between s and T, this time we get

$$\begin{aligned}\varphi(T,y_{s,x}(T))\,\mathrm{e}^{-\int_s^T r(\tau,y_{s,x}(\tau))\,\mathrm{d}\tau}-\varphi(s,x) &= \int_s^T d\left(\varphi(t,y_{s,x}(t))\,\mathrm{e}^{-\int_s^t r(\tau,y_{s,x}(\tau))\,\mathrm{d}\tau}\right)\\ &= \int_s^T \mathrm{e}^{-\int_s^t r(\tau,y_{s,x}(\tau))\,\mathrm{d}\tau}\Big(\varphi_t(t,y_{s,x}(t))\\ &\quad+\nabla_y\varphi(t,y_{s,x}(t))\cdot b(t,y_{s,x}(t))\\ &\quad+\frac{1}{2}\mathrm{Tr}\left[c(t,y_{s,x}(t))\cdot c(t,y_{s,x}(t))^T\right.\\ &\qquad\left.\cdot\,\nabla_y\nabla_y\varphi(t,y_{s,x}(t))\right]\\ &\quad-r(t,y_{s,x}(t))\,\varphi(t,y_{s,x}(t))\Big)\,\mathrm{d}t\\ &\quad+\int_s^T \mathrm{e}^{-\int_s^t r(\tau,y_{s,x}(\tau))\,\mathrm{d}\tau}\nabla_y\varphi(t,y_{s,x}(t))\\ &\quad\cdot c(t,y_{s,x}(t))\cdot\mathrm{d}B(t).\end{aligned}\tag{1.6.3}$$

Taking the conditional expectation $E_{s,x}:=E[\cdot|y(s)=x]$ of both sides in (1.6.3), we get

$$\begin{aligned}&E_{s,x}\varphi(T,y_{s,x}(T))\,\mathrm{e}^{-\int_s^T r(\tau,y_{s,x}(\tau))\,\mathrm{d}\tau}-\varphi(s,x)\\ &=E_{s,x}\int_s^T \mathrm{e}^{-\int_s^t r(\tau,y_{s,x}(\tau))\,\mathrm{d}\tau}\Big(\varphi_t(t,y_{s,x}(t))+\nabla_y\varphi(t,y_{s,x}(t))\cdot b(t,y_{s,x}(t))\\ &\quad+\frac{1}{2}\mathrm{Tr}\left[c(t,y_{s,x}(t))\cdot c(t,y_{s,x}(t))^T\cdot\nabla_y\nabla_y\varphi(t,y_{s,x}(t))\right]\\ &\quad-r(t,y_{s,x}(t))\,\varphi(t,y_{s,x}(t))\Big)\,\mathrm{d}t.\end{aligned}\tag{1.6.4}$$

Let φ be a solution of the second-order PDE

$$\begin{aligned}&\varphi_t(t,y)+\frac{1}{2}\mathrm{Tr}\left[c(t,y)\cdot c(t,y)^T\cdot\nabla_y\nabla_y\varphi(t,y)\right]+b(t,y)\cdot\nabla_y\varphi(t,y)-r(t,y)\varphi(t,y)\\ &=-f(t,y)\end{aligned}\tag{1.6.5}$$

for $t<T$, together with the terminal condition $\varphi(T,y)=g(y)$. Then, combining (1.6.4) and (1.6.5), we get

$$\varphi(s,x)=E_{s,x}\left(g(y_{s,x}(T))\,\mathrm{e}^{-\int_s^T r(\tau,y_{s,x}(\tau))\,\mathrm{d}\tau}+\int_s^T \mathrm{e}^{-\int_s^t r(\tau,y_{s,x}(\tau))\,\mathrm{d}\tau}f(t,y_{s,x}(t))\,\mathrm{d}t\right),\tag{1.6.6}$$

so that the second-order PDE (1.6.5) is solved by φ given by the formula (1.6.6), and conversely, the function φ, given by the formula (1.6.6), is characterized as a solution of the second-order PDE (1.6.5).

Assume that the data c, b, r, and f are all independent of t. Taking the limit $T \to \infty$, we conclude that

$$\varphi(x) = E_{s,x} \int_s^\infty \mathrm{e}^{-\int_s^t r(y_{s,x}(\tau))\,\mathrm{d}\tau} f(y_{s,x}(t))\,\mathrm{d}t = E_{0,x} \int_0^\infty \mathrm{e}^{-\int_0^t r(y_{0,x}(\tau))\,\mathrm{d}\tau} f(y_{0,x}(t))\,\mathrm{d}t \tag{1.6.7}$$

is the entire solution of (1.6.5) for $t \in (-\infty, \infty)$.

Example 1.6.1. (Black–Scholes PDE) Consider a scalar equation, i.e., let $n = 1$, and let

$$\mathrm{d}y(t) = y(t)a\mathrm{d}t + y(t)\sigma\mathrm{d}B(t) \tag{1.6.8}$$

with a, σ constants, $B(t)$ one-dimensional Brownian motion, where a is the constant *appreciation rate*, and where σ is constant *volatility*. Observe (see, e.g., [34]), that (1.6.8) is solved "explicitly" by

$$y(t) = y(s)\mathrm{e}^{\left(a-\frac{\sigma^2}{2}\right)(t-s)+\sigma(B(t)-B(s))} \tag{1.6.9}$$

for $t > s$. Since Brownian motion satisfies $-\infty < B(t) < \infty$ for any $t > s$, the solution $y(t)$ of (1.6.8) beginning in $(0, \infty)$ remains in $(0, \infty)$ for $t > s$. This is important for understanding why PDE (1.6.12) below does not have a boundary condition at $y = 0$.

Setting

$$c = \left(y\sigma\right), b = \{ry\} \tag{1.6.10}$$

in (1.6.6) and (1.6.5), we see that the function

$$\varphi(s,x) = E_{s,x}\left(g(y_{s,x}(T))\mathrm{e}^{-\int_s^T r\mathrm{d}\tau}\right) = \mathrm{e}^{-r(T-s)}E_{s,x}(g(y_{s,x}(T))) \tag{1.6.11}$$

is a solution of the famous Black–Scholes PDE

$$\varphi^{(1,0)}(t,y) + \frac{1}{2}y^2\sigma^2\varphi^{(0,2)}(t,y) + ay\varphi^{(0,1)}(t,y) - r\varphi(t,y) = 0 \tag{1.6.12}$$

in $\{\{t,y\} : t < T, y > 0\}$, together with the terminal condition $\varphi(T,y) = g(y)$. No boundary condition is needed at the boundary $y = 0$, nor is it possible to impose one.

What is the relationship between PDE (1.6.12) and finance? We shall answer this question in detail in subsequent chapters in great generality, but for now, let us just state without explanation that if $y(t)$ is the price of some (perfectly) *tradable* security, to be referred to frequently as the *underlying*, possibly paying an *underlying* dividend at the rate $\mathbb{D}$ (i.e., paying $y(t)\mathbb{D}$ units of currency per year), then any financial contract that pays only the terminal payoff $g(y(T))$ in such an economic environment is priced via (1.6.12), but only after setting $a = r - \mathbb{D}$, and we obtain instead

$$\varphi^{(1,0)}(t,y) + \frac{1}{2}y^2\sigma^2\varphi^{(0,2)}(t,y) + y(r-\mathbb{D})\varphi^{(0,1)}(t,y) - r\varphi(t,y) = 0. \tag{1.6.13}$$

On the other hand, if the underlying is *not* tradable (and $\mathbb{D} = 0$), then (1.6.12) suffices (see (4.18) and its derivation).

Since (1.6.12) implies (1.6.13) by setting $a = r - \mathbb{D}$, we solve only (1.6.12). We do so first in the simplest case in which the terminal condition is given by

$$g(y) = y^\beta \tag{1.6.14}$$

for $\beta \in \mathbb{R}$. (The general solution is given by (1.6.24) below.) Indeed, by substituting $\varphi(t,y) \to h(t)y^\beta$ in (1.6.12) (separating variables), we arrive at the equation

$$y^\beta\left(\left(2r + \beta\left(-\beta\sigma^2 + \sigma^2 - 2a\right)\right)h(t) - 2h'(t)\right) = 0. \tag{1.6.15}$$

We also deduce the terminal condition $y^\beta = \varphi(T,y) = h(T)y^\beta$, implying $h(T) = 1$. This together with (1.6.15) yields a terminal value ODE problem for h, yielding

$$h(t) = \mathrm{e}^{-(T-t)\left(r - a\beta + \frac{1}{2}\beta(1-\beta)\sigma^2\right)}. \tag{1.6.16}$$

Therefore, the (unique) solution of (1.6.12), with the terminal condition (1.6.14), is

$$\varphi(t,y) = \mathrm{e}^{-(T-t)\left(r - a\beta + \frac{1}{2}\beta(1-\beta)\sigma^2\right)}y^\beta, \tag{1.6.17}$$

which in the case $a = r - \mathbb{D}$, i.e., in the case of the Black–Scholes PDE (1.6.13), becomes

$$\varphi(t,y) = \mathrm{e}^{-(T-t)\left(r - \beta(r-\mathbb{D}) + \frac{1}{2}\beta\left((1-\beta)\sigma^2\right)\right)}y^\beta. \tag{1.6.18}$$

Observe that as anticipated, no boundary condition at $y = 0$ was needed, nor is it possible to impose one in the above calculation.

Example 1.6.2. (Zero-coupon bonds and forward contracts.) A (zero-coupon) bond is a financial contract that pays a constant terminal payoff. A forward contract (on an underlying security or commodity) is a financial contract that pays the terminal payoff $y - \mathfrak{k}$, where y is the price of the underlying, and $\mathfrak{k}$ is the *delivery price (or strike)*.

Notice that in the case $\beta = 0$, formula (1.6.18) becomes

$$\varphi(t,y) = \mathrm{e}^{-r(T-t)}, \tag{1.6.19}$$

the price of a bond paying 1 unit of currency, maturing at $t = T$ (under the assumption that the interest rate will remain constant for the remainder of the time till maturity).

In the case $\beta = 1$, formula (1.6.18) becomes

$$\varphi(t,y) = \mathrm{e}^{-\mathbb{D}(T-t)}y, \tag{1.6.20}$$

and by means of a linear combination of (1.6.19) and (1.6.20), we get

$$\varphi(t,y) = \varphi_{\mathfrak{k}}(t,y) = \mathrm{e}^{-\mathbb{D}(T-t)}y - \mathfrak{k}\mathrm{e}^{-r(T-t)}, \tag{1.6.21}$$

the price/value of a *forward contract with delivery price* (strike) $\mathfrak{k}$, a contract that has a terminal payoff $\varphi_{\mathfrak{k}}(T,y) = y - \mathfrak{k}$. Bonds and forward contracts are the simplest financial derivatives, since their payoffs are respectively constant and linear in the underlying. Yet in the case of forward contracts, an additional issue arises, because they are negotiated in such a way that no money changes hands when the contracts are agreed upon. We shall return to this issue shortly.

Example 1.6.3. (Options.) An *option* (on an underlying), more precisely a *call* (*put*), with *strike* k on an underlying security is a financial contract with terminal payoff

$$g(y) = \max[0, y-k] \tag{1.6.22}$$

in the case of *calls*, and

$$g(y) = \max[0, k-y] \tag{1.6.23}$$

in the case of *puts*.

The simplicity of the derivation of the solution (1.6.17) is due to the simplicity of the terminal condition. In the case of options, one has a much more difficult to handle (nondifferentiable) terminal condition (1.6.22) or (1.6.23).

In such a case, no separation of variables is possible, yet an explicit formula is possible to derive – the Black–Scholes–Merton formula. In fact, skipping details, no matter what g is, i.e., no matter what the contract terminal payoff is, the solution of (1.6.12) is

$$\varphi(t,y) = \frac{1}{\sqrt{2\pi}\sqrt{(T-t)\sigma^2}} \mathrm{e}^{-(T-t)\left(\frac{\sigma^2}{2}\left(\frac{1}{2}-\frac{a}{\sigma^2}\right)^2+r\right)} y^{\frac{1}{2}-\frac{a}{\sigma^2}}$$

$$\int_{-\infty}^{\infty} \mathrm{e}^{-\frac{(\xi-\log(y))^2}{2(T-t)\sigma^2}-\xi\left(\frac{1}{2}-\frac{a}{\sigma^2}\right)} g\left(\mathrm{e}^{\xi}\right) \mathrm{d}\xi, \tag{1.6.24}$$

which in the case of calls, i.e., in the case (1.6.22), becomes

$$\varphi(t,y) = \frac{1}{2}\mathrm{e}^{r(t-T)}\left(\mathrm{e}^{a(T-t)}y\left(\operatorname{erf}\left(\left(2\log\left(\frac{y}{k}\right)\right.\right.\right.\right.$$

$$\left.\left.\left. - (t-T)\left(\sigma^2+2a\right)\right)\Big/\left(2\sqrt{2}\sqrt{(T-t)\sigma^2}\right)\right)+1\right)$$

$$+k\left(\operatorname{erfc}\left(\left(2\log\left(\frac{y}{k}\right)\right.\right.\right.$$

$$\left.\left.\left.\left. - (t-T)\left(2a-\sigma^2\right)\right)\Big/\left(2\sqrt{2}\sqrt{(T-t)\sigma^2}\right)\right)-2\right)\right), \tag{1.6.25}$$

while in the case of puts, i.e., in the case (1.6.23), it becomes

$$\varphi(t,y) = \frac{1}{2}e^{r(t-T)}\left(k\operatorname{erfc}\left(\left(2\log\left(\frac{y}{k}\right) -(t-T)\left(2a-\sigma^2\right)\right)\Big/\left(2\sqrt{2}\sqrt{(T-t)\sigma^2}\right)\right)\right.$$
$$-e^{a(T-t)}y\operatorname{erfc}\left(\left(2\log\left(\frac{y}{k}\right)\right.\right.$$
$$\left.\left.\left.-(t-T)\left(\sigma^2+2a\right)\right)\Big/\left(2\sqrt{2}\sqrt{(T-t)\sigma^2}\right)\right)\right), \tag{1.6.26}$$

where

$$\operatorname{erf}(z) := \frac{2}{\sqrt{\pi}}\int_0^z e^{-t^2}\,dt \tag{1.6.27}$$

(the error function) and

$$\operatorname{erfc}(z) := 1 - \operatorname{erf}(z) \tag{1.6.28}$$

(the complementary error function).

We also check whether formulas (1.6.17) and (1.6.24) are consistent. To that end, we calculate the integral in (1.6.24) in the case $g(y) = y^\beta$:

$$\int_{-\infty}^{\infty} e^{-\frac{(\xi-\log(y))^2}{2(T-t)\sigma^2}-\frac{\xi\left(\sigma^2-2a\right)}{2\sigma^2}} g\left(e^\xi\right) d\xi = \int_{-\infty}^{\infty} e^{-\frac{(\xi-\log(y))^2}{2(T-t)\sigma^2}-\frac{\xi\left(\sigma^2-2a\right)}{2\sigma^2}} e^{\xi\beta} d\xi$$
$$= \sqrt{\frac{(T-t)\sigma^2}{y}}\, e^{-\frac{(t-T)\left((2\beta-1)\sigma^2+2a\right)^2}{8\sigma^2}}\sqrt{2\pi}\, y^{\frac{a}{\sigma^2}+\beta}, \tag{1.6.29}$$

and taking (1.6.29) back into (1.6.24), after simplification, formula (1.6.17) is reestablished.

So, $\varphi(t,y)$ given in (1.6.25) or (1.6.26) is the unique $\{y>0\}$-entire solution of (1.6.12), (1.6.22), and (1.6.12), (1.6.23), respectively.

Consider now the (inhomogeneous) Black–Scholes PDE (with nonzero dividend rate $f(t,y)$ units of currency per year)

$$\varphi^{(1,0)}(t,y) + \frac{1}{2}y^2\sigma^2\varphi^{(0,2)}(t,y) + ay\varphi^{(0,1)}(t,y) - r\varphi(t,y) = -f(t,y) \tag{1.6.30}$$

in $\{\{t,y\} : t < T, y > 0\}$, together with the terminal condition $\varphi(T,y) = g(y)$.

As discussed above, the solution of (1.6.30) has a probabilistic representation, i.e., it is characterized as the expected payoff

$$\varphi(s,x) = E_{s,x}\left(g\left(y_{s,x}(T)\right)e^{-r(T-s)} + \int_s^T e^{-r(t-s)} f\left(t, y_{s,x}(t)\right) dt\right). \tag{1.6.31}$$

Since we already have a solution of the homogeneous problem (see (1.6.24), (1.6.25), or (1.6.26)), we can use it via Duhamel's principle to obtain the solution of the inhomogeneous problem (1.6.30). Indeed, in the case of the simplest ODE

$$y'(t) - cy(t) = -f(t), y(T) = g, \tag{1.6.32}$$

for $t < T$, the solution is given by

$$y(t) = \mathrm{e}^{-c(T-t)}g + \int_t^T \mathrm{e}^{-c(\tau-t)}f(\tau)\mathrm{d}\tau = P(T-t)g + \int_t^T P(\tau-t)f(\tau)\mathrm{d}\tau, \tag{1.6.33}$$

where $P(\tau - t)f(\tau) := z(t)$ is the solution of the homogeneous terminal value problem

$$z'(t) - cz(t) = 0, z(\tau) = f(\tau), \tag{1.6.34}$$

for $t < \tau$.

Drawing an analogy between (1.6.30) and (1.6.32), we introduce the linear operator notation

$$\mathfrak{c}\varphi := -\frac{1}{2}y^2\sigma^2\varphi^{(0,2)} - ay\varphi^{(0,1)} + r\varphi, \tag{1.6.35}$$

so that (1.6.30) can be rewritten as $y'(t) - \mathfrak{c}y(t) = -f(t)$. Also the family of solution operators $\{\mathfrak{P}(T-t), t \leq T\}$ for the homogeneous equation $y'(t) - \mathfrak{c}y(t) = 0$ was derived already in (1.6.24), and can now be restated as

$$(\mathfrak{P}(T-t)g)(y) = \frac{1}{\sqrt{2\pi}\sqrt{(T-t)\sigma^2}}\mathrm{e}^{-(T-t)\left(\frac{\sigma^2}{2}\left(\frac{1}{2}-\frac{a}{\sigma^2}\right)^2+r\right)}y^{\frac{1}{2}-\frac{a}{\sigma^2}}$$
$$\int_{-\infty}^{\infty} \mathrm{e}^{-\frac{(\xi-\log(y))^2}{2(T-t)\sigma^2}-\xi\left(\frac{1}{2}-\frac{a}{\sigma^2}\right)} g\left(\mathrm{e}^{\xi}\right)\mathrm{d}\xi. \tag{1.6.36}$$

Observe that the family of solution operators $\{\mathfrak{P}(T-t), t \leq T\}$ forms a semigroup of linear operators, since for $r \leq s \leq t \leq T$, $\mathfrak{P}(s-r) \circ \mathfrak{P}(t-s) = \mathfrak{P}(t-r)$. There exists an extensive body of literature on semigroups of linear (and to a lesser extent also nonlinear) operators (see, e.g., [9], and references therein).

Combining (1.6.33) and (1.6.36), we conclude that the unique solution of (1.6.30), i.e., the probabilistic expression (1.6.31), has the analytic representation (set $\mathfrak{f}(\tau) := f(\tau, \cdot)$)

$$\begin{aligned}\varphi(t,y) &= \left(\mathfrak{P}(T-t)g + \int_t^T \mathfrak{P}(\tau-t)\mathfrak{f}(\tau)\mathrm{d}\tau\right)(y) \\ &= \frac{1}{\sqrt{2\pi}\sqrt{(T-t)\sigma^2}}\mathrm{e}^{-(T-t)\left(\frac{\sigma^2}{2}\left(\frac{1}{2}-\frac{a}{\sigma^2}\right)^2+r\right)}y^{\frac{1}{2}-\frac{a}{\sigma^2}} \\ &\quad \int_{-\infty}^{\infty} \mathrm{e}^{-\frac{(\xi-\log(y))^2}{2(T-t)\sigma^2}-\xi\left(\frac{1}{2}-\frac{a}{\sigma^2}\right)} g\left(\mathrm{e}^{\xi}\right)\mathrm{d}\xi\end{aligned}$$

$$+\int_t^T \frac{1}{\sqrt{2\pi}\sqrt{(\tau-t)\sigma^2}} \mathrm{e}^{-(\tau-t)\left(\frac{\sigma^2}{2}\left(\frac{1}{2}-\frac{a}{\sigma^2}\right)^2+r\right)} y^{\frac{1}{2}-\frac{a}{\sigma^2}}$$

$$\int_{-\infty}^{\infty} \mathrm{e}^{-\frac{(\xi-\log(y))^2}{2(\tau-t)\sigma^2}-\xi\left(\frac{1}{2}-\frac{a}{\sigma^2}\right)} f\left(\tau,\mathrm{e}^{\xi}\right) \mathrm{d}\xi \mathrm{d}\tau \tag{1.6.37}$$

So $\varphi(t,y)$ given in (1.6.37) is the unique $\{y>0\}$-entire solution of (1.6.30).

Example 1.6.4. As mentioned already, forward contracts are entered with no money changing hands, and therefore by negotiating the (forward contract) delivery price $\mathfrak{k}$ such that the corresponding forward contract, at the time the forward contract is agreed upon, say $t=t_0$, has the value zero, i.e., by finding $\mathfrak{k}$ such that $\varphi_{\mathfrak{k}}(t_0, y(t_0)) = 0$. Of course, for $t > t_0$, a contract agreed upon at $t = t_0$ typically has a nonzero value $\varphi_{\mathfrak{k}}(t, y(t))$, $t_0 < t < T$, sometimes positive, sometimes negative, and it is settled at time $t = T$. So at time $t = t_0$, one finds $\mathfrak{k}$ that solves the equation

$$\mathrm{e}^{-\mathbb{D}(T-t)} y - \mathfrak{k}\mathrm{e}^{-r(T-t)} = 0, \tag{1.6.38}$$

i.e., one finds the forward delivery price $\mathfrak{k} = \mathfrak{k}(r,\mathbb{D},T,t)$ for which the price of the forward contract at time $t = t_0$ is equal to zero. Such $\mathfrak{k}$ is called the *forward price* (of the underlying), and in this simple model, it is equal to

$$f(t_0, y) = \mathrm{e}^{(r-\mathbb{D})(T-t_0)} y. \tag{1.6.39}$$

This is a difficult concept: the *forward price* of an underlying is a forward price as of today, taking into the account the convenience or the inconvenience of taking possession of the underlying today or at some given time in the future, and taking into the account the difference in regard to paying for it today or at the time of delivery in the future.

Below we shall discuss a closely related concept, that of the *futures price of an underlying*. The only difference between the two, between the forward price and futures price of an underlying security, is in the ways the corresponding contracts – the forward contract and futures contract – are settled.

Example 1.6.5. (Futures contracts on an underlying, and futures prices of an underlying.) Now imagine that an investor has a long position in a forward contract. At the close of each day, going into the future (until the expiration time $t = T$), he/she sells the contract(s) held, and buys a new one, with the same expiration time. Such a transaction can yield positive or negative cash flow depending on the price movement of the underlying, or in more complicated models, due also to a possible factor movement during the trading day. Such a sequence of forward contracts is equivalent to a single *futures contract*.

Like forward contracts, futures contracts have zero value when agreed upon. Nevertheless, while forward contracts can, and typically do, have nonzero value thereafter, futures contracts, since they are settled at the end of each trading day, and in the idealized world of (stochastic) calculus settled instantaneously, their value/price is always equal to zero. Consequently, futures contract cause (settlement) cash flow, sometimes positive, sometimes negative.

Since futures contracts are settled each trading day, they generate a cash flow during their lifetime, not only at the time of expiration. The generated cash can be deposited to or withdrawn from (depending on whether it is positive or negative) the cash account, generating an income or cost.

If the interest rate r is deterministic, such a settlement rule will not affect the valuation, and under such circumstances, forward prices are equal to the futures prices (see [4]; see also [7]).

Observe that the function f in (1.6.39) (the forward price) does *not* satisfy equation (1.6.13). Instead, it satisfies the equation

$$\varphi^{(1,0)}(t,y)+\frac{1}{2}y^2\sigma^2\varphi^{(0,2)}(t,y)+y(r-\mathbb{D})\varphi^{(0,1)}(t,y)-r\varphi(t,y)=-r\varphi(t,y), \tag{1.6.40}$$

i.e.,

$$\varphi^{(1,0)}(t,y)+\frac{1}{2}y^2\sigma^2\varphi^{(0,2)}(t,y)+y(r-\mathbb{D})\varphi^{(0,1)}(t,y)=0 \tag{1.6.41}$$

for $t<T$, with the terminal condition $\varphi(T,y)=y$. This is a very important observation, and it is the content of a fundamental proposition of Cox et al. [7]:

Proposition 1.6.1 (7, Proposition 7). *The futures price φ is equal to the price of a security that pays dividend $r\varphi$ (per year) and with terminal payoff equal to the underlying y.*

Equation (1.6.40), and therefore also (1.6.41), characterizes the price φ of *a contract with terminal payoff equal to the underlying*, i.e., equal to y, *with the dividend equal to $r\varphi$*. Equivalently, the futures price satisfies the pricing PDE with zero-order term dropped, and with the terminal condition y. In the present situation, the forward price is equal to the futures price, but later we shall see that in the case of stochastic interest rates, that is not the case.

Remark 1.6.1. The rate $\mathbb{D}$ can be positive (if it represents a stock's dividend rate, for example), or negative (if it represents a storage cost rate for a commodity, for example). If, for example, $\mathbb{D}<0, r=0, f(t,y)=\mathrm{e}^{-\mathbb{D}(T-t)}y>y$, then the forward/futures price is higher than the price of the underlying, since otherwise, one could sell the underlying sooner rather than later to save on the storage costs. □

A situation in which the futures price of the underlying is higher than the underlying, i.e., than the *spot price*, i.e., $\varphi(t,y)>y$, or in which the price of a far future delivery is higher than that of a near future delivery, or, at least in this simple model, according to (1.6.39), a situation in which $r>\mathbb{D}$, is called a *contango,* while the reverse, $f(t,y)<y$, i.e., $r<\mathbb{D}$, is called a *backwardation*. Due to near-zero interest rates, the latter case was in effect for the futures on the S&P 500 index (^GSPC) during the summer of 2009, as seen in Fig. 1.3. The plot presents

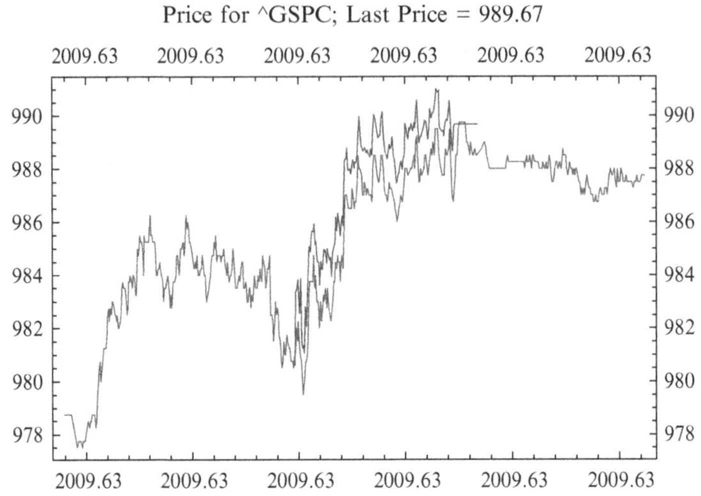

Fig. 1.3 S&P 500 index and its futures

August 18 data for the ^GSPC during regular trading hours, 9:30–16:00, and the data for the futures with September expiration (ESU09.CME), which trade (almost) around the clock.

Exercise 1.6.1. Find the $\{y > 0\}$-entire solution of (1.6.30) for $f(t,y) = \alpha_1 y^\beta$ and $g(y) = \alpha_2 y^\beta$, with and without utilization of the general formula (1.6.37).

Exercise 1.6.2. Assume that $r - \beta\left(a + (\beta - 1)\frac{\sigma^2}{2}\right) > 0$. Find the $\{-\infty < t < \infty\} \times \{y > 0\}$-entire solution of (1.6.30) for $f(t,y) = \alpha_1 y^\beta$, i.e., send $T \to \infty$ in the previous exercise.

1.7 Nonlinear PDEs: Hamilton–Jacobi–Bellman PDEs

Suppose the data in (1.6.1) depend on a feedback control $u : \{t,y\}| \to u(t,y) \in \mathcal{U} \subset \mathbb{R}^l$,

$$\mathrm{d}y^u_{s,x}(t) = b\left(u\left(t, y^u_{s,x}(t)\right), t, y^u_{s,x}(t)\right)\mathrm{d}t + c\left(u\left(t, y^u_{s,x}(t)\right), t, y^u_{s,x}(t)\right) \cdot \mathrm{d}B(t), \quad (1.7.1)$$

for $t > s \in \mathbb{R}$, with an initial condition $y^u_{s,x}(s) = x \in \mathbb{R}^n$. Consider the Hamilton–Jacobi–Bellman (HJB) PDE

$$\max_{u \in \mathcal{U}} \left(\varphi_t(t,y) + \frac{1}{2}\mathrm{Tr}\left[c(u,t,y) \cdot c(u,t,y)^T \cdot \nabla_y \nabla_y \varphi(t,y)\right] \right.$$
$$\left. + \nabla_y \varphi(t,y) \cdot b(u,t,y) - r(u,t,y)\varphi(t,y) + f(u,t,y) \right) = 0 \quad (1.7.2)$$

for $t < T$, together with the terminal condition $\varphi(T,y) = g(y)$.

Assume that the maximum is obtained in (1.7.2). Then (1.7.2) defines a function $u^* : \{t,y\}| \to u^*(t,y) \in \mathcal{U}$ such that φ also satisfies the equation

$$\varphi_t(t,y) + \frac{1}{2}\mathrm{Tr}\left(c(u^*(t,y),t,y)\cdot c(u^*(t,y),t,y)^T \cdot \nabla_y\nabla_y\varphi(t,y)\right) \\ +\nabla_y\varphi(t,y)\cdot b(u^*(t,y),t,y) - r(u^*(t,y),t,y)\,\varphi(t,y) = -f(u^*(t,y),t,y), \quad (1.7.3)$$

and for any other function $u : \{t,y\} \to u(t,y) \in \mathcal{U}$, φ satisfies the inequality

$$\varphi_t(t,y) + \frac{1}{2}\mathrm{Tr}\left(c(u(t,y),t,y)\cdot c(u(t,y),t,y)^T \cdot \nabla_y\nabla_y\varphi(t,y)\right) \\ +\nabla_y\varphi(t,y)\cdot b(u(t,y),t,y) - r(u(t,y),t,y)\varphi(t,y) \le -f(u(t,y),t,y). \quad (1.7.4)$$

Inequality (1.7.4) implies

$$E_{s,x}\varphi\left(T,y^u_{s,x}(T)\right)\mathrm{e}^{-\int_s^T r\left(u\left(\tau,y^u_{s,x}(\tau)\right),\tau,y^u_{s,x}(\tau)\right)\mathrm{d}\tau} - \varphi(s,x) \\ = E_{s,x}\int_s^T \mathrm{e}^{-\int_s^t r\left(u\left(\tau,y^u_{s,x}(\tau)\right),\tau,y^u_{s,x}(\tau)\right)\mathrm{d}\tau}\left(\varphi_t\left(t,y^u_{s,x}(t)\right) + \nabla_y\varphi\left(t,y^u_{s,x}(t)\right)\right. \\ \cdot b\left(u\left(t,y^u_{s,x}(t)\right),t,y^u_{s,x}(t)\right) + \frac{1}{2}\mathrm{Tr}\left[c\left(u\left(t,y^u_{s,x}(t)\right),t,y^u_{s,x}(t)\right)\right. \\ \left.\cdot c\left(u\left(t,y^u_{s,x}(t)\right),t,y^u_{s,x}(t)\right)^T\cdot\nabla_y\nabla_y\varphi\left(t,y^u_{s,x}(t)\right)\right] \\ \left.-r\left(u\left(t,y^u_{s,x}(t)\right),t,y^u_{s,x}(t)\right)\varphi\left(t,y^u_{s,x}(t)\right)\right)\mathrm{d}t \\ \le -E_{s,x}\int_s^T \mathrm{e}^{-\int_s^t r\left(u\left(\tau,y^u_{s,x}(\tau)\right),\tau,y^u_{s,x}(\tau)\right)\mathrm{d}\tau} f\left(u\left(t,y^u_{s,x}(t)\right),t,y^u_{s,x}(t)\right)\mathrm{d}t, \quad (1.7.5)$$

i.e.,

$$\varphi(s,x) \ge E_{s,x}\left(\varphi\left(T,y^u_{s,x}(T)\right)\mathrm{e}^{-\int_s^T r\left(u\left(\tau,y^u_{s,x}(\tau)\right),\tau,y^u_{s,x}(\tau)\right)\mathrm{d}\tau}\right. \\ \left.+\int_s^T \mathrm{e}^{-\int_s^t r\left(u\left(\tau,y^u_{s,x}(\tau)\right),\tau,y^u_{s,x}(\tau)\right)\mathrm{d}\tau} f\left(u\left(t,y^u_{s,x}(t)\right),t,y^u_{s,x}(t)\right)\mathrm{d}t\right), \quad (1.7.6)$$

while (1.7.3) implies

$$E_{s,x}\varphi\left(T,y^{u^*}_{s,x}(T)\right)\mathrm{e}^{-\int_s^T r\left(u^*\left(\tau,y^{u^*}_{s,x}(\tau)\right),\tau,y^{u^*}_{s,x}(\tau)\right)\mathrm{d}\tau} - \varphi(s,x) \\ = E_{s,x}\int_s^T \mathrm{e}^{-\int_s^t r\left(u^*\left(\tau,y^{u^*}_{s,x}(\tau)\right),\tau,y^{u^*}_{s,x}(\tau)\right)\mathrm{d}\tau}\left(\varphi_t\left(t,y^{u^*}_{s,x}(t)\right) + \nabla_y\varphi\left(t,y^{u^*}_{s,x}(t)\right)\right.$$

$$
\begin{aligned}
&\cdot b\left(u^*\left(t,y_{s,x}^{u^*}(t)\right),t,y_{s,x}^{u^*}(t)\right)+\frac{1}{2}\mathrm{Tr}\left(c\left(u^*\left(t,y_{s,x}^{u^*}(t)\right),t,y_{s,x}^{u^*}(t)\right)\right.\\
&\left.\cdot c\left(u^*\left(t,y_{s,x}^{u^*}(t)\right),t,y_{s,x}^{u^*}(t)\right)^T\cdot\nabla_y\nabla_y\varphi\left(t,y_{s,x}^{u^*}(t)\right)\right)\\
&-r\left(u^*\left(t,y_{s,x}^{u^*}(t)\right),t,y_{s,x}^{u^*}(t)\right)\varphi\left(t,y_{s,x}^{u^*}(t)\right)\Big)\,\mathrm{d}t\\
=&-E_{s,x}\int_s^T \mathrm{e}^{-\int_s^t r\left(u^*\left(\tau,y_{s,x}^{u^*}(\tau)\right),\tau,y_{s,x}^{u^*}(\tau)\right)\mathrm{d}\tau} f\left(u^*\left(t,y_{s,x}^{u^*}(t)\right),t,y_{s,x}^{u^*}(t)\right)\mathrm{d}t, \quad (1.7.7)
\end{aligned}
$$

i.e.,

$$
\begin{aligned}
\varphi(s,x)=E_{s,x}\Bigg(&\varphi\left(T,y_{s,x}^{u^*}(T)\right)\mathrm{e}^{-\int_s^T r\left(u^*\left(\tau,y_{s,x}^{u^*}(\tau)\right),\tau,y_{s,x}^{u^*}(\tau)\right)\mathrm{d}\tau}\\
&+\int_s^T \mathrm{e}^{-\int_s^t r\left(u^*\left(\tau,y_{s,x}^{u^*}(\tau)\right),\tau,y_{s,x}^{u^*}(\tau)\right)\mathrm{d}\tau} f\left(u^*\left(t,y_{s,x}^{u^*}(t)\right),t,y_{s,x}^{u^*}(t)\right)\mathrm{d}t\Bigg).
\end{aligned}
\tag{1.7.8}
$$

Putting (1.7.6) and (1.7.8) together, one concludes that

$$
\begin{aligned}
\varphi(s,x)=\max_{u:(t,y)\mapsto u(t,y)\in\mathcal{U}}\Bigg(&E_{s,x}\bigg(g\left(y_{s,x}^{u}(T)\right)\mathrm{e}^{-\int_s^T r\left(u\left(\tau,y_{s,x}^{u}(\tau)\right),\tau,y_{s,x}^{u}(\tau)\right)\mathrm{d}\tau}\\
&+\int_s^T \mathrm{e}^{-\int_s^t r\left(u\left(\tau,y_{s,x}^{u}(\tau)\right),\tau,y_{s,x}^{u}(\tau)\right)\mathrm{d}\tau} f\left(u\left(t,y_{s,x}^{u}(t)\right),t,y_{s,x}^{u}(t)\right)\mathrm{d}t\bigg)\Bigg),
\end{aligned}
\tag{1.7.9}
$$

that is, the solution of the HJB PDE (1.7.2) is the *value function* φ for the stochastic control problem (1.7.9).

Chapter 2
Simple Economies: Complete and Incomplete Markets

A General Multivariable Itô–SDE Framework for Financial Mathematics

2.1 Introduction

Financial contracts are usually traded in financial markets, which are part of a broader economy. An economy is quite a complicated object to comprehend, even more so to attempt to describe mathematically, and the situation is most complicated if the ambition is not just to describe it but also to be able to provide quantitative answers to many relevant financial and economic questions.

We are interested in aspects of the economy, which we shall refer to as (dynamic) economic *factors*, that are quantifiable. Such factors could include (though not all of them will be considered in this work) all stock prices and various market indexes, commodity prices, foreign exchange rates, but also quantities that are not prices, such as the gross national product, money supply, interest rates, commodity reserves, fresh water resources, communication and transportation infrastructure, vulnerability to severe and changing weather patterns. Some of the factors are quite difficult to model, and some of them have a long modeling history, possibly in other scientific disciplines.

Although many factors exist, and although many of them contribute in some way or other to a given economic phenomenon, some more, some less, since models, in order to provide useful answers, have to be tractable, and, moreover, to be tractable they have to be simple enough, for each particular situation we have to find the most relevant factors and ignore the rest.

For example, in recent years a class of financial derivatives was introduced into practice and studied theoretically: the weather derivatives. The purpose of weather derivatives is to bet on or hedge against the financial risks associated with the weather. Now, state-of-the-art meteorology can employ very sophisticated PDEs, coupling Navier–Stokes PDEs with heat diffusion, to model and predict future behavior of the weather. Such equations, even when approximated with their finite-dimensional (stochastic) variants, are very high-dimensional. So, if such equations are to be used for pricing weather derivatives, although conceptually this

S. Stojanovic, *Neutral and Indifference Portfolio Pricing, Hedging and Investing: With Applications in Equity and FX*, DOI 10.1007/978-0-387-71418-9_2,

would fit the framework to be elaborated in this book, one would end up with a very high-dimensional problem, beyond the reach even of today's supercomputing capabilities. Naturally, if that is the case, one has to be less ambitious and look for a list of factors and associated models that ignore some aspects, yet capture the essential parts of the given economic phenomenon.

So, since the subject matter here does not end with the ability to model, and even with the ability to predict with precision future behavior of the modeled quantity, but rather to use such models as an underlying for studying and computing answers in regard to derivative financial contracts, we are confronted with the necessity to deal with simplified models for the constitutive economic quantities and therefore also for the economy. Such simplified models for the economy will be referred to as *simple economies* (to be defined below precisely).

Finally, throughout this book, we shall assume complete information about the factors—all factors (and prices) are observable. Moreover, we assume complete information about all of the model parameters! Of course, such assumptions are unrealistic, and even though we can usually estimate the parameters statistically (see, e.g., Sect. 2.5) and filter dynamic unobservable factors (see, e.g., [34] and references therein), the size of the errors that are necessarily made when such estimations are performed is, unfortunately, not accounted for by the methodology that this book elaborates. That is a consequence of the present stage of development of stochastic control theory under incomplete information, and consequently, the current stage of development of optimal portfolio theory, and consequently, that of pricing and hedging theory.

2.2 Definition of a Simple Economy $\mathfrak{E}$: A Framework for Financial Mathematics

All the major results and all the applications to be developed in this book will be made in the framework to be introduced now. Following [40], we make the following definition.

Definition 2.2.1. A *simple economy* $\mathfrak{E}$ consists of:

(1) A finite (nonempty) set of stochastic *tradables* $S(t) = \{S_1(t), \dots, S_k(t)\}$, or a *market*
(2) A finite (possibly empty) set of dynamic (economic) *factors* $A(t) = \{A_1(t), \dots, A_m(t)\}$
(3) A *money market* that pays interest at the rate, sometimes also called the *short rate*, $r(t, A(t))$

assumed to obey the Itô SDE dynamics:

$$\mathrm{d}S(t) = S(t)\left(a_s(t,A(t)) - \mathbb{D}(t,A(t))\right)\mathrm{d}t + S(t)\sigma_s(t,A(t))\cdot\mathrm{d}B(t), \tag{2.2.1}$$

$$\mathrm{d}A(t) = b(t,A(t))\mathrm{d}t + c(t,A(t))\cdot\mathrm{d}B(t), \tag{2.2.2}$$

where $B(t) = \{B_1(t),\ldots,B_n(t)\}$ is a standard n-dimensional Brownian motion, the vector-valued function $b(t,A)$ is the m-vector of factor drifts, $c(t,A)$ is the $m\times n$ factor-diffusion matrix, $a_s(t,A)$ is the k-vector of *predividend* appreciation rates for the tradables, $\mathbb{D}(t,A)$ is the k-vector of dividend rates of the corresponding assets, and $\sigma_s(t,A)$ is the $k\times n$ volatility matrix. The functions a_s, σ_s, b, c, and r are called *market coefficients* (the information about $\mathbb{D}$ is contained in b and a_s; see (2.2.3) below).

Remark 2.2.1. The factors and tradables typically have nonempty intersection, so that if, for example, factor $A_i(t)$ is also a tradable, denoted by $S_j(t)$, i.e., the ith factor is the jth tradable, then (recall that c_i is the ith row of the matrix c, and similarly $\sigma_{s,j}$ is the jth row of the matrix σ_s)

$$b_i(t,A) = S_j\left(a_{s,j}(t,A) - \mathbb{D}_j(t,A)\right), \tag{2.2.3}$$

$$c_i(t,A) = S_j\sigma_{s,j}(t,A). \tag{2.2.4}$$

In particular, while $a_{s,j}$ is the *predividend* appreciation rate, b_i is the *postdividend* drift coefficient.

Somewhat loosely, we make the following definition.

Definition 2.2.3. The *state space* for a simple economy $\mathfrak{E}$, to be denoted by $\mathcal{A}$, is defined as the set of all possible values for the factors $A(t) = \{A_1(t),\ldots,A_m(t)\}$.

Example 2.2.1. If $A = S = \{S_1\}$, with $S_1(t) = \mathrm{e}^{B_1(t)}$, then $\mathcal{A} = (0,\infty)$.

We also make the following definition.

Definition 2.2.4. A market $S(t) = \{S_1(t),\ldots,S_k(t)\}$ is said to be *nonredundant* if

$$\sigma_s\cdot\sigma_s^{\mathrm{T}} > 0. \tag{2.2.5}$$

Throughout this book we shall always assume market nonredundancy. Nonredundancy outlaws perfectly correlated tradables, meaning that if in some theoretical or practical situations there is market redundancy, i.e., if some of the tradables are perfectly correlated, we either add some small noise, temporarily at least, or designate a nonredundant proxy for the redundant tradables.

On the other hand, factors are allowed to be "redundant." As a matter of fact, we can always start building a model with a large set of factors, e.g., in a way that, say, $S\subset A$; by solving the problem (of calculating the risk premium, pricing, hedging, and/or optimal portfolio strategy), the inessential factors will be set aside automatically. Also, if we start solving the same set of problems with too few factors, it will become transparent quickly that in such a setup there cannot be a solution.

What, then, is the minimal list of factors? That depends on two things. One is the market dynamics postulated in (2.2.1)–(2.2.2). Additionally, when the price of an additional financial contract is being determined, any quantity that affects the contract payoff has to be declared as a factor as well, whether or not it affects the market dynamics. Indeed, while the market dynamics affect the pricing PDE, the payoff structure will affect the terminal condition and, possibly, the right-hand side of the pricing PDE.

Remark 2.2.2. The notation in (2.2.1)–(2.2.2) may still be a bit confusing. For example, $S(t)\sigma_s(t,A(t))$ is by definition obtained by multiplying components of $S(t)$ by the rows of matrix $\sigma_s(t,A(t))$, while $\sigma_s(t,A(t))\cdot \mathrm{d}B(t)$ is the usual matrix–vector product. Observe that vectors, such as $S(t)$, are always understood as one-dimensional arrays, so that they cannot, and need not, be transposed. On the other hand, $\{S(t)\}$ is a "row vector," while $\{S(t)\}^{\mathrm{T}}$ is a "column vector," and they are both two-dimensional arrays, i.e., matrices. It may also be interesting to note that $(S(t)\sigma_s(t,A(t)))\cdot \mathrm{d}B(t) = S(t)\,(\sigma_s(t,A(t))\cdot \mathrm{d}B(t))$.

2.3 Complete and Incomplete Markets

Definition 2.3.1. If $m=0$, the market $S(t)$ is complete. If $m \geq 1$, the market $S(t)$ in a simple economy $\mathfrak{E}$ is said to be *complete* if

$$c\cdot\left(\mathbb{I}_n-\sigma_s^{\mathrm{T}}\cdot\left(\sigma_s\cdot\sigma_s^{\mathrm{T}}\right)^{-1}\cdot\sigma_s\right)\cdot c^{\mathrm{T}}=\mathbb{O}_m, \tag{2.3.1}$$

(where $\mathbb{I}_n$ is the n-identity matrix, $\mathbb{O}_{j\times k}$ is the $j\times k$ zero matrix, and $\mathbb{O}_m{:=}\mathbb{O}_{m\times m}$). Otherwise, the market $S(t)$ is said to be *incomplete*.

More naturally, the first part of Lemma 2.3.1 below should perhaps have been a definition, while Definition 2.3.1 should have been a lemma. They are equivalent, so it is a matter of taste. Our choice was motivated by the structure of the pricing PDEs to be developed in the next chapter.

We shall study both complete and incomplete markets, with emphasis on incomplete markets, while complete markets will be viewed simply as special cases of the incomplete ones. Also, observe that $\sigma_s^{\mathrm{T}}\cdot\left(\sigma_s\cdot\sigma_s^{\mathrm{T}}\right)^{-1}\cdot\sigma_s$ is the projection onto the row space of σ_s. For example, if $\sigma_s=\begin{pmatrix}-2&0&0\\1&0&3\end{pmatrix}$, then $\sigma_s^{\mathrm{T}}\cdot\left(\sigma_s\cdot\sigma_s^{\mathrm{T}}\right)^{-1}\cdot\sigma_s=$ $\begin{pmatrix}1&0&0\\0&0&0\\0&0&1\end{pmatrix}$.

Remark 2.3.1. Observe that $\exists\sigma_s^{-1}\Leftrightarrow\sigma_s^{\mathrm{T}}\cdot\left(\sigma_s\cdot\sigma_s^{\mathrm{T}}\right)^{-1}\cdot\sigma_s=\mathbb{I}_n$, and in such a case, of course, a market is complete. We remark that completeness/incompleteness is determined by the interplay between σ_s and c only. Definition 2.3.1 states, loosely speaking, that *all factor-randomness is tradable* (indeed, see Lemma 2.3.1 below).

Also, we shall see later that if a market is complete, contingent claims have unique prices, while if a market is incomplete, the prices may, and most often do, depend on the investor's risk-aversion.

Lemma 2.3.1. *If $m = 0$, market $S(t)$ is complete. If $m \geq 1$, market $S(t)$ in a simple economy $\mathfrak{E}$ is complete if and only if*

$$\text{SpanOfRows}[c] \subseteq \text{SpanOfRows}[\sigma_s] \tag{2.3.2}$$

("all factor-randomness is tradable"), or equivalently, if and only if

$$c \cdot \left(\mathbb{I}_n - \sigma_s^{\mathrm{T}} \cdot \left(\sigma_s \cdot \sigma_s^{\mathrm{T}}\right)^{-1} \cdot \sigma_s\right) = \mathbb{O}_{m \times n}. \tag{2.3.3}$$

Proof. We first prove the equivalence of conditions (2.3.1) and (2.3.3). Note that since $\sigma_s^{\mathrm{T}} \cdot \left(\sigma_s \cdot \sigma_s^{\mathrm{T}}\right)^{-1} \cdot \sigma_s$ is an orthogonal projection onto the row space of σ_s, $\mathbb{P} := \mathbb{I}_n - \sigma_s^{\mathrm{T}} \cdot \left(\sigma_s \cdot \sigma_s^{\mathrm{T}}\right)^{-1} \cdot \sigma_s$ is an orthogonal projection (onto the orthogonal complement of the row space of σ_s). Therefore, $\mathbb{P}^2 = \mathbb{P} = \mathbb{P}^{\mathrm{T}}$, and (2.3.1) can be rewritten as

$$\begin{aligned} \mathbb{O}_m &= c \cdot \left(\mathbb{I}_n - \sigma_s^{\mathrm{T}} \cdot \left(\sigma_s \cdot \sigma_s^{\mathrm{T}}\right)^{-1} \cdot \sigma_s\right) \cdot c^{\mathrm{T}} = c \cdot \mathbb{P} \cdot c^{\mathrm{T}} = c \cdot \mathbb{P} \cdot \mathbb{P} \cdot c^{\mathrm{T}} \\ &= c \cdot \mathbb{P} \cdot (c \cdot \mathbb{P})^{\mathrm{T}}, \end{aligned} \tag{2.3.4}$$

which holds if and only if $c \cdot \mathbb{P} = \mathbb{O}_{m \times n}$, i.e., if and only if (2.3.3) holds, i.e., if and only if

$$\left(\mathbb{I}_n - \sigma_s^{\mathrm{T}} \cdot \left(\sigma_s \cdot \sigma_s^{\mathrm{T}}\right)^{-1} \cdot \sigma_s\right) \cdot c^{\mathrm{T}} = \mathbb{O}_{n \times m}. \tag{2.3.5}$$

It remains to prove that (2.3.2) is equivalent to (2.3.5). Indeed, (2.3.5) is equivalent to

$$c^{\mathrm{T}} = \sigma_s^{\mathrm{T}} \cdot \left(\sigma_s \cdot \sigma_s^{\mathrm{T}}\right)^{-1} \cdot \sigma_s \cdot c^{\mathrm{T}}, \tag{2.3.6}$$

i.e.,

$$\text{columns}\left[c^{\mathrm{T}}\right] = \sigma_s^{\mathrm{T}} \cdot \left(\sigma_s \cdot \sigma_s^{\mathrm{T}}\right)^{-1} \cdot \sigma_s \cdot \text{columns}\left[c^{\mathrm{T}}\right], \tag{2.3.7}$$

i.e.,

$$\text{rows}[c] = \sigma_s^{\mathrm{T}} \cdot \left(\sigma_s \cdot \sigma_s^{\mathrm{T}}\right)^{-1} \cdot \sigma_s \cdot \text{rows}[c], \tag{2.3.8}$$

i.e.,

$$\text{rows}[c] \subseteq \text{SpanOfRows}\left[\sigma_s\right], \tag{2.3.9}$$

i.e., (2.3.2). □

2.4 Some Market Models: Nonredundancy and Completeness/Incompleteness

2.4.1 Log-Normal (Black–Scholes) Market Model

The following process will be a building block for many models to follow. Consider a simple economy with one tradable $S(t) = \{S_1(t)\}$, and a money market with a deterministic interest rate r. The price of the tradable is assumed to obey the Itô SDE

$$dS_1(t) = S_1(t)((\mathbf{a}_1 - \mathbb{D}_1)\,dt + \mathbf{p}_1 dB_1(t)), \tag{2.4.1}$$

where $\mathbf{a}_1, \mathbf{p}_1, \mathbb{D}_1 \in (-\infty, \infty)$, $\mathbf{p}_1 \neq 0$. This model is referred to as the log-normal (or Black–Scholes) model. Observe that if $\mathbb{D}_1$ represents the underlying dividend yield, then $\mathbb{D}_1 \geq 0$, but if it is, for example, the commodity convenience yield, then it could be positive or negative. Set $A = S = \{S_1\}$. Let the market coefficients be

$$a_s = \{\mathbf{a}_1\}, b = \{S_1(\mathbf{a}_1 - \mathbb{D}_1)\}, \sigma_s = (\mathbf{p}_1), c = (S_1\mathbf{p}_1). \tag{2.4.2}$$

We calculate

$$c \cdot \left(\mathbb{I}_1 - \sigma_s^{\mathrm{T}} \cdot \left(\sigma_s \cdot \sigma_s^{\mathrm{T}}\right)^{-1} \cdot \sigma_s\right) = (0), \tag{2.4.3}$$

$$c \cdot \left(\mathbb{I}_1 - \sigma_s^{\mathrm{T}} \cdot \left(\sigma_s \cdot \sigma_s^{\mathrm{T}}\right)^{-1} \cdot \sigma_s\right) \cdot c^{\mathrm{T}} = (0), \tag{2.4.4}$$

and the market is nonredundant if and only if $\mathbf{p}_1 \neq 0$ and it is complete. By the way, in the case $\mathbf{p}_1 = 0$, the market is deterministic and riskless, and therefore the market and the money market make each other redundant.

We also note that the state space, i.e., the set of all possible values for factors, in the case $A = \{S_1\}$, is equal to $\mathcal{A} = (0, \infty)$.

2.4.2 Stochastic Appreciation Rate

Assume a deterministic interest rate. Consider a simple economy $\mathfrak{E}$ consisting of factors $A = \{S_1, \mathbf{v}\}$ and a tradable $S = \{S_1\}$, with market coefficients

$$a_s = \{\mathbf{v}\}, \sigma_s = (\mathbf{p}_1\ 0), b = \{S_1(\mathbf{v} - \mathbb{D}_1), q_0 + \mathbf{v}q_1\}, c = \begin{pmatrix} S_1\mathbf{p}_1 & 0 \\ w\rho_{2,1} & w\sqrt{1-\rho_{2,1}^2} \end{pmatrix}, \tag{2.4.5}$$

i.e., assume

$$\begin{aligned} \mathrm{d}S_1(t) &= S_1(t)\,(\mathbf{v}(t) - \mathbb{D}_1)\,\mathrm{d}t + S_1(t)\mathbf{p}_1 \mathrm{d}B_1(t), \\ \mathrm{d}\mathbf{v}(t) &= (q_0 + \mathbf{v}(t) q_1)\,\mathrm{d}t + w\mathrm{d}\mathbb{B}_2(t), \end{aligned} \tag{2.4.6}$$

where $\mathrm{d}B_1(t)\mathrm{d}\mathbb{B}_2(t) = \rho_{2,1}\,\mathrm{d}t$. So, a single equity with price S_1 has appreciation rate $\mathbf{v}$ and volatility $\mathbf{p}_1$, where the appreciation rate $\mathbf{v}$ is stochastic, with linear drift $q_0 + \mathbf{v} q_1$ and diffusion w, with instantaneous price/appreciation-rate correlation $\rho_{2,1}$. Then

$$c \cdot \left(\mathbb{I}_2 - \sigma_s^{\mathrm{T}} \cdot \left(\sigma_s \cdot \sigma_s^{\mathrm{T}}\right)^{-1} \cdot \sigma_s\right) = \begin{pmatrix} 0 & 0 \\ 0 & w\sqrt{1-\rho_{2,1}^2} \end{pmatrix}, \tag{2.4.7}$$

$$c \cdot \left(\mathbb{I}_2 - \sigma_s^{\mathrm{T}} \cdot \left(\sigma_s \cdot \sigma_s^{\mathrm{T}}\right)^{-1} \cdot \sigma_s\right) \cdot c^{\mathrm{T}} = \begin{pmatrix} 0 & 0 \\ 0 & w^2\left(1-\rho_{2,1}^2\right) \end{pmatrix}, \tag{2.4.8}$$

so that this market is incomplete if and only if $w^2\left(1-\rho_{2,1}^2\right) \neq 0$. Also observe that the state space for this example is equal to $\mathcal{A} = (0, \infty) \times (-\infty, \infty)$.

2.4.3 *Stochastic Volatility: Heston's Model*

Heston's stochastic volatility model [13] assumes a constant interest rate r, a single tradable $S = \{S_1\}$, and an additional factor $\mathbf{v}(t)$:

$$\begin{aligned} \mathrm{d}S_1(t) &= S_1(t)\,(\mathbf{a}_0 + \lambda_s \mathbf{v}(t) - \mathbb{D}_1)\,\mathrm{d}t + S_1(t)\sqrt{\mathbf{v}(t)}\mathrm{d}B_1(t), \\ \mathrm{d}\mathbf{v}(t) &= (k_v - K_v \mathbf{v}(t))\,\mathrm{d}t + \sigma_v \sqrt{\mathbf{v}(t)}\mathrm{d}\mathbb{B}_2(t), \end{aligned} \tag{2.4.9}$$

where $\mathrm{d}B_1(t)\mathrm{d}\mathbb{B}_2(t) = \rho_{2,1}\mathrm{d}t$, and $\sqrt{\mathbf{v}(t)}$ is the (stochastic) volatility, which makes $\mathbf{v}(t)$ the variance. We have two choices in regard to factors: $A = \{\mathbf{v}\}$ or $A = \{S_1, \mathbf{v}\}$. Indeed, in the first case, if $A = \{\mathbf{v}\}$, then $\mathcal{A} = (0, \infty)$, and the corresponding market coefficients become

$$\begin{aligned} a_s &= \{\mathbf{a}_0 + \lambda_s \mathbf{v}\}, b = \{k_v - K_v \mathbf{v}\}, \sigma_s = \begin{pmatrix} \sqrt{\mathbf{v}} & 0 \end{pmatrix}, \\ c &= \begin{pmatrix} \sigma_v \sqrt{\mathbf{v}} & \rho_{2,1}\, \sigma_v \sqrt{\mathbf{v}} \sqrt{1-\rho_{2,1}^2} \end{pmatrix}, \end{aligned} \tag{2.4.10}$$

and we calculate

$$c \cdot \left(\mathbb{I}_2 - \sigma_s^{\mathrm{T}} \cdot \left(\sigma_s \cdot \sigma_s^{\mathrm{T}}\right)^{-1} \cdot \sigma_s\right) = \begin{pmatrix} 0 & \sqrt{\mathbf{v}}\, \sigma_v \sqrt{1-\rho_{2,1}^2} \end{pmatrix}, \tag{2.4.11}$$

$$c \cdot \left(\mathbb{I}_2 - \sigma_s^{\mathrm{T}} \cdot \left(\sigma_s \cdot \sigma_s^{\mathrm{T}}\right)^{-1} \cdot \sigma_s\right) \cdot c^{\mathrm{T}} = \left(\mathbf{v}\, \sigma_v^2 \left(1-\rho_{2,1}^2\right)\right). \tag{2.4.12}$$

In the second case, if $A = \{S_1, \mathbf{v}\}$, then $\mathcal{A} = (0,\infty)^2$, and the corresponding market coefficients become

$$a_s = \{\mathbf{a}_0 + \lambda_s \mathbf{v}\}, b = \{S_1(\mathbf{a}_0 + \lambda_s \mathbf{v} - \mathbb{D}), k_v - K_v \mathbf{v}\},$$
$$\sigma_s = \left(\sqrt{\mathbf{v}}\, 0\right), c = \begin{pmatrix} S_1\sqrt{\mathbf{v}} & 0 \\ \sigma_v\sqrt{\mathbf{v}}\,\rho_{2,1} & \sigma_v\sqrt{\mathbf{v}}\sqrt{1-\rho_{2,1}^2} \end{pmatrix}, \tag{2.4.13}$$

and this time, we calculate

$$c \cdot \left(\mathbb{I}_2 - \sigma_s^{\mathrm{T}} \cdot \left(\sigma_s \cdot \sigma_s^{\mathrm{T}}\right)^{-1} \cdot \sigma_s\right) = \begin{pmatrix} 0 & 0 \\ 0 & \sqrt{\mathbf{v}}\,\sigma_v\sqrt{1-\rho_{2,1}^2} \end{pmatrix}, \tag{2.4.14}$$

$$c \cdot \left(\mathbb{I}_2 - \sigma_s^{\mathrm{T}} \cdot \left(\sigma_s \cdot \sigma_s^{\mathrm{T}}\right)^{-1} \cdot \sigma_s\right) \cdot c^{\mathrm{T}} = \begin{pmatrix} 0 & 0 \\ 0 & \mathbf{v}\sigma_v^2\left(1-\rho_{2,1}^2\right) \end{pmatrix}. \tag{2.4.15}$$

Either way, this market is incomplete if and only if $\sigma_v^2\left(1-\rho_{2,1}^2\right) \neq 0$.

2.4.4 An Alternative Stochastic Volatility Model

Let us modify Heston's model. Assume a constant interest rate r, a single tradable $S = \{S_1\}$, and an additional factor $\mathbf{v}(t)$:

$$\mathrm{d}S_1(t) = S_1(t)\left(\mathbf{a}_0 + \lambda_s \mathbf{v}(t) - \mathbb{D}_1\right)\mathrm{d}t + S_1(t)\mathbf{v}(t)\mathrm{d}B_1(t),$$
$$\mathrm{d}\mathbf{v}(t) = \left(k_v - K_v \mathbf{v}(t)\right)\mathrm{d}t + \sigma_v \mathbf{v}(t)\mathrm{d}\mathbb{B}_2(t), \tag{2.4.16}$$

where $\mathrm{d}B_1(t)\mathrm{d}\mathbb{B}_2(t) = \rho_{2,1}\mathrm{d}t$, with $k_v, K_v > 0$. This model is realized via the following market coefficients:

Assuming $A = \{\mathbf{v}\}$, we have $\mathcal{A} = (0,\infty)$, and

$$a_s = \{\mathbf{a}_0 + \lambda_s \mathbf{v}\}, \ b = \{k_v - K_v \mathbf{v}\}, \ \sigma_s = (\mathbf{v}0),$$
$$c = \left(\sigma_v \mathbf{v}\rho_{2,1}\sigma_v \mathbf{v}\sqrt{1-\rho_{2,1}^2}\right), \tag{2.4.17}$$

in which case we calculate

$$c \cdot \left(\mathbb{I}_2 - \sigma_s^{\mathrm{T}} \cdot \left(\sigma_s \cdot \sigma_s^{\mathrm{T}}\right)^{-1} \cdot \sigma_s\right) = \left(0 \quad \mathbf{v}\sigma_v\sqrt{1-\rho_{2,1}^2}\right), \tag{2.4.18}$$

$$c \cdot \left(\mathbb{I}_2 - \sigma_s^{\mathrm{T}} \cdot \left(\sigma_s \cdot \sigma_s^{\mathrm{T}}\right)^{-1} \cdot \sigma_s\right) \cdot c^{\mathrm{T}} = \left(\mathbf{v}^2\sigma_v^2\left(1-\rho_{2,1}^2\right)\right), \tag{2.4.19}$$

while in the case $A = \{S_1, \mathbf{v}\}$, we have $\mathcal{A} = (0, \infty)^2$, and

$$a_s = \{\mathbf{a}_0 + \lambda_s \mathbf{v}\}, \; b = \{S_1 (\mathbf{a}_0 + \lambda_s \mathbf{v} - \mathbb{D}), \; k_v - K_v \mathbf{v}\},$$
$$\sigma_s = (\mathbf{v} \quad 0), \; c = \begin{pmatrix} S_1 \mathbf{v} & 0 \\ \sigma_v \mathbf{v} \rho_{2,1} & \sigma_v \mathbf{v} \sqrt{1 - \rho_{2,1}^2} \end{pmatrix}, \tag{2.4.20}$$

in which case we calculate

$$c \cdot \left(\mathbb{I}_2 - \sigma_s^{\mathrm{T}} \cdot \left(\sigma_s \cdot \sigma_s^{\mathrm{T}} \right)^{-1} \cdot \sigma_s \right) = \begin{pmatrix} 0 & 0 \\ 0 & \mathbf{v} \sigma_v \sqrt{1 - \rho_{2,1}^2} \end{pmatrix}, \tag{2.4.21}$$

$$c \cdot \left(\mathbb{I}_2 - \sigma_s^{\mathrm{T}} \cdot \left(\sigma_s \cdot \sigma_s^{\mathrm{T}} \right)^{-1} \cdot \sigma_s \right) \cdot c^{\mathrm{T}} = \begin{pmatrix} 0 & 0 \\ 0 & \mathbf{v}^2 \sigma_v^2 \left(1 - \rho_{2,1}^2 \right) \end{pmatrix}. \tag{2.4.22}$$

So the market is incomplete if and only if $\sigma_v^2 \left(1 - \rho_{2,1}^2 \right) \neq 0$.

2.4.5 *Stochastic Interest Rates: Vasicek Model*

The market completeness can be disturbed in various ways. One of them is by accounting for the randomness of the interest rates (see [5, 7, 12, 16, 23, 32, 43, 46]). Here we embed the very well known Vasicek model [46] in the above simple-economy framework.

Consider an economy with one tradable $S = \{S_1\}$, and a money market with a *stochastic* interest rate $r(t)$, obeying the Itô SDE system

$$\begin{aligned} \mathrm{d}S_1(t) &= S_1(t)((\mathbf{a}_0 + \beta r(t) - \mathbb{D}_1)\,\mathrm{d}t + \mathbf{p}_1 \mathrm{d}B_1(t)), \\ \mathrm{d}r(t) &= (q_0 + q_1 r(t))\,\mathrm{d}t + w \mathrm{d}\mathbb{B}_2(t), \end{aligned} \tag{2.4.23}$$

where $\mathrm{d}B_1(t)\mathrm{d}\mathbb{B}_2(t) = \rho_{2,1}\mathrm{d}t$. We consider two cases.

In the first case, if $A = \{r\}$, we have $\mathcal{A} = (-\infty, \infty)$, and the market coefficients are equal to

$$a_s = \{\mathbf{a}_0 + \beta r\}, \; b = \{q_0 + q_1 r\}, \; \sigma_s = (\mathbf{p}_1 \; 0), \; c = \left(w\rho_{2,1} \; w\sqrt{1 - \rho_{2,1}^2} \right). \tag{2.4.24}$$

We calculate

$$c \cdot \left(\mathbb{I}_2 - \sigma_s^{\mathrm{T}} \cdot \left(\sigma_s \cdot \sigma_s^{\mathrm{T}} \right)^{-1} \cdot \sigma_s \right) = \left(0 \; w\sqrt{1 - \rho_{2,1}^2} \right), \tag{2.4.25}$$

$$c \cdot \left(\mathbb{I}_2 - \sigma_s^{\mathrm{T}} \cdot \left(\sigma_s \cdot \sigma_s^{\mathrm{T}} \right)^{-1} \cdot \sigma_s \right) \cdot c^{\mathrm{T}} = \left(w^2 \left(1 - \rho_{2,1}^2 \right) \right). \tag{2.4.26}$$

In the second case, if $A = \{S_1, r\}$, we have $\mathcal{A} = (0, \infty) \times (-\infty, \infty)$, and the corresponding market coefficients become

$$a_s = \{\mathbf{a}_0 + \beta r\}, \; b = \{S_1(\mathbf{a}_0 + \beta r - \mathbb{D}), \; q_0 + q_1 r\},$$
$$\sigma_s = \begin{pmatrix} \mathbf{p}_1 & 0 \end{pmatrix}, \; c = \begin{pmatrix} S_1 \mathbf{p}_1 & 0 \\ w\rho_{2,1} & w\sqrt{1-\rho_{2,1}^2} \end{pmatrix}, \tag{2.4.27}$$

and this time, we calculate

$$c \cdot \left(\mathbb{I}_2 - \sigma_s^{\mathrm{T}} \cdot \left(\sigma_s \cdot \sigma_s^{\mathrm{T}}\right)^{-1} \cdot \sigma_s\right) = \begin{pmatrix} 0 & 0 \\ 0 & w\sqrt{1-\rho_{2,1}^2} \end{pmatrix}, \tag{2.4.28}$$

$$c \cdot \left(\mathbb{I}_2 - \sigma_s^{\mathrm{T}} \cdot \left(\sigma_s \cdot \sigma_s^{\mathrm{T}}\right)^{-1} \cdot \sigma_s\right) \cdot c^{\mathrm{T}} = \begin{pmatrix} 0 & 0 \\ 0 & w^2\left(1-\rho_{2,1}^2\right) \end{pmatrix}. \tag{2.4.29}$$

So in either case, the conclusion is the same: the market is nonredundant if and only if $\mathbf{p}_1 \neq 0$, and it is complete if and only if $w^2\left(1-\rho_{2,1}^2\right) = 0$.

2.4.6 Stochastic Interest Rates: Cox–Ingersoll–Ross Model

The Cox–Ingersoll–Ross model [7] reads as follows. Consider an economy with one tradable $S(t) = \{S_1(t)\}$, and a money market with a stochastic interest rate $r(t)$, obeying the Itô SDE system

$$\begin{aligned} \mathrm{d}S_1(t) &= S_1(t)((\mathbf{a}_0 + \beta r(t) - \mathbb{D}_1)\,\mathrm{d}t + \mathbf{p}_1 \mathrm{d}B_1(t)), \\ \mathrm{d}r(t) &= (q_0 + q_1 r(t))\,\mathrm{d}t + w\sqrt{r(t)}\mathrm{d}\mathbb{B}_2, \end{aligned} \tag{2.4.30}$$

with $\mathrm{d}B_1(t)\mathrm{d}\mathbb{B}_2(t) = \rho_{2,1}\,\mathrm{d}t$.

If $A = \{r\}$, the state space is equal to $\mathcal{A} = (0, \infty)$, and the corresponding market coefficients are

$$a_s = \{\mathbf{a}_0 + \beta r\}, \; b = \{q_0 + q_1 r\}, \; \sigma_s = \begin{pmatrix} \mathbf{p}_1 & 0 \end{pmatrix}, \; c = \begin{pmatrix} w\sqrt{r}\rho_{2,1} & w\sqrt{r}\sqrt{1-\rho_{2,1}^2} \end{pmatrix}, \tag{2.4.31}$$

and we calculate

$$c \cdot \left(\mathbb{I}_2 - \sigma_s^{\mathrm{T}} \cdot \left(\sigma_s \cdot \sigma_s^{\mathrm{T}}\right)^{-1} \cdot \sigma_s\right) = \begin{pmatrix} 0 & \sqrt{r}w\sqrt{1-\rho_{2,1}^2} \end{pmatrix}, \tag{2.4.32}$$

$$c \cdot \left(\mathbb{I}_2 - \sigma_s^{\mathrm{T}} \cdot \left(\sigma_s \cdot \sigma_s^{\mathrm{T}}\right)^{-1} \cdot \sigma_s\right) \cdot c^{\mathrm{T}} = \left(rw^2\left(1-\rho_{2,1}^2\right)\right), \tag{2.4.33}$$

so that this market is incomplete if and only if $w^2\left(1-\rho_{2,1}^2\right) \neq 0$.

Exercise 2.4.1. Consider the same model by declaring two factors, and derive the same conclusion as above.

2.4.7 Longstaff and Schwartz Interest Rate Model

The Longstaff and Schwartz model [23], which combines stochastic volatility with stochastic interest rates, embedded in the simple-economy framework reads as follows. For example, let $A = \{S_1, x, y\}$, with a tradable $S = \{S_1\}$, and dynamics

$$\begin{aligned} \mathrm{d}S_1(t) &= S_1(t)((\mu x(t) + \theta y(t) - \mathbb{D}_1)\,\mathrm{d}t + \mathbf{p}_1\sqrt{y(t)}\mathrm{d}B_1(t)), \\ \mathrm{d}x(t) &= (\mathbf{a} - \mathbf{b}x(t))\mathrm{d}t + \mathbf{c}\sqrt{x(t)}\mathrm{d}\mathbb{B}_2(t). \\ \mathrm{d}y(t) &= (\mathbf{d} - \mathbf{e}y(t))\mathrm{d}t + \mathbf{f}\sqrt{y(t)}\mathrm{d}\mathbb{B}_3(t), \end{aligned} \tag{2.4.34}$$

where $\mathbf{a}$, $\mathbf{b}$, $\mathbf{c}$, $\mathbf{d}$, $\mathbf{e}$, $\mathbf{f} > 0$, and where $B_1(t)$, $\mathbb{B}_2(t)$, and $\mathbb{B}_3(t)$ are correlated Brownian motions: $\mathrm{d}B_1(t)\mathrm{d}\mathbb{B}_2(t) = \rho_{2,1}\mathrm{d}t$, $\mathrm{d}B_1(t)\mathrm{d}\mathbb{B}_3(t) = \rho_{3,1}\mathrm{d}t$, and $\mathrm{d}\mathbb{B}_2(t)\mathrm{d}\mathbb{B}_3(t) = \rho_{3,2}\,\mathrm{d}t$.

Let the interest rate be equal to $r(t,A) = r(t,S_1,x,y) = r(x,y) = \alpha x + \beta y$, i.e.,

$$r(t) = \alpha x(t) + \beta y(t). \tag{2.4.35}$$

The market coefficients are equal to

$$a_s = \{\mu x + \theta y\},\ b = \{S_1(\mu x + \theta y),\ \mathbf{a} - \mathbf{b}x,\ \mathbf{d} - \mathbf{e}y\},\ \sigma_s = \begin{pmatrix} \mathbf{p}_1\sqrt{y} & 0 & 0 \end{pmatrix},$$

$$c = \begin{pmatrix} \sqrt{y}\ S_1\ \mathbf{p}_1 & 0 & 0 \\ \sqrt{x}\ \mathbf{c}\ \rho_{2,1} & \sqrt{x}\ \mathbf{c}\ \sqrt{1-\rho_{2,1}^2} & 0 \\ \sqrt{y}\ \mathbf{f}\ \rho_{3,1} & \dfrac{\sqrt{y}\ \mathbf{f}\ (\rho_{3,2} - \rho_{2,1}\rho_{3,1})}{\sqrt{1-\rho_{2,1}^2}} & \sqrt{y}\ \mathbf{f}\sqrt{\dfrac{1+2\rho_{2,1}\rho_{3,1}\rho_{3,2} - \left(\rho_{2,1}^2 + \rho_{3,1}^2 + \rho_{3,2}^2\right)}{1-\rho_{2,1}^2}} \end{pmatrix}, \tag{2.4.36}$$

which implies

$$c \cdot \left(\mathbb{I}_3 - \sigma_s^{\mathrm{T}} \cdot \left(\sigma_s \cdot \sigma_s^{\mathrm{T}}\right)^{-1} \cdot \sigma_s\right)$$

$$= \begin{pmatrix} 0 & 0 & 0 \\ 0 & \sqrt{x}\ \mathbf{c}\ \sqrt{1-\rho_{2,1}^2} & 0 \\ 0 & \dfrac{\sqrt{y}\ \mathbf{f}\ (\rho_{3,2} - \rho_{2,1}\rho_{3,1})}{\sqrt{1-\rho_{2,1}^2}} & \sqrt{y}\ \mathbf{f}\sqrt{\dfrac{1+2\rho_{2,1}\rho_{3,1}\rho_{3,2} - \left(\rho_{2,1}^2 + \rho_{3,1}^2 + \rho_{3,2}^2\right)}{1-\rho_{2,1}^2}} \end{pmatrix}, \tag{2.4.37}$$

$$c \cdot \left(\mathbb{I}_3 - \sigma_s^{\mathrm{T}} \cdot \left(\sigma_s \cdot \sigma_s^{\mathrm{T}}\right)^{-1} \cdot \sigma_s\right) \cdot c^{\mathrm{T}}$$

$$= \begin{pmatrix} 0 & 0 & 0 \\ 0 & x\ \mathbf{c}^2\left(1-\rho_{2,1}^2\right) & \sqrt{x}\ \sqrt{y}\ \mathbf{c}\ \mathbf{f}\ (\rho_{3,2}-\rho_{2,1}\rho_{3,1}) \\ 0 & \sqrt{x}\ \sqrt{y}\ \mathbf{c}\ \mathbf{f}(\rho_{3,2}-\rho_{2,1}\rho_{3,1}) & y\ \mathbf{f}^2\left(1-\rho_{3,1}^2\right) \end{pmatrix}, \tag{2.4.38}$$

from which conditions for completeness/incompleteness can easily be deduced. We observe that the state space for this example is equal to $\mathcal{A} = (0, \infty)^3$.

Exercise 2.4.2. Consider the same model by declaring two (appropriately chosen) factors, and derive the same conclusion as above.

2.4.8 Stochastic Interest, Dividend, and Appreciation Rate Model: Part 1

Following [21], the following model, as we shall see later, is surprisingly manageable in spite of its apparent complexity, having three stochastic factors.

Let $S = \{S_1\}$ be the price of a tradable obeying SDE

$$\mathrm{d}S_1(t) = S_1(t)\ (\mathbf{v}(t) + r(t)\ \beta - \mathbb{D}(t))\ \mathrm{d}t + S_1(t)\ \sigma_S \mathrm{d}B_1(t), \tag{2.4.39}$$

where $r(t)$ is the (stochastic) interest rate, $\mathbb{D}(t)$ is the dividend rate, so that $\mathbf{v}(t) + r(t)\beta$ is the predividend appreciation rate. So the factors are $A = \{S_1,\ \mathbf{v},\ r,\ \mathbb{D}\}$, and let $\mathbf{v}(t), r(t),\ \mathbb{D}(t)$ obey SDEs

$$\mathrm{d}\mathbf{v}(t) = \alpha_v\,(\theta_v - \mathbf{v}(t))\,\mathrm{d}t + \sigma_v \mathrm{d}\mathbb{B}_2(t), \tag{2.4.40}$$

$$\mathrm{d}r(t) = \alpha_{\mathbf{r}}\,(\theta_{\mathbf{r}} - r(t))\,\mathrm{d}t + \sigma_{\mathbf{r}} \mathrm{d}\mathbb{B}_3(t), \tag{2.4.41}$$

$$\mathrm{d}\mathbb{D}(t) = \alpha_d\,(\theta_d - \mathbb{D}(t))\,\mathrm{d}t + \sigma_d \mathrm{d}\mathbb{B}_4(t), \tag{2.4.42}$$

where $B_1(t)$, $\mathbb{B}_2(t)$, $\mathbb{B}_3(t)$, and $\mathbb{B}_4(t)$ are correlated Brownian motions.

So the market coefficients are equal to

$$a_s = \{\mathbf{v} + r\beta\},\ \sigma_s = \left(\sigma_1\ 0\ 0\ 0\right), \tag{2.4.43}$$

$$b = \{(\mathbf{v} + r\beta - \mathbb{D})S_1,\ \alpha_v\,(\theta_v - \mathbf{v}),\ \alpha_{\mathbf{r}}\,(\theta_{\mathbf{r}} - r),\ \alpha_d\,(\theta_d - \mathbb{D})\}, \tag{2.4.44}$$

$$c = \begin{pmatrix} S_1\sigma_S & 0 & 0 & 0 \\ \sigma_v\rho_{2,1} & \sigma_v\sqrt{1-\rho_{2,1}^2} & 0 & 0 \\ \sigma_{\mathbf{r}}\rho_{3,1} & \sigma_{\mathbf{r}}\mathbf{s}_{3,2} & \sigma_{\mathbf{r}}\mathbf{s}_{3,3} & 0 \\ \sigma_d\rho_{4,1} & \sigma_d\mathbf{s}_{4,2} & \sigma_d\mathbf{s}_{4,3} & \sigma_d\mathbf{s}_{4,4} \end{pmatrix}, \tag{2.4.45}$$

where $\mathbf{s}_{3,2}$, $\mathbf{s}_{3,3}$, $\mathbf{s}_{4,2}$, $\mathbf{s}_{4,3}$, and $\mathbf{s}_{4,4}$ are given in (A.2.26), (A.2.27), (A.2.29), (A.2.30), and (A.2.31), respectively, which implies

$$\sigma_s \cdot \sigma_s^{\mathrm{T}} = \left(\sigma_S^2\right),$$

$$\mathbb{I}_4 - \sigma_s^{\mathrm{T}} \cdot \left(\sigma_s \cdot \sigma_s^{\mathrm{T}}\right)^{-1} \cdot \sigma_s = \begin{pmatrix} 0 & 0 & 0 & 0 \\ 0 & 1 & 0 & 0 \\ 0 & 0 & 1 & 0 \\ 0 & 0 & 0 & 1 \end{pmatrix}, \tag{2.4.46}$$

and

$$c \cdot \left(\mathbb{I}_4 - \sigma_s^{\mathrm{T}} \cdot \left(\sigma_s \cdot \sigma_s^{\mathrm{T}}\right)^{-1} \cdot \sigma_s\right) \cdot c^{\mathrm{T}}$$

$$= \begin{pmatrix} 0 & 0 & 0 & 0 \\ 0 & \sigma_v^2\left(1-\rho_{2,1}^2\right) & \sigma_v \sigma_{\mathbf{r}} \mathbf{s}_{3,2}\sqrt{1-\rho_{2,1}^2} & \sigma_v \sigma_d \mathbf{s}_{4,2}\sqrt{1-\rho_{2,1}^2} \\ 0 & \sigma_v \sigma_{\mathbf{r}} \mathbf{s}_{3,2}\sqrt{1-\rho_{2,1}^2} & \sigma_{\mathbf{r}}^2 \mathbf{s}_{3,2}^2 + \sigma_{\mathbf{r}}^2 \mathbf{s}_{3,3}^2 & \sigma_d \sigma_{\mathbf{r}} \mathbf{s}_{3,2}\mathbf{s}_{4,2} + \sigma_d \sigma_{\mathbf{r}} \mathbf{s}_{3,3}\mathbf{s}_{4,3} \\ 0 & \sigma_v \sigma_d \mathbf{s}_{4,2}\sqrt{1-\rho_{2,1}^2} & \sigma_d \sigma_{\mathbf{r}} \mathbf{s}_{3,2}\mathbf{s}_{4,2} + \sigma_d \sigma_{\mathbf{r}} \mathbf{s}_{3,3}\mathbf{s}_{4,3} & \sigma_d^2 \mathbf{s}_{4,2}^2 + \sigma_d^2 \mathbf{s}_{4,3}^2 + \sigma_d^2 \mathbf{s}_{4,4}^2 \end{pmatrix}, \tag{2.4.47}$$

from which conditions for completeness/incompleteness can easily be deduced. We observe that the state space for this example is equal to $\mathcal{A} = (0, \infty) \times (-\infty, \infty)^3$.

2.5 Remarks on Data Analysis: Model Fitting

The rest of the chapter is not essential for the reading of this book. It can be skipped altogether, and is provided only to facilitate the practical implementation and empirical verification of the results presented in this book (see, e.g., Example 7.3.2).

2.5.1 *Data Importation and Formatting (Using Mathematica)*

Market data are available, for example at http://finance.yahoo.com/.

Starting at http://finance.yahoo.com, for example, using Get Quotes for ^GSPC, the reader can get historical prices for the S&P 500 index. From there, using the *Download To Spreadsheet* button, one acquires the data in Microsoft Excel Worksheet (.csv) format. If opened and then saved on a computer hard drive (in a file named, say, "SandP500.csv"), the data can be imported into a *Mathematica* session using Import [SandP500.csv,CSV]. Indeed, inside *Mathematica* the data will look like the following (*Warning*: data-providing web sites sometimes change the format of the data; for example, a dash "-" can be changed into slash "/", etc.; in such an event, the code below must be modified):

```
Take[Import["SandP500.csv","CSV"],4]
```

Out[28]=

```
{{Date, Open, High, Low, Close, Volume, Adj Close},
 {2010-09-15, 1119.43, 1126.46, 1114.63, 1125.07,
  3369840000, 1125.07}, {2010-09-14, 1121.16, 1127.36,
  1115.58, 1121.1, 4521050000, 1121.1}, {2010-09-13,
  1113.38, 1123.87, 1113.38, 1121.9,
  4521050000, 1121.9}}
```

The data can then be formatted according to desired specifications. For example, if decimal time (counted in years) and closing prices are extracted from the above data set, then the above, after little *Mathematica* programming

```
MathematicaTime[x_? StringQ,Time_:16]:=
 Module[{y},y=#[[1]]&/@StringPosition[x,"-"];
  Join[ToExpression[StringTake[x,#]]&/@
   {{1,4},{6,7},{9,10}},{Time,0,0}]]

DecimalTime0[x_]:=
 x[[1]]+(FromDate[x]-FromDate[{x[[1]],1,1,0,0,0}])/
   (FromDate[{x[[1]]+1,1,1,0,0,0}]-
    FromDate[{x[[1]],1,1,0,0,0}])//N
DecimalTime[x_]:= DecimalTime0[MathematicaTime[x]]
SandP500Data=Reverse[{DecimalTime[#[[1]]],#[[-1]]}&/@
    Drop[Import[''SandP500.csv'',''CSV''],1]];
```

is transformed into

```
Reverse[Take[SandP500Data, -3]]

{{2009.66, 1028.93},{2009.65, 1030.98}, {2009.65,
 1028.12}}
```

and now it is ready for statistical analysis.

Below we shall use the S&P 500 data to fit the above two models for stochastic volatility. Since volatility is not directly observable in the market, one can estimate the realized volatility quite efficiently using simple statistical estimates (see, e.g., [34]), or one can use ^VIX data, the CBOE (Chicago Board Options Exchange; http://www.cboe.com/) Volatility Index for the S&P 500 index, instead of the estimated volatility (either way making an error). We shall adopt the latter approach here.

One can get the data for ^VIX also from the web site http://finance.yahoo.com, saving it (in a file named, say, "vix.csv"):

```
Take[Import[''vix.csv'', ''CSV''],4]
{{Date, Open, High, Low, Close, Volume, Adj Close},
```

```
{8/28/2009, 24.44, 25.5, 24.28, 24.76, 0, 24.76},
{8/27/2009 , 25.13, 25.89, 24.43, 24.68, 0,24.68},
{8/26/2009, 24.91, 25.56, 24.69, 24.95, 0, 24.95}}
```

Going through the same steps as above for ^GSPC, one arrives at the following (this time also after dividing ^VIX values by 100, since they come in percentage values):

```
VixData=Reverse[{DecimalTime[#[[1]]], #[[5]]/100} &/@
   Drop[Import[''vix.csv'',''CSV''],1]];
Reverse[Take[VixData,-3]]
```

```
{{2009.66,0.2476}, {2009.65,0.2468}, {2009.65,0.2495}}
```

2.5.2 Statistical Analysis

For any of the above models, and the simple-economy framework in general and for any of the derived results that will follow, for such models and results to be useful, the model parameters have somehow to be estimated. Of course, parameter estimation is a whole separate area of research, beyond the scope of this book, but here we just indicate one of the possible approaches to estimation of "static" unobservable parameters (as opposed to dynamic unobservable factors, such as variance $\mathbf{v}(t)$ in the above stochastic volatility models, or some other, more difficult to estimate, parameters suitable for stochastic filtering; see, e.g., [34], and references therein).

Consider two of the above models for stochastic volatility, (2.4.9) and (2.4.16). Set $\mathbb{D}_1 = 0$ in (2.4.9) and (2.4.16). Motivated by

$$\frac{\mathrm{d}B_1(t)}{\sqrt{\mathrm{d}t}}, \frac{\mathrm{d}\mathbb{B}_2(t)}{\sqrt{\mathrm{d}t}} \sim N(0,1) \tag{2.5.1}$$

and

$$\frac{\mathrm{d}B_1(t)}{\sqrt{\mathrm{d}t}}\frac{\mathrm{d}\mathbb{B}_2(t)}{\sqrt{\mathrm{d}t}} = \rho_{2,1}, \tag{2.5.2}$$

equations/models (2.4.9) and (2.4.16) can be rewritten as

$$\begin{aligned}
\frac{\mathrm{d}S_1(t)}{S_1(t)\sqrt{\mathbf{v}(t)}\sqrt{\mathrm{d}t}} - \frac{\mathbf{a}_0 + \lambda_s \mathbf{v}(t)}{\sqrt{\mathbf{v}(t)}}\sqrt{\mathrm{d}t} &= \frac{\mathrm{d}B_1(t)}{\sqrt{\mathrm{d}t}},\\
\frac{\mathrm{d}\mathbf{v}(t)}{\sqrt{\mathbf{v}(t)}\sqrt{\mathrm{d}t}} - \frac{k_v - K_v \mathbf{v}(t)}{\sqrt{\mathbf{v}(t)}}\sqrt{\mathrm{d}t} &= \sigma_v \frac{\mathrm{d}\mathbb{B}_2(t)}{\sqrt{\mathrm{d}t}},
\end{aligned} \tag{2.5.3}$$

and

$$\frac{\mathrm{d}S_1(t)}{S_1(t)\mathbf{v}(t)\sqrt{\mathrm{d}t}} - \frac{\mathbf{a}_0 + \lambda_s \mathbf{v}(t)}{\mathbf{v}(t)}\sqrt{\mathrm{d}t} = \frac{\mathrm{d}B_1(t)}{\sqrt{\mathrm{d}t}},$$

$$\frac{\mathrm{d}\mathbf{v}(t)}{\mathbf{v}(t)\sqrt{\mathrm{d}t}} - \frac{k_v - K_v \mathbf{v}(t)}{\mathbf{v}(t)}\sqrt{\mathrm{d}t} = \sigma_v \frac{\mathrm{d}\mathbb{B}_2(t)}{\sqrt{\mathrm{d}t}},$$

respectively. To estimate parameters $\mathbf{a}_0$ and λ_s, we form quadratic minimizing functionals (assuming that there are n time subintervals, i.e., $n+1$ data points):

$$J_{\text{Heston},1}\left[\mathbf{a}_0, \lambda_s\right] := \sum^{n} \left(\frac{\mathrm{d}S_1(t)}{S_1(t)\sqrt{\mathbf{v}(t)}\sqrt{\mathrm{d}t}} - \frac{\mathbf{a}_0 + \lambda_s \mathbf{v}(t)}{\sqrt{\mathbf{v}(t)}}\sqrt{\mathrm{d}t} \right)^2 \tag{2.5.4}$$

and

$$J_{\text{Non-Heston},1}\left[\mathbf{a}_0, \lambda_s\right] := \sum^{n} \left(\frac{\mathrm{d}S_1(t)}{S_1(t)\mathbf{v}(t)\sqrt{\mathrm{d}t}} - \frac{\mathbf{a}_0 + \lambda_s \mathbf{v}(t)}{\mathbf{v}(t)}\sqrt{\mathrm{d}t} \right)^2 \tag{2.5.5}$$

respectively. Estimating parameters $\mathbf{a}_0$ and λ_s is then reduced to finding minimizers $\mathbf{a}_0^*, \lambda_s^*$, i.e., solving the minimization problems: Find $\mathbf{a}_{H,0}^*,\ \lambda_{H,s}^*$, such that

$$J_{\text{Heston},1}\left[\mathbf{a}_{H,0}^*,\ \lambda_{H,s}^*\right] = \underset{\mathbf{a}_0,\ \lambda_s}{\mathrm{Min}}\, J_{\text{Heston},1}\left[\mathbf{a}_0,\ \lambda_s\right] \tag{2.5.6}$$

in the case of the (2.4.9) model, and $\mathbf{a}_{n-H,0}^*,\ \lambda_{n-H,s}^*$, such that

$$J_{\text{Non-Heston},1}\left[\mathbf{a}_{n-H,0}^*,\ \lambda_{n-H,s}^*\right] = \underset{\mathbf{a}_0, \lambda_s}{\mathrm{Min}}\, J_{\text{Non-Heston},1}\left[\mathbf{a}_0, \lambda_s\right] \tag{2.5.7}$$

in the case of the (2.4.16) model. Similarly, to estimate parameters k_v and K_v, and also as a by-product, the parameter σ_v, we form quadratic functionals

$$J_{\text{Heston},2}\left[k_v, K_v\right] := \sum^{n} \left(\frac{\mathrm{d}\mathbf{v}(t)}{\sqrt{\mathbf{v}(t)}\sqrt{\mathrm{d}t}} - \frac{k_v - K_v \mathbf{v}(t)}{\sqrt{\mathbf{v}(t)}}\sqrt{\mathrm{d}t} \right)^2 \tag{2.5.8}$$

and

$$J_{\text{non-Heston},2}\left[k_v, K_v\right] := \sum^{n} \left(\frac{\mathrm{d}\mathbf{v}(t)}{\mathbf{v}(t)\sqrt{\mathrm{d}t}} - \frac{k_v - K_v \mathbf{v}(t)}{\mathbf{v}(t)}\sqrt{\mathrm{d}t} \right)^2, \tag{2.5.9}$$

respectively. Estimating parameters k_v and K_v is then reduced to finding minimizers k_v^*, K_v^*, i.e., solving the following minimization problems: Find $k_{H,v}^*, K_{H,v}^*$ such that

$$J_{\text{Heston},2}\left[k_{H,v}^*, K_{H,v}^*\right] = \min_{k_v, K_v} J_{\text{Heston},2}\left[k_v, K_v\right] \tag{2.5.10}$$

in the case of the (2.4.9) model, and $k_{n-H,v}^*, K_{n-H,v}^*$ such that

$$J_{\text{non-Heston},2}\left[k_{n-H,v}^*, K_{n-H,v}^*\right] = \min_{k_v^*, K_v^*} J_{\text{non-Heston},2}\left[k_v^*, K_v^*\right] \tag{2.5.11}$$

in the case of the (2.4.16) model. We observe that once the above minimizations (2.5.10) and (2.5.11) are performed, then the estimates for σ_v are obtained as

$$\sigma_{H,v}^* = \sqrt{\frac{1}{n-2} J_{\text{Heston},2}\left[k_{H,v}^*, K_{H,v}^*\right]} \tag{2.5.12}$$

in the case of the (2.4.9) model, and as

$$\sigma_{n-H,v}^* = \sqrt{}\left(\frac{1}{n-2} J_{\text{non-Heston},2}\left[k_{n-H,v}^*, K_{n-H,v}^*\right]\right) \tag{2.5.13}$$

in the case of the (2.4.16) model. Finally, correlation $\rho_{2,1}$ can be estimated using the just-estimated parameters, together with (2.5.2), (2.5.3), and (2.5.4), by means of averaging the expression on the right-hand side of the equation (evaluated on the data)

$$\begin{aligned}\rho_{H,2,1} = \frac{\mathrm{d}B_1(t)}{\sqrt{\mathrm{d}t}} \frac{\mathrm{d}\mathbb{B}_2(t)}{\sqrt{\mathrm{d}t}} = \frac{1}{\sigma_{H,v}^*} \Bigg(&\frac{\mathrm{d}S_1(t)}{S_1(t)\sqrt{\mathbf{v}(t)}\sqrt{\mathrm{d}t}} \\ &- \frac{\mathbf{a}_{H,0}^* + \lambda_{H,s}^* \mathbf{v}(t)}{\sqrt{\mathbf{v}(t)}} \sqrt{\mathrm{d}t} \Bigg) \left(\frac{\mathrm{d}\mathbf{v}(t)}{\sqrt{\mathbf{v}(t)}\sqrt{\mathrm{d}t}} - \frac{k_{H,v}^* - K_{H,v}^* \mathbf{v}(t)}{\sqrt{\mathbf{v}(t)}} \sqrt{\mathrm{d}t} \right)\end{aligned} \tag{2.5.14}$$

in the case of the (2.4.9) model, and

$$\begin{aligned}\rho_{n-H,2,1} &= \frac{\mathrm{d}B_1(t)}{\sqrt{\mathrm{d}t}} \frac{\mathrm{d}\mathbb{B}_2(t)}{\sqrt{\mathrm{d}t}} \\ &= \frac{1}{\sigma_{n-H,v}^*} \left(\frac{\mathrm{d}S_1(t)}{S_1(t)\mathbf{v}(t)\sqrt{\mathrm{d}t}} - \frac{\mathbf{a}_{n-H,0}^* + \lambda_{n-H,s}^* \mathbf{v}(t)}{\mathbf{v}(t)} \sqrt{\mathrm{d}t} \right) \\ &\quad \times \left(\frac{\mathrm{d}\mathbf{v}(t)}{\mathbf{v}(t)\sqrt{\mathrm{d}t}} - \frac{k_{n-H,v}^* - K_{n-H,v}^* \mathbf{v}(t)}{\mathbf{v}(t)} \sqrt{\mathrm{d}t} \right)\end{aligned} \tag{2.5.15}$$

in the case of the (2.4.16) model.

2.5.3 Implementation and Some Empirical Results

To fix ideas, consider the last 1000 data points:

```
n=1000; SP=Take[SandP500Data,-n]; VIX=Take[VixData,-n];
```

Prepare the data:

```
dt = Drop[Transpose[SP][[1]],1]-Drop[Transpose[SP][[1]],-1];
S = Drop[Transpose[SP][[2]],-1];dS=Drop[Transpose[SP][[2]],1]- S;
var = Drop[Transpose[VIX][[2]],-1]^2;
dvar = Drop[Transpose[VIX][[2]]^2,1]- var;
vol = Drop[Transpose[VIX][[2]],-1];
dvol = Drop[Transpose[VIX][[2]],1]- vol;
```

Then implement functionals (2.5.4), (2.5.5), (2.5.8), and (2.5.9):

$$J_{\text{Heston},1}[a0\_,\lambda s\_] = \text{Plus@@}\left(\frac{dS}{S\,\text{Sqrt}[var]\,\text{Sqrt}[dt]} - \frac{a0+\lambda s\,var}{\text{Sqrt}[var]}\text{Sqrt}[dt]\right)^2;$$

$$J_{\text{Heston},2}[kv\_,Kv\_] = \text{Plus@@}\left(\frac{dvar}{\text{Sqrt}[var]\,\text{Sqrt}[dt]} - \frac{kv-Kv\,var}{\text{Sqrt}[var]}\text{Sqrt}[dt]\right)^2;$$

$$J_{\text{Non-Heston},1}[a0\_,\lambda s\_] = \text{Plus@@}\left(\frac{dS}{S\,vol\,\text{Sqrt}[dt]} - \frac{a0+\lambda s\,vol}{vol}\text{Sqrt}[dt]\right)^2;$$

$$J_{\text{Non-Heston},2}[kv\_,Kv\_] = \text{Plus@@}\left(\frac{dvol}{vol\,\text{Sqrt}[dt]} - \frac{kv-Kv\,vol}{vol}\text{Sqrt}[dt]\right)^2;$$

Find their minima and corresponding minimizers:

```
FM1 = FindMinimum[J_Heston,1[a0, λs],{{a0,0},{λs,0}}]
```

{945.405,{a0 → 0.0149222,λs → −0.399536}}

```
FM2 = FindMinimum[J_Heston,2[kv,Kv],{{kv,0},{Kv,0}}]
```

{598.838,{kv → 0.117255,Kv → 1.5123}}

```
FM3 = FindMinimum[J_Non-Heston,1[a0,λs],{{a0,0},{λs,0}}]
```

{945.256,{a0 → 0.0838414, λs → -0.533182}}

```
FM4 = FindMinimum[J_Non-Heston,2[kv,Kv],{{kv,0},{Kv,0}}]
```

{1594.93,{kv → 0.443958, Kv → 1.70071}}

The estimation problem is almost solved. We observe that the λ_s are negative, while the K_v are positive, ensuring mean reversion in all four SDEs.

We pause to check whether the "diffusion parameters" in the first equations in (2.4.9) and (2.4.16) are close to 1 (they are):

```
{Sqrt[FM1[[1]]/(n-2)], Sqrt[FM3[[1]]/(n-2)]}
```

```
{0.973293,0.973216}
```

and then we estimate the "diffusion parameters" in the second equations in (2.4.9) and (2.4.16):

```
{σH,est,σn-H,est} = {Sqrt[FM2[[1]]/(n-2)],Sqrt[FM4[[1]]/(n-2)]}
```

```
{0.774621, 1.26417}
```

Finally, for the purpose of estimating correlations (via (2.5.2), i.e., (2.5.14) and (2.5.15), and for the sake of postestimation analysis/discussion (via (2.5.1), we form the estimation samples using the just-estimated parameters:

```
Sample1 = dS/(S Sqrt[var] Sqrt[dt]) - (a0 + λs var)/Sqrt[var] Sqrt[dt]/.FM1[[2]];

Sample2=dvar/(Sqrt[var] Sqrt[dt]) - (kv - Kv var)/Sqrt[var] Sqrt[dt]/.FM2[[2]];

Sample3=dS/(S vol Sqrt[dt]) - (a0 + λs vol)/vol Sqrt[dt]/.FM3[[2]];

Sample4=dvol/(vol Sqrt[dt]) - (kv - Kv vol)/vol Sqrt[dt]/.FM4[[2]];
```

Then the correlations $\rho_{2,1}$ are estimated (using (2.5.14) and (2.5.15), respectively) as

```
{Mean[1/σH,est Sample1 Sample2], Mean[1/σn-H,est Sample3 Sample4]}
```

```
{-0.690619, -0.815187}
```

Samples 1–4, if the models are correct, should be samples of independent $N(0,1)$ random variables. Are they? To address this question, we consider the corresponding histograms, compared with the $N(0.1)$ histogram:

```
Needs["Histograms`"];

Off[Histogram: :"ltail", Histogram: :"ltail1",
Histogram: : "rtail"];

testcase=
 Histogram[RandomReal[NormalDistribution [0,1],
```

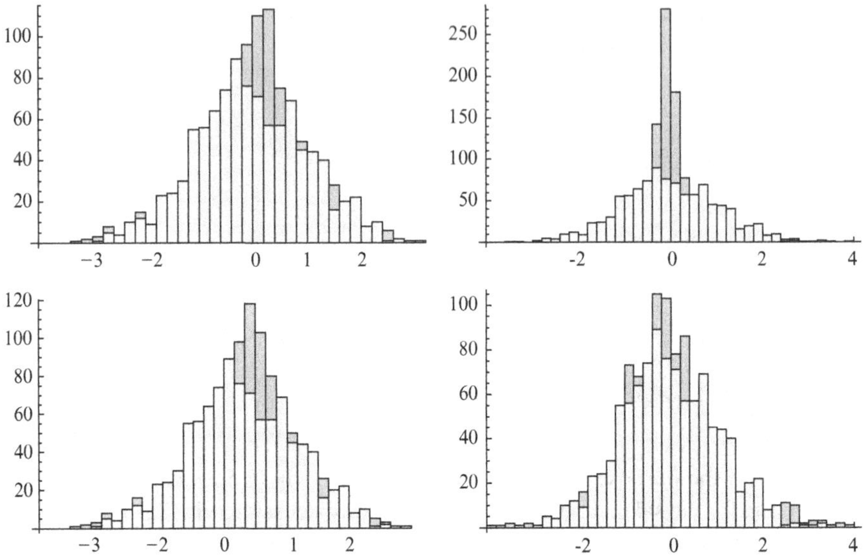

Fig. 2.1 Histogram: Checking for the model randomness normality

```
{Length[Sample1]}],
  HistogramCategories → Range[-4, 4, .2]] /.
 RGBColor[__]→ LightPink; GraphicsArray[
Partition[Show[Histogram[#, HistogramCategories →
  Range[-4,4,.2]],testcase] &/@{Sample1, Sample2,
  Sample3, Sample4}, 2]]
```

In Fig. 2.1 we compare a true $N(0,1)$ sample, with the empirical ones: Samples 1–4, as defined above. Sample 2, which, according to (2.5.1), corresponds to the "Brownian motion" that drives the variance SDE in Heston's model, i.e., the second SDE in (2.4.9), seem to fail the "naked-eye histogram test." The other three samples appear to conform to the $N(0,1)$ distribution somewhat better. The project below encourages much more empirical investigation.

Exercise 2.5.1. Use the above estimation procedure to estimate parameters for both stochastic volatility models (2.4.9) and (2.4.16), for the US market (using ^GSPC and ^VIX data). Use the data for the last 1, 2, and 5 years.

Exercise 2.5.2. Repeat the analysis with Shanghai's SSE Composite Index (000001.SS) data. You may need to estimate the volatility instead of using the volatility index, since there isn't one.

Exercise 2.5.3. Repeat the analysis with the Hong Kong's Hang Seng Index (^HSI) data. You may wish to estimate the volatility instead of using a volatility index.

Exercise 2.5.4. Compare the two stochastic volatility models: which one describes the data better in different markets?

Chapter 3
Investment Portfolio Optimization

Portfolio Optimization without and with Affine Constraints, for General, CRRA, and CARA Utilities

3.1 Introduction

Optimal portfolio theory, by its goal (how to invest in an optimal fashion), by its level of development, and recently also by its implications in other major areas of mathematical finance (pricing and hedging in incomplete markets), is certainly one of its most important areas of research. The essence of finance is investing, and the quantitative framework for optimal investing is optimal portfolio theory.

Modern optimal portfolio theory was initiated by the seminal work of R.C. Merton in the late 1960s and early 1970s (see [25, 26, 28]) as an application and outgrowth of stochastic control theory. Since then, as we shall see in this chapter, many advances were made (see, e.g., [22, 24, 34, 35, 38]). In particular, advances were made in extending Merton's results in the context of non-Log-Normal, yet Markovian markets and, which is very important, under complete market-coefficients information and factor-observability. We remark that due to the limited applicability of the stochastic control "separability principle," portfolio optimization under model-coefficient uncertainty and/or factor nonobservability is a much more difficult and as yet unsettled subject (in the case of nonlogarithmic utilities). Indeed, how much more complicated this problem is can be gleaned, for example, from [34], and much more so from [38]. In this book, however, we shall always assume, when discussing portfolio optimization, and consequently also when discussing pricing and hedging, complete market-coefficient knowledge and factor observability.

We begin with a discussion of the utility of wealth functions, general utilities, CRRA utilities, CARA utilities, and related topics. The rest of the chapter is for the most part a comprehensive study of optimal portfolio theory for simple economies under compete model information and factor observability, and under various kinds of utilities, without and with constraints on the portfolio. Results of this chapter, like most of the results in this book, will be first derived for an arbitrary utility of wealth, and only then specialized to more convenient ones.

S. Stojanovic, *Neutral and Indifference Portfolio Pricing, Hedging and Investing: With Applications in Equity and FX*, DOI 10.1007/978-0-387-71418-9_3,

3.2 Utility of Wealth

Utility theory, utility of money, and expected utility theory are studied quite deeply in economics. Here we shall only introduce and discuss some utility functions that will be used throughout the rest of this work.

Throughout, $X \in (-\infty, \infty)$ will always denote wealth (for better or worse, in some models/situations it is allowed for wealth to be negative), which is the cumulative value of a portfolio, consisting of the values of various investments, cash, or a money market account. The necessity to give a mathematical description of the utility of wealth is both, financial/economic and mathematical. It is financial/economic since, as is well known to economists but also to every investor, the pain experienced from financial loss is much more intense than the pleasure experienced from financial gain. It is mathematical because many of the problems to be considered here are in fact impossible to solve in that they are not well posed in the natural general setting unless one considers utility of wealth as a part of the problem formulation.

Utility of wealth is quantified via a *wealth utility* function, or *utility of wealth*, or just *utility*, denoted by $\psi(X)$, which is a real-valued increasing function, i.e., $\partial\psi(X)/\partial X > 0$. Since there is asymmetry between gains and losses, it is also necessary to assume strict concavity: $\partial^2\psi(X)/\partial X^2 < 0$. The domain of ψ will be denoted by $\mathcal{X} = \mathrm{dom}(\psi)$.

For any wealth utility, one can define the notion of *relative risk-aversion*, to be denoted by γ or $\mathcal{G}$, depending on whether it is considered a parameter or as the (Arrow–Pratt) differential operator

$$\mathcal{G}(\psi)(X) := -X \frac{\partial^2\psi(X)}{\partial X^2} \Big/ \frac{\partial\psi(X)}{\partial X}, \tag{3.2.1}$$

and one can define the notion of *absolute risk-aversion*, to be denoted by ω or Ω, depending on whether it is considered a parameter or as the (Arrow–Pratt) differential operator

$$\Omega(\psi)(X) := -\frac{\partial^2\psi(X)}{\partial X^2} \Big/ \frac{\partial\psi(X)}{\partial X}. \tag{3.2.2}$$

It will also be natural to require that the wealth utility be such that its relative risk-aversion is increasing in X: as wealth increases, the investor becomes more risk-averse (or at least does not become less risk-averse), which translates into the condition $\partial\mathcal{G}(\psi)/\partial X \geq 0$.

Summarizing, in regard to a utility of wealth function we shall require

$$X \in \mathcal{X} = \mathrm{dom}(\psi) \wedge \frac{\partial\psi(X)}{\partial X} > 0 \wedge \frac{\partial^2\psi(X)}{\partial X^2} < 0 \wedge \frac{\partial\mathcal{G}(\psi)}{\partial X} \geq 0,$$

where $\wedge$ is the logical AND. Later, we shall also need the logical OR, to be denoted by $\vee$.

So, for example, solving the ODE

$$\mathcal{G}(\psi)(X) = \gamma, \tag{3.2.3}$$

we get

$$\psi_\gamma(X) = \frac{X^{1-\gamma} c_1}{1-\gamma} + c_2 \tag{3.2.4}$$

for $\gamma \neq 1$, and $\psi_1(X) = c_1 \log(X) + c_2$ in the case $\gamma = 1$, with conditions

$$\left(X \in \mathcal{X} = \operatorname{dom}(\psi_\gamma) \wedge \frac{\partial \psi_\gamma(X)}{\partial X} > 0 \wedge \frac{\partial^2 \psi_\gamma(X)}{\partial X^2} < 0 \wedge \frac{\partial \mathcal{G}(\psi_\gamma)}{\partial X} \geq 0 \right)$$
$$= (X > 0 \wedge c_1 > 0 \wedge \gamma > 0). \tag{3.2.5}$$

We choose $c_1 = 1$ and $c_2 = 0$ in the $\gamma = 1$ case, and $c_1 = 1$ and $c_2 = -\frac{1}{1-\gamma}$ in (3.2.4), obtaining the family of *constant relative risk-aversion* (CRRA) utility functions

$$\psi_\gamma(X) = \frac{X^{1-\gamma} - 1}{1-\gamma} \tag{3.2.6}$$

for $\gamma > 0, \gamma \neq 1$, and $\psi_1(X) = \lim_{\gamma \to 1} \frac{X^{1-\gamma}-1}{1-\gamma} = \log(X)$, i.e.,

$$\psi_\gamma(X) = \begin{cases} \frac{X^{1-\gamma}-1}{1-\gamma}, & \gamma > 0 \wedge \gamma \neq 1, \\ \log(X), & \gamma = 1, \end{cases} \tag{3.2.7}$$

for $X \in \mathcal{X} = (0, \infty)$. We observe, by setting $\gamma = 0$, that $\psi_0(X) = X - 1$. The greater the value of γ, the more risk-averse the CRRA utility functions become.

Alternatively, solving the ODE

$$\Omega(\psi)(X) = \omega, \tag{3.2.8}$$

we get

$$\psi_{C,\omega}(X) = -\frac{e^{-X\omega} c_1}{\omega} + c_2 \tag{3.2.9}$$

for $\omega \neq 0$, and

$$\psi_{C,0}(X) = c_1 X + c_2 \tag{3.2.10}$$

for $\omega = 0$, with conditions

$$\left(X \in \mathcal{X} = \operatorname{dom}(\psi_{C,\omega}) \wedge \frac{\partial \psi_{C,\omega}(X)}{\partial X} > 0 \wedge \frac{\partial^2 \psi_{C,\omega}(X)}{\partial X^2} < 0 \wedge \frac{\partial \mathcal{G}(\psi_{C,\omega})}{\partial X} \geq 0 \right)$$
$$= (c_1 > 0 \wedge \omega > 0). \tag{3.2.11}$$

Imposing also $\psi_0(X) = \psi_{C,0}(X)$, we choose $c_1 = 1$ and $c_2 = -1$ in (3.2.10), and then $c_1 = \mathrm{e}^{\omega}$ and $c_2 = \frac{1}{\omega}$ in (3.2.9), getting the family of *constant absolute* risk-aversion, or CARA, utility functions:

$$\psi_{C,\omega}(X) = \frac{1-\mathrm{e}^{\omega(1-X)}}{\omega} \tag{3.2.12}$$

for $\omega > 0$, for $X \in \mathcal{X} = \mathrm{dom}\,(\psi_{C,\omega}) = (-\infty,\infty)$. Higher the ω, the more risk averse the CARA utility functions are. Notice that

$$\mathcal{G}\,(\psi_{C,\omega})\,(X) = X\omega \tag{3.2.13}$$

Remark 3.2.1. The CRRA and CARA utilities can also be derived or at least approximated from a class of utility functions called HARA (hyperbolic absolute risk-aversion) class. HARA utilities are as tractable as CARA (i.e., a little bit less so than the CRRA). So, precisely, let

$$\psi_{H,\alpha,\gamma,\omega,\eta}(X) := \alpha(\eta+\omega X)^{1-\gamma} \tag{3.2.14}$$

with conditions

$$\begin{aligned}
&\left(X \in \mathcal{X} = \mathrm{dom}\,(\psi_{H,\alpha,\gamma,\omega,\eta}) \wedge \frac{\partial \psi_{H,\alpha,\gamma,\omega,\eta}(X)}{\partial X} > 0 \wedge \frac{\partial^2 \psi_{H,\alpha,\gamma,\omega,\eta}(X)}{\partial X^2}\right.\\
&\qquad \left. < 0 \wedge \frac{\partial \mathcal{G}\,(\psi_{H,\alpha,\gamma,\omega,\eta})}{\partial X} \geq 0\right)\\
&\quad = \left(\omega < 0 \wedge \eta \geq 0 \wedge \alpha < 0 \wedge X < -\frac{\eta}{\omega} \wedge \gamma < 0\right)\\
&\quad \vee \Bigg(\omega > 0 \wedge \eta \geq 0 \wedge\\
&\quad \times \left(\left(\alpha < 0 \wedge X > -\frac{\eta}{\omega} \wedge \gamma > 1\right) \vee \left(\alpha > 0 \wedge X > -\frac{\eta}{\omega} \wedge 0 < \gamma < 1\right)\right)\Bigg).
\end{aligned} \tag{3.2.15}$$

In (3.2.15), the first parameter alternative $(\omega < 0 \wedge \eta \geq 0 \wedge \alpha < 0 \wedge X < -\frac{\eta}{\omega} \wedge \gamma < 0)$ corresponds to the family of HARA utilities capable of approximating the CARA utility. Indeed, setting $\eta = -\gamma \to +\infty$, we obtain

$$\mathcal{G}\,(\psi_{H,\alpha,\gamma,\omega,\eta})\,(X) = \frac{X\gamma\omega}{\eta+X\omega} = \frac{X\gamma\omega}{-\gamma+X\omega} \to -\omega X \tag{3.2.16}$$

for $X < -\eta/\omega = \gamma/\omega$, where $\gamma/\omega > 0$ and $\gamma/\omega \to \infty$. Comparing (3.2.13) and (3.2.16), we see that the parameter ω in (3.2.13), which is the absolute risk-aversion parameter, corresponds to the parameter $-\omega$ in (3.2.16).

The second parameter alternative in (3.2.15), $(\omega > 0 \wedge \eta \geq 0 \wedge ((\alpha < 0 \wedge X > -\frac{\eta}{\omega} \wedge \gamma > 1) \vee (\alpha > 0 \wedge X > -\frac{\eta}{\omega} \wedge 0 < \gamma < 1)))$, corresponds to the family of HARA utilities capable of realizing the CRRA utility (for $X > 0$). Indeed, setting $\eta = 0$, we get

$$\mathcal{G}\left(\psi_{H,\alpha,\gamma,\omega,\eta}\right)(X) = \frac{X\gamma\omega}{\eta + X\omega} = \frac{X\gamma\omega}{X\omega} = \gamma \tag{3.2.17}$$

for $X > -\eta/\omega = 0$. What is not available, unfortunately, is a HARA parameter family that would cover both CRRA and CARA, offering the possibility of "interpolating" between the two of them and combining the relative and absolute risk-aversion investment attitudes.

Remark 3.2.2. To "interpolate" between the CRRA and CARA utilities, we can, for example, consider yet another class of utility functions that is constructed by solving the ODE

$$\mathcal{G}(\psi)(X) = \gamma + X\omega \tag{3.2.18}$$

($\mathcal{G}$ is a nonlinear operator; even though equation (3.2.18) can be rewritten as a linear homogeneous equation, the solution-dependence on the right-hand side of (3.2.18) is nonlinear). We get

$$\psi_{\gamma,\omega}(X) = -X^{1-\gamma}(X\omega)^{\gamma-1}c_1\Gamma(1-\gamma, X\omega) + c_2, \tag{3.2.19}$$

where Γ is the *upper incomplete gamma function*, or *plica function*, given by $\Gamma(a,z) := \int_z^\infty t^{a-1}\mathrm{e}^{-t}\mathrm{d}t$, with conditions

$$\left(X \in \operatorname{dom}\left(\psi_{\gamma,\omega}\right) \wedge \frac{\partial \psi_{\gamma,\omega}(X)}{\partial X} > 0 \wedge \frac{\partial^2 \psi_{\gamma,\omega}(X)}{\partial X^2} < 0 \wedge \frac{\partial \mathcal{G}\left(\psi_{\gamma,\omega}\right)}{\partial X} \geq 0\right)$$
$$= (X > 0 \wedge c_1 > 0 \wedge \omega > 0 \wedge \gamma > 0). \tag{3.2.20}$$

We can choose c_1 and c_2 such that $\psi_{\gamma,\omega}$ is consistent with both ψ_γ and $\psi_{C,\omega}$. Indeed, choosing $c_1 = \mathrm{e}^\omega$ and $c_2 = \mathrm{e}^\omega \omega^{-1+\gamma}\Gamma(1-\gamma,\omega)$, we get

$$\psi_{\gamma,\omega}(X) = \frac{1}{\omega}\left(\mathrm{e}^\omega \omega^\gamma \Gamma(1-\gamma,\omega) - \mathrm{e}^\omega X^{-\gamma}(X\omega)^\gamma \Gamma(1-\gamma, X\omega)\right), \tag{3.2.21}$$

so that

$$\psi_{0,\omega}(X) = \frac{1-\mathrm{e}^{\omega(1-X)}}{\omega} = \psi_{C,\omega}(X). \tag{3.2.22}$$

Summarizing, we have

$$\psi_{\gamma,\omega}(X) = \begin{cases} \frac{1}{\omega}\left(\mathrm{e}^\omega \omega^\gamma \Gamma(1-\gamma,\omega) - \mathrm{e}^\omega X^{-\gamma}(X\omega)^\gamma \Gamma(1-\gamma,X\omega)\right), & \omega > 0, \\ \frac{X^{1-\gamma}-1}{1-\gamma}, & \omega = 0 \wedge \gamma > 0 \wedge \gamma \neq 1, \\ \log(X), & \omega = 0 \wedge \gamma = 1, \end{cases} \tag{3.2.23}$$

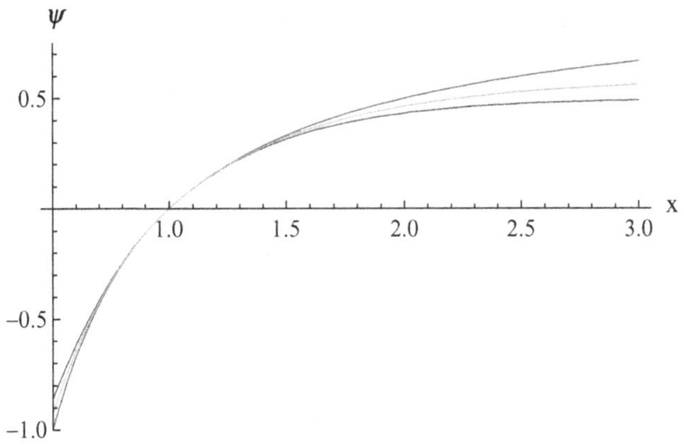

Fig. 3.1 Wealth Utility Functions: CRRA, CARA, and their interpolation

which covers both CRRA and CARA utilities, but also an "interpolation" between the two. In Fig. 3.1, we plot $\psi_2(X) = \psi_{2,0}(X)$, $\psi_{C,2}(X) = \psi_{0,2}(X)$, and $\psi_{1,1}(X)$ in the middle.

Unfortunately, this utility is not as tractable as CRRA, CARA, or HARA, and using it, when possible, would involve numerical computations, i.e., numerical solutions of some complicated PDEs.

3.3 Investment Portfolio Optimization Problems

Consider a simple economy $\mathfrak{E}$, as in Definition 2.2.1, with k tradables $S(t) = \{S_1(t),\ldots,S_k(t)\}$ and m (economic) factors $A(t) = \{A_1(t),\ldots,A_m(t)\}$, and assume the market dynamics (2.2.1)–(2.2.2). Consider also an investor who maintains, by means of continuous trading, a *trading strategy*

$$\Pi(t,X,A) = \{\Pi_1(t,X,A),\ldots,\Pi_k(t,X,A)\}, \tag{3.3.1}$$

where $\Pi_j(t,X,A) \in (-\infty,\infty)$ denotes the monetary value of the investment in the jth considered tradable. If $\Pi_j(t,X,A) > 0$, such a position is referred to as a *long* position, while if $\Pi_j(t,X,A) < 0$, such a position is a *short* position. In addition to portfolio $\Pi(t,X,A)$, the investor has at his/her disposal cash or a money market account for the financing of the trades, and that also pays interest at the rate $r(t,A(t))$. Since $X = X(t)$ denotes the total wealth (e.g., the cumulative value of the brokerage account), the available cash is equal to $X(t) - \sum_{i=1}^{k} \Pi_i(t,X(t),A(t)) \in (-\infty,\infty)$. Denote by $X^{\Pi}(t)$ the wealth at time t corresponding to a particular trading strategy Π. Then, as is well known (see, e.g., [34]),

$$\begin{aligned}\mathrm{d}X^{\Pi}(t) = (\Pi(t,X(t),A(t)) \cdot (a_s(t,A(t)) - r(t,A(t))) \\ + r(t,A(t))X(t))\,\mathrm{d}t + \Pi(t,X(t),A(t)) \cdot \sigma_s(t,A(t)) \cdot \mathrm{d}B(t). \qquad (3.3.2)\end{aligned}$$

This equation will be of fundamental importance for all the results to follow. Equations (2.2.2) and (3.3.2) form a (closed) Markovian stochastic system to be controlled (for general stochastic control theory, we refer to [19]). Various stochastic control problems in regard to the system (2.2.2) and (3.3.2) can be considered, and usually they are interpreted as *portfolio optimization* problems. As mentioned before, such problems arose in the seminal works of R.C. Merton ([25, 26], see also [28]). They can be cast, as we shall do, as problems of maximization of the expected utility of the terminal wealth.

Problem 3.3.1. For a given utility function ψ, find $\varphi(t,X,A)$ and $\Pi^{\star}$ such that

$$\varphi(t,X,A) = \sup_{\Pi} E_{t,X,A}\psi\left(X^{\Pi}(T)\right) = E_{t,X,A}\psi\left(X^{\Pi^{\star}}(T)\right). \qquad (3.3.3)$$

Problem 3.3.2. For a given $(l \times k)$-matrix-valued function $\mu(t,X,A)$, an l-vector-valued function $\xi(t,X,A)$, and a utility function ψ, find $\varphi(t,X,A)$ and $\Pi^{\star}$ such that the *affine constraint on the portfolio*

$$\mu(t,X,A) \cdot \Pi^{\star}(t,X,A) = \xi(t,X,A) \qquad (3.3.4)$$

holds, and such that

$$\varphi(t,X,A) = \sup_{\Pi:\mu\cdot\Pi=\xi} E_{t,X,A}\psi\left(X^{\Pi}(T)\right) = E_{t,X,A}\psi\left(X^{\Pi^{\star}}(T)\right). \qquad (3.3.5)$$

The functions φ in (3.3.3) and (3.3.3) are called *value functions*, while $\Pi^{\star}$ is called the *optimal portfolio strategy*.

Remark 3.3.1. In both Problem 3.3.1 and Problem 3.3.2, it is assumed that $T < \infty$. Later, we shall see whether it is possible, and if so, how, to send $T \to \infty$. In some problems, sending $T \to \infty$ is quite natural. For money managers interested in running a fund indefinitely there are no natural T's when the investment performance is assessed. Does the theory presented for $T < \infty$ apply? We shall see that it does. Also, in pricing theory, where optimal portfolio theory is going to be of fundamental importance, some financial contracts are perpetual – they do not expire. In such problems it is not a simplification but rather the exact problem to send $T \to \infty$. Finally, for many problems in which T is finite but not small, it will be a useful simplification of the problem to send $T \to \infty$.

Remark 3.3.2. A variation of these problems is to consider optimal intertemporal consumption, but for our purposes the above terminal wealth problems will suffice.

Example 3.3.1. To fix ideas, let $k = 2$. Set $\mu(t,X,A) = \begin{pmatrix} 1 & 1 \end{pmatrix}, \xi(t,X,A) = \{X/2\}$. Then the constraint (3.3.4) reads

$$\left\{ \sum_{k=1}^{2} \Pi_k^\star(t,X,A) \right\} = \{\alpha X\} \tag{3.3.6}$$

– the sum of two portfolio investments is equal to the half of the wealth, while the other half is in the cash account. The portfolio is then optimized.

Example 3.3.2. Let $k = 2$. Set $\mu(t,X,A) = \begin{pmatrix} 0 & 1 \end{pmatrix}, \xi(t,X,A) = \{\kappa V(t,X,A)\}$, where $V(t,X,A)$ is the market price of some financial contract. Then the constraint (3.3.4) reads $\{\Pi_2^\star(t,X,A)\} = \{\kappa V(t,X,A)\}$, that is, κ contracts priced at $V(t,X,A)$ are held as a second item in the portfolio. The rest of the portfolio, i.e., the first item in the portfolio, is then optimized.

Remark 3.3.3. We solve optimal portfolio problems in two steps. The first step is to derive the Monge–Ampère-type PDE, which holds no matter what kind of utility function is used, including (3.3.23), for example. The second step, to be done later, is to specialize in the cases of CRRA and CARA utilities.

3.4 Derivation of the Optimal Portfolio Formula and of the Monge–Ampère-type PDE for the Value Function: No Constraints on the Portfolio

The following fully nonlinear differential operator will be used in several important results.

Definition 3.4.1. For a given simple economy $\mathfrak{E}$, let $\mathcal{M}$ always denote the corresponding *differential operator of Monge–Ampère type*, defined by the formula

$$\begin{aligned}\mathcal{M}(\varphi) := {} & \frac{\partial^2\varphi}{\partial X^2}\frac{\partial\varphi}{\partial t} + \frac{1}{2}\frac{\partial^2\varphi}{\partial X^2}\mathrm{Tr}\left(c\cdot c^{\mathrm{T}}\cdot\nabla_A\nabla_A\varphi\right) - \frac{1}{2}\left(\frac{\partial\varphi}{\partial X}\right)^2 (a_s - r)\cdot\left(\sigma_s\cdot\sigma_s^{\mathrm{T}}\right)^{-1} \\ & \cdot(a_s - r) + (b\cdot\nabla_A\varphi)\frac{\partial^2\varphi}{\partial X^2} - \frac{\partial\varphi}{\partial X}(a_s - r)\cdot\left(\sigma_s\cdot\sigma_s^{\mathrm{T}}\right)^{-1}\cdot\sigma_s\cdot c^{\mathrm{T}}\cdot\left(\nabla_A\frac{\partial\varphi}{\partial X}\right) \\ & - \frac{1}{2}\left(\nabla_A\frac{\partial\varphi}{\partial X}\right)\cdot c\cdot\sigma_s^{\mathrm{T}}\cdot\left(\sigma_s\cdot\sigma_s^{\mathrm{T}}\right)^{-1}\cdot\sigma_s\cdot c^{\mathrm{T}}\cdot\left(\nabla_A\frac{\partial\varphi}{\partial X}\right) + rX\frac{\partial^2\varphi}{\partial X^2}\frac{\partial\varphi}{\partial X},\end{aligned} \tag{3.4.1}$$

where $\varphi = \varphi(t,X,A)$ is any sufficiently differentiable function.

Theorem 3.4.1. *The value function $\varphi(t,X,A)$ given in (3.3.3) is the solution of the fully nonlinear PDE*

$$\mathcal{M}(\varphi) = 0 \tag{3.4.2}$$

for $t < T < \infty, X \in \mathcal{X}, A \in \mathcal{A}$, *together with the terminal condition* $\varphi(T,X,A) = \psi(X)$, *while the optimal portfolio strategy, the solution of Problem 3.3.1, is equal to*

$$\Pi^{\star}(t,X,A) = -\left(1 \Big/ \frac{\partial^2 \varphi}{\partial X^2}\right)\left(\frac{\partial \varphi}{\partial X}(a_s - r) + \left(\nabla_A \frac{\partial \varphi}{\partial X}\right)\cdot c \cdot \sigma_s^{\mathrm{T}}\right)\cdot\left(\sigma_s \cdot \sigma_s^{\mathrm{T}}\right)^{-1}. \tag{3.4.3}$$

Proof. To prove (3.4.2) and (3.4.3), we have, from the Itô chain rule (1.5.1),

$$\begin{aligned} \mathrm{d}\varphi = &\left(\frac{\partial \varphi}{\partial t} + \frac{\partial \varphi}{\partial X}\left(\Pi \cdot (a_s - r) + rX\right) + b \cdot \nabla_A \varphi + \frac{1}{2}\frac{\partial^2 \varphi}{\partial X^2}\Pi \cdot \sigma_s \cdot {\sigma_s}^{\mathrm{T}} \cdot \Pi\right. \\ &\left. + \frac{1}{2}\mathrm{Tr}\left(c \cdot c^{\mathrm{T}} \cdot \nabla_A \nabla_A \varphi\right) + \Pi \cdot \sigma_s \cdot c^{\mathrm{T}} \cdot \left(\nabla_A \frac{\partial \varphi}{\partial X}\right)\right)\mathrm{d}t + \frac{\partial \varphi}{\partial X}\Pi \cdot \sigma_s \cdot \mathrm{d}B(t) \\ &+ \nabla_A \varphi \cdot c \cdot \mathrm{d}B(t) = \mathcal{L}\varphi \mathrm{d}t + \frac{\partial \varphi}{\partial X}\Pi \cdot \sigma_s . \mathrm{d}B(t) + \nabla_A \varphi \cdot c \cdot \mathrm{d}B(t), \end{aligned} \tag{3.4.4}$$

where

$$\begin{aligned} \mathcal{L}\varphi := &\frac{\partial \varphi}{\partial t} + \left(\Pi . (a_s - r) + rX\right)\frac{\partial \varphi}{\partial X} + b \cdot \nabla_A \varphi + \frac{1}{2}\Pi \cdot \sigma_s \cdot {\sigma_s}^{\mathrm{T}} \cdot \Pi \frac{\partial^2 \varphi}{\partial X^2} \\ &+ \Pi \cdot \sigma_s \cdot c^{\mathrm{T}} \cdot \left(\nabla_A \frac{\partial \varphi}{\partial X}\right) + \frac{1}{2}\mathrm{Tr}\left(c \cdot c^{\mathrm{T}} \cdot \nabla_A \nabla_A \varphi\right). \end{aligned} \tag{3.4.5}$$

Therefore, as in any stochastic control problem, the associated Hamilton–Jacobi–Bellman (HJB) PDE becomes

$$\begin{aligned} 0 = \max_{\Pi}[\mathcal{L}\varphi] = &\max_{\Pi}\left[\frac{\partial \varphi}{\partial t} + \frac{\partial \varphi}{\partial X}\left(\Pi . (a_s - r) + rX\right) + b \cdot \nabla_A \varphi\right. \\ &\left. + \frac{1}{2}\frac{\partial^2 \varphi}{\partial X^2}\Pi \cdot \sigma_s \cdot {\sigma_s}^{\mathrm{T}} \cdot \Pi + \frac{1}{2}\mathrm{Tr}\left(c \cdot c^{\mathrm{T}} \cdot \nabla_A \nabla_A \varphi\right) + \Pi \cdot \sigma_s \cdot c^{\mathrm{T}} \cdot \left(\nabla_A \frac{\partial \varphi}{\partial X}\right)\right] \\ = &\frac{\partial \varphi}{\partial t} + \frac{\partial \varphi}{\partial X}rX + b \cdot \nabla_A \varphi + \frac{1}{2}\mathrm{Tr}\left(c \cdot c^{\mathrm{T}} \cdot \nabla_A \nabla_A \varphi\right) + \max_{\Pi(t,X,A)} F(\Pi) \end{aligned} \tag{3.4.6}$$

for

$$F(\Pi) := \Pi . (a_s - r)\frac{\partial \varphi}{\partial X} + \frac{1}{2}\Pi \cdot \sigma_s \cdot \sigma_s^{\mathrm{T}} \cdot \Pi \frac{\partial^2 \varphi}{\partial X^2} + \Pi \cdot \sigma_s \cdot c^{\mathrm{T}} \cdot \left(\nabla_A \frac{\partial \varphi}{\partial X}\right) \tag{3.4.7}$$

together with the terminal condition $\varphi(T,X,A) = \psi(X)$. Since $\sigma_s \cdot \sigma_s^{\mathrm{T}} > 0$, and anticipating $\frac{\partial^2 \varphi}{\partial X^2} < 0$, F being quadratic in Π, in order to maximize F, it suffices to solve equation $\nabla_{\Pi} F(\Pi) = 0$. This yields

$$0 = \nabla_{\Pi} F(\Pi) = (a_s - r)\frac{\partial \varphi}{\partial X} + \sigma_s \cdot \sigma_s^{\mathrm{T}} \cdot \Pi \frac{\partial^2 \varphi}{\partial X^2} + \sigma_s \cdot c^{\mathrm{T}} \cdot \left(\nabla_A \frac{\partial \varphi}{\partial X}\right) \tag{3.4.8}$$

and then

$$\sigma_s \cdot \sigma_s^{\mathrm{T}} \cdot \Pi \frac{\partial^2 \varphi}{\partial X^2} = -(a_s - r)\frac{\partial \varphi}{\partial X} - \sigma_s \cdot c^{\mathrm{T}} \cdot \left(\nabla_A \frac{\partial \varphi}{\partial X}\right), \tag{3.4.9}$$

and therefore the optimal portfolio formula (3.4.3) is proved. Setting $\Pi = \Pi^\star$ in (3.4.6), and multiplying (3.4.6) by $\frac{\partial^2 \varphi}{\partial X^2}$, we get

$$\frac{\partial^2 \varphi}{\partial X^2}\left(\frac{\partial \varphi}{\partial t} + (\Pi^\star \cdot (a_s - r) + rX)\frac{\partial \varphi}{\partial X} + b \cdot \nabla_A \varphi + \frac{1}{2}\Pi^\star \cdot \sigma_s \cdot \sigma_s{}^{\mathrm{T}} \cdot \Pi^\star \frac{\partial^2 \varphi}{\partial X^2} + \Pi^\star \cdot \sigma_s \right.$$
$$\left. \cdot c^{\mathrm{T}} \cdot \left(\nabla_A \frac{\partial \varphi}{\partial X}\right) + \frac{1}{2}\mathrm{Tr}\left(c \cdot c^{\mathrm{T}} \cdot \nabla_A \nabla_A \varphi\right)\right) = 0. \tag{3.4.10}$$

□

To write (3.4.10) explicitly, we prepare the formulas

$$\frac{\partial^2 \varphi}{\partial X^2}\frac{\partial \varphi}{\partial X}\Pi^\star \cdot (a_s - r) = -\left(\frac{\partial \varphi}{\partial X}\right)^2 (a_s - r) \cdot \left(\sigma_s \cdot \sigma_s{}^{\mathrm{T}}\right)^{-1} \cdot (a_s - r)$$
$$-\frac{\partial \varphi}{\partial X}\left(\nabla_A \frac{\partial \varphi}{\partial X}\right) \cdot c \cdot \sigma_s{}^{\mathrm{T}} \cdot \left(\sigma_s \cdot \sigma_s{}^{\mathrm{T}}\right)^{-1} \cdot (a_s - r) \tag{3.4.11}$$

and

$$\frac{1}{2}\left(\frac{\partial^2 \varphi}{\partial X^2}\right)^2 \Pi^\star \cdot \sigma_s \cdot \sigma_s{}^{\mathrm{T}} \cdot \Pi^\star = \frac{1}{2}\left(\frac{\partial \varphi}{\partial X}\right)^2 (a_s - r) \cdot \left(\sigma_s \cdot \sigma_s{}^{\mathrm{T}}\right)^{-1} \cdot (a_s - r)$$
$$+\left(\nabla_A \frac{\partial \varphi}{\partial X}\right) \cdot c \cdot \sigma_s{}^{\mathrm{T}} \cdot \left(\sigma_s \cdot \sigma_s{}^{\mathrm{T}}\right)^{-1} \cdot (a_s - r)\frac{\partial \varphi}{\partial X}$$
$$+\frac{1}{2}\left(\nabla_A \frac{\partial \varphi}{\partial X}\right) \cdot c \cdot \sigma_s{}^{\mathrm{T}} \cdot \left(\sigma_s \cdot \sigma_s{}^{\mathrm{T}}\right)^{-1} \cdot \sigma_s \cdot c^{\mathrm{T}} \cdot \left(\nabla_A \frac{\partial \varphi}{\partial X}\right) \tag{3.4.12}$$

and

$$\frac{\partial^2 \varphi}{\partial X^2}\Pi^\star \cdot \sigma_s \cdot c^{\mathrm{T}} \cdot \left(\nabla_A \frac{\partial \varphi}{\partial X}\right) = -\frac{\partial \varphi}{\partial X}(a_s - r) \cdot \left(\sigma_s \cdot \sigma_s{}^{\mathrm{T}}\right)^{-1} \cdot \sigma_s \cdot c^{\mathrm{T}} \cdot \left(\nabla_A \frac{\partial \varphi}{\partial X}\right)$$
$$-\left(\nabla_A \frac{\partial \varphi}{\partial X}\right) \cdot c \cdot \sigma_s{}^{\mathrm{T}} \cdot \left(\sigma_s \cdot \sigma_s{}^{\mathrm{T}}\right)^{-1} \cdot \sigma_s \cdot c^{\mathrm{T}} \cdot \left(\nabla_A \frac{\partial \varphi}{\partial X}\right). \tag{3.4.13}$$

Adding formulas (3.4.11), (3.4.12), and (3.4.13), we get

$$\frac{\partial^2\varphi}{\partial X^2}\frac{\partial\varphi}{\partial X}\Pi^\star\cdot(a_s-r)+\frac{1}{2}\left(\frac{\partial^2\varphi}{\partial X^2}\right)^2\Pi^\star\cdot\sigma_s\cdot{\sigma_s}^{\mathrm{T}}\cdot\Pi^\star+\frac{\partial^2\varphi}{\partial X^2}\Pi^\star\cdot\sigma_s\cdot c^{\mathrm{T}}\cdot\left(\nabla_A\frac{\partial\varphi}{\partial X}\right)$$
$$=-\frac{1}{2}\left(\frac{\partial\varphi}{\partial X}\right)^2(a_s-r)\cdot(\sigma_s\cdot{\sigma_s}^{\mathrm{T}})^{-1}\cdot(a_s-r)-\frac{\partial\varphi}{\partial X}(a_s-r)\cdot(\sigma_s\cdot{\sigma_s}^{\mathrm{T}})^{-1}$$
$$\cdot\sigma_s\cdot c^{\mathrm{T}}\cdot\left(\nabla_A\frac{\partial\varphi}{\partial X}\right)-\frac{1}{2}\left(\nabla_A\frac{\partial\varphi}{\partial X}\right)\cdot c\cdot{\sigma_s}^{\mathrm{T}}\cdot(\sigma_s\cdot{\sigma_s}^{\mathrm{T}})^{-1}\cdot\sigma_s\cdot c^{\mathrm{T}}\cdot\left(\nabla_A\frac{\partial\varphi}{\partial X}\right). \tag{3.4.14}$$

Inserting (3.4.14) into (3.4.10), we prove (3.4.2).

Remark 3.4.1. No boundary conditions (other than the terminal condition at $t=T$) are needed, nor is it possible to impose any, for the PDE (3.4.2), since the PDE holds in $\mathcal{X}\times\mathcal{A}$, so that the solution is then understood as an $(\mathcal{X}\times\mathcal{A})$-entire solution.

Exercise 3.4.1. Consider the log-normal model (2.4.1), with $A=S=\{S_1\}$, and market coefficients (2.4.2). Show that equation (3.4.2) becomes

$$-\frac{1}{2}S_1^2\varphi^{(0,1,1)}(t,X,S_1)^2\mathbf{p}_1^2+\frac{1}{2}S_1^2\varphi^{(0,0,2)}(t,X,S_1)\,\varphi^{(0,2,0)}(t,X,S_1)\,\mathbf{p}_1^2$$
$$-S_1(\mathbf{a}_1-r)\,\varphi^{(0,1,0)}(t,X,S_1)\,\varphi^{(0,1,1)}(t,X,S_1)$$
$$+S_1(\mathbf{a}_1-\mathbb{D}_1)\,\varphi^{(0,0,1)}(t,X,S_1)\,\varphi^{(0,2,0)}(t,X,S_1)$$
$$+rX\varphi^{(0,1,0)}(t,X,S_1)\varphi^{(0,2,0)}(t,X,S_1)+\varphi^{(0,2,0)}(t,X,S_1)\varphi^{(1,0,0)}(t,X,S_1)$$
$$-\frac{(\mathbf{a}_1-r)^2\varphi^{(0,1,0)}(t,X,S_1)^2}{2\mathbf{p}_1^2}=0 \tag{3.4.15}$$

for $t<T, X\in\mathcal{X}, S_1>0$, with the terminal condition $\varphi(T,X,S_1)=\psi(X)$. Then conclude that equation (3.4.16) admits a solution in the form $\varphi(t,X,S_1)=\varphi(t,X)$, and therefore solves

$$-\frac{(\mathbf{a}_1-r)^2\varphi^{(0,1)}(t,X)^2}{2\mathbf{p}_1^2}+rX\varphi^{(0,2)}(t,X)\varphi^{(0,1)}(t,X)+\varphi^{(0,2)}(t,X)\varphi^{(1,0)}(t,X)=0 \tag{3.4.16}$$

for $t<T, X\in\mathcal{X}$, with the terminal condition $\varphi(T,X)=\psi(X)$. Show that the optimal portfolio strategy is equal to

$$\Pi^\star(t,X)=\left\{-\frac{(\mathbf{a}_1-r)\,\varphi^{(0,1)}(t,X)}{\mathbf{p}_1^2\varphi^{(0,2)}(t,X)}\right\}. \tag{3.4.17}$$

Finally, in the case of CRRA utility, i.e., if $\psi(X) = \psi_\gamma(X) = \frac{X^{1-\gamma}-1}{1-\gamma}$, show that

$$\varphi(t,X) = \frac{1}{1-\gamma}\left(X^{1-\gamma}\mathrm{e}^{-(T-t)\left(\frac{1}{2}\frac{\gamma-1}{\gamma}\left(\frac{\mathbf{a}_1-r}{\mathbf{p}_1}\right)^2+r(\gamma-1)\right)}-1\right) \tag{3.4.18}$$

for $t < T, X > 0$, and

$$\Pi^\star(t,X) = \left\{\frac{X(\mathbf{a}_1 - r)}{\gamma \mathbf{p}_1^2}\right\}, \tag{3.4.19}$$

which is Merton's optimal portfolio formula.

Remark 3.4.2. Recall Remark 3.3.2. Can we pass to the limit $T \to \infty$? That is, can we solve the "perpetual" optimal portfolio problem? From Exercise 3.4.1 we can draw a conclusion that is true in general: we cannot pass to the limit $T \to \infty$ in (3.4.18), for if we do, the solution either diverges or converges to zero, but we can pass to the limit $T \to \infty$ in (3.4.19). That seems a trivial fact in (3.4.19), but it is true quite often in general: we cannot pass to the limit $T \to \infty$ in (3.4.2), i.e., we cannot solve this PDE in terms of a time-entire solution, and yet we can *often* pass to the limit $T \to \infty$ in (3.4.3): the optimal portfolio formula will converge. This is nontrivial, and it is not a theorem but merely an observation that *often* holds. We shall return to this point again and again in many examples.

3.5 Derivation of the Optimal Portfolio Formula and of the Monge–Ampère-Type PDE for the Value Function: Affine Constraint on the Portfolio

Solving Problem 3.3.2 is somewhat more difficult. Recall the imposed affine constraint on the portfolio (3.3.4).

Theorem 3.5.1 ([44], cf. [34, 43]). *The value function $\varphi(t,X,A)$ given in (3.3.5) is the solution of the following fully nonlinear PDE, to be referred to as Monge–Ampère-type PDE of constrained portfolio optimization:*

$$\frac{\partial^2\varphi}{\partial X^2}\frac{\partial\varphi}{\partial t} + \frac{1}{2}\frac{\partial^2\varphi}{\partial X^2}\mathrm{Tr}\left(c\cdot c^{\mathrm{T}}\cdot\nabla_A\nabla_A\varphi\right) - \frac{1}{2}\left(\frac{\partial\varphi}{\partial X}\right)^2(a_s-r)\cdot\left(\mathbb{I}_k - \left(\sigma_s\cdot{\sigma_s}^{\mathrm{T}}\right)^{-1}\right.$$
$$\left.\cdot\mu^{\mathrm{T}}\cdot\left(\mu\cdot\left(\sigma_s\cdot{\sigma_s}^{\mathrm{T}}\right)^{-1}\cdot\mu^{\mathrm{T}}\right)^{-1}\cdot\mu\right)\cdot\left(\sigma_s\cdot{\sigma_s}^{\mathrm{T}}\right)^{-1}\cdot(a_s-r) + (b\cdot\nabla_A\varphi)\frac{\partial^2\varphi}{\partial X^2}$$
$$-\frac{\partial\varphi}{\partial X}(a_s-r)\cdot\left(\mathbb{I}_k - \left(\sigma_s\cdot{\sigma_s}^{\mathrm{T}}\right)^{-1}\cdot\mu^{\mathrm{T}}\cdot\left(\mu\cdot\left(\sigma_s\cdot{\sigma_s}^{\mathrm{T}}\right)^{-1}\cdot\mu^{\mathrm{T}}\right)^{-1}\cdot\mu\right)$$
$$\cdot\left(\sigma_s\cdot{\sigma_s}^{\mathrm{T}}\right)^{-1}\cdot\sigma_s\cdot c^{\mathrm{T}}\cdot\left(\nabla_A\frac{\partial\varphi}{\partial X}\right) - \frac{1}{2}\left(\nabla_A\frac{\partial\varphi}{\partial X}\right)\cdot c\cdot{\sigma_s}^{\mathrm{T}}$$

$$
\cdot\left(\mathbb{I}_k-\left(\sigma_s\cdot\sigma_s{}^{\mathrm{T}}\right)^{-1}\cdot\mu^{\mathrm{T}}\cdot\left(\mu\cdot\left(\sigma_s\cdot\sigma_s{}^{\mathrm{T}}\right)^{-1}\cdot\mu^{\mathrm{T}}\right)^{-1}\cdot\mu\right)
$$
$$
\cdot\left(\sigma_s\cdot\sigma_s{}^{\mathrm{T}}\right)^{-1}\cdot\sigma_s\cdot c^{\mathrm{T}}\cdot\left(\nabla_A\frac{\partial\varphi}{\partial X}\right)+rX\frac{\partial^2\varphi}{\partial X^2}\frac{\partial\varphi}{\partial X}+\frac{\partial\varphi}{\partial X}\frac{\partial^2\varphi}{\partial X^2}\xi
$$
$$
\cdot\left(\mu\cdot\left(\sigma_s\cdot\sigma_s{}^{\mathrm{T}}\right)^{-1}\cdot\mu^{\mathrm{T}}\right)^{-1}\cdot\mu\cdot\left(\sigma_s\cdot\sigma_s{}^{\mathrm{T}}\right)^{-1}\cdot(a_s-r)
$$
$$
+\frac{1}{2}\left(\frac{\partial^2\varphi}{\partial X^2}\right)^2\xi\cdot\left(\mu\cdot\left(\sigma_s\cdot\sigma_s{}^{\mathrm{T}}\right)^{-1}\cdot\mu^{\mathrm{T}}\right)^{-1}\cdot\xi+\frac{\partial^2\varphi}{\partial X^2}\xi\cdot\left(\mu\cdot\left(\sigma_s\cdot\sigma_s{}^{\mathrm{T}}\right)^{-1}\cdot\mu^{\mathrm{T}}\right)^{-1}
$$
$$
\cdot\mu\cdot\left(\sigma_s\cdot\sigma_s{}^{\mathrm{T}}\right)^{-1}\cdot\sigma_s\cdot c^{\mathrm{T}}\cdot\left(\nabla_A\frac{\partial\varphi}{\partial X}\right)=0 \tag{3.5.1}
$$

for $t<T<\infty, X\in\mathcal{X}, A\in\mathcal{A}$, together with the terminal condition $\varphi(T,X,A)=\psi(X)$, while the optimal portfolio strategy, the solution of the Problem 3.3.2, is equal to

$$
\Pi^{\star}(t,X,A)=-\left(\frac{\partial\varphi}{\partial X}\Big/\frac{\partial^2\varphi}{\partial X^2}\right)(a_s-r)\cdot\left(\mathbb{I}_k-\left(\sigma_s\cdot\sigma_s{}^{\mathrm{T}}\right)^{-1}\cdot\mu^{\mathrm{T}}\right.
$$
$$
\left.\cdot\left(\mu\cdot\left(\sigma_s\cdot\sigma_s{}^{\mathrm{T}}\right)^{-1}\cdot\mu^{\mathrm{T}}\right)^{-1}\cdot\mu\right)\cdot\left(\sigma_s\cdot\sigma_s{}^{\mathrm{T}}\right)^{-1}-\left(\nabla_A\frac{\partial\varphi}{\partial X}\Big/\frac{\partial^2\varphi}{\partial X^2}\right)\cdot c\cdot\sigma_s{}^{\mathrm{T}}
$$
$$
\cdot\left(\mathbb{I}_k-\left(\sigma_s\cdot\sigma_s{}^{\mathrm{T}}\right)^{-1}\cdot\mu^{\mathrm{T}}\cdot\left(\mu\cdot\left(\sigma_s\cdot\sigma_s{}^{\mathrm{T}}\right)^{-1}\cdot\mu^{\mathrm{T}}\right)^{-1}\cdot\mu\right)\cdot\left(\sigma_s\cdot\sigma_s{}^{\mathrm{T}}\right)^{-1}
$$
$$
+\xi\cdot\left(\mu\cdot\left(\sigma_s\cdot\sigma_s{}^{\mathrm{T}}\right)^{-1}\cdot\mu^{\mathrm{T}}\right)^{-1}\cdot\mu\cdot\left(\sigma_s\cdot\sigma_s{}^{\mathrm{T}}\right)^{-1}. \tag{3.5.2}
$$

Proof. To prove (3.5.1) and (3.5.2), the HJB PDE now reads

$$
0=\max_{\Pi:\mu\cdot\Pi=\xi}[\mathcal{L}\varphi]=\max_{\Pi:\mu\cdot\Pi=\xi}\left[\frac{\partial\varphi}{\partial t}+(\Pi.(a_s-r)+rX)\frac{\partial\varphi}{\partial X}+b\cdot\nabla_A\varphi\right.
$$
$$
\left.+\frac{1}{2}\Pi\cdot\sigma_s\cdot\sigma_s{}^{\mathrm{T}}\cdot\Pi\frac{\partial^2\varphi}{\partial X^2}+\Pi\cdot\sigma_s\cdot c^{\mathrm{T}}\cdot\left(\nabla_A\frac{\partial\varphi}{\partial X}\right)+\frac{1}{2}\mathrm{Tr}\left(c\cdot c^{\mathrm{T}}\cdot\nabla_A\nabla_A\varphi\right)\right] \tag{3.5.3}
$$

with the terminal condition $\varphi(T,X,A)=\psi(X)$, which we rewrite as

$$
\frac{\partial\varphi}{\partial t}+\frac{\partial\varphi}{\partial X}rX+b\cdot\nabla_A\varphi+\frac{1}{2}\mathrm{Tr}\left(c\cdot c^{\mathrm{T}}\cdot\nabla_A\nabla_A\varphi\right)+\max_{\Pi:\mu\cdot\Pi=\xi}F(\Pi)=0, \tag{3.5.4}
$$

where

$$
F(\Pi):=\Pi.(a_s-r)\frac{\partial\varphi}{\partial X}+\frac{1}{2}\Pi\cdot\sigma_s\cdot\sigma_s{}^{\mathrm{T}}\cdot\Pi\frac{\partial^2\varphi}{\partial X^2}+\Pi\cdot\sigma_s\cdot c^{\mathrm{T}}\cdot\left(\nabla_A\frac{\partial\varphi}{\partial X}\right). \tag{3.5.5}
$$

Setting

$$
G(\Pi):=\mu\cdot\Pi=\Pi\cdot\mu^{\mathrm{T}}, \tag{3.5.6}
$$

the affine constraint reads $G(\Pi) = \xi$. Since F is quadratic and G is linear in Π, to find $\Pi^\star$ such that $\max_{G(\Pi)=0} F(\Pi) = F(\Pi^\star)$, we use the Lagrange multiplier method, according to which there exists a vector λ such that $\nabla_\Pi F(\Pi) = \lambda . \nabla_\Pi G(\Pi) = \lambda . \mu = \mu^{\mathrm{T}} \cdot \lambda$, i.e.,

$$\mu^{\mathrm{T}} \cdot \lambda = (a_s - r)\,\frac{\partial \varphi}{\partial X} + \sigma_s \cdot {\sigma_s}^{\mathrm{T}} \cdot \Pi \frac{\partial^2 \varphi}{\partial X^2} + \sigma_s \cdot c^{\mathrm{T}} \cdot \left(\nabla_A \frac{\partial \varphi}{\partial X}\right), \tag{3.5.7}$$

yielding

$$\begin{aligned} \Pi^\star(t,X,A) = & -\left(\frac{\partial \varphi}{\partial X} \Big/ \frac{\partial^2 \varphi}{\partial X^2}\right) (\sigma_s \cdot {\sigma_s}^{\mathrm{T}})^{-1} \cdot (a_s - r) - \left(1 \Big/ \frac{\partial^2 \varphi}{\partial X^2}\right) (\sigma_s \cdot {\sigma_s}^{\mathrm{T}})^{-1} \\ & \cdot \sigma_s \cdot c^{\mathrm{T}} \cdot \left(\nabla_A \frac{\partial \varphi}{\partial X}\right) + \left(1 \Big/ \frac{\partial^2 \varphi}{\partial X^2}\right) (\sigma_s \cdot {\sigma_s}^{\mathrm{T}})^{-1} \cdot \mu^{\mathrm{T}} \cdot \lambda. \end{aligned} \tag{3.5.8}$$

Now, since $G(\Pi^\star) - \xi = 0$, we have

$$\begin{aligned} & \mu \cdot \left(-\left(\frac{\partial \varphi}{\partial X} \Big/ \frac{\partial^2 \varphi}{\partial X^2}\right) (\sigma_s \cdot {\sigma_s}^{\mathrm{T}})^{-1} \cdot (a_s - r) - \left(1 \Big/ \frac{\partial^2 \varphi}{\partial X^2}\right) (\sigma_s \cdot {\sigma_s}^{\mathrm{T}})^{-1} \cdot \sigma_s \cdot c^{\mathrm{T}} \right. \\ & \left. \cdot \left(\nabla_A \frac{\partial \varphi}{\partial X}\right) + \left(1 \Big/ \frac{\partial^2 \varphi}{\partial X^2}\right) (\sigma_s \cdot {\sigma_s}^{\mathrm{T}})^{-1} \cdot \mu^{\mathrm{T}} \cdot \lambda \right) - \xi = 0, \end{aligned} \tag{3.5.9}$$

implying

$$\begin{aligned} \lambda = & \frac{\partial^2 \varphi}{\partial X^2} (\mu \cdot (\sigma_s \cdot {\sigma_s}^{\mathrm{T}})^{-1} \cdot \mu^{\mathrm{T}})^{-1} \cdot \xi + (\mu \cdot (\sigma_s \cdot {\sigma_s}^{\mathrm{T}})^{-1} \cdot \mu^{\mathrm{T}})^{-1} \cdot \mu \cdot (\sigma_s \cdot {\sigma_s}^{\mathrm{T}})^{-1} \\ & \cdot \sigma_s \cdot c^{\mathrm{T}} \cdot \left(\nabla_A \frac{\partial \varphi}{\partial X}\right) + \frac{\partial \varphi}{\partial X} (\mu \cdot (\sigma_s \cdot {\sigma_s}^{\mathrm{T}})^{-1} \cdot \mu^{\mathrm{T}})^{-1} \cdot \mu \cdot (\sigma_s \cdot {\sigma_s}^{\mathrm{T}})^{-1} \cdot (a_s - r), \end{aligned} \tag{3.5.10}$$

from which (3.5.2) follows. Returning to (3.5.4), we get

$$\begin{aligned} & \frac{\partial^2 \varphi}{\partial X^2} \left(\frac{\partial \varphi}{\partial t} + (\Pi^\star \cdot (a_s - r) + rX) \frac{\partial \varphi}{\partial X} + b \cdot \nabla_A \varphi + \frac{1}{2} \Pi^\star \cdot \sigma_s \cdot {\sigma_s}^{\mathrm{T}} \cdot \Pi^\star \frac{\partial^2 \varphi}{\partial X^2} \right. \\ & \left. + \Pi^\star \cdot \sigma_s \cdot c^{\mathrm{T}} \cdot \left(\nabla_A \frac{\partial \varphi}{\partial X}\right) + \frac{1}{2} \mathrm{Tr}\left(c \cdot c^{\mathrm{T}} \cdot \nabla_A \nabla_A \varphi\right)\right) = 0. \end{aligned} \tag{3.5.11}$$

To write this equation explicitly, we prepare formulas

$$\frac{\partial^2\varphi}{\partial X^2}\frac{\partial\varphi}{\partial X}\Pi^\star\cdot(a_s-r)=-\left(\frac{\partial\varphi}{\partial X}\right)^2(a_s-r)\cdot\left(\mathbb{I}_k-(\sigma_s\cdot\sigma_s{}^{\mathrm{T}})^{-1}\cdot\mu^{\mathrm{T}}\right.$$
$$\left.\cdot(\mu\cdot(\sigma_s\cdot\sigma_s{}^{\mathrm{T}})^{-1}\cdot\mu^{\mathrm{T}})^{-1}\cdot\mu\right)\cdot(\sigma_s\cdot\sigma_s{}^{\mathrm{T}})^{-1}\cdot(a_s-r)-\frac{\partial\varphi}{\partial X}\left(\nabla_A\frac{\partial\varphi}{\partial X}\right)\cdot c$$
$$\cdot\sigma_s{}^{\mathrm{T}}\cdot\left(\mathbb{I}_k-(\sigma_s\cdot\sigma_s{}^{\mathrm{T}})^{-1}\cdot\mu^{\mathrm{T}}\cdot(\mu\cdot(\sigma_s\cdot\sigma_s{}^{\mathrm{T}})^{-1}\cdot\mu^{\mathrm{T}})^{-1}\cdot\mu\right)\cdot(\sigma_s\cdot\sigma_s{}^{\mathrm{T}})^{-1}\cdot(a_s-r)$$
$$+\frac{\partial\varphi}{\partial X}\frac{\partial^2\varphi}{\partial X^2}\xi\cdot(\mu\cdot(\sigma_s\cdot\sigma_s{}^{\mathrm{T}})^{-1}\cdot\mu^{\mathrm{T}})^{-1}\cdot\mu\cdot(\sigma_s\cdot\sigma_s{}^{\mathrm{T}})^{-1}\cdot(a_s-r) \tag{3.5.12}$$

and

$$\frac{\partial^2\varphi}{\partial X^2}\Pi^\star\cdot\sigma_s\cdot c^{\mathrm{T}}\cdot\left(\nabla_A\frac{\partial\varphi}{\partial X}\right)=-\frac{\partial\varphi}{\partial X}(a_s-r)\cdot\left(\mathbb{I}_k-(\sigma_s\cdot\sigma_s{}^{\mathrm{T}})^{-1}\cdot\mu^{\mathrm{T}}\right.$$
$$\left.\cdot(\mu\cdot(\sigma_s\cdot\sigma_s{}^{\mathrm{T}})^{-1}\cdot\mu^{\mathrm{T}})^{-1}\cdot\mu\right)\cdot(\sigma_s\cdot\sigma_s{}^{\mathrm{T}})^{-1}\cdot\sigma_s\cdot c^{\mathrm{T}}\cdot\left(\nabla_A\frac{\partial\varphi}{\partial X}\right)$$
$$-\left(\nabla_A\frac{\partial\varphi}{\partial X}\right)\cdot c\cdot\sigma_s{}^{\mathrm{T}}\cdot\left(\mathbb{I}_k-(\sigma_s\cdot\sigma_s{}^{\mathrm{T}})^{-1}\cdot\mu^{\mathrm{T}}\cdot(\mu\cdot(\sigma_s\cdot\sigma_s{}^{\mathrm{T}})^{-1}\cdot\mu^{\mathrm{T}})^{-1}\cdot\mu\right)$$
$$\cdot(\sigma_s\cdot\sigma_s{}^{\mathrm{T}})^{-1}\cdot\sigma_s\cdot c^{\mathrm{T}}\cdot\left(\nabla_A\frac{\partial\varphi}{\partial X}\right)+\frac{\partial^2\varphi}{\partial X^2}\xi\cdot(\mu\cdot(\sigma_s\cdot\sigma_s{}^{\mathrm{T}})^{-1}\cdot\mu^{\mathrm{T}})^{-1}$$
$$\cdot\mu\cdot(\sigma_s\cdot\sigma_s{}^{\mathrm{T}})^{-1}\cdot\sigma_s\cdot c^{\mathrm{T}}\cdot\left(\nabla_A\frac{\partial\varphi}{\partial X}\right). \tag{3.5.13}$$

The last formula that we need requires more work:

$$\frac{1}{2}\left(\frac{\partial^2\varphi}{\partial X^2}\right)^2\Pi^\star\cdot\sigma_s\cdot\sigma_s{}^{\mathrm{T}}\cdot\Pi^\star=\frac{1}{2}\left(\frac{\partial^2\varphi}{\partial X^2}\right)^2\left(-\left(\frac{\partial\varphi}{\partial X}\Big/\frac{\partial^2\varphi}{\partial X^2}\right)(a_s-r)\right.$$
$$\cdot\left(\mathbb{I}_k-(\sigma_s\cdot\sigma_s{}^{\mathrm{T}})^{-1}\cdot\mu^{\mathrm{T}}\cdot(\mu\cdot(\sigma_s\cdot\sigma_s{}^{\mathrm{T}})^{-1}\cdot\mu^{\mathrm{T}})^{-1}\cdot\mu\right)\cdot(\sigma_s\cdot\sigma_s{}^{\mathrm{T}})^{-1}$$
$$-\left(\nabla_A\frac{\partial\varphi}{\partial X}\Big/\frac{\partial^2\varphi}{\partial X^2}\right)\cdot c\cdot\sigma_s{}^{\mathrm{T}}\cdot\left(\mathbb{I}_k-(\sigma_s\cdot\sigma_s{}^{\mathrm{T}})^{-1}\cdot\mu^{\mathrm{T}}\cdot(\mu\cdot(\sigma_s\cdot\sigma_s{}^{\mathrm{T}})^{-1}\cdot\mu^{\mathrm{T}})^{-1}\cdot\mu\right)$$
$$\left.\cdot(\sigma_s\cdot\sigma_s{}^{\mathrm{T}})^{-1}+\xi\cdot(\mu\cdot(\sigma_s\cdot\sigma_s{}^{\mathrm{T}})^{-1}\cdot\mu^{\mathrm{T}})^{-1}\cdot\mu\cdot(\sigma_s\cdot\sigma_s{}^{\mathrm{T}})^{-1}\right)\cdot\sigma_s$$
$$\cdot\sigma_s{}^{\mathrm{T}}\cdot\left(-\left(\frac{\partial\varphi}{\partial X}\Big/\frac{\partial^2\varphi}{\partial X^2}\right)(\sigma_s\cdot\sigma_s{}^{\mathrm{T}})^{-1}\cdot\left(\mathbb{I}_k-\mu^{\mathrm{T}}\cdot(\mu\cdot(\sigma_s\cdot\sigma_s{}^{\mathrm{T}})^{-1}\cdot\mu^{\mathrm{T}})^{-1}\right.\right.$$
$$\left.\cdot\mu\cdot(\sigma_s\cdot\sigma_s{}^{\mathrm{T}})^{-1}\right)\cdot(a_s-r)-(\sigma_s\cdot\sigma_s{}^{\mathrm{T}})^{-1}\cdot\left(\mathbb{I}_k-\mu^{\mathrm{T}}\cdot(\mu\cdot(\sigma_s\cdot\sigma_s{}^{\mathrm{T}})^{-1}\cdot\mu^{\mathrm{T}})^{-1}\right.$$

$$
\begin{aligned}
&\cdot\mu\cdot\left(\sigma_s\cdot{\sigma_s}^{\mathrm{T}}\right)^{-1}\Big)\cdot\sigma_s\cdot c^{\mathrm{T}}\cdot\left(\nabla_A\frac{\partial\varphi}{\partial X}\Big/\frac{\partial^2\varphi}{\partial X^2}\right)+\left(\sigma_s\cdot{\sigma_s}^{\mathrm{T}}\right)^{-1}\cdot\mu^{\mathrm{T}}\\
&\cdot\left(\mu\cdot\left(\sigma_s\cdot{\sigma_s}^{\mathrm{T}}\right)^{-1}\cdot\mu^{\mathrm{T}}\right)^{-1}\cdot\xi\Big)\\
&=\frac{\partial\varphi}{\partial X}(a_s-r)\cdot\left(\mathbb{I}_k-\left(\sigma_s\cdot{\sigma_s}^{\mathrm{T}}\right)^{-1}\cdot\mu^{\mathrm{T}}\cdot\left(\mu\cdot\left(\sigma_s\cdot{\sigma_s}^{\mathrm{T}}\right)^{-1}\cdot\mu^{\mathrm{T}}\right)^{-1}\cdot\mu\right)\\
&\cdot\left(\sigma_s\cdot{\sigma_s}^{\mathrm{T}}\right)^{-1}\cdot\sigma_s\cdot c^{\mathrm{T}}\cdot\left(\nabla_A\frac{\partial\varphi}{\partial X}\right)+\frac{1}{2}\left(\frac{\partial\varphi}{\partial X}\right)^2(a_s-r)\cdot\left(\mathbb{I}_k-\left(\sigma_s\cdot{\sigma_s}^{\mathrm{T}}\right)^{-1}\right.\\
&\cdot\mu^{\mathrm{T}}\cdot\left(\mu\cdot\left(\sigma_s\cdot{\sigma_s}^{\mathrm{T}}\right)^{-1}\cdot\mu^{\mathrm{T}}\right)^{-1}\cdot\mu\big).\left(\sigma_s\cdot{\sigma_s}^{\mathrm{T}}\right)^{-1}\cdot(a_s-r)\\
&+\frac{1}{2}\left(\nabla_A\frac{\partial\varphi}{\partial X}\right)\cdot c\cdot{\sigma_s}^{\mathrm{T}}\cdot\left(\mathbb{I}_k-\left(\sigma_s\cdot{\sigma_s}^{\mathrm{T}}\right)^{-1}\cdot\mu^{\mathrm{T}}\cdot\left(\mu\cdot\left(\sigma_s\cdot{\sigma_s}^{\mathrm{T}}\right)^{-1}\cdot\mu^{\mathrm{T}}\right)^{-1}\cdot\mu\right)\\
&.\left(\sigma_s\cdot{\sigma_s}^{\mathrm{T}}\right)^{-1}\cdot\sigma_s\cdot c^{\mathrm{T}}\cdot\left(\nabla_A\frac{\partial\varphi}{\partial X}\right)+\frac{1}{2}\left(\frac{\partial^2\varphi}{\partial X^2}\right)^2\xi\cdot\left(\mu\cdot\left(\sigma_s\cdot{\sigma_s}^{\mathrm{T}}\right)^{-1}\cdot\mu^{\mathrm{T}}\right)^{-1}\cdot\xi.
\end{aligned}
\tag{3.5.14}
$$

Indeed, to justify (3.5.14), one can think of it as $\frac{1}{2}(a+b+c)^2=ab+ac+bc+\frac{1}{2}a^2+\frac{1}{2}b^2+\frac{1}{2}c^2$, with "mixed" and "quadratic terms" on the right-hand side. Analogously, in (3.5.14), the mixed terms are as follows:

Mixed Term 1

$$
\begin{aligned}
&(a_s-r)\cdot\left(\mathbb{I}_k-\left(\sigma_s\cdot{\sigma_s}^{\mathrm{T}}\right)^{-1}\cdot\mu^{\mathrm{T}}\cdot\left(\mu\cdot\left(\sigma_s\cdot{\sigma_s}^{\mathrm{T}}\right)^{-1}\cdot\mu^{\mathrm{T}}\right)^{-1}\cdot\mu\right).\left(\sigma_s\cdot{\sigma_s}^{\mathrm{T}}\right)^{-1}\cdot\sigma_s\cdot{\sigma_s}^{\mathrm{T}}\\
&\quad.\left(\sigma_s\cdot{\sigma_s}^{\mathrm{T}}\right)^{-1}\cdot\left(\mathbb{I}_k-\mu^{\mathrm{T}}\cdot\left(\mu\cdot\left(\sigma_s\cdot{\sigma_s}^{\mathrm{T}}\right)^{-1}\cdot\mu^{\mathrm{T}}\right)^{-1}\cdot\mu\cdot\left(\sigma_s\cdot{\sigma_s}^{\mathrm{T}}\right)^{-1}\right)\\
&\quad\cdot\sigma_s\cdot c^{\mathrm{T}}\cdot\left(\nabla_A\frac{\partial\varphi}{\partial X}\right)\\
&=(a_s-r)\cdot\left(\mathbb{I}_k-\left(\sigma_s\cdot{\sigma_s}^{\mathrm{T}}\right)^{-1}\cdot\mu^{\mathrm{T}}\cdot\left(\mu\cdot\left(\sigma_s\cdot{\sigma_s}^{\mathrm{T}}\right)^{-1}\cdot\mu^{\mathrm{T}}\right)^{-1}\cdot\mu\right)\cdot\left(\sigma_s\cdot{\sigma_s}^{\mathrm{T}}\right)^{-1}\\
&\quad.\left(\mathbb{I}_k-\mu^{\mathrm{T}}\cdot\left(\mu\cdot\left(\sigma_s\cdot{\sigma_s}^{\mathrm{T}}\right)^{-1}\cdot\mu^{\mathrm{T}}\right)^{-1}\cdot\mu\cdot\left(\sigma_s\cdot{\sigma_s}^{\mathrm{T}}\right)^{-1}\right)\cdot\sigma_s\cdot c^{\mathrm{T}}\cdot\left(\nabla_A\frac{\partial\varphi}{\partial X}\right)\\
&=(a_s-r)\cdot\left(\mathbb{I}_k-\left(\sigma_s\cdot{\sigma_s}^{\mathrm{T}}\right)^{-1}\cdot\mu^{\mathrm{T}}\cdot\left(\mu\cdot\left(\sigma_s\cdot{\sigma_s}^{\mathrm{T}}\right)^{-1}\cdot\mu^{\mathrm{T}}\right)^{-1}\cdot\mu\right)\cdot\left(\sigma_s\cdot{\sigma_s}^{\mathrm{T}}\right)^{-1}\\
&\quad\cdot\sigma_s\cdot c^{\mathrm{T}}\cdot\left(\nabla_A\frac{\partial\varphi}{\partial X}\right),
\end{aligned}
\tag{3.5.15}
$$

since

$$
\begin{aligned}
&\left(\sigma_s\cdot{\sigma_s}^{\mathrm{T}}\right)^{-1}\cdot\mu^{\mathrm{T}}\cdot\left(\mu\cdot\left(\sigma_s\cdot{\sigma_s}^{\mathrm{T}}\right)^{-1}\cdot\mu^{\mathrm{T}}\right)^{-1}\cdot\mu\cdot\left(\sigma_s\cdot{\sigma_s}^{\mathrm{T}}\right)^{-1}\cdot\mu^{\mathrm{T}}\\
&\quad\cdot\left(\mu\cdot\left(\sigma_s\cdot{\sigma_s}^{\mathrm{T}}\right)^{-1}\cdot\mu^{\mathrm{T}}\right)^{-1}\cdot\mu\cdot\left(\sigma_s\cdot{\sigma_s}^{\mathrm{T}}\right)^{-1}\\
&=\left(\sigma_s\cdot{\sigma_s}^{\mathrm{T}}\right)^{-1}\cdot\mu^{\mathrm{T}}\cdot\left(\mu\cdot\left(\sigma_s\cdot{\sigma_s}^{\mathrm{T}}\right)^{-1}\cdot\mu^{\mathrm{T}}\right)^{-1}\cdot\mu\cdot\left(\sigma_s\cdot{\sigma_s}^{\mathrm{T}}\right)^{-1};
\end{aligned}
\tag{3.5.16}
$$

Mixed Term 2

$$
\begin{aligned}
&(a_s - r) \cdot \left(\mathbb{I}_k - \left(\sigma_s \cdot \sigma_s{}^{\mathrm{T}}\right)^{-1} \cdot \mu^{\mathrm{T}} \cdot \left(\mu \cdot \left(\sigma_s \cdot \sigma_s{}^{\mathrm{T}}\right)^{-1} \cdot \mu^{\mathrm{T}}\right)^{-1} \cdot \mu\right) \\
&\qquad \cdot \left(\sigma_s \cdot \sigma_s{}^{\mathrm{T}}\right)^{-1} \cdot \sigma_s \cdot \sigma_s{}^{\mathrm{T}} \cdot \left(\sigma_s \cdot \sigma_s{}^{\mathrm{T}}\right)^{-1} \cdot \mu^{\mathrm{T}} \cdot \left(\mu \cdot \left(\sigma_s \cdot \sigma_s{}^{\mathrm{T}}\right)^{-1} \cdot \mu^{\mathrm{T}}\right)^{-1} \cdot \xi \\
&= (a_s - r) \cdot \left(\mathbb{I}_k - \left(\sigma_s \cdot \sigma_s{}^{\mathrm{T}}\right)^{-1} \cdot \mu^{\mathrm{T}} \cdot \left(\mu \cdot \left(\sigma_s \cdot \sigma_s{}^{\mathrm{T}}\right)^{-1} \cdot \mu^{\mathrm{T}}\right)^{-1} \cdot \mu\right) \\
&\quad \cdot \left(\sigma_s \cdot \sigma_s{}^{\mathrm{T}}\right)^{-1} \cdot \mu^{\mathrm{T}} \cdot \left(\mu \cdot \left(\sigma_s \cdot \sigma_s{}^{\mathrm{T}}\right)^{-1} \cdot \mu^{\mathrm{T}}\right)^{-1} \cdot \xi \\
&= (a_s - r) \cdot \left(\sigma_s \cdot \sigma_s{}^{\mathrm{T}}\right)^{-1} \cdot \mu^{\mathrm{T}} \cdot \left(\mu \cdot \left(\sigma_s \cdot \sigma_s{}^{\mathrm{T}}\right)^{-1} \cdot \mu^{\mathrm{T}}\right)^{-1} \cdot \xi \\
&\quad - (a_s - r) \cdot \left(\sigma_s \cdot \sigma_s{}^{\mathrm{T}}\right)^{-1} \cdot \mu^{\mathrm{T}} \cdot \left(\mu \cdot \left(\sigma_s \cdot \sigma_s{}^{\mathrm{T}}\right)^{-1} \cdot \mu^{\mathrm{T}}\right)^{-1} \cdot \mu \\
&\quad \cdot \left(\sigma_s \cdot \sigma_s{}^{\mathrm{T}}\right)^{-1} \cdot \mu^{\mathrm{T}} \cdot \left(\mu \cdot \left(\sigma_s \cdot \sigma_s{}^{\mathrm{T}}\right)^{-1} \cdot \mu^{\mathrm{T}}\right)^{-1} \cdot \xi \\
&= (a_s - r) \cdot \left(\sigma_s \cdot \sigma_s{}^{\mathrm{T}}\right)^{-1} \cdot \mu^{\mathrm{T}} \cdot \left(\mu \cdot \left(\sigma_s \cdot \sigma_s{}^{\mathrm{T}}\right)^{-1} \cdot \mu^{\mathrm{T}}\right)^{-1} \cdot \xi - (a_s - r) \\
&\quad \cdot \left(\sigma_s \cdot \sigma_s{}^{\mathrm{T}}\right)^{-1} \cdot \mu^{\mathrm{T}} \cdot \left(\mu \cdot \left(\sigma_s \cdot \sigma_s{}^{\mathrm{T}}\right)^{-1} \cdot \mu^{\mathrm{T}}\right)^{-1} \cdot \xi = 0;
\end{aligned}
\tag{3.5.17}
$$

Mixed Term 3

$$
\begin{aligned}
&\left(\nabla_A \frac{\partial \varphi}{\partial X}\right) \cdot c \cdot \sigma_s{}^{\mathrm{T}} \cdot \left(\mathbb{I}_k - \left(\sigma_s \cdot \sigma_s{}^{\mathrm{T}}\right)^{-1} \cdot \mu^{\mathrm{T}} \cdot \left(\mu \cdot \left(\sigma_s \cdot \sigma_s{}^{\mathrm{T}}\right)^{-1} \cdot \mu^{\mathrm{T}}\right)^{-1} \cdot \mu\right) \\
&\qquad \cdot \left(\sigma_s \cdot \sigma_s{}^{\mathrm{T}}\right)^{-1} \cdot \sigma_s \cdot \sigma_s{}^{\mathrm{T}} \cdot \left(\sigma_s \cdot \sigma_s{}^{\mathrm{T}}\right)^{-1} \cdot \mu^{\mathrm{T}} \cdot \left(\mu \cdot \left(\sigma_s \cdot \sigma_s{}^{\mathrm{T}}\right)^{-1} \cdot \mu^{\mathrm{T}}\right)^{-1} \cdot \xi \\
&= \left(\nabla_A \frac{\partial \varphi}{\partial X}\right) \cdot c \cdot \sigma_s{}^{\mathrm{T}} \cdot \left(\mathbb{I}_k - \left(\sigma_s \cdot \sigma_s{}^{\mathrm{T}}\right)^{-1} \cdot \mu^{\mathrm{T}} \cdot \left(\mu \cdot \left(\sigma_s \cdot \sigma_s{}^{\mathrm{T}}\right)^{-1} \cdot \mu^{\mathrm{T}}\right)^{-1} \cdot \mu\right) \\
&\quad \cdot \left(\sigma_s \cdot \sigma_s{}^{\mathrm{T}}\right)^{-1} \cdot \mu^{\mathrm{T}} \cdot \left(\mu \cdot \left(\sigma_s \cdot \sigma_s{}^{\mathrm{T}}\right)^{-1} \cdot \mu^{\mathrm{T}}\right)^{-1} \cdot \xi \\
&= \left(\nabla_A \frac{\partial \varphi}{\partial X}\right) \cdot c \cdot \sigma_s{}^{\mathrm{T}} \cdot \left(\sigma_s \cdot \sigma_s{}^{\mathrm{T}}\right)^{-1} \cdot \mu^{\mathrm{T}} \cdot \left(\mu \cdot \left(\sigma_s \cdot \sigma_s{}^{\mathrm{T}}\right)^{-1} \cdot \mu^{\mathrm{T}}\right)^{-1} \\
&\qquad \cdot \xi - \left(\nabla_A \frac{\partial \varphi}{\partial X}\right) \cdot c \cdot \sigma_s{}^{\mathrm{T}} \cdot \left(\sigma_s . \sigma_s{}^{\mathrm{T}}\right)^{-1} \cdot \mu^{\mathrm{T}} \cdot \left(\mu \cdot \left(\sigma_s \cdot \sigma_s{}^{\mathrm{T}}\right)^{-1} \cdot \mu^{\mathrm{T}}\right)^{-1} \cdot \xi = 0.
\end{aligned}
\tag{3.5.18}
$$

The quadratic terms as these:

Quadratic Term 1

$$
\begin{aligned}
&(a_s - r) \cdot \left(\mathbb{I}_k - \left(\sigma_s \cdot \sigma_s{}^{\mathrm{T}}\right)^{-1} \cdot \mu^{\mathrm{T}} \cdot \left(\mu \cdot \left(\sigma_s \cdot \sigma_s{}^{\mathrm{T}}\right)^{-1} \cdot \mu^{\mathrm{T}}\right)^{-1} \cdot \mu\right) \cdot \left(\sigma_s \cdot \sigma_s{}^{\mathrm{T}}\right)^{-1} \\
&\quad \cdot \sigma_s \cdot \sigma_s{}^{\mathrm{T}} \cdot \left(\sigma_s \cdot \sigma_s{}^{\mathrm{T}}\right)^{-1} \cdot \left(\mathbb{I}_k - \mu^{\mathrm{T}} \cdot \left(\mu \cdot \left(\sigma_s \cdot \sigma_s{}^{\mathrm{T}}\right)^{-1} \cdot \mu^{\mathrm{T}}\right)^{-1}\right. \\
&\quad \left. \cdot \mu \cdot \left(\sigma_s \cdot \sigma_s{}^{\mathrm{T}}\right)^{-1}\right) \cdot (a_s - r) \\
&= (a_s - r) \cdot \left(\mathbb{I}_k - \left(\sigma_s \cdot \sigma_s{}^{\mathrm{T}}\right)^{-1} \cdot \mu^{\mathrm{T}} \cdot \left(\mu \cdot \left(\sigma_s \cdot \sigma_s{}^{\mathrm{T}}\right)^{-1} \cdot \mu^{\mathrm{T}}\right)^{-1} \cdot \mu\right) \cdot \left(\sigma_s \cdot \sigma_s{}^{\mathrm{T}}\right)^{-1} \\
&\quad \cdot \left(\mathbb{I}_k - \mu^{\mathrm{T}} \cdot \left(\mu \cdot \left(\sigma_s \cdot \sigma_s{}^{\mathrm{T}}\right)^{-1} \cdot \mu^{\mathrm{T}}\right)^{-1} \cdot \mu \, . \left(\sigma_s \cdot \sigma_s{}^{\mathrm{T}}\right)^{-1}\right) \cdot (a_s - r)
\end{aligned}
$$

$$= (a_s - r)\cdot\left(\mathbb{I}_k - \left(\sigma_s\cdot\sigma_s^{\mathrm{T}}\right)^{-1}\cdot\mu^{\mathrm{T}}\cdot\left(\mu\cdot\left(\sigma_s\cdot\sigma_s^{\mathrm{T}}\right)^{-1}\cdot\mu^{\mathrm{T}}\right)^{-1}\cdot\mu\right)$$
$$\cdot\left(\sigma_s\cdot\sigma_s^{\mathrm{T}}\right)^{-1}\cdot(a_s - r); \tag{3.5.19}$$

Quadratic Term 2

$$\left(\nabla_A\frac{\partial\varphi}{\partial X}\right)\cdot c\cdot\sigma_s^{\mathrm{T}}\cdot\left(\mathbb{I}_k - \left(\sigma_s\cdot\sigma_s^{\mathrm{T}}\right)^{-1}\cdot\mu^{\mathrm{T}}\cdot\left(\mu\cdot\left(\sigma_s\cdot\sigma_s^{\mathrm{T}}\right)^{-1}\cdot\mu^{\mathrm{T}}\right)^{-1}\cdot\mu\right)$$
$$\cdot\left(\sigma_s\cdot\sigma_s^{\mathrm{T}}\right)^{-1}\cdot\sigma_s\cdot\sigma_s^{\mathrm{T}}\cdot\left(\sigma_s\cdot\sigma_s^{\mathrm{T}}\right)^{-1}\cdot\left(\mathbb{I}_k - \mu^{\mathrm{T}}\cdot\left(\mu\cdot\left(\sigma_s\cdot\sigma_s^{\mathrm{T}}\right)^{-1}\cdot\mu^{\mathrm{T}}\right)^{-1}\cdot\mu\right.$$
$$\left.\cdot\left(\sigma_s\cdot\sigma_s^{\mathrm{T}}\right)^{-1}\right)\cdot\sigma_s\cdot c^{\mathrm{T}}\cdot\left(\nabla_A\frac{\partial\varphi}{\partial X}\right)$$
$$= \left(\nabla_A\frac{\partial\varphi}{\partial X}\right)\cdot c\cdot\sigma_s^{\mathrm{T}}\cdot\left(\mathbb{I}_k - \left(\sigma_s\cdot\sigma_s^{\mathrm{T}}\right)^{-1}\cdot\mu^{\mathrm{T}}\cdot\left(\mu\cdot\left(\sigma_s\cdot\sigma_s^{\mathrm{T}}\right)^{-1}\cdot\mu^{\mathrm{T}}\right)^{-1}\cdot\mu\right)$$
$$\cdot\left(\sigma_s\cdot\sigma_s^{\mathrm{T}}\right)^{-1}\cdot\left(\mathbb{I}_k - \mu^{\mathrm{T}}\cdot\left(\mu\cdot\left(\sigma_s\cdot\sigma_s^{\mathrm{T}}\right)^{-1}\cdot\mu^{\mathrm{T}}\right)^{-1}\cdot\mu\cdot\left(\sigma_s\cdot\sigma_s^{\mathrm{T}}\right)^{-1}\right)$$
$$\cdot\sigma_s\cdot c^{\mathrm{T}}\cdot\left(\nabla_A\frac{\partial\varphi}{\partial X}\right)$$
$$= \left(\nabla_A\frac{\partial\varphi}{\partial X}\right)\cdot c\cdot\sigma_s^{\mathrm{T}}\cdot\left(\mathbb{I}_k - \left(\sigma_s\cdot\sigma_s^{\mathrm{T}}\right)^{-1}\cdot\mu^{\mathrm{T}}\cdot\left(\mu\cdot\left(\sigma_s\cdot\sigma_s^{\mathrm{T}}\right)^{-1}\cdot\mu^{\mathrm{T}}\right)^{-1}\cdot\mu\right)$$
$$\cdot\left(\sigma_s\cdot\sigma_s^{\mathrm{T}}\right)^{-1}\cdot\sigma_s\cdot c^{\mathrm{T}}\cdot\left(\nabla_A\frac{\partial\varphi}{\partial X}\right); \tag{3.5.20}$$

Quadratic Term 3

$$\xi\cdot\left(\mu\cdot\left(\sigma_s\cdot\sigma_s^{\mathrm{T}}\right)^{-1}\cdot\mu^{\mathrm{T}}\right)^{-1}\cdot\mu\cdot\left(\sigma_s\cdot\sigma_s^{\mathrm{T}}\right)^{-1}\cdot\sigma_s\cdot\sigma_s^{\mathrm{T}}\cdot\left(\sigma_s\cdot\sigma_s^{\mathrm{T}}\right)^{-1}\cdot\mu^{\mathrm{T}}$$
$$\cdot\left(\mu\cdot\left(\sigma_s\cdot\sigma_s^{\mathrm{T}}\right)^{-1}\cdot\mu^{\mathrm{T}}\right)^{-1}\cdot\xi$$
$$= \xi\cdot\left(\mu\cdot\left(\sigma_s\cdot\sigma_s^{\mathrm{T}}\right)^{-1}\cdot\mu^{\mathrm{T}}\right)^{-1}\cdot\mu\cdot\left(\sigma_s\cdot\sigma_s^{\mathrm{T}}\right)^{-1}\cdot\mu^{\mathrm{T}}\cdot\left(\mu\cdot\left(\sigma_s\cdot\sigma_s^{\mathrm{T}}\right)^{-1}\cdot\mu^{\mathrm{T}}\right)^{-1}\cdot\xi$$
$$= \xi\cdot\left(\mu\cdot\left(\sigma_s\cdot\sigma_s^{\mathrm{T}}\right)^{-1}\cdot\mu^{\mathrm{T}}\right)^{-1}\cdot\xi. \tag{3.5.21}$$

This ends the justification of formula (3.5.14). Adding up formulas (3.5.12)–(3.5.14) and inserting the result into (3.5.11), we finally get (3.5.1), which completes the proof of the theorem. □

Remark 3.5.1. Equation (3.5.1) can also be written as

$$\mathcal{M}(\varphi) + \frac{1}{2}\left(\frac{\partial\varphi}{\partial X}\right)^2(a_s - r)\cdot\left(\sigma_s\cdot\sigma_s^{\mathrm{T}}\right)^{-1}\cdot\mu^{\mathrm{T}}\cdot\left(\mu\cdot\left(\sigma_s\cdot\sigma_s^{\mathrm{T}}\right)^{-1}\cdot\mu^{\mathrm{T}}\right)^{-1}$$
$$\cdot\mu\cdot\left(\sigma_s\cdot\sigma_s^{\mathrm{T}}\right)^{-1}\cdot(a_s - r) + \frac{\partial\varphi}{\partial X}\left(\nabla_A\frac{\partial\varphi}{\partial X}\right)\cdot c\cdot\sigma_s^{\mathrm{T}}\cdot\left(\sigma_s\cdot\sigma_s^{\mathrm{T}}\right)^{-1}\cdot\mu^{\mathrm{T}}$$

$$\cdot\left(\mu\cdot\left(\sigma_s\cdot\sigma_s{}^{\mathrm{T}}\right)^{-1}\cdot\mu^{\mathrm{T}}\right)^{-1}\cdot\mu\cdot\left(\sigma_s\cdot\sigma_s{}^{\mathrm{T}}\right)^{-1}\cdot(a_s-r)+\frac{\partial\varphi}{\partial X}\frac{\partial^2\varphi}{\partial X^2}\xi$$

$$\cdot\left(\mu\cdot\left(\sigma_s\cdot\sigma_s{}^{\mathrm{T}}\right)^{-1}\cdot\mu^{\mathrm{T}}\right)^{-1}\cdot\mu\cdot\left(\sigma_s\cdot\sigma_s{}^{\mathrm{T}}\right)^{-1}\cdot(a_s-r)+\frac{1}{2}\left(\nabla_A\frac{\partial\varphi}{\partial X}\right)\cdot c$$

$$\cdot\sigma_s{}^{\mathrm{T}}\cdot\left(\sigma_s\cdot\sigma_s{}^{\mathrm{T}}\right)^{-1}\cdot\mu^{\mathrm{T}}\cdot\left(\mu\cdot\left(\sigma_s\cdot\sigma_s{}^{\mathrm{T}}\right)^{-1}\cdot\mu^{\mathrm{T}}\right)^{-1}\cdot\mu\cdot\left(\sigma_s\cdot\sigma_s{}^{\mathrm{T}}\right)^{-1}\cdot\sigma_s\cdot c^{\mathrm{T}}$$

$$\cdot\left(\nabla_A\frac{\partial\varphi}{\partial X}\right)+\frac{1}{2}\left(\frac{\partial^2\varphi}{\partial X^2}\right)^2\xi\cdot\left(\mu\cdot\left(\sigma_s\cdot\sigma_s{}^{\mathrm{T}}\right)^{-1}\cdot\mu^{\mathrm{T}}\right)^{-1}\cdot\xi+\frac{\partial^2\varphi}{\partial X^2}\xi$$

$$\cdot\left(\mu\cdot\left(\sigma_s\cdot\sigma_s{}^{\mathrm{T}}\right)^{-1}\cdot\mu^{\mathrm{T}}\right)^{-1}\cdot\mu\cdot\left(\sigma_s\cdot\sigma_s{}^{\mathrm{T}}\right)^{-1}\cdot\sigma_s\cdot c^{\mathrm{T}}\cdot\left(\nabla_A\frac{\partial\varphi}{\partial X}\right)=0, \quad (3.5.22)$$

where $\mathcal{M}$ is the differential operator of Monge–Ampère type defined in (3.4.1).

Remark 3.5.3. Again, no boundary conditions (other than the terminal condition) are needed, nor is it possible to impose any, for the PDE (3.5.22), since the PDE holds in $\mathcal{X}\times\mathcal{A}$. Also, as before, no condition on the choice of the utility function ψ is needed here.

Remark 3.5.4. An earlier, a much less general version of this theorem was derived in [34].

Exercise 3.5.1. We continue Exercise 3.4.1. Consider the log-normal model (2.4.1), but with two tradables:

$$\begin{aligned}\mathrm{d}\,S_1(t) &= S_1(t)((\mathbf{a}_1-\mathbb{D}_1)\mathrm{d}t+\mathbf{p}_1\mathrm{d}\,B_1(t)),\\ \mathrm{d}\,S_2(t) &= S_2(t)((\mathbf{a}_2-\mathbb{D}_2)\mathrm{d}t+\mathbf{p}_2\mathrm{d}\,\mathbb{B}_2(t)),\end{aligned} \quad (3.5.23)$$

where $\mathbf{a}_i, p_i, \mathbb{D}_i\in\mathbb{R}, \mathbf{p}_i\neq 0$. Set $A=S=\{S_1,S_2\}$, with market coefficients

$$a_s=\{\mathbf{a}_1,\mathbf{a}_2\},b=\{S_1(\mathbf{a}_1-\mathbb{D}_1),S_2(\mathbf{a}_2-\mathbb{D}_2)\},\sigma_s=\begin{pmatrix}\mathbf{p}_1 & 0\\ \mathbf{p}_2\rho_{2,1} & \mathbf{p}_2\sqrt{1-\rho_{2,1}^2}\end{pmatrix},$$

$$c=\begin{pmatrix}S_1\mathbf{p}_1 & 0\\ S_2\mathbf{p}_2\rho_{2,1} & S_2\mathbf{p}_2\sqrt{1-\rho_{2,1}^2}\end{pmatrix}. \quad (3.5.24)$$

First consider the portfolio optimization problem without constraints, and show that equation (3.4.2) reduces to

$$\begin{aligned}&\left(\left((r-\mathbf{a}_2)^2\mathbf{p}_1^2-2(r-\mathbf{a}_1)(r-\mathbf{a}_2)\mathbf{p}_2\rho_{2,1}\mathbf{p}_1+(r-\mathbf{a}_1)^2\mathbf{p}_2^2\right)\varphi^{(0,1)}(t,X)^2\right)\Big/\\ &\quad\left(2\mathbf{p}_1^2\mathbf{p}_2^2\left(\rho_{2,1}^2-1\right)\right)+rX\varphi^{(0,2)}(t,X)\varphi^{(0,1)}(t,X)+\varphi^{(0,2)}(t,X)\varphi^{(1,0)}(t,X)=0\end{aligned} \quad (3.5.25)$$

for $t < T$, with the terminal condition $\varphi(T,X) = \psi(X)$, while the optimal portfolio strategy (3.4.3) becomes

$$\Pi^{\star}(t,X,A) = \Big\{\Big(((\mathbf{a}_1 - r)\mathbf{p}_2 + (r-\mathbf{a}_2)\mathbf{p}_1\rho_{2,1})\varphi^{(0,1)}(t,X)\Big) \Big/ \Big(\mathbf{p}_1^2\mathbf{p}_2\left(\rho_{2,1}^2 - 1\right)\varphi^{(0,2)}(t,X)\Big), \\ \Big(((\mathbf{a}_2 - r)\mathbf{p}_1 + (r-\mathbf{a}_1)\mathbf{p}_2\rho_{2,1})\varphi^{(0,1)}(t,X)\Big) \Big/ \Big(\mathbf{p}_1\mathbf{p}_2^2\left(\rho_{2,1}^2 - 1\right)\varphi^{(0,2)}(t,X)\Big)\Big\}. \tag{3.5.26}$$

Next, in the case of portfolio optimization with affine constraint (3.3.6), show that equation (3.5.22), or equivalently, (3.5.1), reduces to

$$-\frac{(\mathbf{a}_1 - \mathbf{a}_2)^2\varphi^{(0,1)}(t,X)^2}{2\left(\mathbf{p}_1^2 - 2\mathbf{p}_2\rho_{2,1}\mathbf{p}_1 + \mathbf{p}_2^2\right)} + rX\varphi^{(0,2)}(t,X)\varphi^{(0,1)}(t,X) \\ -\Big(\Big(X\alpha\left((r-\mathbf{a}_2)\mathbf{p}_1^2 + (-2r+\mathbf{a}_1+\mathbf{a}_2)\mathbf{p}_2\rho_{2,1}\mathbf{p}_1\right. \\ \left. + (r-\mathbf{a}_1)\mathbf{p}_2^2\right)\Big)\varphi^{(0,2)}(t,X)\varphi^{(0,1)}(t,X)\Big) \Big/ \left(\mathbf{p}_1^2 - 2\mathbf{p}_2\rho_{2,1}\mathbf{p}_1 + \mathbf{p}_2^2\right) \\ +\varphi^{(0,2)}(t,X)\varphi^{(1,0)}(t,X) - \Big(X^2\alpha^2\mathbf{p}_1^2\mathbf{p}_2^2\left(\rho_{2,1}^2 - 1\right)\varphi^{(0,2)}(t,X)^2\Big) \Big/ \\ \left(2\left(\mathbf{p}_1^2 - 2\mathbf{p}_2\rho_{2,1}\mathbf{p}_1 + \mathbf{p}_2^2\right)\right) = 0 \tag{3.5.27}$$

for $t < T$, with terminal condition $\varphi(T,X) = \psi(X)$, while the optimal portfolio strategy (3.5.2) becomes

$$\Pi^{\star}(t,X,A) = \Big\{\Big(-\mathbf{a}_1\varphi^{(0,1)}(t,X) + \mathbf{a}_2\varphi^{(0,1)}(t,X) + X\alpha\mathbf{p}_2(\mathbf{p}_2 - \mathbf{p}_1\rho_{2,1})\varphi^{(0,2)}(t,X)\Big) \Big/ \\ \Big(\left(\mathbf{p}_1^2 - 2\mathbf{p}_2\rho_{2,1}\mathbf{p}_1 + \mathbf{p}_2^2\right)\varphi^{(0,2)}(t,X)\Big), \Big(\mathbf{a}_1\varphi^{(0,1)}(t,X) \\ -\mathbf{a}_2\varphi^{(0,1)}(t,X) + X\alpha\mathbf{p}_1(\mathbf{p}_1 - \mathbf{p}_2\rho_{2,1})\varphi^{(0,2)}(t,X)\Big) \Big/ \\ \Big(\left(\mathbf{p}_1^2 - 2\mathbf{p}_2\rho_{2,1}\mathbf{p}_1 + \mathbf{p}_2^2\right)\varphi^{(0,2)}(t,X)\Big)\Big\}. \tag{3.5.28}$$

Finally, in the case of the CRRA utility, i.e., if $\psi(X) = \psi_\gamma(X) = \frac{X^{1-\gamma}-1}{1-\gamma}$, find the explicit solutions of equations (3.5.25) and (3.5.29), and explicit versions of the optimal portfolio formulas (3.5.26) and (3.5.28).

Remark 3.5.5. Notice that since

$$\sum_{k=1}^{2}\Pi_k^{\star}(t,X,A) = \Big(-\mathbf{a}_1\varphi^{(0,1)}(t,X) + \mathbf{a}_2\varphi^{(0,1)}(t,X) + X\alpha\mathbf{p}_2(\mathbf{p}_2 - \mathbf{p}_1\rho_{2,1})\varphi^{(0,2)} \\ (t,X)\Big) \Big/ \Big(\left(\mathbf{p}_1^2 - 2\mathbf{p}_2\rho_{2,1}\mathbf{p}_1 + \mathbf{p}_2^2\right)\varphi^{(0,2)}(t,X)\Big) + \Big(\mathbf{a}_1\varphi^{(0,1)}(t,X)$$

$$-\mathbf{a}_2\,\varphi^{(0,1)}(t,X)+X\alpha\mathbf{p}_1\,(\mathbf{p}_1-\mathbf{p}_2\rho_{2,1})\,\varphi^{(0,2)}(t,X)\Big)\Big/$$

$$\Big((\mathbf{p}_1^2-2\mathbf{p}_2\rho_{2,1}\mathbf{p}_1+\ \mathbf{p}_2^2)\,\varphi^{(0,2)}(t,X)\Big)$$

$$=\alpha X, \tag{3.5.29}$$

the imposed affine constraint (3.3.6) is indeed satisfied in (3.5.28).

3.6 CRRA Utility Case: No Constraints on the Portfolio

We now specialize to the most convenient (nontrivial) utility function of all, the CRRA utility $\psi_\gamma(X)$, the constant relative risk-aversion utility given in (3.2.7). We remark that log utility is much simpler, even trivial, but not realistic for applications – it is not sufficiently risk-averse; see, e.g., Example 7.3.1 below.

In the case of the CRRA utility (3.2.6), we look for the solutions of equation (3.4.2) in the form

$$\varphi(t,X,A)=\frac{X^{1-\gamma}e^{g_\gamma(t,A)}-1}{1-\gamma} \tag{3.6.1}$$

for $\gamma\neq 1$, and/or alternatively, in the form (see [22])

$$\varphi(t,X,A)=\frac{X^{1-\gamma}f_\gamma(t,A)^\gamma-1}{1-\gamma} \tag{3.6.2}$$

for $\gamma\neq 1$. So, we concurrently prepare two sets of formulas:

$$\frac{\partial\varphi(t,X,A)}{\partial t}=\frac{e^{g_\gamma(t,A)}X^{1-\gamma}}{1-\gamma}\frac{\partial g_\gamma(t,A)}{\partial t}=\frac{X^{1-\gamma}\gamma f_\gamma(t,A)^{\gamma-1}}{1-\gamma}\frac{\partial f_\gamma(t,A)}{\partial t}, \tag{3.6.3}$$

$$\frac{\partial\varphi(t,X,A)}{\partial X}=X^{-\gamma}e^{g_\gamma(t,A)}=X^{-\gamma}f_\gamma(t,A)^\gamma, \tag{3.6.4}$$

$$\frac{\partial^2\varphi(t,X,A)}{\partial X^2}=-X^{-\gamma-1}\gamma e^{g_\gamma(t,A)}=-X^{-\gamma-1}\gamma f_\gamma(t,A)^\gamma, \tag{3.6.5}$$

$$\nabla_A\varphi(t,X,A)=\frac{e^{g_\gamma(t,A)}X^{1-\gamma}}{1-\gamma}\nabla_A g_\gamma(t,A)=\frac{X^{1-\gamma}}{1-\gamma}\gamma f_\gamma(t,A)^{\gamma-1}\nabla_A f_\gamma(t,A), \tag{3.6.6}$$

$$\nabla_A\frac{\partial\varphi(t,X,A)}{\partial X}=e^{g_\gamma(t,A)}X^{-\gamma}\nabla_A g_\gamma(t,A)=X^{-\gamma}\gamma f(t,A)^{\gamma-1}\nabla_A f_\gamma(t,A), \tag{3.6.7}$$

$$
\begin{aligned}
\nabla_A \nabla_A \varphi(t,X,A) &= \frac{X^{1-\gamma}}{1-\gamma} \nabla_A \left(e^{g_\gamma(t,A)} \nabla_A g_\gamma(t,A) \right) \\
&= \frac{X^{1-\gamma}}{1-\gamma} e^{g_\gamma(t,A)} \left(\{\nabla_A g_\gamma(t,A)\}^{\mathrm{T}} \cdot \{\nabla_A g_\gamma(t,A)\} + \nabla_A \nabla_A g_\gamma(t,A) \right) \\
&= -\frac{1}{\gamma-1} X^{1-\gamma} \gamma f_\gamma(t,A)^{\gamma-2} \left((\gamma-1) \{\nabla_A f_\gamma(t,A)\}^{\mathrm{T}} \cdot \{\nabla_A f_\gamma(t,A)\} \right. \\
&\quad \left. + f_\gamma(t,A) \nabla_A \nabla_A f_\gamma(t,A) \right). \qquad (3.6.8)
\end{aligned}
$$

Remark 3.6.1. Observe that since $\nabla_A g_\gamma$ (or $\nabla_A f_\gamma$) is an m-vector, $\{\nabla_A g_\gamma\}$ is a row-vector, i.e., a $1 \times m$ matrix, and consequently, $\{\nabla_A g_\gamma\}^{\mathrm{T}} \cdot \{\nabla_A g_\gamma\}$ is an $m \times m$ matrix. Also observe that

$$
\mathrm{Tr}\left(c \cdot c^{\mathrm{T}} \cdot \left(\{\nabla_A g_\gamma\}^{\mathrm{T}} \cdot \{\nabla_A g_\gamma\} \right) \right) = \nabla_A g_\gamma \cdot c \cdot c^{\mathrm{T}} \cdot \nabla_A g_\gamma. \qquad (3.6.9)
$$

Using (3.6.3)–(3.6.8), we now get (again concurrently two sets of formulas)

$$
\frac{\partial^2 \varphi}{\partial X^2} \frac{\partial \varphi}{\partial t} = -\frac{X^{-2\gamma} \gamma}{1-\gamma} \frac{\partial g_\gamma}{\partial t} e^{2g_\gamma} = -\frac{X^{-2\gamma} \gamma^2 f_\gamma^{2\gamma-1}}{1-\gamma} \frac{\partial f_\gamma}{\partial t}, \qquad (3.6.10)
$$

$$
\begin{aligned}
&-\frac{1}{2} \left(\frac{\partial \varphi}{\partial X} \right)^2 (a_s - r) \cdot \left(\sigma_s \cdot \sigma_s^{\mathrm{T}} \right)^{-1} . (a_s - r) \\
&\qquad = -\frac{1}{2} e^{2g_\gamma} X^{-2\gamma} (a_s - r) \cdot \left(\sigma_s \cdot \sigma_s^{\mathrm{T}} \right)^{-1} . (a_s - r) \\
&\qquad = -\frac{1}{2} X^{-2\gamma} f_\gamma^{2\gamma} (a_s - r) \cdot \left(\sigma_s \cdot \sigma_s^{\mathrm{T}} \right)^{-1} . (a_s - r), \qquad (3.6.11)
\end{aligned}
$$

$$
rX \frac{\partial^2 \varphi}{\partial X^2} \frac{\partial \varphi}{\partial X} = -e^{2g_\gamma} r X^{-2\gamma} \gamma = -r X^{-2\gamma} \gamma f_\gamma^{2\gamma}, \qquad (3.6.12)
$$

$$
b \cdot \nabla_A \varphi \frac{\partial^2 \varphi}{\partial X^2} = -\frac{e^{2g_\gamma} X^{-2\gamma} \gamma b \cdot \nabla_A g_\gamma}{1-\gamma} = -\frac{X^{-2\gamma} \gamma^2 f_\gamma^{2\gamma-1} b \cdot \nabla_A f_\gamma}{1-\gamma}, \qquad (3.6.13)
$$

$$
\begin{aligned}
&-\frac{\partial \varphi}{\partial X} (a_s - r) \cdot \left(\sigma_s \cdot \sigma_s^{\mathrm{T}} \right)^{-1} \cdot \sigma_s \cdot c^{\mathrm{T}} \cdot \left(\nabla_A \frac{\partial \varphi}{\partial X} \right) \\
&= -X^{-2\gamma} e^{2g_\gamma} (a_s - r) \cdot \left(\sigma_s \cdot \sigma_s^{\mathrm{T}} \right)^{-1} \cdot \sigma_s \cdot c^{\mathrm{T}} \cdot \nabla_A g_\gamma \\
&= -X^{-2\gamma} \gamma f_\gamma^{2\gamma-1} (a_s - r) \cdot \left(\sigma_s \cdot \sigma_s^{\mathrm{T}} \right)^{-1} \cdot \sigma_s \cdot c^{\mathrm{T}} \cdot \nabla_A f_\gamma, \qquad (3.6.14)
\end{aligned}
$$

$$
\begin{aligned}
&-\frac{1}{2}\left(\nabla_A \frac{\partial \varphi}{\partial X}\right)\cdot c\cdot \sigma_s{}^{\mathrm{T}}\cdot\left(\sigma_s\cdot\sigma_s{}^{\mathrm{T}}\right)^{-1}\cdot\sigma_s\cdot c^{\mathrm{T}}\cdot\left(\nabla_A \frac{\partial \varphi}{\partial X}\right)\\
&\quad=-\frac{1}{2}\mathrm{e}^{2g_\gamma}X^{-2\gamma}\nabla_A g_\gamma\cdot c\cdot\sigma_s{}^{\mathrm{T}}\cdot\left(\sigma_s\cdot\sigma_s{}^{\mathrm{T}}\right)^{-1}\cdot\sigma_s\cdot c^{\mathrm{T}}\cdot\nabla_A g_\gamma\\
&\quad=-\frac{1}{2}X^{-2\gamma}\gamma^2 f_\gamma{}^{2\gamma-2}\nabla_A f_\gamma\cdot c\cdot\sigma_s{}^{\mathrm{T}}\cdot\left(\sigma_s\cdot\sigma_s{}^{\mathrm{T}}\right)^{-1}\cdot\sigma_s\cdot c^{\mathrm{T}}\cdot\nabla_A f_\gamma,
\end{aligned}
\tag{3.6.15}
$$

and, using also (3.6.9),

$$
\begin{aligned}
&\frac{1}{2}\frac{\partial^2\varphi}{\partial X^2}\mathrm{Tr}\left(c\cdot c^{\mathrm{T}}\cdot\nabla_A\nabla_A\varphi\right)=-\frac{1}{2}\frac{\mathrm{e}^{g_\gamma}X^{-2\gamma}\gamma}{1-\gamma}\mathrm{Tr}\left(c\cdot c^{\mathrm{T}}\cdot\nabla_A\nabla_A\mathrm{e}^{g_\gamma}\right)\\
&\quad=-\frac{1}{2}\frac{\mathrm{e}^{2g_\gamma}X^{-2\gamma}\gamma}{1-\gamma}\left(\nabla_A g_\gamma\cdot c\cdot c^{\mathrm{T}}\cdot\nabla_A g_\gamma+\mathrm{Tr}\left(c\cdot c^{\mathrm{T}}\cdot\nabla_A\nabla_A g_\gamma\right)\right)\\
&\quad=\frac{X^{-2\gamma}\gamma^2 f_\gamma{}^{2\gamma-2}}{2(\gamma-1)}\left((\gamma-1)\nabla_A f_\gamma\cdot c\cdot c^{\mathrm{T}}\cdot\nabla_A f_\gamma+f_\gamma\mathrm{Tr}\left(c\cdot c^{\mathrm{T}}\cdot\nabla_A\nabla_A f_\gamma\right)\right).
\end{aligned}
\tag{3.6.16}
$$

Summing up, we state the following theorem.

Theorem 3.6.1. *In the case of CRRA utility (3.2.6) he value function $\varphi(t,X,A)$ of the optimal investment portfolio problem with no constraints on the portfolio. Problem (3.3.1), given (3.3.3) has the form (3.6.1), where $g_\gamma = g_\gamma(t,A)$ is the solution of the quasilinear PDE*

$$
\begin{aligned}
&\frac{\partial g_\gamma}{\partial t}+\frac{1}{2}Tr\left(c\cdot c^{\mathrm{T}}\cdot\nabla_A\nabla_A g_\gamma\right)+\left(b-\frac{\gamma-1}{\gamma}(a_s-r)\cdot\left(\sigma_s\cdot\sigma_s{}^{\mathrm{T}}\right)^{-1}\cdot\sigma_s\cdot c^{\mathrm{T}}\right)\cdot\nabla_A g_\gamma\\
&\qquad+\frac{1}{2}\nabla_A g_\gamma\cdot c\cdot\left(\mathbb{I}_n-\frac{\gamma-1}{\gamma}\sigma_s{}^{\mathrm{T}}\cdot\left(\sigma_s\cdot\sigma_s{}^{\mathrm{T}}\right)^{-1}\cdot\sigma_s\right)\cdot c^{\mathrm{T}}\cdot\nabla_A g_\gamma\\
&\quad=\frac{\gamma-1}{\gamma}\left(\frac{1}{2}(a_s-r)\cdot\left(\sigma_s\cdot\sigma_s{}^{\mathrm{T}}\right)^{-1}\cdot(a_s-r)+r\gamma\right)
\end{aligned}
\tag{3.6.17}
$$

for $t < T < \infty, A \in \mathcal{A}$, together with the terminal condition $g_\gamma(T,A) = 0$, while the optimal portfolio strategy (3.4.3) is equal to

$$
\Pi_\gamma^\star(t,X,A)=\frac{X}{\gamma}\left(a_s-r+\nabla_A g_\gamma\cdot c\cdot\sigma_s{}^{\mathrm{T}}\right)\cdot\left(\sigma_s\cdot\sigma_s{}^{\mathrm{T}}\right)^{-1}. \tag{3.6.18}
$$

Alternatively, the value function $\varphi(t,X,A)$ has the form (3.6.2), where $f_\gamma = f_\gamma(t,A)$ is the solution of the quasilinear PDE

$$\frac{\partial f_\gamma}{\partial t} + \frac{1}{2}\mathrm{Tr}\left(c\cdot c^{\mathrm{T}}\cdot\nabla_A\nabla_A f_\gamma\right) + \left(b - \frac{\gamma-1}{\gamma}(a_s-r)\cdot\left(\sigma_s\cdot\sigma_s{}^{\mathrm{T}}\right)^{-1}\cdot\sigma_s\cdot c^{\mathrm{T}}\right)\cdot\nabla_A f_\gamma$$
$$+\frac{1}{2}\frac{\gamma-1}{f}\nabla_A f_\gamma\cdot c\cdot\left(\mathbb{I}_n - \sigma_s{}^{\mathrm{T}}\cdot\left(\sigma_s\cdot\sigma_s{}^{\mathrm{T}}\right)^{-1}\cdot\sigma_s\right)\cdot c^{\mathrm{T}}\cdot\nabla_A f_\gamma$$
$$-\frac{\gamma-1}{\gamma^2}\left(\frac{1}{2}(a_s-r)\cdot\left(\sigma_s\cdot\sigma_s{}^{\mathrm{T}}\right)^{-1}\cdot(a_s-r)+r\gamma\right)f_\gamma = 0 \tag{3.6.19}$$

for $t < T < \infty, A\in\mathcal{A}$, together with the terminal condition $f_\gamma(T,A) = 1$, while the optimal portfolio strategy (3.4.3) is equal to

$$\Pi^\star_\gamma(t,X,A) = X\left(\frac{1}{\gamma}(a_s-r) + \frac{\nabla_A f_\gamma}{f_\gamma}\cdot c\cdot\sigma_s{}^{\mathrm{T}}\right)\cdot\left(\sigma_s\cdot\sigma_s{}^{\mathrm{T}}\right)^{-1}. \tag{3.6.20}$$

Proof. What remains to be derived are the optimal portfolio strategies (3.6.18) and (3.6.20). From (3.6.4), (3.6.5), we have □

$$-\frac{\frac{\partial\varphi(t,X,A)}{\partial X}}{\frac{\partial^2\varphi(t,X,A)}{\partial X^2}} = -\frac{e^{g_\gamma(t,A)}X^{-\gamma}}{-e^{g_\gamma(t,A)}X^{-\gamma-1}\gamma} = \frac{X}{\gamma} = -\frac{X^{-\gamma}f_\gamma(t,A)^\gamma}{-X^{-\gamma-1}\gamma f_\gamma(t,A)^\gamma}, \tag{3.6.21}$$

while from (3.6.4), (3.6.7) we obtain

$$-\frac{\nabla_A\frac{\partial\varphi(t,X,A)}{\partial X}}{\frac{\partial^2\varphi(t,X,A)}{\partial X^2}} = -\frac{e^{g_\gamma(t,A)}X^{-\gamma}\nabla_A g_\gamma(t,A)}{-e^{g_\gamma(t,A)}X^{-\gamma-1}\gamma} = \frac{X}{\gamma}\nabla_A g_\gamma(t,A), \tag{3.6.22}$$

$$-\frac{\nabla_A\frac{\partial\varphi(t,X,A)}{\partial X}}{\frac{\partial^2\varphi(t,X,A)}{\partial X^2}} = -\frac{X^{-\gamma}\gamma f(t,A)^{\gamma-1}\nabla_A f_\gamma(t,A)}{-X^{-\gamma-1}\gamma f_\gamma(t,A)^\gamma} = \frac{X}{f_\gamma(t,A)}\nabla_A f_\gamma(t,A), \tag{3.6.23}$$

which, together with (3.4.3), imply (3.6.18) and (3.6.20).

Remark 3.6.2. The optimal portfolio formula (3.6.18) provides a correct intuition about $\Pi^\star_\gamma$ as $\gamma\to\infty$, while formula (3.6.20) may be misleading in that regard, since quite generally, we have

$$\frac{\nabla_A f_\gamma}{f_\gamma} = \frac{\nabla_A g_\gamma}{\gamma} \longrightarrow 0 \tag{3.6.24}$$

as $\gamma\to\infty$.

Remark 3.6.3. Notice that if the market is complete, i.e., (2.3.1) holds, then equation (3.6.24) reduces to a linear equation:

$$\frac{\partial f_\gamma}{\partial t} + \frac{1}{2}\mathrm{Tr}\left(c \cdot c^{\mathrm{T}} \cdot \nabla_A \nabla_A f_\gamma\right) + \left(b - \frac{\gamma-1}{\gamma}(a_s - r)\cdot\left(\sigma_s \cdot \sigma_s{}^{\mathrm{T}}\right)^{-1} \cdot \sigma_s \cdot c^{\mathrm{T}}\right) \cdot \nabla_A f_\gamma$$
$$-\frac{\gamma-1}{\gamma^2}\left(\frac{1}{2}(a_s - r)\cdot\left(\sigma_s \cdot \sigma_s{}^{\mathrm{T}}\right)^{-1}\cdot(a_s - r) + r\gamma\right) f_\gamma = 0 \qquad (3.6.25)$$

for $t < T < \infty, A \in \mathcal{A}$, with the terminal condition $f_\gamma(T,A) = 1$.

Remark 3.6.4. Alternatively, equation (3.6.19) follows from (3.6.17) (or vice versa), by setting $g_\gamma = \gamma \log\left(f_\gamma\right)$ and using the formulas

$$\frac{\partial g_\gamma}{\partial t} = \frac{\gamma}{f_\gamma}\frac{\partial f_\gamma}{\partial t}, \qquad (3.6.26)$$

$$\nabla_A g_\gamma = \frac{\gamma}{f_\gamma}\nabla_A f_\gamma, \qquad (3.6.27)$$

$$\nabla_A \nabla_A g_\gamma = \nabla_A\left(\frac{\gamma}{f_\gamma}\nabla_A f_\gamma\right) = -\frac{\gamma}{f_\gamma{}^2}\left\{\nabla_A f_\gamma\right\}^{\mathrm{T}} \cdot \left\{\nabla_A f_\gamma\right\} + \frac{\gamma}{f_\gamma}\nabla_A \nabla_A f_\gamma, \qquad (3.6.28)$$

$$\frac{1}{2}\mathrm{Tr}\left(c\cdot c^{\mathrm{T}}\cdot\nabla_A\nabla_A g_\gamma\right) = \frac{1}{2}\mathrm{Tr}\left(c\cdot c^{\mathrm{T}}\cdot\left(-\frac{\gamma}{f_\gamma{}^2}\left\{\nabla_A f_\gamma\right\}^{\mathrm{T}}\cdot\left\{\nabla_A f_\gamma\right\} + \frac{\gamma}{f_\gamma}\nabla_A\nabla_A f_\gamma\right)\right)$$
$$= -\frac{1}{2}\frac{\gamma}{f_\gamma{}^2}\nabla_A f_\gamma \cdot c \cdot c^{\mathrm{T}} \cdot \nabla_A f_\gamma + \frac{1}{2}\frac{\gamma}{f_\gamma}\mathrm{Tr}\left(c\cdot c^{\mathrm{T}}\cdot\nabla_A\nabla_A f_\gamma\right). \qquad (3.6.29)$$

Remark 3.6.5. As mentioned before, the log utility, i.e., the $\gamma = 1$ case, is trivial, since equations (3.6.17) and (3.6.19) solved by $g_\gamma = 0$ and $f_\gamma = 1$, respectively.

Example 3.6.1. (CRRA optimal portfolio under stochastic appreciation rate.) To illustrate the conclusion of Theorem 3.6.1 consider a simple economy $\mathfrak{E}$ consisting of factors $A = \{S_1, \mathbf{v}\}$, a tradable $S = \{S_1\}$ with market coefficients (2.4.5) and a constant interest rate r. Assume also the mean reversion for the appreciation rate, i.e., assume $q_1 < 0$.

Under such conditions, the PDE (3.6.17) becomes (for $g = g_\gamma$)

$$\left(q_0 + \mathbf{v}q_1 - \frac{w(\mathbf{v}-r)(\gamma-1)\rho_{2,1}}{\gamma\mathbf{p}_1}\right) g^{(0,0,1)}(t,S_1,\mathbf{v})$$
$$+\left(S_1\left(\mathbf{v}-\mathbb{D}_1\right) - \frac{(\mathbf{v}-r)(\gamma-1)S_1}{\gamma}\right) g^{(0,1,0)}(t,S_1,\mathbf{v})$$
$$+\frac{1}{2\gamma}\left(w^2\left(\gamma-(\gamma-1)\rho_{2,1}^2\right) g^{(0,0,1)}(t,S_1,\mathbf{v})^2 + 2wS_1\mathbf{p}_1\rho_{2,1}\right.$$
$$\left. g^{(0,1,0)}(t,S_1,\mathbf{v})\, g^{(0,0,1)}(t,S_1,\mathbf{v}) + S_1^2\mathbf{p}_1^2 g^{(0,1,0)}(t,S_1,\mathbf{v})^2\right)$$

$$+\frac{1}{2}\left(g^{(0,0,2)}(t,S_1,\mathbf{v})w^2+S_1\mathbf{p}_1\left(2w\rho_{2,1}g^{(0,1,1)}(t,S_1,\mathbf{v})+S_1\mathbf{p}_1 g^{(0,2,0)}(t,S_1,\mathbf{v})\right)\right)$$
$$+g^{(1,0,0)}(t,S_1,\mathbf{v})=\frac{\gamma-1}{\gamma}\left(\frac{1}{2}\left(\frac{\mathbf{v}-r}{\mathbf{p}_1}\right)^2+r\gamma\right) \tag{3.6.30}$$

for $t < T$, with the terminal condition $g(T,S_1,\mathbf{v}) = 0$. We look for the solution of (3.6.30) in the form $g(t,S_1,\mathbf{v}) = g(t,\mathbf{v})$, and therefore, equation (3.6.30) simplifies to

$$\frac{\left(\gamma-(\gamma-1)\rho_{2,1}^2\right)g^{(0,1)}(t,\mathbf{v})^2w^2}{2\gamma}+\frac{1}{2}g^{(0,2)}(t,\mathbf{v})w^2$$
$$+\left(q_0+\mathbf{v}q_1-\frac{w(\mathbf{v}-r)(\gamma-1)\rho_{2,1}}{\gamma\mathbf{p}_1}\right)g^{(0,1)}(t,\mathbf{v})+g^{(1,0)}(t,\mathbf{v})$$
$$=\frac{\gamma-1}{\gamma}\left(\frac{1}{2}\left(\frac{\mathbf{v}-r}{\mathbf{p}_1}\right)^2+r\gamma\right) \tag{3.6.31}$$

for $t < T$, with the terminal condition $g(T,\mathbf{v}) = 0$. We look for the solution of (3.6.31) in the form $g(t,\mathbf{v}) = \mathbb{F}_{2,\gamma}(t)\mathbf{v}^2+\mathbb{F}_{1,\gamma}(t)\mathbf{v}+\mathbb{F}_{0,\gamma}(t)$, and therefore, equation (3.6.30) simplifies to

$$\frac{1}{2\gamma}\left(\gamma-(\gamma-1)\rho_{2,1}^2\right)\left(\mathbb{F}_{1,\gamma}(t)+2\mathbf{v}\mathbb{F}_{2,\gamma}(t)\right)^2w^2+\mathbb{F}_{2,\gamma}(t)w^2+\left(q_0+\mathbf{v}q_1\right.$$
$$\left.-\frac{w(\mathbf{v}-r)(\gamma-1)\rho_{2,1}}{\gamma\mathbf{p}_1}\right)\left(\mathbb{F}_{1,\gamma}(t)+2\mathbf{v}\mathbb{F}_{2,\gamma}(t)\right)+\mathbb{F}'_{0,\gamma}(t)+\mathbf{v}\mathbb{F}'_{1,\gamma}(t)+\mathbf{v}^2\mathbb{F}'_{2,\gamma}(t)$$
$$=\frac{\gamma-1}{\gamma}\left(\frac{1}{2}\left(\frac{\mathbf{v}-r}{\mathbf{p}_1}\right)^2+r\gamma\right) \tag{3.6.32}$$

for any $\mathbf{v}$, for $t < T$, with the terminal conditions $\mathbb{F}_{2,\gamma}(T) = \mathbb{F}_{1,\gamma}(T) = \mathbb{F}_{0,\gamma}(T) = 0$. Equation (3.6.32) implies the three ODEs

$$\frac{(1-\gamma)r^2}{2\gamma\mathbf{p}_1^2}+(1-\gamma)r+\frac{w(\gamma-1)\rho_{2,1}\mathbb{F}_{1,\gamma}(t)r}{\gamma\mathbf{p}_1}+\frac{w^2\left(\gamma-(\gamma-1)\rho_{2,1}^2\right)\mathbb{F}_{1,\gamma}(t)^2}{2\gamma}$$
$$+q_0\mathbb{F}_{1,\gamma}(t)+w^2\mathbb{F}_{2,\gamma}(t)+\mathbb{F}_{0,\gamma}{}'(t)=0, \tag{3.6.33}$$

$$\frac{2\left(\gamma-(\gamma-1)\rho_{2,1}^2\right)\mathbb{F}_{1,\gamma}(t)\mathbb{F}_{2,\gamma}(t)w^2}{\gamma}+\frac{2r(\gamma-1)\rho_{2,1}\mathbb{F}_{2,\gamma}(t)w}{\gamma\mathbf{p}_1}$$
$$-\frac{(\gamma-1)\rho_{2,1}\mathbb{F}_{1,\gamma}(t)w}{\gamma\mathbf{p}_1}+q_1\mathbb{F}_{1,\gamma}(t)+2q_0\mathbb{F}_{2,\gamma}(t)+\mathbb{F}_{1,\gamma}{}'(t)-\frac{r(1-\gamma)}{\gamma\mathbf{p}_1^2}=0, \tag{3.6.34}$$

and

$$\frac{2w^2\left(\gamma-(\gamma-1)\rho_{2,1}^2\right)\mathbb{F}_{2,\gamma}(t)^2}{\gamma}+2q_1\mathbb{F}_{2,\gamma}(t)-\frac{2w(\gamma-1)\rho_{2,1}\mathbb{F}_{2,\gamma}(t)}{\gamma\mathbf{p}_1}+\mathbb{F}'_{2,\gamma}(t)$$
$$+\frac{1-\gamma}{2\gamma\mathbf{p}_1^2}=0 \tag{3.6.35}$$

for $t<T$, with the terminal conditions $\mathbb{F}_{2,\gamma}(T)=\mathbb{F}_{1,\gamma}(T)=\mathbb{F}_{0,\gamma}(T)=0$. Since not g_γ but only $\nabla_A g_\gamma$ is needed to compute the optimal portfolio strategy (3.6.18), and since $\nabla_A g_\gamma=\nabla_{\{S_1,\mathbf{v}\}}\left(\mathbb{F}_{2,\gamma}(t)\mathbf{v}^2+\mathbb{F}_{1,\gamma}(t)\mathbf{v}+\mathbb{F}_{0,\gamma}(t)\right)=\left\{0,\mathbb{F}_{1,\gamma}(t)+2\mathbf{v}\mathbb{F}_{2,\gamma}(t)\right\}$, equation (3.6.33) is not needed. Indeed, since

$$(a_s-r)\cdot\left(\sigma_s\cdot\sigma_s{}^{\mathrm{T}}\right)^{-1}=\left\{\frac{\mathbf{v}-r}{\mathbf{p}_1^2}\right\} \tag{3.6.36}$$

and

$$\nabla_A g_\gamma\cdot c\cdot\sigma_s{}^{\mathrm{T}}\cdot\left(\sigma_s\cdot\sigma_s{}^{\mathrm{T}}\right)^{-1}=\left\{0,\mathbb{F}_{1,\gamma}(t)+2\mathbf{v}\mathbb{F}_{2,\gamma}(t)\right\}\cdot\begin{pmatrix}S_1\\ \frac{w\rho_{2,1}}{\mathbf{p}_1}\end{pmatrix}$$
$$=\left\{\frac{w\rho_{2,1}\left(\mathbb{F}_{1,\gamma}(t)+2\mathbf{v}\mathbb{F}_{2,\gamma}(t)\right)}{\mathbf{p}_1}\right\}, \tag{3.6.37}$$

we have

$$\Pi_\gamma^\star(t,X,A)=\frac{X}{\gamma}\left\{\frac{\mathbf{v}-r}{\mathbf{p}_1^2}+\frac{w\rho_{2,1}}{\mathbf{p}_1}\left(\mathbb{F}_{1,\gamma}(t)+2\mathbf{v}\mathbb{F}_{2,\gamma}(t)\right)\right\}. \tag{3.6.38}$$

Remark 3.6.6. What happens when $\gamma\to\infty$, i.e., what would be an "infinitely conservative" portfolio strategy? Equation (3.6.17) seems more suitable to answer that question than (3.6.19), since the γ-dependent expressions in (3.6.17) all have the form $(\gamma-1)/\gamma$, and $(\gamma-1)/\gamma\to 1$, as $\gamma\to\infty$. Indeed, passing to the limit $\gamma\to\infty$ in (3.6.34) and (3.6.35), we conclude that $\nabla_A g_\gamma=\left\{0,\mathbb{F}_{1,\gamma}(t)+2\mathbf{v}\mathbb{F}_{2,\gamma}(t)\right\}$ converges as $\gamma\to\infty$, and consequently $\nabla_A f_\gamma/f_\gamma=\nabla_A g_\gamma/\gamma\to 0$ as $\gamma\to\infty$. So, the optimal portfolio formula (3.6.20) is somewhat misleading, since γ does not appear explicitly in the second term.

Remark 3.6.7. Recall Remark 3.4.2. Here we can see explicitly what happens in general. Equations (3.6.34) and (3.6.35), due to the presence of the zero-order (damping) term, have a steady-state solutions (i.e., we can pass to the limit $T\to\infty$ for them, or equivalently, $t\to-\infty$; say $\mathbb{A}_1:=\lim_{t\to-\infty}\mathbb{F}_{1,\gamma}(t)$ and $\mathbb{A}_2:=\lim_{t\to-\infty}\mathbb{F}_{2,\gamma}(t)$). The steady-state solutions are, moreover, characterized as solutions of the algebraic equations (by means of setting derivative terms to zero)

$$0=\frac{2w^2\left(\gamma-(\gamma-1)\rho_{2,1}^2\right)\mathbb{A}_2^2}{\gamma}+2q_1\mathbb{A}_2-\frac{2w(\gamma-1)\rho_{2,1}\mathbb{A}_2}{\gamma\mathbf{p}_1}+\frac{1-\gamma}{2\gamma\mathbf{p}_1^2} \tag{3.6.39}$$

and

$$\frac{2\mathbb{A}_1\mathbb{A}_2\left(\gamma-(\gamma-1)\rho_{2,1}^2\right)w^2}{\gamma}+\frac{2r(\gamma-1)\mathbb{A}_2\rho_{2,1}w}{\gamma\mathbf{p}_1}-\frac{(\gamma-1)\mathbb{A}_1\rho_{2,1}w}{\gamma\mathbf{p}_1}$$

$$+q_1\mathbb{A}_1+2q_0\mathbb{A}_2-\frac{r(1-\gamma)}{\gamma\mathbf{p}_1^2}=0. \tag{3.6.40}$$

On the other hand, we cannot pass to the limit $T \to \infty$, or equivalently, $t \to -\infty$, in equation (3.6.33), since that equation has no zero-order term. Indeed, assuming the convergence for $\mathbb{F}_{1,\gamma}(t)$ and $\mathbb{F}_{2,\gamma}(t)$, in the long run, equation (3.6.33) is like the equation

$$\mathrm{const}+\mathbb{F}_{0,\gamma}'(t)=0, \tag{3.6.41}$$

which cannot have a steady state. Since all three functions $\mathbb{F}_{0,\gamma}(t)$, $\mathbb{F}_{1,\gamma}(t)$, and $\mathbb{F}_{2,\gamma}(t)$ constitute the value function, i.e., the solution of the equation (3.6.17), yet only two, $\mathbb{F}_{1,\gamma}(t)$ and $\mathbb{F}_{2,\gamma}(t)$, constitute the optimal portfolio strategy (3.6.18), the optimal portfolio can and does converge to a steady state, as well as the corresponding derivative expressions of the value function, while the value function itself does not converge to a steady state.

3.7 CRRA Utility Case: Affine Constraints on the Portfolio

In the case of portfolio optimization under CRRA utility and affine constraints on the portfolio, we look for the solutions of the equation (3.5.22), again in the form (3.6.1). Yet for this derivation to work out, beyond what is already in (3.5.22), the constraint (3.3.4) has to be consistent with (3.6.1), i.e., with the CRRA utility. So a CRRA suitable form of the affine constraint is

$$\mu(t,A)\cdot\Pi^\star(t,X,A)=X\eta(t,A), \tag{3.7.1}$$

i.e., assuming

$$\begin{aligned}\mu(t,X,A)&=\mu(t,A),\\ \xi(t,X,A)&=X\eta(t,A),\end{aligned} \tag{3.7.2}$$

in (3.3.4). Recall formulas (3.6.3)–(3.6.9), and then prepare the formulas for all the (new) terms in equation (3.5.1) (already taking into the account the new form of the affine constraint):

$$-\frac{1}{2}\left(\frac{\partial\varphi}{\partial X}\right)^2(a_s-r)\cdot\left(\mathbb{I}_k-\left(\sigma_s\cdot\sigma_s{}^{\mathrm{T}}\right)^{-1}\cdot\mu^{\mathrm{T}}\cdot\left(\mu\cdot\left(\sigma_s\cdot\sigma_s{}^{\mathrm{T}}\right)^{-1}\cdot\mu^{\mathrm{T}}\right)^{-1}\cdot\mu\right)$$

$$\cdot\left(\sigma_s\cdot\sigma_s{}^{\mathrm{T}}\right)^{-1}\cdot(a_s-r)$$

$$= -\frac{1}{2}e^{2g\gamma}X^{-2\gamma}(a_s-r)\cdot\left(\mathbb{I}_k-\left(\sigma_s\cdot\sigma_s{}^{\mathrm{T}}\right)^{-1}\cdot\mu^{\mathrm{T}}\cdot\left(\mu\cdot\left(\sigma_s\cdot\sigma_s{}^{\mathrm{T}}\right)^{-1}\cdot\mu^{\mathrm{T}}\right)^{-1}\cdot\mu\right)$$
$$\cdot\left(\sigma_s\cdot\sigma_s{}^{\mathrm{T}}\right)^{-1}\cdot(a_s-r), \tag{3.7.3}$$

$$\frac{\partial\varphi}{\partial X}\frac{\partial^2\varphi}{\partial X^2}X\left(\eta\cdot\left(\mu\cdot\left(\sigma_s\cdot\sigma_s{}^{\mathrm{T}}\right)^{-1}\cdot\mu^{\mathrm{T}}\right)^{-1}\cdot\mu\cdot\left(\sigma_s\cdot\sigma_s{}^{\mathrm{T}}\right)^{-1}\cdot(a_s-r)+r\right)$$
$$= -e^{2g\gamma}X^{-2\gamma}\gamma\left(\eta.\left(\mu\cdot\left(\sigma_s\cdot\sigma_s{}^{\mathrm{T}}\right)^{-1}\cdot\mu^{\mathrm{T}}\right)^{-1}\cdot\mu\cdot\left(\sigma_s\cdot\sigma_s{}^{\mathrm{T}}\right)^{-1}\cdot(a_s-r)+r\right), \tag{3.7.4}$$

$$-\frac{\partial\varphi}{\partial X}\left(\nabla_A\frac{\partial\varphi}{\partial X}\right)\cdot c\cdot\sigma_s{}^{\mathrm{T}}\cdot\left(\mathbb{I}_k-\left(\sigma_s\cdot\sigma_s{}^{\mathrm{T}}\right)^{-1}\cdot\mu^{\mathrm{T}}\cdot\left(\mu\cdot\left(\sigma_s\cdot\sigma_s{}^{\mathrm{T}}\right)^{-1}\cdot\mu^{\mathrm{T}}\right)^{-1}\cdot\mu\right)$$
$$\cdot\left(\sigma_s\cdot\sigma_s{}^{\mathrm{T}}\right)^{-1}\cdot(a_s-r)$$
$$= -\frac{\partial\varphi}{\partial X}(a_s-r)\cdot\left(\sigma_s\cdot\sigma_s{}^{\mathrm{T}}\right)^{-1}\cdot\left(\mathbb{I}_k-\left(\sigma_s\cdot\sigma_s{}^{\mathrm{T}}\right)^{-1}\cdot\mu^{\mathrm{T}}\cdot\right.$$
$$\left.\left(\mu\cdot\left(\sigma_s\cdot\sigma_s{}^{\mathrm{T}}\right)^{-1}\cdot\mu^{\mathrm{T}}\right)^{-1}\cdot\mu\right)^{\mathrm{T}}\cdot\sigma_s\cdot c^{\mathrm{T}}\cdot\left(\nabla_A\frac{\partial\varphi}{\partial X}\right)$$
$$= -X^{-2\gamma}e^{2g\gamma}(a_s-r)\cdot\left(\sigma_s\cdot\sigma_s{}^{\mathrm{T}}\right)^{-1}\cdot\left(\mathbb{I}_k-\mu^{\mathrm{T}}\cdot\left(\mu\cdot\left(\sigma_s\cdot\sigma_s{}^{\mathrm{T}}\right)^{-1}\cdot\mu^{\mathrm{T}}\right)^{-1}\right.$$
$$\left.\cdot\mu\cdot\left(\sigma_s\cdot\sigma_s{}^{\mathrm{T}}\right)^{-1}\right)\cdot\sigma_s\cdot c^{\mathrm{T}}\cdot\nabla_A g_\gamma, \tag{3.7.5}$$

$$-\frac{1}{2}\left(\nabla_A\frac{\partial\varphi}{\partial X}\right)\cdot c\cdot\sigma_s{}^{\mathrm{T}}\cdot\left(\sigma_s\cdot\sigma_s{}^{\mathrm{T}}\right)^{-1}\cdot\left(\mathbb{I}_k-\mu^{\mathrm{T}}\cdot\left(\mu\cdot\left(\sigma_s\cdot\sigma_s{}^{\mathrm{T}}\right)^{-1}\cdot\mu^{\mathrm{T}}\right)^{-1}\right.$$
$$\left.\cdot\mu\cdot\left(\sigma_s\cdot\sigma_s{}^{\mathrm{T}}\right)^{-1}\right)\cdot\sigma_s\cdot c^{\mathrm{T}}\cdot\left(\nabla_A\frac{\partial\varphi}{\partial X}\right)$$
$$= -\frac{1}{2}e^{2g\gamma}X^{-2\gamma}\nabla_A g_\gamma\cdot c\cdot\sigma_s{}^{\mathrm{T}}\cdot\left(\sigma_s\cdot\sigma_s{}^{\mathrm{T}}\right)^{-1}\cdot\left(\mathbb{I}_k-\mu^{\mathrm{T}}\cdot\left(\mu\cdot\left(\sigma_s\cdot\sigma_s{}^{\mathrm{T}}\right)^{-1}\cdot\mu^{\mathrm{T}}\right)^{-1}\right.$$
$$\left.\cdot\mu\cdot\left(\sigma_s\cdot\sigma_s{}^{\mathrm{T}}\right)^{-1}\right)\cdot\sigma_s\cdot c^{\mathrm{T}}\cdot\nabla_A g_\gamma, \tag{3.7.6}$$

$$\frac{1}{2}\left(\frac{\partial^2\varphi}{\partial X^2}\right)^2 X\eta.\left(\mu\cdot\left(\sigma_s\cdot\sigma_s{}^{\mathrm{T}}\right)^{-1}\cdot\mu^{\mathrm{T}}\right)^{-1}\cdot\eta X$$
$$= \frac{1}{2}e^{2g\gamma}X^{-2\gamma}\gamma^2\eta.\left(\mu\cdot\left(\sigma_s\cdot\sigma_s{}^{\mathrm{T}}\right)^{-1}\cdot\mu^{\mathrm{T}}\right)^{-1}\cdot\eta, \tag{3.7.7}$$

$$\frac{\partial^2\varphi}{\partial X^2}X\eta.\left(\mu\cdot\left(\sigma_s\cdot\sigma_s{}^{\mathrm{T}}\right)^{-1}\cdot\mu^{\mathrm{T}}\right)^{-1}\cdot\mu\cdot\left(\sigma_s\cdot\sigma_s{}^{\mathrm{T}}\right)^{-1}\cdot\sigma_s\cdot c^{\mathrm{T}}\cdot\left(\nabla_A\frac{\partial\varphi}{\partial X}\right)$$
$$= -e^{2g\gamma}X^{-2\gamma}\gamma\eta\cdot\left(\mu\cdot\left(\sigma_s\cdot\sigma_s{}^{\mathrm{T}}\right)^{-1}\cdot\mu^{\mathrm{T}}\right)^{-1}\cdot\mu\cdot\left(\sigma_s\cdot\sigma_s{}^{\mathrm{T}}\right)^{-1}\cdot\sigma_s\cdot c^{\mathrm{T}}\cdot\nabla_A g_\gamma. \tag{3.7.8}$$

We therefore have the following.

Theorem 3.7.1. *In the case of the CRRA utility (3.2.6), the value function $\varphi(t,X,A)$ of the optimal investment portfolio problem with affine constraints on the portfolio, Problem 3.3.2, given in (3.3.5),has the form (3.6.1), where $g_\gamma = g_\gamma(t,A)$ is the solution of the following quasilinear PDE:*

$$\frac{\partial g_\gamma}{\partial t} + \frac{1}{2}Tr\left(c\cdot c^{\mathrm{T}}\cdot\nabla_A\nabla_A g_\gamma\right) + \left(b - \frac{\gamma-1}{\gamma}(a_s - r)\cdot\left(\sigma_s\cdot\sigma_s^{\mathrm{T}}\right)^{-1}\cdot\left(\mathbb{I}_k - \mu^{\mathrm{T}}\cdot(\mu\right.\right.$$
$$\left.\cdot\left(\sigma_s\cdot\sigma_s^{\mathrm{T}}\right)^{-1}\cdot\mu^{\mathrm{T}}\right)^{-1}\cdot\mu\cdot\left(\sigma_s\cdot\sigma_s^{\mathrm{T}}\right)^{-1}\Big)\cdot\sigma_s\cdot c^{\mathrm{T}} - (\gamma-1)\eta.\left(\mu\cdot\left(\sigma_s\cdot\sigma_s^{\mathrm{T}}\right)^{-1}\cdot\right.$$
$$\left.\mu^{\mathrm{T}}\right)^{-1}\cdot\mu\cdot\left(\sigma_s\cdot\sigma_s^{\mathrm{T}}\right)^{-1}\cdot\sigma_s\cdot c^{\mathrm{T}}\Big)\cdot\nabla_A g_\gamma + \frac{1}{2}\nabla_A g_\gamma\cdot c\cdot\left(\mathbb{I}_n - \frac{\gamma-1}{\gamma}\sigma_s^{\mathrm{T}}\cdot\left(\sigma_s\cdot\sigma_s^{\mathrm{T}}\right)^{-1}\right.$$
$$\left.\cdot\left(\mathbb{I}_k - \mu^{\mathrm{T}}\cdot\left(\mu\cdot\left(\sigma_s\cdot\sigma_s^{\mathrm{T}}\right)^{-1}\cdot\mu^{\mathrm{T}}\right)^{-1}\cdot\mu\cdot\left(\sigma_s\cdot\sigma_s^{\mathrm{T}}\right)^{-1}\right)\cdot\sigma_s\right)\cdot c^{\mathrm{T}}\cdot\nabla_A g_\gamma$$
$$= \frac{1}{2}\frac{\gamma-1}{\gamma}(a_s - r)\cdot\left(\mathbb{I}_k - \left(\sigma_s\cdot\sigma_s^{\mathrm{T}}\right)^{-1}\cdot\mu^{\mathrm{T}}\cdot\left(\mu\cdot\left(\sigma_s\cdot\sigma_s^{\mathrm{T}}\right)^{-1}\cdot\mu^{\mathrm{T}}\right)^{-1}\cdot\mu\right)\cdot$$
$$\left(\sigma_s\cdot\sigma_s^{\mathrm{T}}\right)^{-1}\cdot(a_s - r) + (\gamma-1)\left(\eta.\left(\mu\cdot\left(\sigma_s\cdot\sigma_s^{\mathrm{T}}\right)^{-1}\cdot\mu^{\mathrm{T}}\right)^{-1}\cdot\mu\cdot\left(\sigma_s\cdot\sigma_s^{\mathrm{T}}\right)^{-1}\right.$$
$$\left.\cdot(a_s - r) + r\right) - \frac{1}{2}\gamma(\gamma-1)\eta.\left(\mu\cdot\left(\sigma_s\cdot\sigma_s^{\mathrm{T}}\right)^{-1}\cdot\mu^{\mathrm{T}}\right)^{-1}\cdot\eta \tag{3.7.9}$$

for $t < T < \infty, A \in \mathcal{A}$, with terminal condition $g_\gamma(T,A) = 0$. Finally, the optimal portfolio strategy (3.5.2) simplifies to

$$\Pi_\gamma^\star(t,X,A) = X\left(\frac{1}{\gamma}\left(a_s - r + \nabla_A g_\gamma\cdot c\cdot\sigma_s^{\mathrm{T}}\right)\cdot\left(\mathbb{I}_k - \left(\sigma_s\cdot\sigma_s^{\mathrm{T}}\right)^{-1}\cdot\mu^{\mathrm{T}}\right.\right.$$
$$\left.\cdot\left(\mu\cdot\left(\sigma_s\cdot\sigma_s^{\mathrm{T}}\right)^{-1}\cdot\mu^{\mathrm{T}}\right)^{-1}\cdot\mu\right)\cdot\left(\sigma_s\cdot\sigma_s^{\mathrm{T}}\right)^{-1} + \eta$$
$$\left.\cdot\left(\mu\cdot\left(\sigma_s\cdot\sigma_s^{\mathrm{T}}\right)^{-1}\cdot\mu^{\mathrm{T}}\right)^{-1}\cdot\mu\cdot\left(\sigma_s\cdot\sigma_s^{\mathrm{T}}\right)^{-1}\right), \tag{3.7.10}$$

and in particular,

$$\Pi_1^\star(t,X,A) = X\left((a_s - r)\cdot\left(\mathbb{I}_k - \left(\sigma_s\cdot\sigma_s^{\mathrm{T}}\right)^{-1}\cdot\mu^{\mathrm{T}}\cdot\left(\mu\cdot\left(\sigma_s\cdot\sigma_s^{\mathrm{T}}\right)^{-1}\cdot\mu^{\mathrm{T}}\right)^{-1}\cdot\mu\right)\right.$$
$$\left.\cdot\left(\sigma_s\cdot\sigma_s^{\mathrm{T}}\right)^{-1} + \eta.\left(\mu\cdot\left(\sigma_s\cdot\sigma_s^{\mathrm{T}}\right)^{-1}\cdot\mu^{\mathrm{T}}\right)^{-1}\cdot\mu\cdot\left(\sigma_s\cdot\sigma_s^{\mathrm{T}}\right)^{-1}\right) \tag{3.7.11}$$

and

$$\Pi_\infty^\star(t,X,A) = X\eta.\left(\mu\cdot\left(\sigma_s\cdot\sigma_s^{\mathrm{T}}\right)^{-1}\cdot\mu^{\mathrm{T}}\right)^{-1}\cdot\mu\cdot\left(\sigma_s\cdot\sigma_s^{\mathrm{T}}\right)^{-1}, \tag{3.7.12}$$

provided $\nabla_A g_\gamma/\gamma \to 0$ as $\gamma \to \infty$.

Remark 3.7.1. Again, the log utility, i.e., the $\gamma = 1$ case, is trivial, since equation (3.7.9) is solved by $g_\gamma = 0$.

Remark 3.7.2. An alternative, more direct way (without assuming (3.7.2)) to derive formula (3.7.12) is by means of wealth-variance minimization. Indeed, from the wealth evolution equation (3.3.2) we compute the wealth variance as

$$\begin{aligned}\mathfrak{W}^{\Pi}(t) := & \lim_{dt \to 0} \frac{\left(dX^{\Pi}(t)\right)^2}{dt} \\ = & \, \Pi(t,X(t),A(t)) \cdot \sigma_s(t,A(t)) \cdot \sigma_s(t,A(t))^{\mathrm{T}} \cdot \Pi(t,X(t),A(t)) \geq 0. \quad (3.7.13)\end{aligned}$$

Then, for any t, X, and A, the problem is to find $\Pi(t,X,A)$ that minimizes the wealth variance, i.e., the quadratic function $f(\Pi) := \frac{1}{2}\Pi \cdot \sigma_s(t,A) \cdot \sigma_s(t,A)^{\mathrm{T}} \cdot \Pi$. This problem is trivial (solved by $\Pi = 0$) unless there is a constraint on the portfolio, such as

$$g(\Pi) := \Pi.\mu(t,X,A)^{\mathrm{T}} - \xi(t,X,A) = \mu(t,X,A) \cdot \Pi - \xi(t,X,A) = 0. \quad (3.7.14)$$

Again, we can apply the method of Lagrange multipliers, according to which there exists an l-vector $\lambda = \lambda(t,X,A)$ such that $\sigma \cdot \sigma^{\mathrm{T}} \cdot \Pi = \nabla_{\Pi} f(\Pi) = \lambda.\nabla_{\Pi} g(\Pi) = \lambda.\mu$, which implies

$$\sigma_s \cdot \sigma_s^{\mathrm{T}} \cdot \Pi = \lambda.\mu = \mu^{\mathrm{T}} \cdot \lambda, \quad (3.7.15)$$

and consequently

$$\Pi^{\star} = \left(\sigma_s \cdot \sigma_s^{\mathrm{T}}\right)^{-1} \cdot \mu^{\mathrm{T}} \cdot \lambda, \quad (3.7.16)$$

with $\lambda = \lambda(t,X,A)$ still unknown. Using the affine constraint $\mu \cdot \Pi^{\star} = \xi$, we get $\mu \cdot \left(\sigma_s \cdot \sigma_s^{\mathrm{T}}\right)^{-1} \cdot \mu^{\mathrm{T}} \cdot \lambda = \xi$, and therefore, since it is assumed that the $k \times k$ matrix $\mu \cdot \left(\sigma \cdot \sigma^{\mathrm{T}}\right)^{-1} \cdot \mu^{\mathrm{T}}$ is invertible,

$$\lambda = \left(\mu \cdot \left(\sigma_s \cdot \sigma_s^{\mathrm{T}}\right)^{-1} \cdot \mu^{\mathrm{T}}\right)^{-1} \cdot \xi(t,X,A), \quad (3.7.17)$$

from which

$$\Pi^{\star}(t,X,A) = \left(\sigma_s \cdot \sigma_s^{\mathrm{T}}\right)^{-1} \cdot \mu^{\mathrm{T}} \cdot \left(\mu \cdot \left(\sigma_s \cdot \sigma_s^{\mathrm{T}}\right)^{-1} \cdot \mu^{\mathrm{T}}\right)^{-1} \cdot \xi(t,X,A) \quad (3.7.18)$$

follows.

Exercise 3.7.1. Show that if the wealth-variance minimization portfolio strategy $\Pi_{\infty}^{\star}(t,X,A)$ is applied, then the realized wealth volatility is equal to

$$\frac{1}{X}\sqrt{\left(\Pi_{\infty}^{\star}(t,X,A) \cdot \sigma_s \cdot \sigma_s^{\mathrm{T}} \cdot \Pi_{\infty}^{\star}(t,X,A)\right)} = \sqrt{\eta \cdot \left(\mu \cdot \left(\sigma_s \cdot \sigma_s^{\mathrm{T}}\right)^{-1} \cdot \mu^{\mathrm{T}}\right)^{-1} \cdot \eta}. \quad (3.7.19)$$

Exercise 3.7.2. Set $g_\gamma = \gamma \log\left(f_\gamma\right)$ in (3.7.9) and (3.7.10) and derive the corresponding PDE and portfolio strategy, respectively. Recall Remark 3.6.2. Does the equation characterizing f_γ again become linear in the complete market case?

3.8 CARA Utility Case: No Constraints on the Portfolio – Deterministic Interest Rates

The CARA utility (3.2.12) is somewhat less tractable then the CRRA utility (3.2.7), since the derivation below will require that the interest rates be deterministic. Moreover, the derived results appear less compact than in the case of the CRRA utility.

Set

$$\varphi(t,X,A) = \frac{1-\mathrm{e}^{\omega\left(1-X\mathrm{e}^{\int_t^{\mathrm{T}} r(\tau)\,\mathrm{d}\tau}\right)}\mathrm{e}^{g_\omega(t,A)}}{\omega} = \frac{1}{\omega}\left(1-\mathrm{e}^{-\mathrm{e}^{\int_t^{\mathrm{T}} r(\tau)\,\mathrm{d}\tau}X\omega+\omega+g_\omega(t,A)}\right) \tag{3.8.1}$$

for $\omega > 0$. Then

$$\varphi(T,X,A) = \frac{1-\mathrm{e}^{-(X-1)\omega+g_\omega(T,A)}}{\omega} = \frac{1-\mathrm{e}^{\omega(1-X)}}{\omega} = \psi_{C,\omega}(X), \tag{3.8.2}$$

provided

$$g_\omega(T,A) = 0. \tag{3.8.3}$$

We now prepare formulas

$$\frac{\partial\varphi(t,X,A)}{\partial t} = -\frac{1}{\omega}\mathrm{e}^{-\mathrm{e}^{\int_t^{\mathrm{T}} r(\tau)\,\mathrm{d}\tau}X\omega+\omega+g_\omega(t,A)}\left(\mathrm{e}^{\int_t^{\mathrm{T}} r(\tau)\,\mathrm{d}\tau}X\omega r(t)+\frac{\partial g_\omega(t,A)}{\partial t}\right), \tag{3.8.4}$$

$$\frac{\partial\varphi(t,X,A)}{\partial X} = \mathrm{e}^{-\mathrm{e}^{\int_t^{\mathrm{T}} r(\tau)\,\mathrm{d}\tau}X\omega+\omega+\int_t^{\mathrm{T}} r(\tau)\,\mathrm{d}\tau+g_\omega(t,A)}, \tag{3.8.5}$$

$$\frac{\partial^2\varphi(t,X,A)}{\partial X^2} = -\mathrm{e}^{-\mathrm{e}^{\int_t^{\mathrm{T}} r(\tau)\,\mathrm{d}\tau}X\omega+\omega+2\int_t^{\mathrm{T}} r(\tau)\,\mathrm{d}\tau+g_\omega(t,A)}\omega, \tag{3.8.6}$$

$$\nabla_A\varphi(t,X,A) = -\frac{1}{\omega}\mathrm{e}^{-\mathrm{e}^{\int_t^{\mathrm{T}} r(\tau)\,\mathrm{d}\tau}X\omega+\omega+g_\omega(t,A)}\nabla_A g_\omega(t,A), \tag{3.8.7}$$

$$\nabla_A\frac{\partial\varphi(t,X,A)}{\partial X} = \mathrm{e}^{-\mathrm{e}^{\int_t^{\mathrm{T}} r(\tau)\,\mathrm{d}\tau}X\omega+\omega+\int_t^{\mathrm{T}} r(\tau)\,\mathrm{d}\tau+g_\omega(t,A)}\nabla_A g_\omega(t,A), \tag{3.8.8}$$

$$\nabla_A \nabla_A \varphi(t,X,A) = -\frac{1}{\omega} \mathrm{e}^{-\mathrm{e}^{\int_t^{\mathrm{T}} r(\tau)\,\mathrm{d}\tau} X\omega + \omega + g_\omega(t,A)}$$
$$\left(\{\nabla_A g_\omega(t,A)\}^{\mathrm{T}} \cdot \{\nabla_A g_\omega(t,A)\} + \nabla_A \nabla_A g_\omega(t,A) \right), \tag{3.8.9}$$

and then

$$\begin{aligned} \frac{\partial^2 \varphi}{\partial X^2} \frac{\partial \varphi}{\partial t} &= \left(-\mathrm{e}^{-\mathrm{e}^{\int_t^{\mathrm{T}} r(\tau)\,\mathrm{d}\tau} X\omega + \omega + 2\int_t^{\mathrm{T}} r(\tau)\,\mathrm{d}\tau + g_\omega(t,A)} \omega \right) \\ &\times \left(-\frac{1}{\omega} \mathrm{e}^{-\mathrm{e}^{\int_t^{\mathrm{T}} r(\tau)\,\mathrm{d}\tau} X\omega + \omega + g_\omega(t,A)} \left(\mathrm{e}^{\int_t^{\mathrm{T}} r(\tau)\,\mathrm{d}\tau} X\omega r(t) + \frac{\partial g_\omega(t,A)}{\partial t} \right) \right) \\ &= \mathrm{e}^{2\left(-\mathrm{e}^{\int_t^{\mathrm{T}} r(\tau)\,\mathrm{d}\tau} X\omega + \omega + \int_t^{\mathrm{T}} r(\tau)\,\mathrm{d}\tau + g_\omega(t,A)\right)} \left(\mathrm{e}^{\int_t^{\mathrm{T}} r(\tau)\,\mathrm{d}\tau} X\omega r(t) + \frac{\partial g_\omega(t,A)}{\partial t} \right) \end{aligned} \tag{3.8.10}$$

and

$$\begin{aligned} &-\frac{1}{2} \left(\frac{\partial \varphi}{\partial X} \right)^2 (a_s - r) \cdot \left(\sigma_s \cdot \sigma_s^{\mathrm{T}} \right)^{-1} \cdot (a_s - r) \\ &= -\frac{1}{2} \left(\mathrm{e}^{-\mathrm{e}^{\int_t^{\mathrm{T}} r(\tau)\,\mathrm{d}\tau} X\omega + \omega + \int_t^{\mathrm{T}} r(\tau)\,\mathrm{d}\tau + g_\omega(t,A)} \right)^2 (a_s - r) \cdot \left(\sigma_s \cdot \sigma_s^{\mathrm{T}} \right)^{-1} \cdot (a_s - r) \\ &= -\frac{1}{2} \mathrm{e}^{2\left(-\mathrm{e}^{\int_t^{\mathrm{T}} r(\tau)\,\mathrm{d}\tau} X\omega + \omega + \int_t^{\mathrm{T}} r(\tau)\,\mathrm{d}\tau + g_\omega(t,A)\right)} (a_s - r) \cdot \left(\sigma_s \cdot \sigma_s^{\mathrm{T}} \right)^{-1} \cdot (a_s - r) \end{aligned} \tag{3.8.11}$$

and

$$\begin{aligned} rX \frac{\partial^2 \varphi}{\partial X^2} \frac{\partial \varphi}{\partial X} &= rX \left(-\mathrm{e}^{-\mathrm{e}^{\int_t^{\mathrm{T}} r(\tau)\,\mathrm{d}\tau} X\omega + \omega + 2\int_t^{\mathrm{T}} r(\tau)\,\mathrm{d}\tau + g_\omega(t,A)} \omega \right) \\ &\quad \times \mathrm{e}^{-\mathrm{e}^{\int_t^{\mathrm{T}} r(\tau)\,\mathrm{d}\tau} X\omega + \omega + \int_t^{\mathrm{T}} r(\tau)\,\mathrm{d}\tau + g_\omega(t,A)} \\ &= -\mathrm{e}^{3\int_t^{\mathrm{T}} r(\tau)\,\mathrm{d}\tau + 2\left(-\mathrm{e}^{\int_t^{\mathrm{T}} r(\tau)\,\mathrm{d}\tau} X\omega + \omega + g_\omega(t,A)\right)} rX\omega \end{aligned} \tag{3.8.12}$$

and

$$\begin{aligned} b \cdot \nabla_A \varphi \frac{\partial^2 \varphi}{\partial X^2} &= b \cdot \left(-\frac{1}{\omega} \mathrm{e}^{-\mathrm{e}^{\int_t^{\mathrm{T}} r(\tau)\,\mathrm{d}\tau} X\omega + \omega + g_\omega(t,A)} \nabla_A g_\omega(t,A) \right) \\ &\times \left(-\mathrm{e}^{-\mathrm{e}^{\int_t^{\mathrm{T}} r(\tau)\,\mathrm{d}\tau} X\omega + \omega + 2\int_t^{\mathrm{T}} r(\tau)\,\mathrm{d}\tau + g_\omega(t,A)} \omega \right) \\ &= \mathrm{e}^{2\left(-\mathrm{e}^{\int_t^{\mathrm{T}} r(\tau)\,\mathrm{d}\tau} X\omega + \omega + \int_t^{\mathrm{T}} r(\tau)\,\mathrm{d}\tau + g_\omega(t,A)\right)} b \cdot \nabla_A g_\omega(t,A) \end{aligned} \tag{3.8.13}$$

and

$$-\frac{\partial\varphi}{\partial X}(a_s-r)\cdot\left(\sigma_s\cdot\sigma_s^{\mathrm{T}}\right)^{-1}\cdot\sigma_s\cdot c^{\mathrm{T}}\cdot\left(\nabla_A\frac{\partial\varphi}{\partial X}\right)$$

$$=-\mathrm{e}^{-\mathrm{e}^{\int_t^{\mathrm{T}}r(\tau)\,\mathrm{d}\tau}X\omega+\omega+\int_t^{\mathrm{T}}r(\tau)\,\mathrm{d}\tau+g_\omega(t,A)}(a_s-r)\cdot\left(\sigma_s\cdot\sigma_s^{\mathrm{T}}\right)^{-1}\cdot\sigma_s\cdot c^{\mathrm{T}}$$

$$\cdot\left(\mathrm{e}^{-\mathrm{e}^{\int_t^{\mathrm{T}}r(\tau)\,\mathrm{d}\tau}X\omega+\omega+\int_t^{\mathrm{T}}r(\tau)\,\mathrm{d}\tau+g_\omega(t,A)}\nabla_A g_\omega(t,A)\right)$$

$$=-\mathrm{e}^{2\left(-\mathrm{e}^{\int_t^{\mathrm{T}}r(\tau)\,\mathrm{d}\tau}X\omega+\omega+\int_t^{\mathrm{T}}r(\tau)\,\mathrm{d}\tau+g_\omega(t,A)\right)}(a_s-r)$$

$$\cdot\left(\sigma_s\cdot\sigma_s^{\mathrm{T}}\right)^{-1}\cdot\sigma_s\cdot c^{\mathrm{T}}\cdot\nabla_A g_\omega(t,A) \qquad (3.8.14)$$

and

$$-\frac{1}{2}\left(\nabla_A\frac{\partial\varphi}{\partial X}\right)\cdot c\cdot\sigma_s^{\mathrm{T}}\cdot\left(\sigma_s\cdot\sigma_s^{\mathrm{T}}\right)^{-1}\cdot\sigma_s\cdot c^{\mathrm{T}}\cdot\left(\nabla_A\frac{\partial\varphi}{\partial X}\right)$$

$$=-\frac{1}{2}\left(\mathrm{e}^{-\mathrm{e}^{\int_t^{\mathrm{T}}r(\tau)\,\mathrm{d}\tau}X\omega+\omega+\int_t^{\mathrm{T}}r(\tau)\,\mathrm{d}\tau+g_\omega(t,A)}\nabla_A g_\omega(t,A)\right)\cdot c\cdot\sigma_s^{\mathrm{T}}\cdot\left(\sigma_s\cdot\sigma_s^{\mathrm{T}}\right)^{-1}$$

$$\cdot\sigma_s\cdot c^{\mathrm{T}}\cdot\left(\mathrm{e}^{-\mathrm{e}^{\int_t^{\mathrm{T}}r(\tau)\,\mathrm{d}\tau}X\omega+\omega+\int_t^{\mathrm{T}}r(\tau)\,\mathrm{d}\tau+g_\omega(t,A)}\nabla_A g_\omega(t,A)\right)$$

$$=-\frac{1}{2}\mathrm{e}^{2\left(-\mathrm{e}^{\int_t^{\mathrm{T}}r(\tau)\,\mathrm{d}\tau}X\omega+\omega+\int_t^{\mathrm{T}}r(\tau)\,\mathrm{d}\tau+g_\omega(t,A)\right)}$$

$$\times\nabla_A g_\omega(t,A)\cdot c\cdot\sigma_s^{\mathrm{T}}\cdot\left(\sigma_s\cdot\sigma_s^{\mathrm{T}}\right)^{-1}\cdot\sigma_s\cdot c^{\mathrm{T}}\cdot\nabla_A g_\omega(t,A) \qquad (3.8.15)$$

and, using also (3.6.9),

$$\frac{1}{2}\frac{\partial^2\varphi}{\partial X^2}\mathrm{Tr}\left(c\cdot c^{\mathrm{T}}\cdot\nabla_A\nabla_A\varphi\right)$$

$$=\frac{1}{2}\left(-\mathrm{e}^{-\mathrm{e}^{\int_t^{\mathrm{T}}r(\tau)\,\mathrm{d}\tau}X\omega+\omega+2\int_t^{\mathrm{T}}r(\tau)\,\mathrm{d}\tau+g_\omega(t,A)}\omega\right)$$

$$\times\mathrm{Tr}\left(c\cdot c^{\mathrm{T}}\cdot\left(-\frac{1}{\omega}\mathrm{e}^{-\mathrm{e}^{\int_t^{\mathrm{T}}r(\tau)\,\mathrm{d}\tau}X\omega+\omega+g_\omega(t,A)}\right.\right.$$

$$\left.\left.\left(\{\nabla_A g_\omega(t,A)\}^{\mathrm{T}}\cdot\{\nabla_A g_\omega(t,A)\}+\nabla_A\nabla_A g_\omega(t,A)\right)\right)\right)$$

$$=\frac{1}{2}\mathrm{e}^{2\left(-\mathrm{e}^{\int_t^{\mathrm{T}}r(\tau)\,\mathrm{d}\tau}X\omega+\omega+\int_t^{\mathrm{T}}r(\tau)\,\mathrm{d}\tau+g_\omega(t,A)\right)}$$

$$\times \mathrm{Tr}\left(c\cdot c^{\mathrm{T}}\cdot\left(\{\nabla_A g_\omega(t,A)\}^{\mathrm{T}}\cdot\{\nabla_A g_\omega(t,A)\}+\nabla_A\nabla_A g_\omega(t,A)\right)\right)$$

$$=\frac{1}{2}\mathrm{e}^{2\left(-\mathrm{e}^{\int_t^{\mathrm{T}} r(\tau)\,\mathrm{d}\tau}X\omega+\omega+\int_t^{\mathrm{T}} r(\tau)\,\mathrm{d}\tau+g_\omega(t,A)\right)}\left(\nabla_A g_\omega(t,A)c\cdot c^{\mathrm{T}}\cdot\nabla_A g_\omega(t,A)\right.$$
$$\left.+\mathrm{Tr}\left(c\cdot c^{\mathrm{T}}\cdot\nabla_A\nabla_A g_\omega(t,A)\right)\right) \tag{3.8.16}$$

yielding the following theorem.

Theorem 3.8.1. *In the case of the CARA utility (3.2.9), and if interest rates are deterministic, the value function $\varphi(t,X,A)$ of the optimal investment portfolio problem with no constraints on the portfolio, Problem 3.3.1, given in (3.3.3), has the form (3.8.1), where $g=g(t,A)$ is the solution of the quasilinear PDE*

$$\frac{\partial g(t,A)}{\partial t}+\frac{1}{2}Tr\left(c\cdot c^{\mathrm{T}}\cdot\nabla_A\nabla_A g(t,A)\right)+\left(b-(a_s-r)\cdot\left(\sigma_s\cdot\sigma_s^{\mathrm{T}}\right)^{-1}\cdot\sigma_s\cdot c^{\mathrm{T}}\right)$$
$$\cdot\nabla_A g(t,A)+\frac{1}{2}\nabla_A g(t,A)\cdot c\cdot\left(\mathbb{I}_n-\sigma_s^{\mathrm{T}}\cdot\left(\sigma_s\cdot\sigma_s^{\mathrm{T}}\right)^{-1}\cdot\sigma_s\right)\cdot c^{\mathrm{T}}\cdot\nabla_A g(t,A)$$
$$=\frac{1}{2}(a_s-r)\cdot\left(\sigma_s\cdot\sigma_s^{\mathrm{T}}\right)^{-1}\cdot(a_s-r) \tag{3.8.17}$$

for $t<T<\infty, A\in\mathcal{A}$, together with the terminal condition $g(T,A)=0$, while the optimal portfolio strategy (3.4.3) is equal to

$$\Pi_\omega^\star(t,A)=\frac{1}{\omega}\mathrm{e}^{-\int_t^{\mathrm{T}} r(\tau)\,\mathrm{d}\tau}\left(a_s-r+\nabla_A g\cdot c\cdot\sigma_s^{\mathrm{T}}\right)\cdot\left(\sigma_s\cdot\sigma_s^{\mathrm{T}}\right)^{-1}. \tag{3.8.18}$$

Proof. What remains to be proved is the optimal portfolio formula (3.8.18). To that end, we have

$$\Pi_\omega^\star(t,X,A)=-\left(1\Big/\frac{\partial^2\varphi}{\partial X^2}\right)\left(\frac{\partial\varphi}{\partial X}(a_s-r)+\left(\nabla_A\frac{\partial\varphi}{\partial X}\right)\cdot c\cdot\sigma_s^{\mathrm{T}}\right)\cdot\left(\sigma_s\cdot\sigma_s^{\mathrm{T}}\right)^{-1}$$
$$=\left(\frac{\mathrm{e}^{-\int_t^{\mathrm{T}} r(\tau)\,\mathrm{d}\tau}}{\omega}(a_s-r)+\frac{\mathrm{e}^{-\int_t^{\mathrm{T}} r(\tau)\,\mathrm{d}\tau}\nabla_A g(t,A)}{\omega}c\cdot\sigma_s^{\mathrm{T}}\right)\cdot\left(\sigma_s\cdot\sigma_s^{\mathrm{T}}\right)^{-1}, \tag{3.8.19}$$

from which (3.8.18) follows. □

Remark 3.8.1. Observe that equation (3.8.17), and therefore also its solution g, does not depend on the parameter ω, the absolute risk-aversion parameter. The (absolute) risk-aversion parameter ω affects the optimal portfolio formula (3.8.18).

3.9 CARA Utility Case: Affine Constraints on the Portfolio – Deterministic Interest Rates

In the case of an optimal portfolio under affine constraints, we need now a constraint that is compatible with the CARA utility, as opposed to the one used in Sect. 3.7, which was suitable for the CRRA utility. A constraint suitable for the CARA utility reads as follows (cf. (3.7.1)):

$$\mu(t,A)\cdot\Pi^{\star}(t,X,A)=\xi(t,A), \tag{3.9.1}$$

i.e., $\mu(t,X,A)=\mu(t,A)$ and $\xi(t,X,A)=\xi(t,A)$ in (3.3.4). We look for the solutions of equation (3.5.22), again in the form (3.8.1). Recall formulas (3.8.4)–(3.8.9), and then prepare the formulas for all the (new) terms in equation (3.5.1):

$$\begin{aligned}
&-\frac{1}{2}\left(\frac{\partial\varphi}{\partial X}\right)^2(a_s-r)\cdot\left(\mathbb{I}_k-\left(\sigma_s\cdot\sigma_s^{\mathrm{T}}\right)^{-1}\cdot\mu^{\mathrm{T}}\cdot\left(\mu\cdot\left(\sigma_s\cdot\sigma_s^{\mathrm{T}}\right)^{-1}\cdot\mu^{\mathrm{T}}\right)^{-1}\cdot\mu\right)\\
&\quad\cdot\left(\sigma_s\cdot\sigma_s^{\mathrm{T}}\right)^{-1}\cdot(a_s-r)=-\frac{1}{2}\left(\mathrm{e}^{-\mathrm{e}^{\int_t^{\mathrm{T}}r(\tau)\,\mathrm{d}\tau}X\omega+\omega+\int_t^{\mathrm{T}}r(\tau)\,\mathrm{d}\tau+g_{\omega}(t,A)}\right)^2(a_s-r)\\
&\quad\cdot\left(\mathbb{I}_k-\left(\sigma_s\cdot\sigma_s^{\mathrm{T}}\right)^{-1}\cdot\mu^{\mathrm{T}}\cdot\left(\mu\cdot\left(\sigma_s\cdot\sigma_s^{\mathrm{T}}\right)^{-1}\cdot\mu^{\mathrm{T}}\right)^{-1}\cdot\mu\right)\cdot\left(\sigma_s\cdot\sigma_s^{\mathrm{T}}\right)^{-1}\cdot(a_s-r)\\
&=-\frac{1}{2}\mathrm{e}^{2\left(-\mathrm{e}^{\int_t^{\mathrm{T}}r(\tau)\,\mathrm{d}\tau}X\omega+\omega+\int_t^{\mathrm{T}}r(\tau)\,\mathrm{d}\tau+g_{\omega}(t,A)\right)}(a_s-r)\cdot\left(\mathbb{I}_k-\left(\sigma_s\cdot\sigma_s^{\mathrm{T}}\right)^{-1}\cdot\mu^{\mathrm{T}}\right.\\
&\quad\left.\cdot\left(\mu\cdot\left(\sigma_s\cdot\sigma_s^{\mathrm{T}}\right)^{-1}\cdot\mu^{\mathrm{T}}\right)^{-1}\cdot\mu\right)\cdot\left(\sigma_s\cdot\sigma_s^{\mathrm{T}}\right)^{-1}\cdot(a_s-r)s
\end{aligned}\tag{3.9.2}$$

and

$$\begin{aligned}
&-\frac{\partial\varphi}{\partial X}(a_s-r)\cdot\left(\mathbb{I}_k-\left(\sigma_s\cdot\sigma_s^{\mathrm{T}}\right)^{-1}\cdot\mu^{\mathrm{T}}\cdot\left(\mu\cdot\left(\sigma_s\cdot\sigma_s^{\mathrm{T}}\right)^{-1}\cdot\mu^{\mathrm{T}}\right)^{-1}\cdot\mu\right)\\
&\quad\cdot\left(\sigma_s\cdot\sigma_s^{\mathrm{T}}\right)^{-1}\cdot\sigma_s\cdot c^{\mathrm{T}}\cdot\left(\nabla_A\frac{\partial\varphi}{\partial X}\right)\\
&=-\left(\mathrm{e}^{-\mathrm{e}^{\int_t^{\mathrm{T}}r(\tau)\,\mathrm{d}\tau}X\omega+\omega+\int_t^{\mathrm{T}}r(\tau)\,\mathrm{d}\tau+g_{\omega}(t,A)}\right)(a_s-r)\cdot\left(\mathbb{I}_k-\left(\sigma_s\cdot\sigma_s^{\mathrm{T}}\right)^{-1}\right.\\
&\quad\left.\cdot\mu^{\mathrm{T}}\cdot\left(\mu\cdot\left(\sigma_s\cdot\sigma_s^{\mathrm{T}}\right)^{-1}\cdot\mu^{\mathrm{T}}\right)^{-1}\cdot\mu\right)\cdot\left(\sigma_s\cdot\sigma_s^{\mathrm{T}}\right)^{-1}\cdot\sigma_s\cdot c^{\mathrm{T}}\\
&\quad\cdot\left(\mathrm{e}^{-\mathrm{e}^{\int_t^{\mathrm{T}}r(\tau)\,\mathrm{d}\tau}X\omega+\omega+\int_t^{\mathrm{T}}r(\tau)\,\mathrm{d}\tau+g_{\omega}(t,A)}\nabla_A g_{\omega}(t,A)\right)\\
&=-\mathrm{e}^{2\left(-\mathrm{e}^{\int_t^{\mathrm{T}}r(\tau)\,\mathrm{d}\tau}X\omega+\omega+\int_t^{\mathrm{T}}r(\tau)\,\mathrm{d}\tau+g_{\omega}(t,A)\right)}(a_s-r)\cdot\left(\mathbb{I}_k-\left(\sigma_s\cdot\sigma_s^{\mathrm{T}}\right)^{-1}\cdot\mu^{\mathrm{T}}\right.\\
&\quad\left.\cdot\left(\mu\cdot\left(\sigma_s\cdot\sigma_s^{\mathrm{T}}\right)^{-1}\cdot\mu^{\mathrm{T}}\right)^{-1}\cdot\mu\right)\cdot\left(\sigma_s\cdot\sigma_s^{\mathrm{T}}\right)^{-1}\cdot\sigma_s\cdot c^{\mathrm{T}}\cdot\nabla_A g_{\omega}(t,A)
\end{aligned}\tag{3.9.3}$$

and

$$
\begin{aligned}
&-\frac{1}{2}\left(\nabla_A \frac{\partial \varphi}{\partial X}\right)\cdot c\cdot\sigma_s^{\mathrm{T}}\cdot\left(\mathbb{I}_k-\left(\sigma_s\cdot\sigma_s^{\mathrm{T}}\right)^{-1}\cdot\mu^{\mathrm{T}}\cdot\left(\mu\cdot\left(\sigma_s\cdot\sigma_s^{\mathrm{T}}\right)^{-1}\cdot\mu^{\mathrm{T}}\right)^{-1}\cdot\mu\right)\\
&\qquad\cdot\left(\sigma_s\cdot\sigma_s^{\mathrm{T}}\right)^{-1}\cdot\sigma_s.c^{\mathrm{T}}\cdot\left(\nabla_A\frac{\partial\varphi}{\partial X}\right)\\
&=-\frac{1}{2}\left(\mathrm{e}^{-\mathrm{e}^{\int_t^{\mathrm{T}} r(\tau)\,\mathrm{d}\tau}X\omega+\omega+\int_t^{\mathrm{T}} r(\tau)\,\mathrm{d}\tau+g_\omega(t,A)}\nabla_A g_\omega(t,A)\right)\cdot c\cdot\sigma_s^{\mathrm{T}}\\
&\quad\cdot\left(\mathbb{I}_k-\left(\sigma_s\cdot\sigma_s^{\mathrm{T}}\right)^{-1}\cdot\mu^{\mathrm{T}}\cdot\left(\mu\cdot\left(\sigma_s\cdot\sigma_s^{\mathrm{T}}\right)^{-1}\cdot\mu^{\mathrm{T}}\right)^{-1}\cdot\mu\right)\cdot\left(\sigma_s\cdot\sigma_s^{\mathrm{T}}\right)^{-1}\cdot\sigma_s\\
&\quad\cdot c^{\mathrm{T}}\cdot\left(\mathrm{e}^{-\mathrm{e}^{\int_t^{\mathrm{T}} r(\tau)\,\mathrm{d}\tau}X\omega+\omega+\int_t^{\mathrm{T}} r(\tau)\,\mathrm{d}\tau+g_\omega(t,A)}\nabla_A g_\omega(t,A)\right)\\
&=-\frac{1}{2}\mathrm{e}^{2\left(-\mathrm{e}^{\int_t^{\mathrm{T}} r(\tau)\,\mathrm{d}\tau}X\omega+\omega+\int_t^{\mathrm{T}} r(\tau)\,\mathrm{d}\tau+g_\omega(t,A)\right)}\nabla_A g_\omega(t,A)\cdot c\cdot\sigma_s^{\mathrm{T}}\cdot\left(\mathbb{I}_k-\left(\sigma_s\cdot\sigma_s^{\mathrm{T}}\right)^{-1}\right.\\
&\quad\left.\cdot\mu^{\mathrm{T}}\cdot\left(\mu\cdot\left(\sigma_s\cdot\sigma_s^{\mathrm{T}}\right)^{-1}\cdot\mu^{\mathrm{T}}\right)^{-1}\cdot\mu\right)\cdot\left(\sigma_s\cdot\sigma_s^{\mathrm{T}}\right)^{-1}\cdot\sigma_s\cdot c^{\mathrm{T}}\cdot\nabla_A g_\omega(t,A)
\end{aligned}
\tag{3.9.4}
$$

and

$$
\begin{aligned}
&\frac{\partial\varphi}{\partial X}\frac{\partial^2\varphi}{\partial X^2}\xi\cdot\left(\mu\cdot\left(\sigma_s\cdot\sigma_s^{\mathrm{T}}\right)^{-1}\cdot\mu^{\mathrm{T}}\right)^{-1}\cdot\mu\cdot\left(\sigma_s\cdot\sigma_s^{\mathrm{T}}\right)^{-1}\cdot(a_s-r)\\
&\quad=\mathrm{e}^{-\mathrm{e}^{\int_t^{\mathrm{T}} r(\tau)\,\mathrm{d}\tau}X\omega+\omega+\int_t^{\mathrm{T}} r(\tau)\,\mathrm{d}\tau+g_\omega(t,A)}\\
&\quad\times\left(-\mathrm{e}^{-\mathrm{e}^{\int_t^{\mathrm{T}} r(\tau)\,\mathrm{d}\tau}X\omega+\omega+2\int_t^{\mathrm{T}} r(\tau)\,\mathrm{d}\tau+g_\omega(t,A)}\mu\right)\xi\cdot\left(\mu\cdot\left(\sigma_s\cdot\sigma_s^{\mathrm{T}}\right)^{-1}\cdot\mu^{\mathrm{T}}\right)^{-1}\\
&\quad\cdot\mu\cdot\left(\sigma_s\cdot\sigma_s^{\mathrm{T}}\right)^{-1}.(a_s-r)=-\mathrm{e}^{3\int_t^{\mathrm{T}} r(\tau)\,\mathrm{d}\tau+2\left(-\mathrm{e}^{\int_t^{\mathrm{T}} r(\tau)\,\mathrm{d}\tau}X\omega+\omega+g_\omega(t,A)\right)}\\
&\quad\times\mu\xi\cdot\left(\mu\cdot\left(\sigma_s\cdot\sigma_s^{\mathrm{T}}\right)^{-1}\cdot\mu^{\mathrm{T}}\right)^{-1}\cdot\mu\cdot\left(\sigma_s\cdot\sigma_s^{\mathrm{T}}\right)^{-1}\cdot(a_s-r)
\end{aligned}
\tag{3.9.5}
$$

and

$$
\begin{aligned}
&\frac{1}{2}\left(\frac{\partial^2\varphi}{\partial X^2}\right)^2\xi\cdot\left(\mu\cdot\left(\sigma_s\cdot\sigma_s^{\mathrm{T}}\right)^{-1}\cdot\mu^{\mathrm{T}}\right)^{-1}\cdot\xi=\frac{1}{2}\left(-\mathrm{e}^{-\mathrm{e}^{\int_t^{\mathrm{T}} r(\tau)\,\mathrm{d}\tau}X\omega+\omega+2\int_t^{\mathrm{T}} r(\tau)\,\mathrm{d}\tau+g_\omega(t,A)}\omega\right)^2\\
&\times\xi\cdot\left(\mu\cdot\left(\sigma_s\cdot\sigma_s^{\mathrm{T}}\right)^{-1}\cdot\mu^{\mathrm{T}}\right)^{-1}\cdot\xi\\
&=\frac{1}{2}\mathrm{e}^{2\left(-\mathrm{e}^{\int_t^{\mathrm{T}} r(\tau)\,\mathrm{d}\tau}X\omega+\omega+2\int_t^{\mathrm{T}} r(\tau)\,\mathrm{d}\tau+g_\omega(t,A)\right)}\omega^2\xi\cdot\left(\mu\cdot\left(\sigma_s\cdot\sigma_s^{\mathrm{T}}\right)^{-1}\cdot\mu^{\mathrm{T}}\right)^{-1}\cdot\xi
\end{aligned}
\tag{3.9.6}
$$

and

$$\frac{\partial^2\varphi}{\partial X^2}\xi\cdot\left(\mu\cdot\left(\sigma_s\cdot\sigma_s^{\mathrm{T}}\right)^{-1}\cdot\mu^{\mathrm{T}}\right)^{-1}\cdot\mu\cdot\left(\sigma_s\cdot\sigma_s^{\mathrm{T}}\right)^{-1}\cdot\sigma_s\cdot c^{\mathrm{T}}\cdot\left(\nabla_A\frac{\partial\varphi}{\partial X}\right)$$

$$=-\mathrm{e}^{-\mathrm{e}^{\int_t^{\mathrm{T}}r(\tau)\,\mathrm{d}\tau}X\omega+\omega+2\int_t^{\mathrm{T}}r(\tau)\,\mathrm{d}\tau+g_\omega(t,A)}\omega\xi\cdot\left(\mu\cdot\left(\sigma_s\cdot\sigma_s^{\mathrm{T}}\right)^{-1}\cdot\mu^{\mathrm{T}}\right)^{-1}$$

$$\cdot\mu\cdot\left(\sigma_s\cdot\sigma_s^{\mathrm{T}}\right)^{-1}\cdot\sigma_s\cdot c^{\mathrm{T}}\cdot\left(\mathrm{e}^{-\mathrm{e}^{\int_t^{\mathrm{T}}r(\tau)\,\mathrm{d}\tau}X\omega+\omega+\int_t^{\mathrm{T}}r(\tau)\,\mathrm{d}\tau+g_\omega(t,A)}\nabla_A g_\omega(t,A)\right)$$

$$=-\mathrm{e}^{3\int_t^{\mathrm{T}}r(\tau)\,\mathrm{d}\tau+2\left(-\mathrm{e}^{\int_t^{\mathrm{T}}r(\tau)\,\mathrm{d}\tau}X\omega+\omega+g_\omega(t,A)\right)}\omega\xi\cdot\left(\mu\cdot\left(\sigma_s\cdot\sigma_s^{\mathrm{T}}\right)^{-1}\cdot\mu^{\mathrm{T}}\right)^{-1}$$

$$\cdot\mu\cdot\left(\sigma_s\cdot\sigma_s^{\mathrm{T}}\right)^{-1}\cdot\sigma_s.c^{\mathrm{T}}\cdot\nabla_A g_\omega(t,A), \tag{3.9.7}$$

yielding the following theorem.

Theorem 3.9.1. *In the case of the CARA utility (3.2.9),* the value function$\varphi(t,X,A)$ *of the optimal investment portfolio problem with affine constraints on the portfolio, Problem 3.3.2, given in (3.3.5), has the form (3.8.1), where* $g_\omega = g_\omega(t,A)$ *is the solution of the following quasilinear PDE:*

$$\frac{\partial g_\omega(t,A)}{\partial t}+\frac{1}{2}Tr\left(c\cdot c^{\mathrm{T}}\cdot\nabla_A\nabla_A g_\omega(t,A)\right)+\left(b-(a_s-r)\cdot\left(\mathbb{I}_k-\left(\sigma_s\cdot\sigma_s^{\mathrm{T}}\right)^{-1}\cdot\mu^{\mathrm{T}}\right.\right.$$

$$\left.\cdot\left(\mu\cdot\left(\sigma_s\cdot\sigma_s^{\mathrm{T}}\right)^{-1}\cdot\mu^{\mathrm{T}}\right)^{-1}\cdot\mu\right)\cdot\left(\sigma_s\cdot\sigma_s^{\mathrm{T}}\right)^{-1}\cdot\sigma_s\cdot c^{\mathrm{T}}-\mathrm{e}^{\int_t^{\mathrm{T}}r(\tau)\,\mathrm{d}\tau}\omega\xi$$

$$\left.\cdot\left(\mu\cdot\left(\sigma_s\cdot\sigma_s^{\mathrm{T}}\right)^{-1}\cdot\mu^{\mathrm{T}}\right)^{-1}\cdot\mu\cdot\left(\sigma_s\cdot\sigma_s^{\mathrm{T}}\right)^{-1}\cdot\sigma_s\cdot c^{\mathrm{T}}\right)\cdot\nabla_A g_\omega(t,A)$$

$$+\frac{1}{2}\nabla_A g_\omega(t,A)\cdot c\cdot\left(\mathbb{I}_n-\sigma_s^{\mathrm{T}}\cdot\left(\mathbb{I}_k-\left(\sigma_s\cdot\sigma_s^{\mathrm{T}}\right)^{-1}\cdot\mu^{\mathrm{T}}\cdot\left(\mu\cdot\left(\sigma_s\cdot\sigma_s^{\mathrm{T}}\right)^{-1}\right.\right.\right.$$

$$\left.\left.\left.\cdot\mu^{\mathrm{T}}\right)^{-1}\cdot\mu\right)\cdot\left(\sigma_s\cdot\sigma_s^{\mathrm{T}}\right)^{-1}\cdot\sigma_s\right)\cdot c^{\mathrm{T}}\cdot\nabla_A g_\omega(t,A)$$

$$=\frac{1}{2}(a_s-r)\cdot\left(\mathbb{I}_k-\left(\sigma_s\cdot\sigma_s^{\mathrm{T}}\right)^{-1}\cdot\mu^{\mathrm{T}}\cdot\left(\mu\cdot\left(\sigma_s\cdot\sigma_s^{\mathrm{T}}\right)^{-1}\cdot\mu^{\mathrm{T}}\right)^{-1}\cdot\mu\right)$$

$$\cdot\left(\sigma_s\cdot\sigma_s^{\mathrm{T}}\right)^{-1}\cdot(a_s-r)+\mathrm{e}^{\int_t^{\mathrm{T}}r(\tau)\,\mathrm{d}\tau}\mu\xi\cdot\left(\mu\cdot\left(\sigma_s\cdot\sigma_s^{\mathrm{T}}\right)^{-1}\cdot\mu^{\mathrm{T}}\right)^{-1}$$

$$\cdot\mu\cdot\left(\sigma_s\cdot\sigma_s^{\mathrm{T}}\right)^{-1}\cdot(a_s-r)-\frac{1}{2}\mathrm{e}^{2\int_t^{\mathrm{T}}r(\tau)\,\mathrm{d}\tau}\omega^2\xi\cdot\left(\mu\cdot\left(\sigma_s\cdot\sigma_s^{\mathrm{T}}\right)^{-1}\cdot\mu^{\mathrm{T}}\right)^{-1}\cdot\xi \tag{3.9.8}$$

for $t < T < \infty, A \in \mathcal{A}$, with the terminal condition $g_\omega(T,A) = 0$. The optimal portfolio strategy (3.5.2) simplifies to

$$\begin{aligned}\Pi^\star_\omega(t,A) &= \frac{\mathrm{e}^{-\int_t^T r(\tau)\,\mathrm{d}\tau}}{\omega}\left(a_s - r + \nabla_A g_\omega(t,A)\cdot c\cdot \sigma_s^{\mathrm{T}}\right)\cdot\left(\mathbb{I}_k - \left(\sigma_s\cdot\sigma_s^{\mathrm{T}}\right)^{-1}\right.\\&\quad \left.\cdot\mu^{\mathrm{T}}\cdot\left(\mu\cdot\left(\sigma_s\cdot\sigma_s^{\mathrm{T}}\right)^{-1}\cdot\mu^{\mathrm{T}}\right)^{-1}\cdot\mu\right)\cdot\left(\sigma_s\cdot\sigma_s^{\mathrm{T}}\right)^{-1}\\&\quad +\xi\cdot\left(\mu\cdot\left(\sigma_s\cdot\sigma_s^{\mathrm{T}}\right)^{-1}\cdot\mu^{\mathrm{T}}\right)^{-1}\cdot\mu\cdot\left(\sigma_s\cdot\sigma_s^{\mathrm{T}}\right)^{-1}\end{aligned} \tag{3.9.9}$$

and

$$\Pi^\star_\infty(t,A) = \lim_{\omega\to\infty}\Pi^\star_\omega[t,X,A] = \xi\cdot\left(\mu\cdot\left(\sigma_s\cdot\sigma_s^{\mathrm{T}}\right)^{-1}\cdot\mu^{\mathrm{T}}\right)^{-1}\cdot\mu\cdot\left(\sigma_s\cdot\sigma_s^{\mathrm{T}}\right)^{-1}, \tag{3.9.10}$$

provided $\nabla_A g_\omega/\,\omega \to 0$, as $\omega\to\infty$.

Proof. It remains to prove the optimal portfolio formula (3.9.9). To that end, using (3.8.1),

$$\begin{aligned}\Pi^\star_\omega(t,X,A) &= -\left(\frac{\partial\varphi}{\partial X}\Big/\frac{\partial^2\varphi}{\partial X^2}\right)(a_s - r)\cdot\left(\mathbb{I}_k - \left(\sigma_s\cdot\sigma_s^{\mathrm{T}}\right)^{-1}\cdot\mu^{\mathrm{T}}\right.\\&\quad\left.\cdot\left(\mu\cdot\left(\sigma_s\cdot\sigma_s^{\mathrm{T}}\right)^{-1}\cdot\mu^{\mathrm{T}}\right)^{-1}\cdot\mu\right)\cdot\left(\sigma_s\cdot\sigma_s^{\mathrm{T}}\right)^{-1} - \left(\nabla_A\frac{\partial\varphi}{\partial X}\Big/\frac{\partial^2\varphi}{\partial X^2}\right)\\&\quad\cdot c\cdot\sigma_s^{\mathrm{T}}\cdot\left(\mathbb{I}_k - \left(\sigma_s\cdot\sigma_s^{\mathrm{T}}\right)^{-1}\cdot\mu^{\mathrm{T}}\cdot\left(\mu\cdot\left(\sigma_s\cdot\sigma_s^{\mathrm{T}}\right)^{-1}\cdot\mu^{\mathrm{T}}\right)^{-1}\cdot\mu\right)\\&\quad\cdot\left(\sigma_s\cdot\sigma_s^{\mathrm{T}}\right)^{-1} + \xi\cdot\left(\mu\cdot\left(\sigma_s\cdot\sigma_s^{\mathrm{T}}\right)^{-1}\cdot\mu^{\mathrm{T}}\right)^{-1}\cdot\mu\cdot\left(\sigma_s\cdot\sigma_s^{\mathrm{T}}\right)^{-1}\\&= \frac{\mathrm{e}^{-\int_t^T r(\tau)\,\mathrm{d}\tau}}{\omega}(a_s - r)\cdot\left(\mathbb{I}_k - \left(\sigma_s\cdot\sigma_s^{\mathrm{T}}\right)^{-1}\cdot\mu^{\mathrm{T}}\cdot\left(\mu\cdot\left(\sigma_s\cdot\sigma_s^{\mathrm{T}}\right)^{-1}\right.\right.\\&\quad\left.\left.\cdot\mu^{\mathrm{T}}\right)^{-1}\cdot\mu\right)\cdot\left(\sigma_s\cdot\sigma_s^{\mathrm{T}}\right)^{-1} + \frac{\mathrm{e}^{-\int_t^T r(\tau)\,\mathrm{d}\tau}}{\omega}\nabla_A g_\omega(t,A)\cdot c\cdot\sigma_s^{\mathrm{T}}\\&\quad\cdot\left(\mathbb{I}_k - \left(\sigma_s\cdot\sigma_s^{\mathrm{T}}\right)^{-1}\cdot\mu^{\mathrm{T}}\cdot\left(\mu\cdot\left(\sigma_s\cdot\sigma_s^{\mathrm{T}}\right)^{-1}\cdot\mu^{\mathrm{T}}\right)^{-1}\cdot\mu\right)\cdot\left(\sigma_s\cdot\sigma_s^{\mathrm{T}}\right)^{-1}\\&\quad+\xi\cdot\left(\mu\cdot\left(\sigma_s\cdot\sigma_s^{\mathrm{T}}\right)^{-1}\cdot\mu^{\mathrm{T}}\right)^{-1}\cdot\mu\cdot\left(\sigma_s\cdot\sigma_s^{\mathrm{T}}\right)^{-1},\end{aligned} \tag{3.9.11}$$

and formula (3.9.9) follows. □

3.10 Case Study: Stochastic Volatility

We showcase some of the results of this chapter by applying them to the model discussed in Sect. 2.4.3. The results presented in this section draw on results from [42].

As in Sect. 2.4.3, we assume the stochastic volatility model (2.4.9), with $S = \{S_1\}$, $A = \{S_1, \mathbf{v}\}$, and market coefficients (2.4.13). We consider the CRRA utility of wealth. Equation (3.6.17) becomes

$$
\begin{aligned}
& g_\gamma^{(1,0,0)}(t,S_1,\mathbf{v}) + \frac{1}{2}\mathbf{v}\left(g_\gamma^{(0,2,0)}(t,S_1,\mathbf{v})S_1^2 + 2\sigma_v\rho_{2,1}g_\gamma^{(0,1,1)}(t,S_1,\mathbf{v})S_1\right. \\
& \left. +\sigma_v^2 g_\gamma^{(0,0,2)}(t,S_1,\mathbf{v})\right) + \left(k_v - \mathbf{v}K_v - \frac{(\gamma-1)(\mathbf{a}_0+\mathbf{v}\lambda_s - r)\sigma_v\rho_{2,1}}{\gamma}\right) \\
& g_\gamma^{(0,0,1)}(t,S_1,\mathbf{v}) + \left(S_1(\mathbf{a}_0+\mathbf{v}\lambda_s - \mathbb{D}) - \frac{(\gamma-1)S_1(\mathbf{a}_0+\mathbf{v}\lambda_s - r)}{\gamma}\right) \\
& g_\gamma^{(0,1,0)}(t,S_1,\mathbf{v}) + \frac{1}{2\gamma}\mathbf{v}\left(\sigma_v^2(\gamma-(\gamma-1)\rho_{2,1}^2)g_\gamma^{(0,0,1)}(t,S_1,\mathbf{v})^2\right. \\
& \left. +2S_1\sigma_v\rho_{2,1}g_\gamma^{(0,1,0)}(t,S_1,\mathbf{v})g_\gamma^{(0,0,1)}(t,S_1,\mathbf{v}) + S_1^2 g_\gamma^{(0,1,0)}(t,S_1,\mathbf{v})^2\right) \\
& = \frac{\gamma-1}{\gamma}\left(\frac{(\mathbf{a}_0+\mathbf{v}\lambda_s - r)^2}{2\mathbf{v}} + r\gamma\right) \qquad (3.10.1)
\end{aligned}
$$

for $t < T < \infty, S_1 > 0, \mathbf{v} > 0$, with the terminal condition $g_\gamma(T,S_1,\mathbf{v}) = 0$. We are looking for the solution in the form $g_\gamma(t,S_1,\mathbf{v}) = g_\gamma(t,\mathbf{v})$, and the above PDE reduces to

$$
\begin{aligned}
& g_\gamma^{(1,0)}(t,\mathbf{v}) + \frac{1}{2}\mathbf{v}g_\gamma^{(0,2)}(t,\mathbf{v})\sigma_v^2 + \frac{\mathbf{v}\left(\gamma-(\gamma-1)\rho_{2,1}^2\right)g_\gamma^{(0,1)}(t,\mathbf{v})^2\sigma_v^2}{2\gamma} \\
& + \left(k_v - \mathbf{v}K_v - \frac{(\gamma-1)(\mathbf{a}_0+\mathbf{v}\lambda_s - r)\sigma_v\rho_{2,1}}{\gamma}\right)g_\gamma^{(0,1)}(t,\mathbf{v}) \\
& = \frac{\gamma-1}{\gamma}\left(\frac{(\mathbf{a}_0+\mathbf{v}\lambda_s - r)^2}{2\mathbf{v}} + r\gamma\right) \qquad (3.10.2)
\end{aligned}
$$

for $t < T < \infty, \mathbf{v} > 0$, with the terminal condition $g_\gamma(T,\mathbf{v}) = 0$. There are two cases to consider: whether or not $\mathbf{a}_0 = r$. The assumption $\mathbf{a}_0 = r$ makes the model much simpler (yet, unfortunately, more rigid for practical implementation), since the right-hand side of (3.10.2) becomes linear in $\mathbf{v}$,

$$
\frac{\gamma-1}{\gamma}\left(\frac{(\mathbf{a}_0+\mathbf{v}\lambda_s - r)^2}{2\mathbf{v}} + r\gamma\right) = \frac{\gamma-1}{\gamma}\left(\frac{\mathbf{v}\lambda_s^2}{2} + r\gamma\right),
$$

and therefore, one can find (cf. [22]) the solution g_γ linear in $\mathbf{v}$:

$$g_\gamma(t,\mathbf{v}):=\mathbb{A}_0(t)+\mathbf{v}\mathbb{A}_1(t). \tag{3.10.3}$$

Indeed, plugging (3.10.3) into (3.10.2), we get

$$\begin{aligned}&\frac{\mathbf{v}\sigma_v^2\left(\gamma-(\gamma-1)\rho_{2,1}^2\right)\mathbb{A}_1(t)^2}{2\gamma}+\left(k_v-\mathbf{v}K_v-\frac{\mathbf{v}(\gamma-1)\lambda_s\sigma_v\rho_{2,1}}{\gamma}\right)\mathbb{A}_1(t)\\&+\mathbb{A}_0'(t)+\mathbf{v}\mathbb{A}_1'(t)=\frac{\gamma-1}{\gamma}\left(\frac{\mathbf{v}\lambda_s^2}{2}+r\gamma\right)\end{aligned} \tag{3.10.4}$$

for any $\mathbf{v}>0$. Therefore, two ODEs characterizing $\mathbb{A}_0(t)$ and $\mathbb{A}_1(t)$ are

$$\mathbb{A}_0'(t)+k_v\mathbb{A}_1(t)=r(\gamma-1) \tag{3.10.5}$$

and

$$\begin{aligned}&\mathbb{A}_1'(t)+\left(\frac{(1-\gamma)\sigma_v\rho_{2,1}\lambda_s}{\gamma}-K_v\right)\mathbb{A}_1(t)+\frac{\sigma_v^2\left(\gamma-(\gamma-1)\rho_{2,1}^2\right)\mathbb{A}_1(t)^2}{2\gamma}\\&=\frac{\gamma-1}{\gamma}\frac{\lambda_s^2}{2}\end{aligned} \tag{3.10.6}$$

for $t<T$, with the terminal conditions $\mathbb{A}_0(T)=\mathbb{A}_1(T)=0$. A solution of (3.10.6) is

$$\begin{aligned}\mathbb{A}_1(t)=&-\frac{1}{\sigma_v^2\left((\gamma-1)\rho_{2,1}^2-\gamma\right)}\Bigg(\gamma K_v+(\gamma-1)\lambda_s\sigma_v\rho_{2,1}\\&+\sqrt{\alpha}\tan\Bigg(\frac{1}{2\sqrt{\alpha}}\Big((t-T)\gamma K_v^2+2(t-T)(\gamma-1)\lambda_s\sigma_v\rho_{2,1}K_v\\&+(t-T)(\gamma-1)\lambda_s^2\sigma_v^2-2\sqrt{\alpha}\tan^{-1}\left(\frac{1}{\sqrt{\alpha}}(\gamma K_v+(\gamma-1)\lambda_s\sigma_v\rho_{2,1})\right)\Big)\Bigg)\Bigg)\end{aligned} \tag{3.10.7}$$

for

$$\alpha=-\gamma\left(\gamma K_v^2+(\gamma-1)\lambda_s\sigma_v\left(2\rho_{2,1}K_v+\lambda_s\sigma_v\right)\right), \tag{3.10.8}$$

and once (3.10.6) is solved, $\mathbb{A}_0(t)$ can be found by integrating (3.10.5). That is unnecessary for the sake of portfolio optimization, since $\nabla_A g_\gamma$ is all that is needed, and $\nabla_A g_\gamma = \nabla_{\{S_1,\mathbf{v}\}} g_\gamma = \{0, \mathbb{A}_1(t)\}$. Consequently, applying (3.6.18), we get, under the crucial assumption $\mathbf{a}_0 = r$, the optimal portfolio formula

$$\begin{aligned}
\Pi_\gamma^\star(t,X,A) &= \frac{X}{\gamma}\left(a_s - r + \nabla_A g_\gamma \cdot c \cdot \sigma_s^{\mathrm{T}}\right).\left(\sigma_s \cdot \sigma_s^{\mathrm{T}}\right)^{-1} \\
&= \left\{\frac{1}{\mathbf{v}\gamma} X \left(\mathbf{a}_0 + \mathbf{v}\lambda_s - r + \sqrt{\mathbf{v}}\left(\sqrt{\mathbf{v}}\sigma_v \rho_{2,1} g_\gamma^{(0,0,1)}(t,S_1,\mathbf{v})\right.\right.\right. \\
&\quad \left.\left.\left. + \sqrt{\mathbf{v}} S_1 g_\gamma^{(0,1,0)}(t,S_1,\mathbf{v})\right)\right)\right\} \\
&= \left\{\frac{1}{\mathbf{v}\gamma} X \left(\mathbf{a}_0 + \mathbf{v}\lambda_s - r + \mathbf{v}\sigma_v \rho_{2,1} g_\gamma^{(0,1)}(t,\mathbf{v})\right)\right\} \\
&= \left\{\frac{X\left(\lambda_s + \sigma_v \rho_{2,1} g_\gamma^{(0,1)}(t,\mathbf{v})\right)}{\gamma}\right\} = \left\{\frac{X\left(\lambda_s + \sigma_v \rho_{2,1} \mathbb{A}_1(t)\right)}{\gamma}\right\}.
\end{aligned} \tag{3.10.9}$$

What if $\mathbf{a}_0 \neq r$? In that case, an explicit solution of (3.10.2) is unknown – only a numerical solution is available. Nevertheless, in that case, following [37], we can still find the "perpetual optimal portfolio formula" – the one corresponding to $\lim_{t\to-\infty} \nabla_A g_\gamma$. Indeed, differentiating (3.10.2) with respect to $\mathbf{v}$, we get

$$\begin{aligned}
&\frac{\left(\gamma - (\gamma-1)\rho_{2,1}^2\right) g_\gamma^{(0,1)}(t,\mathbf{v})^2 \sigma_v^2}{2\gamma} + \frac{1}{2} g_\gamma^{(0,2)}(t,\mathbf{v})\sigma_v^2 + \frac{1}{\gamma}\mathbf{v}\left(\gamma - (\gamma-1)\rho_{2,1}^2\right) \\
&g_\gamma^{(0,1)}(t,\mathbf{v}) g_\gamma^{(0,2)}(t,\mathbf{v})\sigma_v^2 + \frac{1}{2}\mathbf{v} g_\gamma^{(0,3)}(t,\mathbf{v})\sigma_v^2 + \left(-K_v - \frac{(\gamma-1)\lambda_s \sigma_v \rho_{2,1}}{\gamma}\right) \\
&g_\gamma^{(0,1)}(t,\mathbf{v}) + \left(k_v - \mathbf{v}K_v - \frac{1}{\gamma}(\gamma-1)\left(\mathbf{a}_0 + \mathbf{v}\lambda_s - r\right)\sigma_v \rho_{2,1}\right) g_\gamma^{(0,2)}(t,\mathbf{v}) \\
&+ g_\gamma^{(1,1)}(t,\mathbf{v}) = \frac{\gamma-1}{\gamma}\left(\frac{\lambda_s\left(\mathbf{a}_0 + \mathbf{v}\lambda_s - r\right)}{\mathbf{v}} - \frac{\left(\mathbf{a}_0 + \mathbf{v}\lambda_s - r\right)^2}{2\mathbf{v}^2}\right),
\end{aligned} \tag{3.10.10}$$

and since we anticipate a steady-state solution $h(\mathbf{v}) := \lim_{t\to-\infty} g_\gamma^{(0,1)}(t,\mathbf{v})$, we make the substitution

$$g_\gamma(t,\mathbf{v}) \to \int h(\mathbf{v})\,\mathrm{d}\mathbf{v} \tag{3.10.11}$$

in (3.10.10), obtaining

$$\frac{h(\mathbf{v})^2\left(\gamma-(\gamma-1)\rho_{2,1}^2\right)\sigma_v^2}{2\gamma}+\frac{1}{2}h'(\mathbf{v})\sigma_v^2+\frac{\mathbf{v}h(\mathbf{v})\left(\gamma-(\gamma-1)\rho_{2,1}^2\right)h'(\mathbf{v})\sigma_v^2}{\gamma}$$
$$+\frac{1}{2}\mathbf{v}h''(\mathbf{v})\sigma_v^2\left(-K_v-\frac{(\gamma-1)\lambda_s\sigma_v\rho_{2,1}}{\gamma}\right)h(\mathbf{v})$$
$$+\left(k_v-\mathbf{v}K_v-\frac{1}{\gamma}(\gamma-1)\left(\mathbf{a}_0+\mathbf{v}\lambda_s-r\right)\sigma_v\rho_{2,1}\right)$$
$$h'(\mathbf{v})=\frac{\gamma-1}{\gamma}\left(\frac{\lambda_s\left(\mathbf{a}_0+\mathbf{v}\lambda_s-r\right)}{\mathbf{v}}-\frac{\left(\mathbf{a}_0+\mathbf{v}\lambda_s-r\right)^2}{2\mathbf{v}^2}\right)$$
$$=\frac{1}{2}\frac{\gamma-1}{\gamma}\left(\lambda_s^2-\frac{\left(\mathbf{a}_0-r\right)^2}{\mathbf{v}^2}\right) \tag{3.10.12}$$

for any $\mathbf{v}>0$. We need to find an entire solution of (3.10.12). While in the case $\mathbf{a}_0-r$, equation (3.10.12) is easy to solve, since it admits a constant solution $h(\mathbf{v})=\mathbb{H}_0$, characterized as a solution of the algebraic equation

$$\frac{\mathbb{H}_0^2\left(\gamma-(\gamma-1)\rho_{2,1}^2\right)\sigma_v^2}{2\gamma}-\left(K_v+\frac{(\gamma-1)\lambda_s\sigma_v\rho_{2,1}}{\gamma}\right)\mathbb{H}_0=\frac{(\gamma-1)\lambda_s^2}{2\gamma}, \tag{3.10.13}$$

that is not the case when $\mathbf{a}_0\neq r$. Even though there are zero-, first-, and second-order derivatives of h as well as multiplicative terms on the left-hand side of (3.10.12) we can find an entire solution of (3.10.12) in the simple form

$$h(\mathbf{v})=\mathbb{H}_0+\frac{\mathbb{H}_{-1}}{\mathbf{v}}. \tag{3.10.14}$$

Indeed, plugging (3.10.14) into (3.10.12), we get

$$\frac{\mathbb{H}_{-1}\sigma_v^2}{2\mathbf{v}^2}+\frac{1}{2\gamma}\left(\frac{\mathbb{H}_{-1}}{\mathbf{v}}+\mathbb{H}_0\right)^2(\gamma-(\gamma-1)\rho_{2,1}^2)\sigma_v^2-\frac{1}{\mathbf{v}\gamma}\mathbb{H}_{-1}\left(\frac{\mathbb{H}_{-1}}{\mathbf{v}}+\mathbb{H}_0\right)$$
$$\left(\gamma-(\gamma-1)\rho_{2,1}^2\right)\sigma_v^2+\left(-K_v-\frac{(\gamma-1)\lambda_s\sigma_v\rho_{2,1}}{\gamma}\right)\left(\frac{\mathbb{H}_{-1}}{\mathbf{v}}+\mathbb{H}_0\right)$$
$$-\frac{1}{\mathbf{v}^2}\mathbb{H}_{-1}\left(k_v-\mathbf{v}K_v-\frac{1}{\gamma}(\gamma-1)\left(-r+\mathbf{a}_0+\mathbf{v}\lambda_s\right)\sigma_v\rho_{2,1}\right)$$
$$=\frac{\gamma-1}{2\gamma}\left(\lambda_s^2-\frac{\left(\mathbf{a}_0-r\right)^2}{\mathbf{v}^2}\right), \tag{3.10.15}$$

which simplifies to

$$\frac{1}{\mathbf{v}^2}\left(\frac{1}{2}(\mathbf{a}_0-r)^2\left(1-\frac{1}{\gamma}\right)+\mathbb{H}_{-1}\left(\frac{1}{2\gamma}\sigma_v(\gamma\sigma_v-2(\gamma-1)(r-\mathbf{a}_0)\rho_{2,1})-k_v\right)\right.$$
$$\left.+\,\mathbb{H}_{-1}^2\frac{\sigma_v^2}{2}\left(\rho_{2,1}^2\left(1-\frac{1}{\gamma}\right)-1\right)\right)-\frac{1}{2\gamma}((\gamma-1)\lambda_s^2+2\mathbb{H}_0(\gamma K_v+(\gamma-1)$$
$$\sigma_v\rho_{2,1}\lambda_s)+\mathbb{H}_0^2\sigma_v^2((\gamma-1)\rho_{2,1}^2-\gamma))=0, \tag{3.10.16}$$

and therefore, we obtain two quadratic equations characterizing $\mathbb{H}_{-1}$ and $\mathbb{H}_0$:

$$\mathbb{H}_{-1}^2\frac{\sigma_v^2}{2}\left(1-\frac{\gamma-1}{\gamma}\rho_{2,1}^2\right)-\mathbb{H}_{-1}\left(\frac{\sigma_v^2}{2}+\frac{\gamma-1}{\gamma}\sigma_v\,(\mathbf{a}_0-r)\,\rho_{2,1}-k_v\right)-\frac{1}{2}\frac{\gamma-1}{\gamma}$$
$$(\mathbf{a}_0-r)^2=0, \tag{3.10.17}$$

$$\mathbb{H}_0^2\frac{\sigma_v^2}{2}\left(1-\frac{\gamma-1}{\gamma}\rho_{2,1}^2\right)-\mathbb{H}_0\left(K_v+\frac{\gamma-1}{\gamma}\sigma_v\rho_{2,1}\lambda_s\right)-\frac{1}{2}\frac{\gamma-1}{\gamma}\lambda_s^2=0. \tag{3.10.18}$$

Once $\mathbb{H}_{-1}$ and $\mathbb{H}_0$ have been found, then the "perpetual optimal portfolio formula" becomes

$$\begin{aligned}\Pi_\gamma^\star(t,X,A) &= \left\{\frac{1}{\mathbf{v}\gamma}X\left(\mathbf{a}_0+\mathbf{v}\lambda_s-r+\mathbf{v}\sigma_v\rho_{2,1}g_\gamma^{(0,1)}(t,\mathbf{v})\right)\right\}\\
&\underset{t\to-\infty}{\longrightarrow}\left\{\frac{1}{\mathbf{v}\gamma}X\left(\mathbf{a}_0+\mathbf{v}\lambda_s-r+\mathbf{v}\sigma_v\rho_{2,1}\lim_{t\to-\infty}g_\gamma^{(0,1)}(t,\mathbf{v})\right)\right\}\\
&= \left\{\frac{1}{\mathbf{v}\gamma}X\left(\mathbf{a}_0+\mathbf{v}\lambda_s-r+\mathbf{v}\sigma_v\rho_{2,1}h(\mathbf{v})\right)\right\}\\
&= \left\{\frac{1}{\mathbf{v}\gamma}X\left(\mathbf{a}_0+\mathbf{v}\lambda_s-r+\mathbf{v}\sigma_v\rho_{2,1}\left(\mathbb{H}_0+\frac{\mathbb{H}_{-1}}{\mathbf{v}}\right)\right)\right\}.\end{aligned} \tag{3.10.19}$$

Proposition 3.10.1. *For Heston's stochastic volatility model (2.4.9), with $S=\{S_1\}$, $A=\{S_1,\mathbf{v}\}$, and market coefficients (2.4.13), the perpetual optimal portfolio formula for the CRRA utility of wealth with risk-aversion γ becomes*

$$\Pi_\gamma^\star(t,X,A)=\left\{\frac{X}{\gamma}\left(\frac{\mathbf{a}_0-r}{\mathbf{v}}+\lambda_s+\sigma_v\rho_{2,1}\left(\mathbb{H}_0+\frac{\mathbb{H}_{-1}}{\mathbf{v}}\right)\right)\right\}, \tag{3.10.20}$$

where $\mathbb{H}_{-1}$ and $\mathbb{H}_0$ are found as the solution of the quadratic system (3.10.17)–(3.10.18).

Exercise 3.10.1. Modify the results of this section to show that in the case $\mathbf{a}_0 = r$ and CARA utility of wealth, applying (3.10.17) instead of (3.6.17), we get

$$g(t,\mathbf{v}) = \mathbb{A}_0(t) + \mathbf{v}\mathbb{A}_1(t), \tag{3.10.21}$$

where

$$\mathbb{A}_1(t) = -\left(K_v + \lambda_s\sigma_v\rho_{2,1} + \sqrt{\alpha}\tan\left(\left((t-T)K_v^2 + 2(t-T)\lambda_s\sigma_v\rho_{2,1}K_v \right.\right.\right.$$
$$\left.\left.\left. +(t-T)\lambda_s^2\sigma_v^2 - 2\sqrt{\alpha}\tan^{-1}\left(\frac{K_v + \lambda_s\sigma_v\rho_{2,1}}{\sqrt{\alpha}}\right)\right)\Big/\left(2\sqrt{\alpha}\right)\right)\right)\Big/\left(\sigma_v^2\left(\rho_{2,1}^2 - 1\right)\right) \tag{3.10.22}$$

with

$$\alpha = -K_v^2 - 2\lambda_s\sigma_v\rho_{2,1}K_v - \lambda_s^2\sigma_v^2. \tag{3.10.23}$$

Furthermore, using (3.8.18) instead of (3.6.18), calculate the optimal portfolio formula.

Exercise 3.10.2. Modify the results of this section to show that in the case $\mathbf{a}_0 \neq r$ and CARA utility of wealth, if $h(\mathbf{v}) := \lim_{t\to-\infty} g^{(0,1)}(t,\mathbf{v})$, then (3.10.14) holds with

$$\mathbb{H}_{-1}^2\left(\frac{1}{2}\sigma_v^2\left(1-\rho_{2,1}^2\right)\right) - \mathbb{H}_{-1}\left(\frac{1}{2}\sigma_v^2 - k_v + (\mathbf{a}_0 - r)\sigma_v\rho_{2,1}\right) - \frac{(\mathbf{a}_0 - r)^2}{2} = 0 \tag{3.10.24}$$

and

$$\mathbb{H}_0^2\left(\frac{1}{2}\sigma_v^2\left(1-\rho_{2,1}^2\right)\right) - \mathbb{H}_0\left(\sigma_v\rho_{2,1}\lambda_s + K_v\right) - \frac{1}{2}\lambda_s^2 = 0. \tag{3.10.25}$$

Remark 3.10.1. Equations (3.10.24) and (3.10.25) are obtained by sending $\gamma \to \infty$ in equations (3.10.17) and (3.10.18) espectively.

3.11 Case Study: Stochastic Interest Rates

We showcase some of the results of this chapter also by applying them to the model discussed in Sect. 2.4.5. The CARA utility results are inapplicable now, since the interest rate is stochastic. We shall concentrate on the results from Sect. 3.6. The results presented in this section draw on results from [43] and [16].

As in Sect. 2.4.5, we assume the stochastic interest rate model (2.4.23), with $S = \{S_1\}, A = \{S_1, r\}$, and market coefficients (2.4.27). We are looking for the solution of (3.6.17) in the form $g_\gamma(t,S_1,r) = g_\gamma(t,r)$, and equation (3.6.17) becomes

$$\left(q_0 + rq_1 - \frac{w(\gamma-1)(\mathbf{a}_0+\beta r - r)\rho_{2,1}}{\gamma \mathbf{p}_1}\right) g^{(0,1)}(t,r) + \frac{1}{2}\left(\left(1-\frac{\gamma-1}{\gamma}\right)\rho_{2,1}^2\right.$$

$$\left. g^{(0,1)}(t,r)w^2 + \left(1-\rho_{2,1}^2\right) g^{(0,1)}(t,r)w^2\right) g^{(0,1)}(t,r) + \frac{1}{2}w^2 g^{(0,2)}(t,r) + g^{(1,0)}(t,r)$$

$$= \frac{\gamma-1}{\gamma}\left(\frac{(\beta r - r + \mathbf{a}_0)^2}{2\mathbf{p}_1^2} + r\gamma\right) \tag{3.11.1}$$

for $t < T < \infty, r \in (-\infty,\infty)$, with the terminal condition $g_\gamma(T,r) = 0$.

We look for the solution in the form $g_\gamma(t,r) = \mathbb{G}_2(t)r^2 + \mathbb{G}_1(t)r + \mathbb{G}_0(t)$, obtaining

$$\mathbb{G}_2'(t)r^2 + \mathbb{G}_1'(t)r + w^2\mathbb{G}_2(t) + \left(q_0 + rq_1 - \frac{w(\gamma-1)(\beta r - r + \mathbf{a}_0)\rho_{2,1}}{\gamma \mathbf{p}_1}\right)(\mathbb{G}_1(t)$$

$$+2r\mathbb{G}_2(t)) + \frac{1}{2}(\mathbb{G}_1(t) + 2r\mathbb{G}_2(t))\left(\left(1-\frac{\gamma-1}{\gamma}\right)\rho_{2,1}^2(\mathbb{G}_1(t)\right.$$

$$\left. +2r\mathbb{G}_2(t))w^2 + (1-\rho_{2,1}^2)(\mathbb{G}_1(t) + 2r\mathbb{G}_2(t))w^2\right) + \mathbb{G}_0'(t)$$

$$= \frac{\gamma-1}{\gamma}\left(\frac{(\mathbf{a}_0 + \beta r - r)^2}{2\mathbf{p}_1^2} + r\gamma\right) \tag{3.11.2}$$

for $t < T$, with the terminal condition $\mathbb{G}_0(T) = \mathbb{G}_1(T) = \mathbb{G}_2(T) = 0$, for any $r \in (-\infty,\infty)$, therefore arriving at three ODEs characterizing $\mathbb{G}_0(t), \mathbb{G}_1(t), \mathbb{G}_2(t)$:

$$\mathbb{G}_2'(t) + \frac{2w^2\left(\gamma - (\gamma-1)\rho_{2,1}^2\right)\mathbb{G}_2(t)^2}{\gamma} + \frac{1}{\gamma \mathbf{p}_1}2\left(\gamma q_1 \mathbf{p}_1 + w(\gamma-1)(1-\beta)\rho_{2,1}\right)$$

$$\mathbb{G}_2(t) = \frac{(\beta-1)^2(\gamma-1)}{2\gamma \mathbf{p}_1^2}, \tag{3.11.3}$$

$$\mathbb{G}_1'(t) + 2w^2\left(1 - \frac{\gamma-1}{\gamma}\rho_{2,1}^2\right)\mathbb{G}_1(t)\mathbb{G}_2(t) + \left(q_1 + \frac{w(1-\beta)(\gamma-1)\rho_{2,1}}{\gamma \mathbf{p}_1}\right)\mathbb{G}_1(t)$$

$$+2\left(q_0 - \frac{w(\gamma-1)\mathbf{a}_0\rho_{2,1}}{\gamma \mathbf{p}_1}\right)\mathbb{G}_2(t) = \frac{\gamma-1}{\gamma}\left(\gamma + \frac{(\beta-1)\mathbf{a}_0}{\mathbf{p}_1^2}\right), \tag{3.11.4}$$

$$\mathbb{G}_0'(t) + \left(q_0 - \frac{w(\gamma-1)\mathbf{a}_0\rho_{2,1}}{\gamma\mathbf{p}_1}\right)\mathbb{G}_1(t) + \frac{w^2\left(\gamma-(\gamma-1)\rho_{2,1}^2\right)\mathbb{G}_1(t)^2}{2\gamma} + w^2\mathbb{G}_2(t)$$
$$= \frac{1}{2}\frac{\gamma-1}{\gamma}\frac{\mathbf{a}_0^2}{\mathbf{p}_1^2} \tag{3.11.5}$$

for $t < T$, with the terminal conditions $\mathbb{G}_0(T) = \mathbb{G}_1(T) = \mathbb{G}_2(T) = 0$.

An important (for its simplicity) and instructive special case, solved in [43], is $\beta = 1$. In such a case, the above system simplifies substantially to

$$\mathbb{G}_2(t) = 0, \tag{3.11.6}$$

$$\mathbb{G}_1'(t) + q_1\mathbb{G}_1(t) = \gamma - 1, \tag{3.11.7}$$

$$\mathbb{G}_0'(t) + \left(q_0 - \frac{w(\gamma-1)\mathbf{a}_0\rho_{2,1}}{\gamma\mathbf{p}_1}\right)\mathbb{G}_1(t) + \frac{w^2\left(\gamma-(\gamma-1)\rho_{2,1}^2\right)\mathbb{G}_1(t)^2}{2\gamma}$$
$$= \frac{1}{2}\frac{\gamma-1}{\gamma}\frac{\mathbf{a}_0^2}{\mathbf{p}_1^2} \tag{3.11.8}$$

for $t < T$, with the terminal conditions $\mathbb{G}_0(T) = \mathbb{G}_1(T) = 0$. Obviously, (3.11.7) is solved by

$$\mathbb{G}_1(t) = (\gamma-1)\frac{1-\mathrm{e}^{(T-t)q_1}}{q_1} \tag{3.11.9}$$

($\mathbb{G}_0(t)$ is not needed). Returning to the general case $\beta \neq 1$, to find $\nabla_A g_\gamma(t,r)$, we need to solve (3.11.3)–(3.11.4). To do so is possible, either explicitly or numerically, but instead we shall confine ourselves to finding only its "perpetual" version, i.e., $\lim_{t\to-\infty}\nabla_A g_\gamma(t,r)$. To that end, we differentiate (3.11.1) with respect to r, obtaining

$$\left(q_1 - \frac{w(\beta-1)(\gamma-1)\rho_{2,1}}{\gamma\mathbf{p}_1}\right)g^{(0,1)}(t,r) + \left(q_0 + rq_1 - \frac{1}{\gamma\mathbf{p}_1}w(\gamma-1)\left(\beta r - r\right.\right.$$
$$\left.\left.+\mathbf{a}_0\right)\rho_{2,1}\right)g^{(0,2)}(t,r) + \frac{1}{\gamma}w^2\left(\gamma-(\gamma-1)\rho_{2,1}^2\right)g^{(0,1)}(t,r)g^{(0,2)}(t,r)$$
$$+\frac{1}{2}w^2 g^{(0,3)}(t,r) + g^{(1,1)}(t,r) = \frac{(\gamma-1)}{\gamma}\left(\gamma + \frac{(\beta-1)(\beta r - r + \mathbf{a}_0)}{\mathbf{p}_1^2}\right).$$

Since we anticipate the existence of $h(r) := \lim_{t\to-\infty} g^{(0,1)}(t,r)$, we make the substitution

$$g(t,r) \to \int h(r)\,dr, \tag{3.11.10}$$

obtaining

$$\frac{h(r)\left(\gamma-(\gamma-1)\rho_{2,1}^2\right)h'(r)w^2}{\gamma}+\frac{1}{2}h''(r)w^2+h(r)\left(q_1-\frac{w(\beta-1)(\gamma-1)\rho_{2,1}}{\gamma\mathbf{p}_1}\right)$$
$$+\left(q_0+rq_1-\frac{1}{\gamma\mathbf{p}_1}w(\gamma-1)(\beta r-r+\mathbf{a}_0)\rho_{2,1}\right)h'(r)$$
$$=\frac{(\gamma-1)}{\gamma}\left(\gamma+\frac{(\beta-1)(\beta r-r+\mathbf{a}_0)}{\mathbf{p}_1^2}\right) \tag{3.11.11}$$

for $-\infty<r<\infty$. We are looking for the entire solution of (3.11.11) in the form $h(r)=\mathbb{H}_0+\mathbb{H}_1 r$, and therefore we obtain

$$\frac{\mathbb{H}_1(\mathbb{H}_0+r\mathbb{H}_1)\left(\gamma-(\gamma-1)\rho_{2,1}^2\right)w^2}{\gamma}+(\mathbb{H}_0+r\mathbb{H}_1)\left(q_1-\frac{w(\beta-1)(\gamma-1)\rho_{2,1}}{\gamma\mathbf{p}_1}\right)$$
$$+\mathbb{H}_1\left(q_0+rq_1-\frac{1}{\gamma\mathbf{p}_1}w(\gamma-1)((\beta-1)r+\mathbf{a}_0)\rho_{2,1}\right)$$
$$=\frac{(\gamma-1)}{\gamma}\left(\gamma+\frac{(\beta-1)((\beta-1)r+\mathbf{a}_0)}{\mathbf{p}_1^2}\right) \tag{3.11.12}$$

for $-\infty<r<\infty$, yielding two algebraic equations. One of them is quadratic,

$$\frac{w^2\left(\gamma-(\gamma-1)\rho_{2,1}^2\right)}{\gamma}\mathbb{H}_1^2+\frac{1}{\gamma\mathbf{p}_1}2\left(\gamma q_1\mathbf{p}_1-w(\beta-1)(\gamma-1)\rho_{2,1}\right)\mathbb{H}_1$$
$$-\frac{(\beta-1)^2(\gamma-1)}{\gamma\mathbf{p}_1^2}=0, \tag{3.11.13}$$

and once that one is solved (a nontrivial issue is the choice of the proper solution among the two solutions of the quadratic equation, which, for example, can be settled by comparison with the solution of the above ODEs: $\mathbb{H}_0=\lim_{t\to-\infty}\mathbb{G}_1(t)$, $\mathbb{H}_1=2\lim_{t\to-\infty}\mathbb{G}_2(t)$). The other is linear,

$$\left(q_1+\frac{1}{\gamma\mathbf{p}_1}w\left(w\mathbb{H}_1\mathbf{p}_1\left(\gamma-(\gamma-1)\rho_{2,1}^2\right)-(\beta-1)(\gamma-1)\rho_{2,1}\right)\right)\mathbb{H}_0-\gamma+q_0\mathbb{H}_1$$
$$-\frac{1}{\gamma\mathbf{p}_1^2}(\gamma-1)\mathbf{a}_0\left(\beta-1+w\mathbb{H}_1\mathbf{p}_1\rho_{2,1}\right)+1=0, \tag{3.11.14}$$

characterizing $\mathbb{H}_0$ and $\mathbb{H}_1$. If $\beta=1$, equation (3.11.13) is solved by

$$\mathbb{H}_1=0, \tag{3.11.15}$$

and then equation (3.11.14) by

$$\mathbb{H}_0 = \frac{\gamma - 1}{q_1}. \tag{3.11.16}$$

Also observe that if $\gamma = 1$, then for any β, $\mathbb{H}_0 = \mathbb{H}_1 = 0$. Summarizing, we have the following.

Proposition 3.11.1. *For the stochastic interest rate model (2.4.23), with* $S = \{S_1\}$, $A = \{S_1, r\}$, *and market coefficients (2.4.27), the perpetual optimal portfolio formula for the CRRA utility of wealth with risk-aversion* γ *is*

$$\Pi^\star_\gamma(t,X,A) = \left\{\frac{X}{\gamma}\left(\frac{\mathbf{a}_0 + (\beta-1)r}{\mathbf{p}_1^2} + \frac{w\left(\mathbb{H}_0 + r\mathbb{H}_1\right)\rho_{2,1}}{\mathbf{p}_1}\right)\right\}, \tag{3.11.17}$$

where $\mathbb{H}_0$ *and* $\mathbb{H}_1$ *are found as the solution of the system (3.11.13)–(3.11.14).*

Proof. We have

$$\begin{aligned}\Pi^\star_\gamma(t,X,A) &= \left\{\frac{1}{\gamma\mathbf{p}_1^2}X(\beta r - r + \mathbf{a}_0 + \mathbf{p}_1(w\rho_{2,1}g^{(0,0,1)}(t,S_1,r)\right.\\ &\qquad \left. + S_1\mathbf{p}_1 g^{(0,1,0)}(t,S_1,r)))\right\}\\ &= \left\{\frac{X\left(\beta r - r + \mathbf{a}_0 + wh(r)\mathbf{p}_1\rho_{2,1}\right)}{\gamma\mathbf{p}_1^2}\right\}\\ &= \left\{\frac{1}{\gamma\mathbf{p}_1^2}X\left(\beta r - r + \mathbf{a}_0 + w\left(\mathbb{H}_0 + r\mathbb{H}_1\right)\mathbf{p}_1\rho_{2,1}\right)\right\}. \end{aligned} \tag{3.11.18}$$

□

3.12 Stochastic Interest, Dividend, and Appreciation Rate Model: Part 2

Continuing Sect. 2.4.8, following [21], we solve the CRRA optimal portfolio problem for the market model (2.4.39)–(2.4.42).

By setting $g_\gamma(t, S_1, \mathbf{v}, r, \mathbb{D}) = g(t, \mathbf{v}, r)$, equation (3.6.17) becomes

$$\begin{aligned}&\left(\alpha_\mathbf{r}\left(\theta_\mathbf{r} - r\right) - \frac{(\beta r - r + \mathbf{v})(\gamma - 1)\sigma_\mathbf{r}\rho_{3,1}}{\gamma\sigma_S}\right) g^{(0,0,1)}(t,\mathbf{v},r)\\ &\quad + \left(\alpha_v\left(\theta_v - \mathbf{v}\right) - \frac{(\beta r - r + \mathbf{v})(\gamma - 1)\sigma_v\rho_{2,1}}{\gamma\sigma_S}\right) g^{(0,1,0)}(t,\mathbf{v},r)\end{aligned}$$

$$+\frac{1}{2}\left(g^{(0,1,0)}(t,\mathbf{v},r)\left(\left(1-\frac{\gamma-1}{\gamma}\right)\sigma_v\rho_{2,1}(\sigma_{\mathbf{r}}\rho_{3,1}g^{(0,0,1)}(t,\mathbf{v},r)\right.\right.$$

$$+\sigma_v\rho_{2,1}g^{(0,1,0)}(t,\mathbf{v},r))+\sigma_v\sqrt{1-\rho_{2,1}^2}((\sigma_{\mathbf{r}}(\rho_{3,2}-\rho_{2,1}\rho_{3,1})g^{(0,0,1)}(t,\mathbf{v},.r))\Big/$$

$$\left.\left.\left(\sqrt{1-\rho_{2,1}^2}\right)+\sigma_v\sqrt{1-\rho_{2,1}^2}g^{(0,1,0)}(t,\mathbf{v},r))\right)\right)$$

$$+g^{(0,0,1)}(t,\mathbf{v},r)\left(\left(-\rho_{3,1}^2+\frac{(\rho_{3,2}-\rho_{2,1}\rho_{3,1})^2}{\rho_{2,1}^2-1}+1\right)g^{(0,0,1)}(t,\mathbf{v},r)\sigma_{\mathbf{r}}^2\right.$$

$$+\left(1-\frac{\gamma-1}{\gamma}\right)\rho_{3,1}\left(\sigma_{\mathbf{r}}\rho_{3,1}g^{(0,0,1)}(t,\mathbf{v},r)+\sigma_v\rho_{2,1}g^{(0,1,0)}(t,\mathbf{v},r)\right)\sigma_{\mathbf{r}}$$

$$+\frac{1}{\sqrt{1-\rho_{2,1}^2}}\sigma_{\mathbf{r}}(\rho_{3,2}-\rho_{2,1}\rho_{3,1})((\sigma_{\mathbf{r}}(\rho_{3,2}-\rho_{2,1}\rho_{3,1})$$

$$\left.\left.g^{(0,0,1)}(t,\mathbf{v},r))/\left(\sqrt{1-\rho_{2,1}^2}\right)+\sigma_v\sqrt{1-\rho_{2,1}^2}g^{(0,1,0)}(t,\mathbf{v},r))\right)\right)$$

$$+\frac{1}{2}\left(\left(\rho_{3,1}^2\sigma_{\mathbf{r}}^2+\frac{(\rho_{3,2}-\rho_{2,1}\rho_{3,1})^2\sigma_{\mathbf{r}}^2}{1-\rho_{2,1}^2}+\right.\right.$$

$$\left.\left(-\rho_{3,1}^2+\frac{(\rho_{3,2}-\rho_{2,1}\rho_{3,1})^2}{\rho_{2,1}^2-1}+1\right)\sigma_{\mathbf{r}}^2\right)g^{(0,0,2)}(t,\mathbf{v},r)$$

$$+2(\sigma_v\sigma_{\mathbf{r}}$$

$$\rho_{2,1}\rho_{3,1}+\sigma_v\sigma_{\mathbf{r}}(\rho_{3,2}-\rho_{2,1}\rho_{3,1}))g^{(0,1,1)}(t,\mathbf{v},r)+(\rho_{2,1}^2\sigma_v^2$$

$$\left.+(1-\rho_{2,1}^2)\sigma_v^2)g^{(0,2,0)}(t,\mathbf{v},r)\right)+g^{(1,0,0)}(t,\mathbf{v},r)=\frac{(\gamma-1)\left(\frac{(\beta r-r+\mathbf{v})^2}{2\sigma_S^2}+r\gamma\right)}{\gamma} \tag{3.12.1}$$

for $t<T<\infty, -\infty<\mathbf{v}<\infty, -\infty<r<\infty$, with the terminal condition $g(T,\mathbf{v},r)=0$. We look for the solution in the form

$$g(t,\mathbf{v},r)=\mathbb{H}_5(t)r^2+\mathbf{v}\mathbb{H}_3(t)r+\mathbb{H}_4(t)r+\mathbb{H}_0(t)+\mathbf{v}\mathbb{H}_1(t)+\mathbf{v}^2\mathbb{H}_2(t), \tag{3.12.2}$$

obtaining six ODEs characterizing the functions $\mathbb{H}_0(t),\mathbb{H}_1(t),\mathbb{H}_2(t),\mathbb{H}_3(t),\mathbb{H}_4(t)$, and $\mathbb{H}_5(t)$:

$$\frac{1}{2}\left(1-\frac{\gamma-1}{\gamma}\right)\sigma_v^2\rho_{2,1}^2\mathbb{H}_1(t)^2+\frac{1}{2}\sigma_v^2\left(1-\rho_{2,1}^2\right)\mathbb{H}_1(t)^2+\alpha_v\theta_v\mathbb{H}_1(t)+\left(1-\frac{\gamma-1}{\gamma}\right)$$

$$\sigma_v\sigma_{\mathbf{r}}\rho_{2,1}\rho_{3,1}\mathbb{H}_4(t)\mathbb{H}_1(t)+\sigma_v\sigma_{\mathbf{r}}(\rho_{3,2}-\rho_{2,1}\rho_{3,1})\mathbb{H}_4(t)\mathbb{H}_1(t)$$

$$+\frac{1}{2}\left(1-\frac{\gamma-1}{\gamma}\right)\sigma_{\mathbf{r}}^2\rho_{3,1}^2\mathbb{H}_4(t)^2+\frac{\sigma_{\mathbf{r}}^2(\rho_{3,2}-\rho_{2,1}\rho_{3,1})^2\mathbb{H}_4(t)^2}{2\left(1-\rho_{2,1}^2\right)}$$

$$+\frac{1}{2}\sigma_{\mathbf{r}}^2\left(-\rho_{3,1}^2+\frac{(\rho_{3,2}-\rho_{2,1}\rho_{3,1})^2}{\rho_{2,1}^2-1}+1\right)\mathbb{H}_4(t)^2+\alpha_{\mathbf{r}}\theta_{\mathbf{r}}\mathbb{H}_4(t)$$

$$+\frac{1}{2}\left(2\left(\rho_{2,1}^2\sigma_v^2+(1-\rho_{2,1}^2)\,\sigma_v^2\right)\mathbb{H}_2(t)+2\left(\sigma_v\sigma_{\mathbf{r}}\rho_{2,1}\rho_{3,1}\right.\right.$$

$$\left.+\sigma_v\sigma_{\mathbf{r}}(\rho_{3,2}-\rho_{2,1}\rho_{3,1})\right)\mathbb{H}_3(t)+2\left(\rho_{3,1}^2\sigma_{\mathbf{r}}^2+\frac{(\rho_{3,2}-\rho_{2,1}\rho_{3,1})^2\sigma_{\mathbf{r}}^2}{1-\rho_{2,1}^2}\right.$$

$$\left.\left.+\left(-\rho_{3,1}^2+\frac{(\rho_{3,2}-\rho_{2,1}\rho_{3,1})^2}{\rho_{2,1}^2-1}+1\right)\sigma_{\mathbf{r}}^2\right)\mathbb{H}_5(t)\right)+\mathbb{H}_0'(t)=0, \tag{3.12.3}$$

$$\left(1-\frac{\gamma-1}{\gamma}\right)\rho_{2,1}^2\mathbb{H}_1(t)\mathbb{H}_3(t)\sigma_v^2+\left(1-\rho_{2,1}^2\right)\mathbb{H}_1(t)\mathbb{H}_3(t)\sigma_v^2+\frac{(\gamma-1)\rho_{2,1}\mathbb{H}_1(t)\sigma_v}{\gamma\sigma_S}$$

$$+\left(1-\frac{\gamma-1}{\gamma}\right)\sigma_{\mathbf{r}}\rho_{2,1}\rho_{3,1}\mathbb{H}_3(t)\mathbb{H}_4(t)\sigma_v+\sigma_{\mathbf{r}}(\rho_{3,2}-\rho_{2,1}\rho_{3,1})\mathbb{H}_3(t)\mathbb{H}_4(t)\sigma_v$$

$$+2\left(1-\frac{\gamma-1}{\gamma}\right)\sigma_{\mathbf{r}}\rho_{2,1}\rho_{3,1}\mathbb{H}_1(t)\mathbb{H}_5(t)\sigma_v+2\sigma_{\mathbf{r}}(\rho_{3,2}-\rho_{2,1}\rho_{3,1})\mathbb{H}_1(t)\mathbb{H}_5(t)\sigma_v$$

$$-\gamma+\alpha_v\theta_v\mathbb{H}_3(t)-\alpha_{\mathbf{r}}\mathbb{H}_4(t)+\frac{(\gamma-1)\sigma_{\mathbf{r}}\rho_{3,1}\mathbb{H}_4(t)}{\gamma\sigma_S}+2\alpha_{\mathbf{r}}\theta_{\mathbf{r}}\mathbb{H}_5(t)$$

$$+2\left(1-\frac{\gamma-1}{\gamma}\right)\sigma_{\mathbf{r}}^2\rho_{3,1}^2\mathbb{H}_4(t)\mathbb{H}_5(t)+\frac{1}{1-\rho_{2,1}^2}2\sigma_{\mathbf{r}}^2(\rho_{3,2}-\rho_{2,1}\rho_{3,1})^2\mathbb{H}_4(t)\mathbb{H}_5(t)$$

$$+2\sigma_{\mathbf{r}}^2\left(-\rho_{3,1}^2+\frac{(\rho_{3,2}-\rho_{2,1}\rho_{3,1})^2}{\rho_{2,1}^2-1}+1\right)\mathbb{H}_4(t)\mathbb{H}_5(t)+\mathbb{H}_4'(t)+1=0, \tag{3.12.4}$$

$$\frac{1}{2}\left(1-\frac{\gamma-1}{\gamma}\right)\sigma_v^2\rho_{2,1}^2\mathbb{H}_3(t)^2+\frac{1}{2}\sigma_v^2\left(1-\rho_{2,1}^2\right)\mathbb{H}_3(t)^2+\frac{(\gamma-1)\sigma_v\rho_{2,1}\mathbb{H}_3(t)}{\gamma\sigma_S}$$

$$+2\left(1-\frac{\gamma-1}{\gamma}\right)\sigma_v\sigma_{\mathbf{r}}\rho_{2,1}\rho_{3,1}\mathbb{H}_5(t)\mathbb{H}_3(t)+2\sigma_v\sigma_{\mathbf{r}}(\rho_{3,2}-\rho_{2,1}\rho_{3,1})\mathbb{H}_5(t)\mathbb{H}_3(t)$$

$$+2\left(1-\frac{\gamma-1}{\gamma}\right)\sigma_{\mathbf{r}}^2\rho_{3,1}^2\mathbb{H}_5(t)^2+\frac{2\sigma_{\mathbf{r}}^2(\rho_{3,2}-\rho_{2,1}\rho_{3,1})^2\mathbb{H}_5(t)^2}{1-\rho_{2,1}^2}+2\sigma_{\mathbf{r}}^2$$

$$\left(-\rho_{3,1}^2+\frac{(\rho_{3,2}-\rho_{2,1}\rho_{3,1})^2}{\rho_{2,1}^2-1}+1\right)\mathbb{H}_5(t)^2-2\alpha_{\mathbf{r}}\mathbb{H}_5(t)+\frac{2(\gamma-1)\sigma_{\mathbf{r}}\rho_{3,1}\mathbb{H}_5(t)}{\gamma\sigma_S}$$

$$+\mathbb{H}_5'(t)+\frac{1-\gamma}{2\gamma\sigma_S^2}=0, \tag{3.12.5}$$

$$
\begin{aligned}
&2\left(1-\frac{\gamma-1}{\gamma}\right)\rho_{2,1}^2\mathbb{H}_1(t)\mathbb{H}_2(t)\sigma_v^2+2\left(1-\rho_{2,1}^2\right)\mathbb{H}_1(t)\mathbb{H}_2(t)\sigma_v^2\\
&+\frac{(1-\gamma)\rho_{2,1}\mathbb{H}_1(t)\sigma_v}{\gamma\sigma_S}+\left(1-\frac{\gamma-1}{\gamma}\right)\sigma_{\mathbf{r}}\rho_{2,1}\rho_{3,1}\mathbb{H}_1(t)\mathbb{H}_3(t)\sigma_v\\
&+\sigma_{\mathbf{r}}\left(\rho_{3,2}-\rho_{2,1}\rho_{3,1}\right)\mathbb{H}_1(t)\mathbb{H}_3(t)\sigma_v+2\left(1-\frac{\gamma-1}{\gamma}\right)\sigma_{\mathbf{r}}\rho_{2,1}\rho_{3,1}\mathbb{H}_2(t)\mathbb{H}_4(t)\sigma_v\\
&+2\sigma_{\mathbf{r}}\left(\rho_{3,2}-\rho_{2,1}\rho_{3,1}\right)\mathbb{H}_2(t)\mathbb{H}_4(t)\sigma_v-\alpha_v\mathbb{H}_1(t)+2\alpha_v\theta_v\mathbb{H}_2(t)+\alpha_{\mathbf{r}}\theta_{\mathbf{r}}\mathbb{H}_3(t)\\
&+\frac{(1-\gamma)\sigma_{\mathbf{r}}\rho_{3,1}\mathbb{H}_4(t)}{\gamma\sigma_S}+\left(1-\frac{\gamma-1}{\gamma}\right)\sigma_{\mathbf{r}}^2\rho_{3,1}^2\mathbb{H}_3(t)\mathbb{H}_4(t)\\
&+\frac{\sigma_{\mathbf{r}}^2\left(\rho_{3,2}-\rho_{2,1}\rho_{3,1}\right)^2\mathbb{H}_3(t)\mathbb{H}_4(t)}{1-\rho_{2,1}^2}+\sigma_{\mathbf{r}}^2\left(-\rho_{3,1}^2+\frac{\left(\rho_{3,2}-\rho_{2,1}\rho_{3,1}\right)^2}{\rho_{2,1}^2-1}+1\right)\\
&\times\mathbb{H}_3(t)\mathbb{H}_4(t)+\mathbb{H}_1'(t)=0,
\end{aligned}
\tag{3.12.6}
$$

$$
\begin{aligned}
&2\left(1-\frac{\gamma-1}{\gamma}\right)\rho_{2,1}^2\mathbb{H}_2(t)\mathbb{H}_3(t)\sigma_v^2+2\left(1-\rho_{2,1}^2\right)\mathbb{H}_2(t)\mathbb{H}_3(t)\sigma_v^2\\
&+\left(1-\frac{\gamma-1}{\gamma}\right)\sigma_{\mathbf{r}}\rho_{2,1}\rho_{3,1}\mathbb{H}_3(t)^2\sigma_v+\sigma_{\mathbf{r}}\left(\rho_{3,2}-\rho_{2,1}\rho_{3,1}\right)\mathbb{H}_3(t)^2\sigma_v\\
&+\frac{2(\gamma-1)\rho_{2,1}\mathbb{H}_2(t)\sigma_v}{\gamma\sigma_S}+\frac{(1-\gamma)\rho_{2,1}\mathbb{H}_3(t)\sigma_v}{\gamma\sigma_S}+4\left(1-\frac{\gamma-1}{\gamma}\right)\sigma_{\mathbf{r}}\rho_{2,1}\rho_{3,1}\\
&\times\mathbb{H}_2(t)\mathbb{H}_5(t)\sigma_v+4\sigma_{\mathbf{r}}\left(\rho_{3,2}-\rho_{2,1}\rho_{3,1}\right)\mathbb{H}_2(t)\mathbb{H}_5(t)\sigma_v-\alpha_v\mathbb{H}_3(t)-\alpha_{\mathbf{r}}\mathbb{H}_3(t)\\
&+\frac{(\gamma-1)\sigma_{\mathbf{r}}\rho_{3,1}\mathbb{H}_3(t)}{\gamma\sigma_S}+2\left(1-\frac{\gamma-1}{\gamma}\right)\sigma_{\mathbf{r}}^2\rho_{3,1}^2\mathbb{H}_3(t)\mathbb{H}_5(t)\\
&+\frac{1}{1-\rho_{2,1}^2}2\sigma_{\mathbf{r}}^2\left(\rho_{3,2}-\rho_{2,1}\rho_{3,1}\right)^2\mathbb{H}_3(t)\mathbb{H}_5(t)\\
&+2\sigma_{\mathbf{r}}^2\left(-\rho_{3,1}^2+\frac{\left(\rho_{3,2}-\rho_{2,1}\rho_{3,1}\right)^2}{\rho_{2,1}^2-1}+1\right)\mathbb{H}_3(t)\mathbb{H}_5(t)+\mathbb{H}_3'(t)\\
&-\frac{2(\gamma-1)\sigma_{\mathbf{r}}\rho_{3,1}\mathbb{H}_5(t)}{\gamma\sigma_S}+\frac{\gamma-1}{\gamma\sigma_S^2}=0,
\end{aligned}
\tag{3.12.7}
$$

$$
\begin{aligned}
&2\left(1-\frac{\gamma-1}{\gamma}\right)\sigma_v^2\rho_{2,1}^2\mathbb{H}_2(t)^2+2\sigma_v^2\left(1-\rho_{2,1}^2\right)\mathbb{H}_2(t)^2-2\alpha_v\mathbb{H}_2(t)+2\left(1-\frac{\gamma-1}{\gamma}\right)\\
&\sigma_v\sigma_{\mathbf{r}}\rho_{2,1}\rho_{3,1}\mathbb{H}_3(t)\mathbb{H}_2(t)+2\sigma_v\sigma_{\mathbf{r}}\left(\rho_{3,2}-\rho_{2,1}\rho_{3,1}\right)\mathbb{H}_3(t)\mathbb{H}_2(t)\\
&-\frac{2(\gamma-1)\sigma_v\rho_{2,1}\mathbb{H}_2(t)}{\gamma\sigma_S}+\frac{1}{2}\left(1-\frac{\gamma-1}{\gamma}\right)\sigma_{\mathbf{r}}^2\rho_{3,1}^2\mathbb{H}_3(t)^2
\end{aligned}
$$

$$+\frac{\sigma_{\mathbf{r}}^2\left(\rho_{3,2}-\rho_{2,1}\rho_{3,1}\right)^2\mathbb{H}_3(t)^2}{2\left(1-\rho_{2,1}^2\right)}+\frac{1}{2}\sigma_{\mathbf{r}}^2\left(-\rho_{3,1}^2+\frac{\left(\rho_{3,2}-\rho_{2,1}\rho_{3,1}\right)^2}{\rho_{2,1}^2-1}+1\right)\mathbb{H}_3(t)^2$$

$$+\frac{(1-\gamma)\sigma_{\mathbf{r}}\rho_{3,1}\mathbb{H}_3(t)}{\gamma\sigma_S}+\mathbb{H}_2'(t)+\frac{1-\gamma}{2\gamma\sigma_S^2}=0, \qquad (3.12.8)$$

for $t < T < \infty$, with the terminal condition $\mathbb{H}_i(T) = 0$, for $i = 0, 1, \ldots, 5$. Since $\mathbb{H}_0(t)$ is not needed for ∇g_γ, and since $\mathbb{H}_0(t)$ does not appear in equations (3.12.4)–(3.12.8), only the ODE system (3.12.4)–(3.12.8) needs to be solved, which is simple enough to do using numerical ODE solvers (for example, in *Mathematica* using NDSolve). Summarizing, we have the following proposition.

Proposition 3.12.1. *For the stochastic interest, dividend, and appreciation rate model (2.4.39)–(2.4.42), the optimal portfolio formula for the CRRA utility of wealth with risk-aversion γ is*

$$\Pi_\gamma^\star(t,X,A)=\left\{\frac{X}{\gamma}\left(\frac{\mathbf{v}+r(\beta-1)}{\sigma_S^2}+\frac{1}{\sigma_S}\left(\sigma_v\rho_{2,1}\left(\mathbb{H}_1(t)+2\mathbf{v}\mathbb{H}_2(t)+r\mathbb{H}_3(t)\right)\right.\right.\right.$$

$$\left.\left.\left.+\sigma_{\mathbf{r}}\rho_{3,1}\left(\mathbf{v}\mathbb{H}_3(t)+\mathbb{H}_4(t)+2r\mathbb{H}_5(t)\right)\right)\right)\right\}, \qquad (3.12.9)$$

where $\mathbb{H}_i(t)$, $i = 1, \ldots, 5$, are characterized as the solution of the system of the five ODEs (3.12.4)–(3.12.8).

Chapter 4
Pricing: Neutral and Indifference

Portfolio Optimization Approach to Pricing: Neutral and Indifference Pricing of Portfolios of Financial Contracts in Incomplete Markets

4.1 Introduction

There is probably no more important concept nor more important issue in financial dealings than the prices at which deals are made.

Since the seminal work of Black and Scholes [2] there have been many works in pricing of derivative financial contracts. Black and Scholes's main idea was to eliminate randomness from the time evolution of a financial portfolio consisting of a financial contract that is being priced, an underlying security that is actively traded for the purpose of such randomness elimination, i.e., *hedging*, and a cash or money market account that is used for financing such trades. Once randomness is eliminated, once risk is taken out of investing, then the remaining financial structure must, by the force of the market, behave, i.e., appreciate, like a riskless cash or money market account. This principle yields the derivation of the Black–Scholes PDE (1.6.12), and thereby, the determination of the Black–Scholes prices (see [2, 27]; see also [14] and references given there).

Yet it is only for simple underlying models that risk-elimination is possible. In more complicated models, as in reality, risk-elimination is only partially achievable. Thus, following the Black and Scholes argument, one can derive a pricing PDE modulo an undetermined first-order coefficient. Consequently, at first, studies did not concentrate on a mathematical determination of the risk premium, but rather on what could be deduced *assuming* a certain risk premium – what can be concluded from the derived pricing equation, even though (at least) one of the coefficients was unknown.

Theoretically and practically, it became necessary to develop a pricing framework and a methodology that implicitly were able to determine the unknown coefficient in the pricing PDE, i.e., able to price the remaining risk, i.e., to find the so-called *risk premium*. Not one, but two successful frameworks for determination of risk premium were found: neutral and indifference pricing frameworks. The former asks the following question:

S. Stojanovic, *Neutral and Indifference Portfolio Pricing, Hedging and Investing: With Applications in Equity and FX*, DOI 10.1007/978-0-387-71418-9_4,

Question 4.1.1. Assuming a certain type of the investor's risk-aversion, what should the price of an asset (or a portfolio of assets) be so that it is optimal to hold a specific (and in particular, *no*) investment position in it?

The latter asks the following question:

Question 4.1.2. Assuming a certain type of investor's risk-aversion, what should the price of an asset be so that having a specific position or not having one is irrelevant (indifferent) from the point of view of investing in the rest of the market?

These two pricing frameworks, neutral and indifference, are not equivalent, they are different, obviously, from the point of view of their financial interpretation, i.e., financial meaning, and consequently, from the point of view of their applicability. They are also quite different, as it turns out, and to be seen below, methodologically, since the former can handle the problem of pricing individual components of a portfolio of assets (see (4.7.2)), while the latter can handle (at least from the point of view of mathematical proof) only the problem of pricing of a single-type asset. It was nevertheless obvious how to conjecture the result in the vector case, i.e., in the case of pricing individual components of a portfolio of assets.

Our methodology for solving all of those questions is by and large analytical—it is based on portfolio optimization (as presented in Chap. 3 in an analytic fashion), and differential equations [36, 40]. As pointed out already, this was made possible by the discovery of what we call the "fundamental matrix of derivatives pricing and hedging" ([40]; see Theorem 4.6.1 below). So far, the more established approach to pricing in incomplete markets was quite a bit different in methodology, relying heavily on abstract probability and "equivalent martingale measures" (see, e.g., [6, 18, 30, 31, 33]). Which approach is better? That is for the reader to decide, and the answer probably will depend quite a bit on personal preferences between differential equations on one side and abstract probability theory on the other (which, by the way, could be one of the readings of the joke at the opening of this book). Still, what was achieved here and maybe not elsewhere is a very general, yet very clean and streamlined methodology, from model building, (and as we are going to see in this chapter) to risk premium determination and price calculation according to different pricing methodologies, and furthermore, (as it is going to be elaborated in Chap. 5) to various hedging methodologies.

But first, we shall need to formalize mathematically Questions 4.1.1 and 4.1.2 using and building on the framework of simple economies presented in Chap. 2, and definitions and results of the portfolio optimization theory presented in Chap. 3

4.2 Definition of the Extended, i.e., Auxiliary Simple Economy $\mathfrak{E}_a$-*A* Framework for Pricing and Hedging

Following [40], consider a simple economy $\mathfrak{E}$, as in Definition 2.2.1, with k tradables $S(t) = \{S_1(t), \dots, S_k(t)\}$, with m factors $A(t) = \{A_1(t), \dots, A_m(t)\}$, and assume the market dynamics (2.2.1)–(2.2.2). Additionally, consider an extended, or *auxiliary*, simple economy, denoted by $\mathfrak{E}_a$, and such that $\mathfrak{E}_a \supset \mathfrak{E}$, having l additional tradables, with the vector of prices $V(t) = \{V_1(t), \ldots, V_l(t)\}$, which is to be determined. Then the extended economy $\mathfrak{E}_a$ has the same set of factors $A(t)$, but an extended set of tradables $\{S(t), V(t)\}$. More formally, we make the following definition.

Definition 4.2.1. *The auxiliary simple economy* $\mathfrak{E}_a$ consists of

(1) A set of extended *tradables*, or an extended market,

$$S_a(t) = \{S(t), V(t)\} = \{S_1(t), \ldots, S_k(t), V_1(t), \ldots, V_l(t)\} \tag{4.2.1}$$

(2) A finite (possibly empty) set of dynamic (economic) *factors* $A(t) = \{A_1(t), \ldots, A_m(t)\}$
(3) A money market that pays interest at the *rate* $r(t, A(t))$

Here $S(t)$ and $A(t)$ are assumed to obey Itô SDE dynamics (2.2.1)–(2.2.2); $V(t) = \{V_1(t), \ldots, V_l(t)\}$ is assumed to be a set of $l < \infty$ dividend and terminal payoff European contingent claims (financial contracts) expiring at time $T \leq \infty$; dividend payoffs (or liabilities, if negative) per each contract are assumed to be in the amount of

$$\mathcal{D}(t) = \mathcal{D} = \{\mathcal{D}_1, \ldots, \mathcal{D}_l\} = \{\mathcal{D}_1(t, A(t)), \ldots, \mathcal{D}_l(t, A(t))\} \tag{4.2.2}$$

units of currency (such as dollars) per year, for $t < T$; the terminal payoffs (or liabilities, if negative) per each contract are assumed to be in the amount of

$$V(T) = \{V_1(T), \ldots, V_l(T)\} = \{\upsilon_1(A(T)), \ldots, \upsilon_l(A(T))\} = \upsilon(A(T)). \tag{4.2.3}$$

The market coefficients of the auxiliary economy $\mathfrak{E}_a$ will be denoted by a_a, σ_a, b, c (see (4.5.4) below).

Remark 4.2.1. If $T = \infty$, i.e., if the contracts never expire (or mature), the contracts are said to be *perpetual*, and the terminal payoff (4.2.3) is then not imposed, i.e., it is irrelevant.

Remark 4.2.2. Above, $\mathcal{D} = \{\mathcal{D}_1, \ldots, \mathcal{D}_l\}$ is the dividend payoff *per year*. This is different from the way dividend *rates* $\mathbb{D}$ are accounted for in (2.2.1); they are accounted as "percentages" of the security prices (per year), while dividend $\mathcal{D}$ is accounted in units of currency (per year).

Remark 4.2.3. The contract terminal payoffs as well as the dividend payoffs are functions of factors A, and not of tradables S, so that if they are functions of a tradable underlying, as most often is the case, then such a tradable has to be declared as a factor as well, in which case necessarily $A \cap S \neq \emptyset$. This, of course, is not a restriction of any kind.

Remark 4.2.4. The tradables in the extended economy are $\{S(t), V(t)\}$, with the same factors $A(t)$ as in $\mathfrak{E}$. For the end results it will be necessary only that the original market be nonredundant, i.e., that $\sigma_s \cdot \sigma_s^{\mathrm{T}} > 0$. Typically, the extended market is redundant, i.e., one has only that $\sigma_{\mathrm{a}} \cdot \sigma_{\mathrm{a}}^{\mathrm{T}} \geq 0$ (see (4.6.1) below), since portfolios of contracts can contain many tradables. This may present only circumventable, technical issues, but is irrelevant from the point of view of the substantive end results.

4.3 Definition of Neutral and Indifference Pricing Problems; Hypothesis 4.3.1

Consider a self-financing portfolio strategy $\Pi(t,X,A)=\{\Pi_1,\ldots,\Pi_k,\Pi_{k+1},\ldots,\Pi_{k+l}\}$ in the auxiliary economy $\mathfrak{E}_{\mathrm{a}}$, where $\{\Pi_1,\ldots,\Pi_k\}$ is the cash value of the investments into the market $S \subset \mathfrak{E}$, and $\{\Pi_{k+1},\ldots,\Pi_{k+l}\}$ are the cash values of investments in the considered contracts.

For a given utility function ψ, let $\Pi_\psi^\star(t,X,A) = \{\Pi_{\psi,1}^\star,\ldots,\Pi_{\psi,k}^\star,\Pi_{\psi,k+1}^\star,\ldots,\Pi_{\psi,k+l}^\star\}$ denote the solution of Merton's optimal portfolio problem:

Problem 4.3.1. For a given utility function ψ, find $\Pi_\psi^\star$ such that

$$E_{t,X,A}\psi\left(X^{\Pi_\psi^\star}(T)\right) = \sup_{\Pi=\{\Pi_1,\ldots,\Pi_k,\Pi_{k+1},\ldots,\Pi_{k+l}\}} E_{t,X,A}\psi\left(X^{\Pi}(T)\right). \tag{4.3.1}$$

Due to its current importance, we make more precise a notion that we introduced previously:

Definition 4.3.1. A market *position*, or *portfolio* (of financial contracts), is any l-vector, denoted by $\kappa = \kappa(t,X,A) = \{\kappa_1(t,X,A),\ldots,\kappa_l(t,X,A)\}$, with κ_j representing the number of financial contracts of type j held, i.e., bought, if $\kappa_j > 0$, or shorted, if $\kappa_j < 0$, by an investor (or by a financial institution); $\kappa_j = 0$ is allowed and is also meaningful (and also self-explanatory).

Therefore, the vector κ quantifies the *portfolio of financial contracts* to be priced or, later, to be hedged.

Definition 4.3.2 (cf. [15]; see also [40,44,45]). For a given utility function $\psi(X)$, a market position $\kappa(t,X,A)$, dividend payoff $\mathcal{D}(t,A)$, and terminal payoff $\upsilon(A)$, the ψ-κ- *neutral* price (per single contract) $V_{\psi,\kappa}(t,X,A) = \{V_{\psi,\kappa,1}(t,X,A),\ldots, V_{\psi,\kappa,l}(t,X,A)\}$ is defined as a solution of the equation

$$\Pi_\psi^\star(t,X,A) = \{*, \kappa(t,X,A)V(t,X,A)\} \tag{4.3.2}$$

for $t < T$, where $*$ is an arbitrary expression, or, equivalently, a solution of the system of equations

$$\Pi_{\psi,k+j}^\star(t,X,A) = \kappa_j(t,X,A)V_{\psi,\kappa,j}(t,X,A) \tag{4.3.3}$$

for $t < T$, for $j = 1,\dots,l$, where $\Pi^{\star}_{\psi}$ is the solution of Problem 4.3.1, with the terminal condition

$$V_{\psi,\kappa,j}(T,X,A) = \upsilon_j(A) \tag{4.3.4}$$

for $j = 1,\dots,l$. If there is no confusion in regard to which utility function is used, then the ψ-κ- *neutral* price is also referred to as the κ- *neutral* price. If there is no confusion in regard to which portfolio is being priced, the κ-neutral price is also referred to as the *neutral* price. If $\psi(X) = \psi_\gamma(X)$ given in (3.2.7), i.e., if ψ is the CRRA utility, then the ψ_γ-κ-neutral price will be referred to as the γ-κ-neutral price or just the γ-neutral price.

Problem 4.3.2 (Neutral pricing). For a given utility function $\psi(X)$, market position $\kappa(t,X,A)$, dividend payoff $\mathcal{D}(t,A)$, and terminal payoff $\upsilon(A)$, find the ψ-κ-*neutral* price.

Remark 4.3.1. According to the condition (4.3.2), Definition 2.3.1 asks what the price of assets in the portfolio κ should be so that such a position is in fact an optimal position when paired with investing into the rest of the market. This might be the one of the most important problems in finance, since it connects its two most fundamental concepts: prices and investing.

On the other hand, for a given utility function ψ, let $\varphi_{\psi,c}(t,X,A)$ be the value function and let $\Pi^{\star}_{\psi,c}(t,X,A) = \{\Pi^{\star}_{\psi,c,1},\dots,\Pi^{\star}_{\psi,c,k},\Pi^{\star}_{\psi,c,k+1},\dots,\Pi^{\star}_{\psi,c,k+l}\}$ denote the solution of the optimal portfolio problem with the affine constraints on the portfolio

$$\mu.\Pi(t,X,A) = \xi(t,X,A) \tag{4.3.5}$$

with

$$\mu = \left(\mathbb{O}_{l\times k} \; \mathbb{I}_l \right) \tag{4.3.6}$$

and

$$\begin{aligned}\xi(t,X,A) &= V(t,X,A)\kappa(t,X,A) \\ &= \{V_1(t,X,A)\kappa_1(t,X,A),\dots,V_l(t,X,A)\kappa_l(t,X,A)\}\end{aligned} \tag{4.3.7}$$

(amounting to the constraint $\{\Pi_{k+1},\dots,\Pi_{k+l}\} = \{V_1\kappa_1,\dots,V_l\kappa_l\}$), so that

$$\varphi_{\psi,c}(t,X,A) = \sup_{\Pi:\mu.\Pi=\xi} E_{t,X,A}\psi\left(X^{\Pi}(T)\right) = E_{t,X,A}\psi\left(X^{\Pi^{\star}_{\psi,c}}(T)\right). \tag{4.3.8}$$

Problem 4.3.3. For a given utility function ψ and market position $\kappa(t,X,A)$, find $\varphi_{\psi,c}(t,X,A)$.

Definition 4.3.6 (cf. [1, 17, 30, 44]). For a given utility function $\psi(X)$, market position $\kappa(t,X,A)$, dividend payoff $\mathcal{D}(t,A)$, and terminal payoff $\upsilon(A)$, the ψ-κ- *indifference* price (per single contract) $V_{\psi,\iota,\kappa}(t,X,A) = \big\{V_{\psi,\iota,\kappa,1}(t,X,A),\dots, V_{\psi,\iota,\kappa,l}(t,X,A)\big\}$ is defined as a solution of the equation

$$\varphi_{\psi,c}(t,X,A) = \varphi(t,X,A) \tag{4.3.9}$$

for $t \leq T$, where $\varphi(t,X,A)$ is given by (3.3.3), i.e., it is the value function for Problem 3.3.1, and $\varphi_{\psi,c}(t,X,A)$ is the solution of Problem 4.3.3. If there is no confusion in regard to which utility function is used, then the ψ-κ-indifference price is also referred to as the κ-indifference price. If there is no confusion in regard to which portfolio is being priced, then the κ-indifference price is also referred to as the *indifference* price.

Problem 4.3.4 (Indifference pricing). For a given utility function $\psi(X)$, market position $\kappa(t,X,A)$, dividend payoff $\mathcal{D}(t,A)$, and terminal payoff $\upsilon(A)$, find the ψ-κ-*indifference*price.

Remark 4.3.2. More explicitly, (4.3.9) reads

$$\begin{aligned} &\sup_{\Pi=\{\Pi_1,\ldots,\Pi_k,\Pi_{k+1},\ldots,\Pi_{k+l}\}:\{\Pi_{k+1},\ldots,\Pi_{k+l}\} = \{V_1\kappa_1,\ldots,V_l\kappa_l\}} E_{t,X,A}\psi\left(X^{\Pi}(T)\right) \\ &= \sup_{\Pi=\{\Pi_1,\ldots,\Pi_k\}} E_{t,X,A}\psi\left(X^{\Pi}(T)\right), \end{aligned} \tag{4.3.10}$$

so that the indifference pricing definition asks what the price of assets $V = \{V_1,\ldots,V_l\}$ in the portfolio κ should be so that having them or not having them is irrelevant (indifferent), i.e., yields the same maximized expected utility of the terminal wealth, from the point of view of investing in the rest of the market.

Remark 4.3.3. There is an important methodological difference between Problems 4.3.2 and 4.3.4, since condition (4.3.2), i.e., equation (4.3.2), is a vector equation, while condition (4.3.9), i.e., equation (4.3.9), i.e., equation (4.3.10), is only a scalar equation. The vector condition (4.3.2) enables us to prove the pricing system in the case of neutral pricing in the scalar case as well as in the vector case, i.e., in the case of pricing a whole portfolio of contracts. The scalar nature of (4.3.9) will amount to the inability to *prove* the pricing system in the vector case in the case of indifference pricing – proof is available only in the scalar case ($l = 1$). Nevertheless, it is obvious how to conjecture such an equation.

Remark 4.3.4. There is an important financial difference between Problems 4.3.2 and 4.3.4 and that is emphasized in this chapter's introduction as Questions 4.1.1 and 4.1.2 Those two different kinds of pricing methodologies will generally produce different prices.

It will be important to make the following assumption.

Hypothesis 1 *The contract price vector V is a function of time t and of (economic) factors A only, i.e., $V = V(t,A)$, i.e., the price V does not depend on the investor's wealth X.*

Remark 4.3.5. Hypothesis 1 can be circumvented for some of the results that follow by declaring wealth X as an additional factor, i.e., by, for example, appending A with X, and, drawing from the wealth evolution equation (3.3.2), when written for an appropriate optimal portfolio strategy, appending b and c in the appropriate manner (see Sect. 4.16 below).

4.4 Further Vector Calculus Notation

As we are pricing portfolios, i.e., seeking pricing systems of PDEs, we shall need some more vector calculus notation, some of which is not widely used.

Assuming Hypothesis 1, for the vector of derivative prices $V(t,A) = V = \{V_1,\ldots,V_l\}$, the gradient ∇_A will be understood as the gradient with respect to A, i.e., $\nabla_A V_j = \{\partial V_j/\partial A_1,\ldots,\partial V_j/\partial A_m\}$, for $j = 1,\ldots,l$, while

$$\begin{aligned}\nabla_A V &= \nabla_A \{V_1,\ldots,V_l\} := \{\nabla_A V_1,\ldots,\nabla_A V_l\}\\ &= \left\{\left\{\frac{\partial V_1}{\partial A_1},\ldots,\frac{\partial V_1}{\partial A_m}\right\},\ldots,\left\{\frac{\partial V_l}{\partial A_1},\ldots,\frac{\partial V_l}{\partial A_m}\right\}\right\}\\ &= \begin{pmatrix}\frac{\partial V_1}{\partial A_1} & \cdots & \frac{\partial V_1}{\partial A_m}\\ & \cdots & \\ \frac{\partial V_l}{\partial A_1} & \cdots & \frac{\partial V_l}{\partial A_m}\end{pmatrix}.\end{aligned} \tag{4.4.1}$$

Similarly, $\nabla_A \nabla_A V_j$ is the Hessian matrix for $j = 1,\ldots,l$,

$$\begin{aligned}\nabla_A \nabla_A V_j &= \nabla_A \left\{\frac{\partial V_j}{\partial A_1},\ldots,\frac{\partial V_j}{\partial A_m}\right\} = \left\{\nabla_A \frac{\partial V_j}{\partial A_1},\ldots,\nabla_A \frac{\partial V_j}{\partial A_m}\right\}\\ &= \left\{\left\{\frac{\partial^2 V_j}{\partial A_1\,\partial A_1},\ldots,\frac{\partial^2 V_j}{\partial A_1\,\partial A_m}\right\},\ldots,\left\{\frac{\partial^2 V_j}{\partial A_m\,\partial A_1},\ldots,\frac{\partial^2 V_j}{\partial A_m\,\partial A_m}\right\}\right\}\\ &= \begin{pmatrix}\frac{\partial^2 V_j}{\partial A_1\,\partial A_1} & \cdots & \frac{\partial^2 V_j}{\partial A_1\,\partial A_m}\\ & \cdots & \\ \frac{\partial^2 V_j}{\partial A_m\,\partial A_1} & \cdots & \frac{\partial^2 V_j}{\partial A_m\,\partial A_m}\end{pmatrix},\end{aligned} \tag{4.4.2}$$

while $\nabla_A \nabla_A V$ is a vector of Hessian matrices (a 3-dimensional array, i.e., rank-3 tensor)

$$\nabla_A \nabla_A V = \nabla_A \nabla_A \{V_1,\ldots,V_l\} = \nabla_A \{\nabla_A V_1,\ldots,\nabla_A V_l\} = \{\nabla_A \nabla_A V_1,\ldots,\nabla_A \nabla_A V_l\}. \tag{4.4.3}$$

As before, we shall denote by $\mathrm{Tr}(A_j)$ the usual trace of a matrix A_j, while

$$\mathrm{tr}(\{A_1,\ldots,A_l\}) := \{\mathrm{Tr}(A_1),\ldots,\mathrm{Tr}(A_l)\} \tag{4.4.4}$$

(which is different from the usual definition of the trace of a tensor of rank 3). For example,

$$\mathrm{tr}(\nabla_A \nabla_A V) = \{\Delta V_1,\ldots,\Delta V_l\}, \tag{4.4.5}$$

where Δ is the usual Laplacian: $\Delta V_j = \sum_{i=1}^{m} \partial^2 V_j / \partial A_i^2$. Furthermore, a matrix can, by definition, be multiplied (denoted by apposition; as opposed to the "dot" product, denoted by "$\cdot$") or divided by a vector rowwise. For example ($1/V := \{1/V_1, \ldots, 1/V_l.\}$),

$$
\begin{aligned}
\frac{\nabla_A V}{V} &= \frac{\{\nabla_A V_1, \ldots, \nabla_A V_l\}}{\{V_1, \ldots, V_l\}} := \left\{ \frac{\nabla_A V_1}{V_1}, \ldots, \frac{\nabla_A V_l}{V_l} \right\} \\
&= \begin{pmatrix} \frac{1}{V_1}\frac{\partial V_1}{\partial A_1} & \cdots & \frac{1}{V_1}\frac{\partial V_1}{\partial A_m} \\ & \cdots & \\ \frac{1}{V_l}\frac{\partial V_l}{\partial A_1} & \cdots & \frac{1}{V_l}\frac{\partial V_l}{\partial A_m} \end{pmatrix} = \begin{pmatrix} \frac{1}{V_1} & \cdots & 0 \\ & \cdots & \\ 0 & \cdots & \frac{1}{V_l} \end{pmatrix} \cdot \begin{pmatrix} \frac{\partial V_1}{\partial A_1} & \cdots & \frac{\partial V_1}{\partial A_m} \\ & \cdots & \\ \frac{\partial V_l}{\partial A_1} & \cdots & \frac{\partial V_l}{\partial A_m} \end{pmatrix} \\
&= \left(\frac{1}{V} \mathbb{I}_l \right) \cdot \nabla_A V. \qquad (4.4.6)
\end{aligned}
$$

We shall use $\nabla_A V/V$ and its transpose $(\nabla_A V/V)^{\mathrm{T}} = (\nabla_A V)^{\mathrm{T}} \cdot (\frac{1}{V}\mathbb{I}_l)$ quite often. In (4.4.6), "dot", i.e., "$\cdot$"denotes the usual matrix–matrix product, already mentioned. Similarly, we shall need (see, e.g., (4.5.1) and (4.5.2) below) a product of a rank-3 tensor α (say $2 \times 2 \times 2$) and a matrix β (say 2×2), defined as

$$
\begin{aligned}
\alpha.\beta &= \left\{ \begin{pmatrix} \alpha_{1,1,1} & \alpha_{1,1,2} \\ \alpha_{1,2,1} & \alpha_{1,2,2} \end{pmatrix}, \begin{pmatrix} \alpha_{2,1,1} & \alpha_{2,1,2} \\ \alpha_{2,2,1} & \alpha_{2,2,2} \end{pmatrix} \right\} \cdot \begin{pmatrix} \beta_{1,1} & \beta_{1,2} \\ \beta_{2,1} & \beta_{2,2} \end{pmatrix} \\
&:= \left\{ \begin{pmatrix} \beta_{1,1}\alpha_{1,1,1} + \beta_{2,1}\alpha_{1,1,2} & \beta_{1,2}\alpha_{1,1,1} + \beta_{2,2}\alpha_{1,1,2} \\ \beta_{1,1}\alpha_{1,2,1} + \beta_{2,1}\alpha_{1,2,2} & \beta_{1,2}\alpha_{1,2,1} + \beta_{2,2}\alpha_{1,2,2} \end{pmatrix}, \right. \\
&\qquad \left. \begin{pmatrix} \beta_{1,1}\alpha_{2,1,1} + \beta_{2,1}\alpha_{2,1,2} & \beta_{1,2}\alpha_{2,1,1} + \beta_{2,2}\alpha_{2,1,2} \\ \beta_{1,1}\alpha_{2,2,1} + \beta_{2,1}\alpha_{2,2,2} & \beta_{1,2}\alpha_{2,2,1} + \beta_{2,2}\alpha_{2,2,2} \end{pmatrix} \right\},
\end{aligned}
$$

or equivalently, $(\alpha.\beta)_{i,j,k} := \sum_q \alpha_{i,j,q}\beta_{q,k})$.

4.5 Market Coefficients for the Auxiliary Simple Economy $\mathfrak{E}_{\mathrm{a}}$

Of course, $\partial V/\partial t := \{\partial V_1/\partial t, \ldots, \partial V_l/\partial t\}$, and assuming Hypothesis 1, from (1.5.1), (2.2.1), and (2.2.2), applying quite a few of the above notions, the *vector* Itô chain rule for $V(t,A(t))$ can be written as

$$
\begin{aligned}
\mathrm{d}V(t,A(t)) = &\left(\frac{\partial V(t,A(t))}{\partial t} + \frac{1}{2}\operatorname{tr}\left(\nabla_A \nabla_A V(t,A(t)) \cdot c(t,A(t)) \cdot c(t,A(t))^{\mathrm{T}} \right) \right. \\
&\left. + \nabla_A V(t,A(t)) \cdot b(t,A(t)) \right) \mathrm{d}t + \nabla_A V(t,A(t)) \cdot c(t,A(t)) \cdot \mathrm{d}B(t),
\end{aligned}
\qquad (4.5.1)
$$

implying

$$\frac{\mathrm{d}V}{V} = \frac{1}{V}\left(\partial V/\partial t + \frac{1}{2}\mathrm{tr}\left(\nabla_A\nabla_A V\cdot c\cdot c^{\mathrm{T}}\right) + \nabla_A V.b\right)\mathrm{d}t + \frac{\nabla_A V}{V}\cdot c\cdot \mathrm{d}B(t). \quad (4.5.2)$$

From (4.5.2) we can read off the market coefficients for the auxiliary economy $\mathfrak{E}_{\mathrm{a}}$. The (vector) appreciation rate *before dividends* for considered contracts (with (vector) price $V(t,A(t))$) is equal to

$$\frac{1}{V}\left(\partial V/\partial t + \frac{1}{2}\mathrm{tr}\left(\nabla_A\nabla_A Vc\cdot c^{\mathrm{T}}\right) + \nabla_A V.b + \mathcal{D}\right), \quad (4.5.3)$$

and consequently, the market coefficients for the auxiliary economy $\mathfrak{E}_{\mathrm{a}}$ are

$$a_{\mathrm{a}} = \left\{a_s, \frac{1}{V}\left(\partial V/\partial t + \frac{1}{2}\mathrm{tr}\left(\nabla_A\nabla_A V\cdot c\cdot c^{\mathrm{T}}\right) + \nabla_A V.b + \mathcal{D}\right)\right\}, \quad (4.5.4)$$

the *predividend appreciation rate*, and

$$\sigma_{\mathrm{a}} = \begin{pmatrix} \sigma_s \\ \frac{\nabla_A V}{V}\cdot c \end{pmatrix}, \quad (4.5.5)$$

the volatility matrix. Of course, since there is no change in the factors, market coefficients b and c remain unchanged.

4.6 The Fundamental Matrix of Derivative Pricing and Hedging

For everything that follows, neutral and indifference pricing, as well as hedging, conservative and relaxed, the inverse of the matrix

$$\sigma_{\mathrm{a}}\cdot\sigma_{\mathrm{a}}^{\mathrm{T}} = \begin{pmatrix} \sigma_s\cdot\sigma_s^{\mathrm{T}} & \sigma_s\cdot c^{\mathrm{T}}\cdot\left(\frac{\nabla_A V}{V}\right)^{\mathrm{T}} \\ \frac{\nabla_A V}{V}\cdot c\cdot\sigma_s^{\mathrm{T}} & \frac{\nabla_A V}{V}\cdot c\cdot c^{\mathrm{T}}\cdot\left(\frac{\nabla_A V}{V}\right)^{\mathrm{T}} \end{pmatrix}, \quad (4.6.1)$$

the matrix $\left(\sigma_{\mathrm{a}}\cdot\sigma_{\mathrm{a}}^{\mathrm{T}}\right)^{-1}$, is key. That should not be surprising, at least in the case of pricing, considering the definition of both neutral and indifference pricing and the fact that such a matrix inverse is present in all of the relevant optimal portfolio formulas and in the PDEs characterizing the various value functions. So, considering its above-stated importance, we make the following definition.

Definition 4.6.1. The matrix $(\sigma_{\mathrm{a}}\cdot\sigma_{\mathrm{a}}^{\mathrm{T}})^{-1}$ will be referred to as the *fundamental matrix of derivative pricing and hedging*, or if the context is unambiguous, the *fundamental matrix.*

Remark 4.6.1. The fundamental matrix of derivative pricing and hedging (4.6.1), suitable for pricing and hedging of portfolios of contracts, was found and utilized by the author in [40], and previously in [39] for the case $l = 1$, i.e., in the case of pricing and hedging of a single type contract.

Definition 4.6.2. Define the matrix $\mathfrak{X}$ by

$$\mathfrak{X} := \nabla_A V \cdot c \cdot \left(\mathbb{I}_n - \sigma_s^{\mathrm{T}} \cdot \left(\sigma_s \cdot \sigma_s^{\mathrm{T}}\right)^{-1} \cdot \sigma_s\right) \cdot c^{\mathrm{T}} \cdot (\nabla_A V)^{\mathrm{T}}. \tag{4.6.2}$$

Theorem 4.6.1 (Stojanovic [40]). *Suppose the matrix $\mathfrak{X}$ given in (4.6.2) is invertible. Then $\sigma_{\mathrm{a}} \cdot \sigma_{\mathrm{a}}^{\mathrm{T}}$ is invertible, and moreover, its inverse $(\sigma_{\mathrm{a}} \cdot \sigma_{\mathrm{a}}^{\mathrm{T}})^{-1}$ is given by the explicit formula*

$$\left(\sigma_{\mathrm{a}} \cdot \sigma_{\mathrm{a}}^{\mathrm{T}}\right)^{-1} = \begin{pmatrix} \mathfrak{S}_{1,1} & \mathfrak{S}_{1,2} \\ \mathfrak{S}_{2,1} & \mathfrak{S}_{2,2} \end{pmatrix} \tag{4.6.3}$$

with

$$\begin{aligned}
\mathfrak{S}_{1,1} &= \left(\sigma_s \cdot \sigma_s^{\mathrm{T}}\right)^{-1} + \left(\sigma_s \cdot \sigma_s^{\mathrm{T}}\right)^{-1} \cdot \sigma_s \cdot c^{\mathrm{T}} \cdot (\nabla_A V)^{\mathrm{T}} \cdot \mathfrak{X}^{-1} \cdot \nabla_A V \cdot c \cdot \sigma_s^{\mathrm{T}} \cdot \left(\sigma_s \cdot \sigma_s^{\mathrm{T}}\right)^{-1}, \\
\mathfrak{S}_{1,2} &= -\left(\sigma_s \cdot \sigma_s^{\mathrm{T}}\right)^{-1} \cdot \sigma_s \cdot c^{\mathrm{T}} \cdot (\nabla_A V)^{\mathrm{T}} \cdot \mathfrak{X}^{-1} \cdot (V\mathbb{I}_l), \\
\mathfrak{S}_{2,1} &= -(V\mathbb{I}_l) \cdot \mathfrak{X}^{-1} \cdot \nabla_A V \cdot c \cdot \sigma_s^{\mathrm{T}} \cdot \left(\sigma_s \cdot \sigma_s^{\mathrm{T}}\right)^{-1}, \\
\mathfrak{S}_{2,2} &= (V\mathbb{I}_l) \cdot \mathfrak{X}^{-1} \cdot (V\mathbb{I}_l).
\end{aligned} \tag{4.6.4}$$

Proof. As is usually the case, it is much easier to check whether a given matrix is an inverse than to find the inverse of a given matrix.

Indeed, observe that $\mathfrak{S}_{1,1}$ and $\mathfrak{S}_{2,2}$ are symmetric matrices, while $(\mathfrak{S}_{1,2})^{\mathrm{T}} = \mathfrak{S}_{2,1}$, so that the right side of (4.6.3) and $\sigma_{\mathrm{a}} \cdot \sigma_{\mathrm{a}}^{\mathrm{T}}$ are symmetric matrices. Therefore, to prove (4.6.3)–(4.6.4), it suffices to prove (the third identity in)

$$\begin{aligned}
&\sigma_{\mathrm{a}} \cdot \sigma_{\mathrm{a}}^{\mathrm{T}} \cdot \begin{pmatrix} \mathfrak{S}_{1,1} & \mathfrak{S}_{1,2} \\ \mathfrak{S}_{2,1} & \mathfrak{S}_{2,2} \end{pmatrix} \\
&= \begin{pmatrix} \sigma_s \cdot \sigma_s^{\mathrm{T}} & \sigma_s \cdot c^{\mathrm{T}} \cdot \left(\frac{\nabla_A V}{V}\right)^{\mathrm{T}} \\ \frac{\nabla_A V}{V} \cdot c \cdot \sigma_s^{\mathrm{T}} & \frac{\nabla_A V}{V} \cdot c \cdot c^{\mathrm{T}} \cdot \left(\frac{\nabla_A V}{V}\right)^{\mathrm{T}} \end{pmatrix} \cdot \begin{pmatrix} \mathfrak{S}_{1,1} & \mathfrak{S}_{1,2} \\ \mathfrak{S}_{2,1} & \mathfrak{S}_{2,2} \end{pmatrix} \\
&= \begin{pmatrix} \sigma_s \cdot \sigma_s^{\mathrm{T}} \cdot \mathfrak{S}_{1,1} + \sigma_s \cdot c^{\mathrm{T}} \cdot \left(\frac{\nabla_A V}{V}\right)^{\mathrm{T}} \cdot \mathfrak{S}_{2,1} & \sigma_s \cdot \sigma_s^{\mathrm{T}} \cdot \mathfrak{S}_{1,2} + \sigma_s \cdot c^{\mathrm{T}} \cdot \left(\frac{\nabla_A V}{V}\right)^{\mathrm{T}} \cdot \mathfrak{S}_{2,2} \\ \frac{\nabla_A V}{V} \cdot c \cdot \sigma_s^{\mathrm{T}} \cdot \mathfrak{S}_{1,1} + \frac{\nabla_A V}{V} \cdot c \cdot c^{\mathrm{T}} \cdot \left(\frac{\nabla_A V}{V}\right)^{\mathrm{T}} \cdot \mathfrak{S}_{2,1} & \frac{\nabla_A V}{V} \cdot c \cdot \sigma_s^{\mathrm{T}} \cdot \mathfrak{S}_{1,2} + \frac{\nabla_A V}{V} \cdot c \cdot c^{\mathrm{T}} \cdot \left(\frac{\nabla_A V}{V}\right)^{\mathrm{T}} \cdot \mathfrak{S}_{2,2} \end{pmatrix} \\
&\stackrel{?}{=} \begin{pmatrix} \mathbb{I}_n & \mathbb{O}_{n\times l} \\ \mathbb{O}_{l\times n} & \mathbb{I}_l \end{pmatrix} = \mathbb{I}_{n+l}.
\end{aligned} \tag{4.6.5}$$

To that end, observing that $\left(\frac{\nabla_A V}{V}\right)^{\mathrm{T}} \cdot (V\mathbb{I}_l) = (\nabla_A V)^{\mathrm{T}}$, we compute

$$\begin{aligned}
&\sigma_s \cdot \sigma_s^{\mathrm{T}} \cdot \mathfrak{S}_{1,1} + \sigma_s \cdot c^{\mathrm{T}} \cdot \left(\frac{\nabla_A V}{V}\right)^{\mathrm{T}} \cdot \mathfrak{S}_{2,1} = \sigma_s \cdot \sigma_s^{\mathrm{T}} \cdot \left(\sigma_s \cdot \sigma_s^{\mathrm{T}}\right)^{-1} \\
&\quad + \sigma_s \cdot \sigma_s^{\mathrm{T}} \cdot \left(\sigma_s \cdot \sigma_s^{\mathrm{T}}\right)^{-1} \cdot \sigma_s \cdot c^{\mathrm{T}} \cdot (\nabla_A V)^{\mathrm{T}} \cdot \mathfrak{X}^{-1} \cdot \nabla_A V \cdot c \cdot \sigma_s^{\mathrm{T}} \cdot \left(\sigma_s \cdot \sigma_s^{\mathrm{T}}\right)^{-1} \\
&\quad - \sigma_s \cdot c^{\mathrm{T}} \cdot \left(\frac{\nabla_A V}{V}\right)^{\mathrm{T}} \cdot (V\mathbb{I}_l) \cdot \mathfrak{X}^{-1} \cdot \nabla_A V \cdot c \cdot \sigma_s^{\mathrm{T}} \cdot \left(\sigma_s \cdot \sigma_s^{\mathrm{T}}\right)^{-1} \\
&= \mathbb{I}_n + \sigma_s \cdot c^{\mathrm{T}} \cdot (\nabla_A V)^{\mathrm{T}} \cdot \mathfrak{X}^{-1} \cdot \nabla_A V \cdot c \cdot \sigma_s^{\mathrm{T}} \cdot \left(\sigma_s \cdot \sigma_s^{\mathrm{T}}\right)^{-1} \\
&\quad - \sigma_s \cdot c^{\mathrm{T}} \cdot (\nabla_A V)^{\mathrm{T}} \cdot \mathfrak{X}^{-1} \cdot \nabla_A V \cdot c \cdot \sigma_s^{\mathrm{T}} \cdot \left(\sigma_s \cdot \sigma_s^{\mathrm{T}}\right)^{-1} = \mathbb{I}_n
\end{aligned} \tag{4.6.6}$$

and

$$\begin{aligned}
&\sigma_s \cdot \sigma_s^{\mathrm{T}} \cdot \mathfrak{S}_{1,2} + \sigma_s \cdot c^{\mathrm{T}} \cdot \left(\frac{\nabla_A V}{V}\right)^{\mathrm{T}} \cdot \mathfrak{S}_{2,2} \\
&= -\sigma_s \cdot \sigma_s^{\mathrm{T}} \cdot \left(\sigma_s \cdot \sigma_s^{\mathrm{T}}\right)^{-1} \cdot \sigma_s \cdot c^{\mathrm{T}} \cdot (\nabla_A V)^{\mathrm{T}} \cdot \mathfrak{X}^{-1} \cdot (V\mathbb{I}_l) + \sigma_s \cdot c^{\mathrm{T}} \cdot \left(\frac{\nabla_A V}{V}\right)^{\mathrm{T}} \\
&\quad \cdot (V\mathbb{I}_l) \cdot \mathfrak{X}^{-1} \cdot (V\mathbb{I}_l) \\
&= -\sigma_s \cdot c^{\mathrm{T}} \cdot (\nabla_A V)^{\mathrm{T}} \cdot \mathfrak{X}^{-1} \cdot (V\mathbb{I}_l) + \sigma_s \cdot c^{\mathrm{T}} \cdot (\nabla_A V)^{\mathrm{T}} \cdot \mathfrak{X}^{-1} \cdot (V\mathbb{I}_l) \\
&= \mathbb{O}_{n\times l}
\end{aligned} \tag{4.6.7}$$

and

$$\begin{aligned}
&\frac{\nabla_A V}{V} \cdot c \cdot \sigma_s^{\mathrm{T}} \cdot \mathfrak{S}_{1,1} + \frac{\nabla_A V}{V} \cdot c \cdot c^{\mathrm{T}} \cdot \left(\frac{\nabla_A V}{V}\right)^{\mathrm{T}} \cdot \mathfrak{S}_{2,1} \\
&= \frac{\nabla_A V}{V} \cdot c \cdot \sigma_s^{\mathrm{T}} \cdot \left(\sigma_s \cdot \sigma_s^{\mathrm{T}}\right)^{-1} + \frac{\nabla_A V}{V} \cdot c \cdot \sigma_s^{\mathrm{T}} \cdot \left(\sigma_s \cdot \sigma_s^{\mathrm{T}}\right)^{-1} \cdot \sigma_s \cdot c^{\mathrm{T}} \cdot (\nabla_A V)^{\mathrm{T}} \cdot \mathfrak{X}^{-1} \\
&\quad \cdot \nabla_A V \cdot c \cdot \sigma_s^{\mathrm{T}} \cdot \left(\sigma_s \cdot \sigma_s^{\mathrm{T}}\right)^{-1} - \frac{\nabla_A V}{V} \cdot c \cdot c^{\mathrm{T}} \cdot (\nabla_A V)^{\mathrm{T}} \cdot \mathfrak{X}^{-1} \cdot \nabla_A V \cdot c \cdot \sigma_s^{\mathrm{T}} \cdot \left(\sigma_s \cdot \sigma_s^{\mathrm{T}}\right)^{-1} \\
&= \frac{\nabla_A V}{V} \cdot c \cdot \sigma_s^{\mathrm{T}} \cdot \left(\sigma_s \cdot \sigma_s^{\mathrm{T}}\right)^{-1} + \frac{\nabla_A V}{V} \cdot c \cdot \left(\sigma_s^{\mathrm{T}} \cdot \left(\sigma_s \cdot \sigma_s^{\mathrm{T}}\right)^{-1} \cdot \sigma_s - \mathbb{I}_n\right) \cdot c^{\mathrm{T}} \cdot (\nabla_A V)^{\mathrm{T}} \\
&\quad \cdot \mathfrak{X}^{-1} \cdot \nabla_A V \cdot c \cdot \sigma_s^{\mathrm{T}} \cdot \left(\sigma_s \cdot \sigma_s^{\mathrm{T}}\right)^{-1} \\
&= \frac{\nabla_A V}{V} \cdot c \cdot \sigma_s^{\mathrm{T}} \cdot \left(\sigma_s \cdot \sigma_s^{\mathrm{T}}\right)^{-1} - \frac{\nabla_A V}{V} \cdot c \cdot \sigma_s^{\mathrm{T}} \cdot \left(\sigma_s \cdot \sigma_s^{\mathrm{T}}\right)^{-1} \\
&= \mathbb{O}_{l\times n}
\end{aligned} \tag{4.6.8}$$

as well as

$$
\begin{aligned}
&\frac{\nabla_A V}{V}\cdot c\cdot\sigma_s^{\mathrm{T}}\cdot\mathfrak{S}_{1,2}+\frac{\nabla_A V}{V}\cdot c\cdot c^{\mathrm{T}}\cdot\left(\frac{\nabla_A V}{V}\right)^{\mathrm{T}}\cdot\mathfrak{S}_{2,2}\\
&=-\frac{\nabla_A V}{V}\cdot c\cdot\sigma_s^{\mathrm{T}}\cdot\left(\sigma_s\cdot\sigma_s^{\mathrm{T}}\right)^{-1}\cdot\sigma_s\cdot c^{\mathrm{T}}\cdot(\nabla_A V)^{\mathrm{T}}\cdot\mathfrak{X}^{-1}\cdot(V\mathbb{I}_l)\\
&\quad+\frac{\nabla_A V}{V}\cdot c\cdot c^{\mathrm{T}}\cdot\left(\frac{\nabla_A V}{V}\right)^{\mathrm{T}}\cdot(V\mathbb{I}_l)\cdot\mathfrak{X}^{-1}\cdot(V\mathbb{I}_l) \qquad (4.6.9)\\
&=\frac{\nabla_A V}{V}\cdot c\cdot\left(\mathbb{I}_n-\sigma_s^{\mathrm{T}}\cdot\left(\sigma_s\cdot\sigma_s^{\mathrm{T}}\right)^{-1}\cdot\sigma_s\right)\cdot c^{\mathrm{T}}\cdot(\nabla_A V)^{\mathrm{T}}\cdot\mathfrak{X}^{-1}\cdot(V\mathbb{I}_l)\\
&=\mathbb{I}_l, \qquad (4.6.10)
\end{aligned}
$$

and (4.6.3)–(4.6.4) is proved. □

4.7 Any-Utility Neutral Pricing

We are now in a position to solve Problem 4.3.2.

Theorem 4.7.1. *Assuming Hypothesis 1, in the above framework, for a position $\kappa = \kappa(t,X,A)$, and for a utility of wealth $\psi = \psi(X)$, the ψ-κ-neutral price (per single contract) $V = V_{\psi,\kappa} = \{V_{\psi,\kappa,1}(t,A),\ldots,V_{\psi,\kappa,l}(t,A)\}$ is the solution of the following coupled pricing system of PDEs:*

$$
\mathcal{M}(\varphi)-\frac{1}{2}\left(\frac{\partial^2\varphi}{\partial X^2}\right)^2\kappa.\nabla_A V\cdot c\cdot\left(\mathbb{I}_n-\sigma_s^{\mathrm{T}}\cdot\left(\sigma_s\cdot\sigma_s^{\mathrm{T}}\right)^{-1}\cdot\sigma_s\right)\cdot c^{\mathrm{T}}\cdot(\nabla_A V)^{\mathrm{T}}\cdot\kappa=0 \qquad (4.7.1)
$$

for $t < T$, together with the terminal condition $\varphi(T,X,A) = \psi(X)$, where the operator $\mathcal{M}$ is defined in (3.4.1), and

$$
\begin{aligned}
&\frac{\partial V}{\partial t}+\frac{1}{2}\mathrm{tr}\left(\nabla_A\nabla_A V\cdot c\cdot c^{\mathrm{T}}\right)+\nabla_A V.\left(b-(a_s-r)\cdot\left(\sigma_s\cdot\sigma_s^{\mathrm{T}}\right)^{-1}\cdot\sigma_s\cdot c^{\mathrm{T}}\right)\\
&\quad+\left(\nabla_A\frac{\partial\varphi}{\partial X}/\frac{\partial\varphi}{\partial X}\right)\cdot c\cdot\left(\mathbb{I}_n-\sigma_s^{\mathrm{T}}\cdot\left(\sigma_s\cdot\sigma_s^{\mathrm{T}}\right)^{-1}\cdot\sigma_s\right)\cdot c^{\mathrm{T}}\cdot(\nabla_A V)^{\mathrm{T}}\\
&\quad+\left(\frac{\partial^2\varphi}{\partial X^2}/\frac{\partial\varphi}{\partial X}\right)\nabla_A V\cdot c\cdot\left(\mathbb{I}_n-\sigma_s^{\mathrm{T}}\cdot\left(\sigma_s\cdot\sigma_s^{\mathrm{T}}\right)^{-1}\cdot\sigma_s\right)\cdot c^{\mathrm{T}}\cdot(\nabla_A V)^{\mathrm{T}}\cdot\kappa-rV\\
&=-\mathcal{D} \qquad (4.7.2)
\end{aligned}
$$

for $t < T$, together with the terminal condition $V(T,A) = \upsilon(A)$.

Remark 4.7.1. If $\kappa \neq 0$, then (4.7.1)–(4.7.2) is a coupled system. If $\kappa = 0$, then (4.7.1) becomes (3.4.2), and can be solved first, after which (4.7.2) is solved.

Proof of the theorem. We derive (4.7.1) first. Set

$$\mathcal{Y} := \left(\mathbb{I}_n - \sigma_s^{\mathrm{T}} \cdot \left(\sigma_s \cdot \sigma_s^{\mathrm{T}}\right)^{-1} \cdot \sigma_s\right) \cdot c^{\mathrm{T}} \cdot (\nabla V)^{\mathrm{T}} \cdot \mathfrak{X}^{-1} \cdot \nabla V \cdot c \cdot \left(\mathbb{I}_n - \sigma_s^{\mathrm{T}} \cdot \left(\sigma_s \cdot \sigma_s^{\mathrm{T}}\right)^{-1} \cdot \sigma_s\right) \tag{4.7.3}$$

and

$$\mathcal{Z} := \frac{\partial V}{\partial t} + \frac{1}{2}\mathrm{tr}\left(\nabla_A \nabla_A V \cdot c \cdot c^{\mathrm{T}}\right) + \nabla_A V.b - rV + \mathcal{D}. \tag{4.7.4}$$

Assuming (4.7.2), we have

$$\mathcal{Z} = -\left(\frac{\partial^2 \varphi}{\partial X^2} \Big/ \frac{\partial \varphi}{\partial X}\right) \kappa \cdot \mathfrak{X} + \nabla_A V \cdot c \cdot \sigma_s^{\mathrm{T}} \cdot \left(\sigma_s \cdot \sigma_s^{\mathrm{T}}\right)^{-1} \cdot (a_s - r) - \nabla_A V \cdot c \cdot \left(\mathbb{I}_n - \sigma_s^{\mathrm{T}} \cdot \left(\sigma_s \cdot \sigma_s^{\mathrm{T}}\right)^{-1} \cdot \sigma_s\right) \cdot c^{\mathrm{T}} \cdot \left(\nabla_A \frac{\partial \varphi}{\partial X}\right) \Big/ \frac{\partial \varphi}{\partial X}, \tag{4.7.5}$$

and, as is going to be useful,

$$(a_s - r) \cdot \left(\sigma_s \cdot \sigma_s^{\mathrm{T}}\right)^{-1} \cdot \sigma_s \cdot c^{\mathrm{T}} \cdot (\nabla_A V)^{\mathrm{T}} - \mathcal{Z} = \left(\frac{\partial^2 \varphi}{\partial X^2} \Big/ \frac{\partial \varphi}{\partial X}\right) \kappa.\mathfrak{X} + \left(\nabla_A \frac{\partial \varphi}{\partial X} \Big/ \frac{\partial \varphi}{\partial X}\right) \cdot c \cdot \left(\mathbb{I}_n - \sigma_s^{\mathrm{T}} \cdot \left(\sigma_s \cdot \sigma_s^{\mathrm{T}}\right)^{-1} \cdot \sigma_s\right) \cdot c^{\mathrm{T}} \cdot (\nabla_A V)^{\mathrm{T}}. \tag{4.7.6}$$

Remark 4.7.2. The three lemmas below were first found and used by the author, in the context of a single contract ($l = 1$) and zero-contract position ($\kappa = 0$) in [39]. That result is then extended to the case $l > 1, \kappa = 0$ in [44], and finally to the case $l > 1, \kappa \neq 0$ in [45].

Lemma 4.7.1. *For any function* $\varphi = \varphi(t, X, A)$,

$$-\frac{1}{2}\left(\nabla_A \frac{\partial \varphi}{\partial X}\right) \cdot c \cdot \sigma_{\mathrm{a}}^{\mathrm{T}} \cdot \left(\sigma_{\mathrm{a}} \cdot \sigma_{\mathrm{a}}^{\mathrm{T}}\right)^{-1} \cdot \sigma_{\mathrm{a}} \cdot c^{\mathrm{T}} \cdot \left(\nabla_A \frac{\partial \varphi}{\partial X}\right) = -\frac{1}{2}\left(\nabla_A \frac{\partial \varphi}{\partial X}\right) \cdot c \cdot \sigma_s^{\mathrm{T}} \cdot \left(\sigma_s \cdot \sigma_s^{\mathrm{T}}\right)^{-1} \cdot \sigma_s \cdot c^{\mathrm{T}} \cdot \left(\nabla_A \frac{\partial \varphi}{\partial X}\right) - \frac{1}{2}\left(\nabla_A \frac{\partial \varphi}{\partial X}\right) \cdot c \cdot \mathcal{Y} \cdot c^{\mathrm{T}} \cdot \left(\nabla_A \frac{\partial \varphi}{\partial X}\right). \tag{4.7.7}$$

Lemma 4.7.2. *Let V be the solution of (4.7.2). Then for any function $\varphi = \varphi(t,X,A)$,*

$$-\frac{1}{2}\left(\frac{\partial\varphi}{\partial X}\right)^2 (a_\mathrm{a}-r)\cdot\left(\sigma_\mathrm{a}\cdot\sigma_\mathrm{a}^\mathrm{T}\right)^{-1}\cdot(a_\mathrm{a}-r)$$

$$=-\frac{1}{2}\left(\frac{\partial\varphi}{\partial X}\right)^2 (a_s-r)\cdot\left(\sigma_s\cdot\sigma_s^\mathrm{T}\right)^{-1}\cdot(a_s-r)-\frac{1}{2}\left(\nabla_A\frac{\partial\varphi}{\partial X}\right)\cdot c\cdot\mathcal{Y}\cdot c^\mathrm{T}$$

$$\cdot\left(\nabla_A\frac{\partial\varphi}{\partial X}\right)-\frac{1}{2}\left(\frac{\partial^2\varphi}{\partial X^2}\right)^2\kappa.\mathfrak{X}\cdot\kappa-\frac{\partial^2\varphi}{\partial X^2}\left(\nabla_A\frac{\partial\varphi}{\partial X}\right)$$

$$\cdot c\cdot\left(\mathbb{I}_n-\sigma_s^\mathrm{T}\cdot\left(\sigma_s\cdot\sigma_s^\mathrm{T}\right)^{-1}\cdot\sigma_s\right)\cdot c^\mathrm{T}\cdot(\nabla_A V)^\mathrm{T}\cdot\kappa. \tag{4.7.8}$$

Lemma 4.7.3. *Let V be the solution of (4.7.2). Then for any function $\varphi = \varphi(t,X,A)$,*

$$-\frac{\partial\varphi}{\partial X}(a_\mathrm{a}-r)\cdot\left(\sigma_\mathrm{a}\cdot\sigma_\mathrm{a}^\mathrm{T}\right)^{-1}\cdot\sigma_\mathrm{a}\cdot c^\mathrm{T}\cdot\left(\nabla_A\frac{\partial\varphi}{\partial X}\right)$$

$$=-\frac{\partial\varphi}{\partial X}(a_s-r)\cdot\left(\sigma_s\cdot\sigma_s^\mathrm{T}\right)^{-1}\cdot\sigma_s\cdot c^\mathrm{T}\cdot\left(\nabla_A\frac{\partial\varphi}{\partial X}\right)+\left(\nabla_A\frac{\partial\varphi}{\partial X}\right)\cdot c\cdot\mathcal{Y}\cdot c^\mathrm{T}$$

$$\cdot\left(\nabla_A\frac{\partial\varphi}{\partial X}\right)+\frac{\partial^2\varphi}{\partial X^2}\left(\nabla_A\frac{\partial\varphi}{\partial X}\right)\cdot c\cdot\left(\mathbb{I}_n-\sigma_s^\mathrm{T}\cdot\left(\sigma_s\cdot\sigma_s^\mathrm{T}\right)^{-1}\cdot\sigma_s\right)$$

$$\cdot c^\mathrm{T}\cdot(\nabla_A V)^\mathrm{T}\cdot\kappa. \tag{4.7.9}$$

Adding up formulas (4.7.7), (4.7.8), and (4.7.9), we get a crucial result:

Corollary 4.7.1. *Let V be the solution of (4.7.2). Then for any function $\varphi = \varphi(t,X,A)$,*

$$\frac{\partial^2\varphi}{\partial X^2}\frac{\partial\varphi}{\partial t}+\frac{1}{2}\frac{\partial^2\varphi}{\partial X^2}\mathrm{Tr}\left(c\cdot c^\mathrm{T}\cdot\nabla_A\nabla_A\varphi\right)-\frac{1}{2}\left(\frac{\partial\varphi}{\partial X}\right)^2(a_\mathrm{a}-r)\cdot\left(\sigma_\mathrm{a}\cdot\sigma_\mathrm{a}^\mathrm{T}\right)^{-1}\cdot(a_\mathrm{a}-r)$$

$$+(b\cdot\nabla_A\varphi)\frac{\partial^2\varphi}{\partial X^2}-\frac{\partial\varphi}{\partial X}(a_\mathrm{a}-r)\cdot\left(\sigma_\mathrm{a}\cdot\sigma_\mathrm{a}^\mathrm{T}\right)^{-1}\cdot\sigma_\mathrm{a}\cdot c^\mathrm{T}\cdot\left(\nabla_A\frac{\partial\varphi}{\partial X}\right)$$

$$-\frac{1}{2}\left(\nabla_A\frac{\partial\varphi}{\partial X}\right)\cdot c\cdot\sigma_\mathrm{a}^\mathrm{T}\cdot\left(\sigma_\mathrm{a}\cdot\sigma_\mathrm{a}^\mathrm{T}\right)^{-1}\cdot\sigma_\mathrm{a}\cdot c^\mathrm{T}\cdot\left(\nabla_A\frac{\partial\varphi}{\partial X}\right)+rX\frac{\partial^2\varphi}{\partial X^2}\frac{\partial\varphi}{\partial X}$$

$$=\mathcal{M}(\varphi)-\frac{1}{2}\left(\frac{\partial^2\varphi}{\partial X^2}\right)^2\kappa.\nabla_A V\cdot c\cdot\left(\mathbb{I}_n-\sigma_s^\mathrm{T}\cdot\left(\sigma_s\cdot\sigma_s^\mathrm{T}\right)^{-1}\cdot\sigma_s\right)\cdot c^\mathrm{T}\cdot(\nabla_A V)^\mathrm{T}\cdot\kappa, \tag{4.7.10}$$

where the operator $\mathcal{M}$ is defined in (3.4.1). In particular, (4.7.10) implies that in the case $\kappa = 0$, for the sake of neutral pricing, it suffices to solve (3.4.2).

Proof of Lemma 1. We calculate

$$\sigma_{\mathrm{a}}^{\mathrm{T}}\cdot\left(\sigma_{\mathrm{a}}\cdot\sigma_{\mathrm{a}}^{\mathrm{T}}\right)^{-1}\cdot\sigma_{\mathrm{a}} = \left(\sigma_s^{\mathrm{T}}\ c^{\mathrm{T}}\cdot\left(\tfrac{\nabla_A V}{V}\right)^{\mathrm{T}}\right)\cdot\begin{pmatrix}\mathfrak{S}_{1,1} & \mathfrak{S}_{1,2}\\ \mathfrak{S}_{2,1} & \mathfrak{S}_{2,2}\end{pmatrix}\cdot\begin{pmatrix}\sigma_s\\ \frac{\nabla_A V}{V}\cdot c\end{pmatrix}$$

$$= \sigma_s^{\mathrm{T}}\cdot\mathfrak{S}_{1,1}\cdot\sigma_s + 2c^{\mathrm{T}}\cdot\left(\frac{\nabla_A V}{V}\right)^{\mathrm{T}}\cdot\mathfrak{S}_{2,1}\cdot\sigma_s + c^{\mathrm{T}}\cdot\left(\frac{\nabla_A V}{V}\right)^{\mathrm{T}}\cdot\mathfrak{S}_{2,2}\cdot\frac{\nabla_A V}{V}\cdot c$$

$$= \sigma_s^{\mathrm{T}}\cdot\left(\left(\sigma_s\cdot\sigma_s^{\mathrm{T}}\right)^{-1} + \left(\sigma_s\cdot\sigma_s^{\mathrm{T}}\right)^{-1}\cdot\sigma_s\cdot c^{\mathrm{T}}\cdot(\nabla_A V)^{\mathrm{T}}\cdot\mathfrak{X}^{-1}\cdot\nabla_A V\right.$$

$$\left.\cdot c\cdot\sigma_s^{\mathrm{T}}\cdot\left(\sigma_s\cdot\sigma_s^{\mathrm{T}}\right)^{-1}\right)\cdot\sigma_s - 2c^{\mathrm{T}}\cdot\left(\frac{\nabla_A V}{V}\right)^{\mathrm{T}}\cdot(V\mathbb{I}_l)\cdot\mathfrak{X}^{-1}\cdot\nabla_A V$$

$$\cdot c\cdot\sigma_s^{\mathrm{T}}\cdot\left(\sigma_s\cdot\sigma_s^{\mathrm{T}}\right)^{-1}\cdot\sigma_s + c^{\mathrm{T}}\cdot\left(\frac{\nabla_A V}{V}\right)^{\mathrm{T}}\cdot(V\mathbb{I}_l)\cdot\mathfrak{X}^{-1}\cdot(V\mathbb{I}_l)\cdot\frac{\nabla_A V}{V}\cdot c$$

$$= \sigma_s^{\mathrm{T}}\cdot\left(\sigma_s\cdot\sigma_s^{\mathrm{T}}\right)^{-1}\cdot\sigma_s + \sigma_s^{\mathrm{T}}\cdot\left(\sigma_s\cdot\sigma_s^{\mathrm{T}}\right)^{-1}\cdot\sigma_s\cdot c^{\mathrm{T}}\cdot(\nabla_A V)^{\mathrm{T}}\cdot\mathfrak{X}^{-1}$$

$$\cdot\nabla_A V\cdot c\cdot\sigma_s^{\mathrm{T}}\cdot\left(\sigma_s\cdot\sigma_s^{\mathrm{T}}\right)^{-1}\cdot\sigma_s - 2c^{\mathrm{T}}\cdot(\nabla_A V)^{\mathrm{T}}\cdot\mathfrak{X}^{-1}\cdot\nabla_A V$$

$$\cdot c\cdot\sigma_s^{\mathrm{T}}\cdot\left(\sigma_s\cdot\sigma_s^{\mathrm{T}}\right)^{-1}\cdot\sigma_s + c^{\mathrm{T}}\cdot(\nabla_A V)^{\mathrm{T}}\cdot\mathfrak{X}^{-1}\cdot\nabla_A V\cdot c$$

$$= \sigma_s^{\mathrm{T}}\cdot\left(\sigma_s\cdot\sigma_s^{\mathrm{T}}\right)^{-1}\cdot\sigma_s + \left(\mathbb{I}_n - \sigma_s^{\mathrm{T}}\cdot\left(\sigma_s\cdot\sigma_s^{\mathrm{T}}\right)^{-1}\cdot\sigma_s\right)\cdot c^{\mathrm{T}}\cdot(\nabla_A V)^{\mathrm{T}}$$

$$\cdot\mathfrak{X}^{-1}\cdot\nabla_A V\cdot c\cdot\left(\mathbb{I}_n - \sigma_s^{\mathrm{T}}\cdot\left(\sigma_s\cdot\sigma_s^{\mathrm{T}}\right)^{-1}\cdot\sigma_s\right),$$

and therefore

$$\sigma_{\mathrm{a}}^{\mathrm{T}}\cdot\left(\sigma_{\mathrm{a}}\cdot\sigma_{\mathrm{a}}^{\mathrm{T}}\right)^{-1}\cdot\sigma_{\mathrm{a}} = \sigma_s^{\mathrm{T}}\cdot\left(\sigma_s\cdot\sigma_s^{\mathrm{T}}\right)^{-1}\cdot\sigma_s + \mathcal{Y}\cdot \tag{4.7.11}$$

Premultiplying (4.7.11) by $-\frac{1}{2}\left(\nabla_A\frac{\partial\varphi}{\partial X}\right)\cdot c$ and postmultiplying it by $c^{\mathrm{T}}\cdot\left(\nabla_A\frac{\partial\varphi}{\partial X}\right)$, we arrive at (4.7.7).

Proof of Lemma 2. Recall (4.7.4). We have

$$a_{\mathrm{a}} - r = \left\{a_s - r, \frac{1}{V}\left(\frac{\partial V}{\partial t} + \frac{1}{2}\mathrm{tr}\left(\nabla_A\nabla_A V\cdot c\cdot c^{\mathrm{T}}\right) + \nabla_A V.b - rV + \mathcal{D}\right)\right\}$$

$$= \left\{a_s - r, \frac{1}{V}\mathcal{Z}\right\}.$$

Therefore

$$
\begin{aligned}
&(a_{\mathrm{a}}-r)\cdot\left(\sigma_{\mathrm{a}}\cdot\sigma_{\mathrm{a}}^{\mathrm{T}}\right)^{-1}\cdot(a_{\mathrm{a}}-r)\\
&=\left\{a_s-r,\frac{1}{V}\mathcal{Z}\right\}\cdot\begin{pmatrix}\mathfrak{S}_{1,1} & \mathfrak{S}_{1,2}\\ \mathfrak{S}_{2,1} & \mathfrak{S}_{2,2}\end{pmatrix}\cdot\left\{a_s-r,\frac{1}{V}\mathcal{Z}\right\}\\
&=\left\{(a_s-r)\cdot\mathfrak{S}_{1,1}+\frac{1}{V}\mathcal{Z}\cdot\mathfrak{S}_{2,1},(a_s-r)\cdot\mathfrak{S}_{1,2}+\frac{1}{V}\mathcal{Z}\cdot\mathfrak{S}_{2,2}\right\}\cdot\left\{a_s-r,\frac{1}{V}\mathcal{Z}\right\}\\
&=(a_s-r)\cdot\mathfrak{S}_{1,1}\cdot(a_s-r)+\frac{2}{V}\mathcal{Z}\cdot\mathfrak{S}_{2,1}\cdot(a_s-r)+\frac{1}{V^2}\mathcal{Z}\cdot\mathfrak{S}_{2,2}\cdot\mathcal{Z},
\end{aligned}
$$

and then

$$
\begin{aligned}
&-\frac{1}{2}\left(\frac{\partial\varphi}{\partial X}\right)^2(a_{\mathrm{a}}-r)\cdot\left(\sigma_{\mathrm{a}}\cdot\sigma_{\mathrm{a}}^{\mathrm{T}}\right)^{-1}\cdot(a_{\mathrm{a}}-r)\\
&=-\frac{1}{2}\left(\frac{\partial\varphi}{\partial X}\right)^2\left((a_s-r)\cdot\mathfrak{S}_{1,1}\cdot(a_s-r)+\frac{2}{V}\mathcal{Z}\cdot\mathfrak{S}_{2,1}\cdot(a_s-r)+\frac{1}{V^2}\mathcal{Z}\cdot\mathfrak{S}_{2,2}\cdot\mathcal{Z}\right)\\
&=-\frac{1}{2}\left(\frac{\partial\varphi}{\partial X}\right)^2\Big((a_s-r)\cdot\Big(\left(\sigma_s\cdot\sigma_s^{\mathrm{T}}\right)^{-1}+\left(\sigma_s\cdot\sigma_s^{\mathrm{T}}\right)^{-1}\cdot\sigma_s\cdot c^{\mathrm{T}}\cdot(\nabla_A V)^{\mathrm{T}}\\
&\qquad\cdot\mathfrak{X}^{-1}\cdot\nabla_A V\cdot c\cdot\sigma_s^{\mathrm{T}}\cdot\left(\sigma_s\cdot\sigma_s^{\mathrm{T}}\right)^{-1}\Big)\cdot(a_s-r)-2\mathcal{Z}\cdot\mathfrak{X}^{-1}\\
&\qquad\cdot\nabla_A V\cdot c\cdot\sigma_s^{\mathrm{T}}\cdot\left(\sigma_s\cdot\sigma_s^{\mathrm{T}}\right)^{-1}\cdot(a_s-r)+\mathcal{Z}\cdot\mathfrak{X}^{-1}\cdot\mathcal{Z}\Big)\\
&=-\frac{1}{2}\left(\frac{\partial\varphi}{\partial X}\right)^2(a_s-r)\cdot\left(\sigma_s\cdot\sigma_s^{\mathrm{T}}\right)^{-1}\cdot(a_s-r)\\
&\quad-\frac{1}{2}\left(\frac{\partial\varphi}{\partial X}\right)^2\left((a_s-r)\cdot\left(\sigma_s\cdot\sigma_s^{\mathrm{T}}\right)^{-1}\cdot\sigma_s\cdot c^{\mathrm{T}}\cdot(\nabla_A V)^{\mathrm{T}}-\mathcal{Z}\right)\\
&\quad.\mathfrak{X}^{-1}\cdot\left((a_s-r)\cdot\left(\sigma_s\cdot\sigma_s^{\mathrm{T}}\right)^{-1}\cdot\sigma_s\cdot c^{\mathrm{T}}\cdot(\nabla_A V)^{\mathrm{T}}-\mathcal{Z}\right).
\end{aligned}
$$

Recall (4.7.6). We have

$$
\begin{aligned}
&-\frac{1}{2}\left(\frac{\partial\varphi}{\partial X}\right)^2(a_{\mathrm{a}}-r)\cdot\left(\sigma_{\mathrm{a}}\cdot\sigma_{\mathrm{a}}^{\mathrm{T}}\right)^{-1}\cdot(a_{\mathrm{a}}-r)\\
&=-\frac{1}{2}\left(\frac{\partial\varphi}{\partial X}\right)^2\Bigg((a_s-r)\cdot\left(\sigma_s\cdot\sigma_s^{\mathrm{T}}\right)^{-1}\cdot(a_s-r)+\left(\left(\frac{\partial^2\varphi}{\partial X^2}\Big/\frac{\partial\varphi}{\partial X}\right)\kappa.\mathfrak{X}\right.\\
&\qquad+\left(\nabla_A\frac{\partial\varphi}{\partial X}\frac{\partial\varphi}{\partial X}\right)\cdot c\cdot\left(\mathbb{I}_n-\sigma_s^{\mathrm{T}}\cdot\left(\sigma_s\cdot\sigma_s^{\mathrm{T}}\right)^{-1}\cdot\sigma_s\right)
\end{aligned}
$$

$$\cdot c^{\mathrm{T}} \cdot (\nabla_A V)^{\mathrm{T}}\Big) . \mathfrak{X}^{-1} \cdot \left(\left(\frac{\partial^2 \varphi}{\partial X^2} / \frac{\partial \varphi}{\partial X}\right) \kappa \cdot \mathfrak{X} + \left(\nabla_A \frac{\partial \varphi}{\partial X} / \frac{\partial \varphi}{\partial X}\right)\right.$$

$$\left.\cdot c \cdot \left(\mathbb{I}_n - \sigma_s^{\mathrm{T}} \cdot \left(\sigma_s \cdot \sigma_s^{\mathrm{T}}\right)^{-1} \cdot \sigma_s\right) \cdot c^{\mathrm{T}} \cdot (\nabla_A V)^{\mathrm{T}}\right)\Big)$$

$$= -\frac{1}{2}\left(\frac{\partial \varphi}{\partial X}\right)^2 (a_s - r) \cdot \left(\sigma_s \cdot \sigma_s^{\mathrm{T}}\right)^{-1} \cdot (a_s - r)$$

$$-\frac{1}{2}\left(\nabla_A \frac{\partial \varphi}{\partial X}\right) \cdot c \cdot \mathcal{Y} \cdot c^{\mathrm{T}} \cdot \left(\nabla_A \frac{\partial \varphi}{\partial X}\right) - \frac{1}{2}\left(\frac{\partial^2 \varphi}{\partial X^2}\right)^2 \kappa . \mathfrak{X} \cdot \kappa$$

$$-\frac{\partial^2 \varphi}{\partial X^2}\left(\nabla_A \frac{\partial \varphi}{\partial X}\right) \cdot c \cdot \left(\mathbb{I}_n - \sigma_s^{\mathrm{T}} \cdot \left(\sigma_s \cdot \sigma_s^{\mathrm{T}}\right)^{-1} \cdot \sigma_s\right) \cdot c^{\mathrm{T}} \cdot (\nabla_A V)^{\mathrm{T}} \cdot \kappa,$$

completing the proof of formula (4.7.8).

Proof of Lemma 3. Similarly, we have

$$(a_{\mathrm{a}} - r) \cdot \left(\sigma_{\mathrm{a}} \cdot \sigma_{\mathrm{a}}^{\mathrm{T}}\right)^{-1} \cdot \sigma_{\mathrm{a}}$$

$$= \left\{a_s - r, \frac{1}{V}\mathcal{Z}\right\} \cdot \begin{pmatrix} \mathfrak{S}_{1,1} & \mathfrak{S}_{1,2} \\ \mathfrak{S}_{2,1} & \mathfrak{S}_{2,2} \end{pmatrix} \cdot \begin{pmatrix} \sigma_s \\ \frac{\nabla V}{V} \cdot c \end{pmatrix}$$

$$= (a_s - r) \cdot \mathfrak{S}_{1,1} \cdot \sigma_s + \frac{1}{V}\mathcal{Z} \cdot \mathfrak{S}_{2,1} \cdot \sigma_s + (a_s - r) \cdot \mathfrak{S}_{1,2} \cdot \frac{\nabla V}{V} \cdot c + \frac{1}{V}\mathcal{Z} \cdot \mathfrak{S}_{2,2} \cdot \frac{\nabla V}{V} \cdot c$$

$$= (a_s - r) \cdot \left(\sigma_s \cdot \sigma_s^{\mathrm{T}}\right)^{-1} \cdot \sigma_s + \left((a_s - r) \cdot \left(\sigma_s \cdot \sigma_s^{\mathrm{T}}\right)^{-1} \cdot \sigma_s \cdot c^{\mathrm{T}} \cdot (\nabla_A V)^{\mathrm{T}} - \mathcal{Z}\right) \cdot$$

$$\mathfrak{X}^{-1} \cdot \nabla V \cdot c^{\mathrm{T}} \cdot \left(\sigma_s^{\mathrm{T}} \cdot \left(\sigma_s \cdot \sigma_s^{\mathrm{T}}\right)^{-1} \cdot \sigma_s - \mathbb{I}_n\right)$$

$$= (a_s - r) \cdot \left(\sigma_s \cdot \sigma_s^{\mathrm{T}}\right)^{-1} \cdot \sigma_s + \left(\left(\frac{\partial^2 \varphi}{\partial X^2} \Big/ \frac{\partial \varphi}{\partial X}\right) \kappa . \mathfrak{X} + \left(\nabla_A \frac{\partial \varphi}{\partial X} \Big/ \frac{\partial \varphi}{\partial X}\right) \cdot c \cdot \left(\mathbb{I}_n - \sigma_s^{\mathrm{T}}\right.\right.$$

$$\left.\left.\cdot \left(\sigma_s \cdot \sigma_s^{\mathrm{T}}\right)^{-1} \cdot \sigma_s\right) \cdot c^{\mathrm{T}} \cdot (\nabla_A V)^{\mathrm{T}}\right) . \mathfrak{X}^{-1} \cdot \nabla_A V \cdot c^{\mathrm{T}} \cdot \left(\sigma_s^{\mathrm{T}} \cdot \left(\sigma_s \cdot \sigma_s^{\mathrm{T}}\right)^{-1} \cdot \sigma_s - \mathbb{I}_n\right)$$

$$= (a_s - r) \cdot \left(\sigma_s \cdot \sigma_s^{\mathrm{T}}\right)^{-1} \cdot \sigma_s + \left(\nabla_A V \cdot c \cdot \left(\mathbb{I}_n - \sigma_s^{\mathrm{T}} \cdot \left(\sigma_s \cdot \sigma_s^{\mathrm{T}}\right)^{-1} \cdot \sigma_s\right)\right.$$

$$\left.\cdot c^{\mathrm{T}} \cdot \left(\nabla_A \frac{\partial \varphi}{\partial X}\right) \Big/ \frac{\partial \varphi}{\partial X}\right) . \mathfrak{X}^{-1} \cdot \nabla_A V \cdot c^{\mathrm{T}} \cdot \left(\sigma_s^{\mathrm{T}} \cdot \left(\sigma_s \cdot \sigma_s^{\mathrm{T}}\right)^{-1} \cdot \sigma_s - \mathbb{I}_n\right)$$

$$+ \left(\frac{\partial^2 \varphi}{\partial X^2} \Big/ \frac{\partial \varphi}{\partial X}\right) \kappa . \nabla_A V \cdot c^{\mathrm{T}} \cdot \left(\sigma_s^{\mathrm{T}} \cdot \left(\sigma_s \cdot \sigma_s^{\mathrm{T}}\right)^{-1} \cdot \sigma_s - \mathbb{I}_n\right),$$

and therefore

$$
\begin{aligned}
&-\frac{\partial\varphi}{\partial X}(a_{\mathrm{a}}-r)\cdot\left(\sigma_{\mathrm{a}}\cdot\sigma_{\mathrm{a}}^{\mathrm{T}}\right)^{-1}\cdot\sigma_{\mathrm{a}}\cdot c^{\mathrm{T}}\cdot\left(\nabla_A\frac{\partial\varphi}{\partial X}\right)\\
&=-\frac{\partial\varphi}{\partial X}\Bigg((a_s-r)\cdot\left(\sigma_s\cdot\sigma_s^{\mathrm{T}}\right)^{-1}\cdot\sigma_s+\Bigg(\nabla_A V\cdot c\cdot\left(\mathbb{I}_n-\sigma_s^{\mathrm{T}}\cdot\left(\sigma_s\cdot\sigma_s^{\mathrm{T}}\right)^{-1}\cdot\sigma_s\right)\\
&\qquad\cdot c^{\mathrm{T}}\cdot\left(\nabla_A\frac{\partial\varphi}{\partial X}\right)\Big/\frac{\partial\varphi}{\partial X}\Bigg)\cdot\mathfrak{X}^{-1}\cdot\nabla_A V\cdot c^{\mathrm{T}}\cdot\left(\sigma_s^{\mathrm{T}}\cdot\left(\sigma_s\cdot\sigma_s^{\mathrm{T}}\right)^{-1}\cdot\sigma_s-\mathbb{I}_n\right)\\
&\qquad+\left(\frac{\partial^2\varphi}{\partial X^2}\Big/\frac{\partial\varphi}{\partial X}\right)\kappa.\nabla_A V\cdot c^{\mathrm{T}}\cdot\left(\sigma_s^{\mathrm{T}}\cdot\left(\sigma_s\cdot\sigma_s^{\mathrm{T}}\right)^{-1}\cdot\sigma_s-\mathbb{I}_n\right)\Bigg)\cdot c^{\mathrm{T}}\cdot\left(\nabla_A\frac{\partial\varphi}{\partial X}\right)\\
&=-\frac{\partial\varphi}{\partial X}(a_s-r)\cdot\left(\sigma_s\cdot\sigma_s^{\mathrm{T}}\right)^{-1}\cdot\sigma_s\cdot c^{\mathrm{T}}\cdot\left(\nabla_A\frac{\partial\varphi}{\partial X}\right)+\left(\nabla_A\frac{\partial\varphi}{\partial X}\right)\cdot c\cdot\mathcal{Y}\cdot c^{\mathrm{T}}\cdot\left(\nabla_A\frac{\partial\varphi}{\partial X}\right)\\
&\qquad+\frac{\partial^2\varphi}{\partial X^2}\kappa\cdot\nabla_A V\cdot c^{\mathrm{T}}\cdot\left(\mathbb{I}_n-\sigma_s^{\mathrm{T}}\cdot\left(\sigma_s\cdot\sigma_s^{\mathrm{T}}\right)^{-1}\cdot\sigma_s\right)\cdot c^{\mathrm{T}}\cdot\left(\nabla_A\frac{\partial\varphi}{\partial X}\right),
\end{aligned}
$$

completing the proof of formula (4.7.9), which also completes the proof of Lemmas 4.7.1, 4.7.2, and 4.7.3 leading to Corollary 4.7.1 and ultimately the proof of (4.7.1).

Now we prove (4.7.2). To that end, we compute the optimal portfolio $\Pi^{\star}_{\psi}(t,X,A)$ in (4.3.2). Using the fundamental matrix (4.6.3)–(4.6.4), we get

$$
\begin{aligned}
&c\cdot\sigma_{\mathrm{a}}^{\mathrm{T}}\cdot\left(\sigma_{\mathrm{a}}\cdot\sigma_{\mathrm{a}}^{\mathrm{T}}\right)^{-1}\\
&=c\cdot\left(\sigma_s^{\mathrm{T}}\;\;c^{\mathrm{T}}\cdot\left(\tfrac{\nabla_A V}{V}\right)^{\mathrm{T}}\right)\cdot\left(\sigma_{\mathrm{a}}\cdot\sigma_{\mathrm{a}}^{\mathrm{T}}\right)^{-1}\\
&=\left(c\cdot\sigma_s^{\mathrm{T}}\;\;c\cdot c^{\mathrm{T}}\cdot\left(\tfrac{\nabla_A V}{V}\right)^{\mathrm{T}}\right)\cdot\left(\sigma_{\mathrm{a}}\cdot\sigma_{\mathrm{a}}^{\mathrm{T}}\right)^{-1}\\
&=\left(c\cdot\sigma_s^{\mathrm{T}}\;\;c\cdot c^{\mathrm{T}}\cdot\left(\tfrac{\nabla_A V}{V}\right)^{\mathrm{T}}\right)\cdot\begin{pmatrix}* & -\left(\sigma_s\cdot\sigma_s^{\mathrm{T}}\right)^{-1}\cdot\sigma_s\cdot c^{\mathrm{T}}\cdot\left(\nabla_A V\right)^{\mathrm{T}}\cdot\mathfrak{X}^{-1}\cdot\left(V\mathbb{I}_l\right)\\ * & \left(V\mathbb{I}_l\right)\cdot\mathfrak{X}^{-1}\cdot\left(V\mathbb{I}_l\right)\end{pmatrix}\\
&=\begin{pmatrix}* & -c\cdot\sigma_s^{\mathrm{T}}\cdot\left(\sigma_s\cdot\sigma_s^{\mathrm{T}}\right)^{-1}\cdot\sigma_s\cdot c^{\mathrm{T}}\cdot\left(\nabla_A V\right)^{\mathrm{T}}\cdot\mathfrak{X}^{-1}\cdot\left(V\mathbb{I}_l\right)+c\cdot c^{\mathrm{T}}\cdot\left(\tfrac{\nabla_A V}{V}\right)^{\mathrm{T}}\\ & \cdot\left(V\mathbb{I}_l\right)\cdot\mathfrak{X}^{-1}\cdot\left(V\mathbb{I}_l\right)\end{pmatrix}\\
&=\left(*\;\;c\cdot\left(\mathbb{I}_n-\sigma_s^{\mathrm{T}}\cdot\left(\sigma_s\cdot\sigma_s^{\mathrm{T}}\right)^{-1}\cdot\sigma_s\right)\cdot c^{\mathrm{T}}\cdot\left(\nabla_A V\right)^{\mathrm{T}}\cdot\mathfrak{X}^{-1}\cdot\left(V\mathbb{I}_l\right)\right)
\end{aligned}
\tag{4.7.12}
$$

and also

$$
\begin{aligned}
&(a_{\mathrm{a}}-r)\cdot\left(\sigma_{\mathrm{a}}\cdot\sigma_{\mathrm{a}}^{\mathrm{T}}\right)^{-1}\\
&=\left\{a_s-r,\frac{1}{V}\left(\frac{\partial V}{\partial t}+\frac{1}{2}\mathrm{tr}\left(\nabla_A\nabla_A V\cdot c\cdot c^{\mathrm{T}}\right)+\nabla_A V.b+\mathcal{D}\right)-r\right\}
\end{aligned}
$$

$$\cdot\begin{pmatrix} * - \left(\sigma_s\cdot\sigma_s^{\mathrm{T}}\right)^{-1}\cdot\sigma_s\cdot c^{\mathrm{T}}\cdot(\nabla_A V)^{\mathrm{T}}\cdot\mathfrak{X}^{-1}\cdot(V\mathbb{I}_l) \\ * \qquad (V\mathbb{I}_l)\cdot\mathfrak{X}^{-1}\cdot(V\mathbb{I}_l)\end{pmatrix}$$

$$= \Big\{*, -(a_s-r)\cdot\left(\sigma_s\cdot\sigma_s^{\mathrm{T}}\right)^{-1}\cdot\sigma_s\cdot c^{\mathrm{T}}\cdot(\nabla_A V)^{\mathrm{T}}\cdot\mathfrak{X}^{-1}\cdot(V\mathbb{I}_l)$$

$$+\frac{1}{V}\left(\frac{\partial V}{\partial t}+\frac{1}{2}\mathrm{tr}\left(\nabla_A\nabla_A V\cdot c\cdot c^{\mathrm{T}}\right)+\nabla_A V.b+\mathcal{D}-rV\right)\cdot(V\mathbb{I}_l)\cdot\mathfrak{X}^{-1}\cdot(V\mathbb{I}_l)\Big\}$$

$$= \Big\{*, \Big(-(a_s-r)\cdot\left(\sigma_s\cdot\sigma_s^{\mathrm{T}}\right)^{-1}\cdot\sigma_s\cdot c^{\mathrm{T}}\cdot(\nabla_A V)^{\mathrm{T}}+\frac{\partial V}{\partial t}+\frac{1}{2}\mathrm{tr}\left(\nabla_A\nabla_A V\cdot c\cdot c^{\mathrm{T}}\right)$$

$$+\nabla_A V.b - rV+\mathcal{D}\Big)\cdot\mathfrak{X}^{-1}\cdot(V\mathbb{I}_l)\Big\}. \tag{4.7.13}$$

Therefore, we can transform (3.4.22) into

$$\Pi_\psi^\star(t,X,A) = -\left(1\Big/\frac{\partial^2\varphi}{\partial X^2}\right)\left(\frac{\partial\varphi}{\partial X}(a_{\mathrm{a}}-r)+\left(\nabla_A\frac{\partial\varphi}{\partial X}\right)\cdot c\cdot\sigma_{\mathrm{a}}^{\mathrm{T}}\right)\cdot\left(\sigma_{\mathrm{a}}\cdot\sigma_{\mathrm{a}}^{\mathrm{T}}\right)^{-1}$$

$$= -\frac{1}{\frac{\partial^2\varphi}{\partial X^2}}\left(\frac{\partial\varphi}{\partial X}(a_{\mathrm{a}}-r)\cdot\left(\sigma_{\mathrm{a}}\cdot\sigma_{\mathrm{a}}^{\mathrm{T}}\right)^{-1}+\left(\nabla_A\frac{\partial\varphi}{\partial X}\right)\cdot c\cdot\sigma_{\mathrm{a}}^{\mathrm{T}}\cdot\left(\sigma_{\mathrm{a}}\cdot\sigma_{\mathrm{a}}^{\mathrm{T}}\right)^{-1}\right)$$

$$= -\frac{1}{\frac{\partial^2\varphi}{\partial X^2}}\Big(\frac{\partial\varphi}{\partial X}\Big\{*, \Big(-(a_s-r)\cdot\left(\sigma_s\cdot\sigma_s^{\mathrm{T}}\right)^{-1}\cdot\sigma_s\cdot c^{\mathrm{T}}\cdot(\nabla_A V)^{\mathrm{T}}+\frac{\partial V}{\partial t}$$

$$+\frac{1}{2}\mathrm{tr}\left(\nabla_A\nabla_A V\cdot c\cdot c^{\mathrm{T}}\right)+\nabla_A V.b-rV+\mathcal{D}\Big)\cdot\mathfrak{X}^{-1}\cdot(V\mathbb{I}_l)\Big\}$$

$$+\left(\nabla_A\frac{\partial\varphi}{\partial X}\right)\cdot\Big(*\; c\cdot\left(\mathbb{I}_n-\sigma_s^{\mathrm{T}}\cdot\left(\sigma_s\cdot\sigma_s^{\mathrm{T}}\right)^{-1}\cdot\sigma_s\right)\cdot c^{\mathrm{T}}\cdot(\nabla_A V)^{\mathrm{T}}$$

$$\cdot\mathfrak{X}^{-1}\cdot(V\mathbb{I}_l)\Big)\Big)$$

$$= -\frac{1}{\frac{\partial^2\varphi}{\partial X^2}}\Big\{*, \Big(\frac{\partial\varphi}{\partial X}\Big(-(a_s-r)\cdot\left(\sigma_s\cdot\sigma_s^{\mathrm{T}}\right)^{-1}\cdot\sigma_s\cdot c^{\mathrm{T}}\cdot(\nabla_A V)^{\mathrm{T}}+\frac{\partial V}{\partial t}$$

$$+\frac{1}{2}\mathrm{tr}\left(\nabla_A\nabla_A V\cdot c\cdot c^{\mathrm{T}}\right)+\nabla_A V.b-rV+\mathcal{D}\Big)+\left(\nabla_A\frac{\partial\varphi}{\partial X}\right)$$

$$\cdot c\cdot\left(\mathbb{I}_n-\sigma_s^{\mathrm{T}}\cdot\left(\sigma_s\cdot\sigma_s^{\mathrm{T}}\right)^{-1}\cdot\sigma_s\right)\cdot c^{\mathrm{T}}\cdot(\nabla_A V)^{\mathrm{T}}\Big)\cdot\mathfrak{X}^{-1}\cdot(V\mathbb{I}_l)\Big\}$$

and therefore

$$\Pi_\psi^\star(t,X,A) = \Big\{*, \Big(-\left(\frac{\partial\varphi}{\partial X}\Big/\frac{\partial^2\varphi}{\partial X^2}\right)\Big(-(a_s-r)\cdot\left(\sigma_s\cdot\sigma_s^{\mathrm{T}}\right)^{-1}\cdot\sigma_s\cdot c^{\mathrm{T}}\cdot(\nabla_A V)^{\mathrm{T}}$$

$$+\frac{\partial V}{\partial t}+\frac{1}{2}\mathrm{tr}\left(\nabla_A\nabla_A V\cdot c\cdot c^{\mathrm{T}}\right)+\nabla_A V.b-rV+\mathcal{D}\Big)$$

$$-\left(\left(\nabla_A \frac{\partial \varphi}{\partial X}\right) \Big/ \frac{\partial^2 \varphi}{\partial X^2}\right) \cdot c \cdot \left(\mathbb{I}_n - \sigma_s^{\mathrm{T}} \cdot \left(\sigma_s \cdot \sigma_s^{\mathrm{T}}\right)^{-1} \cdot \sigma_s\right)$$

$$\cdot c^{\mathrm{T}} \cdot (\nabla_A V)^{\mathrm{T}} \Big) . \mathfrak{X}^{-1} \cdot (V \mathbb{I}_l) \Big\}, \tag{4.7.14}$$

which, by means of (4.3.2), implies

$$\left(-\left(\frac{\partial \varphi}{\partial X} \Big/ \frac{\partial^2 \varphi}{\partial X^2}\right)\left(-(a_s - r) \cdot \left(\sigma_s \cdot \sigma_s^{\mathrm{T}}\right)^{-1} \cdot \sigma_s \cdot c^{\mathrm{T}} \cdot (\nabla_A V)^{\mathrm{T}}\right.\right.$$

$$\left. + \frac{\partial V}{\partial t} + \frac{1}{2} \operatorname{tr}\left(\nabla_A \nabla_A V \cdot c \cdot c^{\mathrm{T}}\right) + \nabla_A V . b - rV + \mathcal{D}\right) - \left(\left(\nabla_A \frac{\partial \varphi}{\partial X}\right) \Big/ \frac{\partial^2 \varphi}{\partial X^2}\right)$$

$$\left. \cdot c \cdot \left(\mathbb{I}_n - \sigma_s^{\mathrm{T}} \cdot \left(\sigma_s \cdot \sigma_s^{\mathrm{T}}\right)^{-1} \cdot \sigma_s\right) \cdot c^{\mathrm{T}} \cdot (\nabla_A V)^{\mathrm{T}} \right) \cdot \mathfrak{X}^{-1} \cdot (V \mathbb{I}_l)$$

$$= \kappa V, \tag{4.7.15}$$

and therefore (4.7.2) follows, which completes the proof of Theorem 4.9.1. As a corollary of Theorem 4.9.1 we have the following.

Theorem 4.7.2. *Assuming Hypothesis 1, in the above framework, for a position* $\kappa = 0$, *and for any utility of wealth* $\psi = \psi(X)$, *the* ψ-$(\kappa = 0)$*-neutral price (per single contract)* $V = V_{\psi,\kappa} = \left\{V_{\psi,\kappa,1}(t,A), \ldots, V_{\psi,\kappa,l}(t,A)\right\}$ *is the solution of the following uncoupled pricing system of PDEs:*

$$\mathcal{M}(\varphi) = 0 \tag{4.7.16}$$

for $t < T$, *together with the terminal condition* $\varphi(T,X,A) = \psi(X)$, *where the operator* $\mathcal{M}$ *is defined in 3.4.1, and*

$$\frac{\partial V}{\partial t} + \frac{1}{2} \operatorname{tr}\left(\nabla_A \nabla_A V \cdot c \cdot c^{\mathrm{T}}\right) + \nabla_A V . \left(b - (a_s - r) \cdot \left(\sigma_s \cdot \sigma_s^{\mathrm{T}}\right)^{-1} \cdot \sigma_s \cdot c^{\mathrm{T}}\right)$$

$$+ \left(\nabla_A \frac{\partial \varphi}{\partial X} \Big/ \frac{\partial \varphi}{\partial X}\right) \cdot c \cdot \left(\mathbb{I}_n - \sigma_s^{\mathrm{T}} \cdot \left(\sigma_s \cdot \sigma_s^{\mathrm{T}}\right)^{-1} \cdot \sigma_s\right) \cdot c^{\mathrm{T}} \cdot (\nabla_A V)^{\mathrm{T}} - rV$$

$$= -\mathcal{D} \tag{4.7.17}$$

for $t < T$, *together with the terminal condition* $V(T,A) = \upsilon(A)$.

So while the pricing system in the case $\kappa \neq 0$ *is genuinely coupled, in the case* $\kappa = 0$ *it is completely uncoupled (system (4.7.17) is itself also uncoupled).*

Corollary 4.7.3. *If a market is complete, then for any position* $\kappa = \kappa(t,X,A)$ *and any utility of wealth* $\psi = \psi(X)$, *the* ψ-κ*- neutral price (per single contract)*

$V = V_{\psi,\kappa}(t,A) = \{V_{\psi,\kappa,1}(t,A),\dots,V_{\psi,\kappa,l}(t,A)\}$ *is the solution of the following pricing system of PDEs:*

$$\frac{\partial V}{\partial t} + \frac{1}{2}\mathrm{tr}\left(\nabla_A\nabla_A V\cdot c\cdot c^{\mathrm{T}}\right) + \nabla_A V.\left(b-(a_s-r)\cdot\left(\sigma_s\cdot\sigma_s^{\mathrm{T}}\right)^{-1}\cdot\sigma_s\cdot c^{\mathrm{T}}\right) - rV = -\mathcal{D} \tag{4.7.18}$$

for $t < T$, together with the terminal condition $V(T,A) = \upsilon(A)$.

Proof. Since the market is complete (see Definition 2.3.1), we have

$$c\cdot\left(\mathbb{I}_n - \sigma_s^{\mathrm{T}}\cdot\left(\sigma_s\cdot\sigma_s^{\mathrm{T}}\right)^{-1}\cdot\sigma_s\right)\cdot c^{\mathrm{T}} = \mathbb{O}_m. \tag{4.7.19}$$

□

4.8 Any-Utility Indifference Pricing

We are also in a position to solve Problem 4.3.3.

Remark 4.8.1. Theorem 4.8.1 below is stated for the general case $l \geq 1$, even though it is proved only for the case $l = 1$; furthermore, in the rest of the book it will be occasionally ignored that in the case $l \geq 2$, Theorem 4.8.1 is only a conjecture, and sometimes we may use it in the case $l \geq 2$ without further warning.

Theorem 4.8.1. *Assuming Hypothesis 1, in the above framework (in the case $l = 1$ and conjecturallyin the case $l \geq 2$) for a position $\kappa = \kappa(t,X,A)$, and for the utility of wealth $\psi = \psi(X)$, the ψ-κ- indifferenceprice (per single contract) $V = V_{\psi,\iota,\kappa} = \{V_{\psi,\iota,\kappa,1}(t,A),\dots,V_{\psi,\iota,\kappa,l}(t,A)\}$ is the solution of the following uncoupled pricing system of PDEs:*

$$\mathcal{M}(\varphi) = 0 \tag{4.8.1}$$

for $t < T$, together with the terminal condition $\varphi(T,X,A) = \psi(X)$, where the operator $\mathcal{M}$ is defined in (3.4.1), and

$$\begin{aligned}
&\frac{\partial V}{\partial t} + \frac{1}{2}\mathrm{tr}\left(\nabla_A\nabla_A V\cdot c\cdot c^{\mathrm{T}}\right) + \nabla_A V.\left(b-(a_s-r)\cdot\left(\sigma_s\cdot\sigma_s^{\mathrm{T}}\right)^{-1}\cdot\sigma_s\cdot c^{\mathrm{T}}\right)\\
&\quad+\left(\nabla_A\frac{\partial\varphi}{\partial X}/\frac{\partial\varphi}{\partial X}\right)\cdot c\cdot\left(\mathbb{I}_n-\sigma_s^{\mathrm{T}}\cdot\left(\sigma_s\cdot\sigma_s^{\mathrm{T}}\right)^{-1}\cdot\sigma_s\right)\cdot c^{\mathrm{T}}\cdot\left(\nabla_A V\right)^{\mathrm{T}}\\
&\quad+\frac{1}{2}\left(\frac{\partial^2\varphi}{\partial X^2}/\frac{\partial\varphi}{\partial X}\right)\nabla_A V\cdot c\cdot\left(\mathbb{I}_n-\sigma_s^{\mathrm{T}}\cdot\left(\sigma_s\cdot\sigma_s^{\mathrm{T}}\right)^{-1}\cdot\sigma_s\right)\cdot c^{\mathrm{T}}\cdot\left(\nabla_A V\right)^{\mathrm{T}}\cdot\kappa - rV\\
&= -\mathcal{D}
\end{aligned} \tag{4.8.2}$$

for $t < T$, together with the terminal condition $V(T,A) = \upsilon(A)$.

Proof. Using (4.6.3)–(4.6.4) and (4.3.6)–(4.3.7), we first derive the formula

$$\mu.\left(\sigma_{\mathrm{a}}\cdot\sigma_{\mathrm{a}}^{\mathrm{T}}\right)^{-1}\cdot\mu^{\mathrm{T}} = \left(\mathbb{O}_{l\times k}\ \mathbb{I}_l\right)\cdot\begin{pmatrix}\mathfrak{S}_{1,1} & \mathfrak{S}_{1,2}\\ \mathfrak{S}_{2,1} & \mathfrak{S}_{2,2}\end{pmatrix}\cdot\begin{pmatrix}\mathbb{O}_{k\times l}\\ \mathbb{I}_l\end{pmatrix} = \mathfrak{S}_{2,2}, \tag{4.8.3}$$

and then

$$\left(\mu.\left(\sigma_a\cdot\sigma_a^T\right)^{-1}\cdot\mu^T\right)^{-1} = (\mathfrak{S}_{2,2})^{-1} = \left((V\mathbb{I}_l)\cdot\mathfrak{X}^{-1}\cdot(V\mathbb{I}_l)\right)^{-1}$$
$$= \left(\frac{1}{V}\mathbb{I}_l\right)\cdot\mathfrak{X}\cdot\left(\frac{1}{V}\mathbb{I}_l\right) \tag{4.8.4}$$

and

$$\mu^T\cdot\left(\mu\cdot\left(\sigma_a\cdot\sigma_a^T\right)^{-1}\cdot\mu^T\right)^{-1} = \begin{pmatrix}\mathbb{O}_{k\times l}\\ \mathbb{I}_l\end{pmatrix}\cdot\left(\frac{1}{V}\mathbb{I}_l\right)\cdot\mathfrak{X}\cdot\left(\frac{1}{V}\mathbb{I}_l\right)$$
$$= \begin{pmatrix}\mathbb{O}_{k\times l}\\ \left(\frac{1}{V}\mathbb{I}_l\right)\cdot\mathfrak{X}\cdot\left(\frac{1}{V}\mathbb{I}_l\right)\end{pmatrix} \tag{4.8.5}$$

and

$$\left(\sigma_a\cdot\sigma_a^T\right)^{-1}\cdot\mu^T\cdot\left(\mu\cdot\left(\sigma_a\cdot\sigma_a^T\right)^{-1}\cdot\mu^T\right)^{-1} = \begin{pmatrix}\mathfrak{S}_{1,1} & \mathfrak{S}_{1,2}\\ \mathfrak{S}_{2,1} & \mathfrak{S}_{2,2}\end{pmatrix}\cdot\begin{pmatrix}\mathbb{O}_{k\times l}\\ \left(\frac{1}{V}\mathbb{I}_l\right)\cdot\mathfrak{X}\cdot\left(\frac{1}{V}\mathbb{I}_l\right)\end{pmatrix}$$
$$= \begin{pmatrix}\mathfrak{S}_{1,2}\cdot\left(\frac{1}{V}\mathbb{I}_l\right)\cdot\mathfrak{X}\cdot\left(\frac{1}{V}\mathbb{I}_l\right)\\ \mathfrak{S}_{2,2}\cdot\left(\frac{1}{V}\mathbb{I}_l\right)\cdot\mathfrak{X}\cdot\left(\frac{1}{V}\mathbb{I}_l\right)\end{pmatrix}, \tag{4.8.6}$$

where

$$\mathfrak{S}_{1,2}\cdot\left(\frac{1}{V}\mathbb{I}_l\right)\cdot\mathfrak{X}\cdot\left(\frac{1}{V}\mathbb{I}_l\right) = -\left(\sigma_s\cdot\sigma_s^T\right)^{-1}\cdot\sigma_s\cdot c^T\cdot(\nabla_A V)^T\cdot\mathfrak{X}^{-1}\cdot(V\mathbb{I}_l)$$
$$\cdot\left(\frac{1}{V}\mathbb{I}_l\right)\cdot\mathfrak{X}\cdot\left(\frac{1}{V}\mathbb{I}_l\right)$$
$$= -\left(\sigma_s\cdot\sigma_s^T\right)^{-1}\cdot\sigma_s\cdot c^T\cdot(\nabla_A V)^T\cdot\mathfrak{X}^{-1}\cdot\mathfrak{X}\cdot\left(\frac{1}{V}\mathbb{I}_l\right)$$
$$= -\left(\sigma_s\cdot\sigma_s^T\right)^{-1}\cdot\sigma_s\cdot c^T\cdot(\nabla_A V)^T\cdot\left(\frac{1}{V}\mathbb{I}_l\right) \tag{4.8.7}$$

and

$$\mathfrak{S}_{2,2}\cdot\left(\frac{1}{V}\mathbb{I}_l\right)\cdot\mathfrak{X}\cdot\left(\frac{1}{V}\mathbb{I}_l\right) = (V\mathbb{I}_l)\cdot\mathfrak{X}^{-1}\cdot(V\mathbb{I}_l)\cdot\left(\frac{1}{V}\mathbb{I}_l\right)\cdot\mathfrak{X}\cdot\left(\frac{1}{V}\mathbb{I}_l\right)$$
$$= (V\mathbb{I}_l)\cdot\mathfrak{X}^{-1}\cdot\mathfrak{X}\cdot\left(\frac{1}{V}\mathbb{I}_l\right) = (V\mathbb{I}_l)\cdot\left(\frac{1}{V}\mathbb{I}_l\right) = \mathbb{I}_l, \tag{4.8.8}$$

so that

$$\left(\sigma_a\cdot\sigma_a^T\right)^{-1}\cdot\mu^T\cdot\left(\mu\cdot\left(\sigma_a\cdot\sigma_a^T\right)^{-1}\cdot\mu^T\right)^{-1} = \begin{pmatrix}-\left(\sigma_s\cdot\sigma_s^T\right)^{-1}\cdot\sigma_s\cdot c^T\cdot(\nabla_A V)^T\cdot\left(\frac{1}{V}\mathbb{I}_l\right)\\ \mathbb{I}_l\end{pmatrix} \tag{4.8.9}$$

and

$$\left(\sigma_a\cdot\sigma_a^T\right)^{-1}\cdot\mu^T\cdot\left(\mu\cdot\left(\sigma_a\cdot\sigma_a^T\right)^{-1}\cdot\mu^T\right)^{-1}\cdot\mu=\begin{pmatrix}-\left(\sigma_s\cdot\sigma_s^T\right)^{-1}\cdot\sigma_s\cdot c^T\cdot(\nabla_A V)^T\cdot\left(\frac{1}{V}\mathbb{I}_l\right)\\ \mathbb{I}_l\end{pmatrix}\cdot$$

$$\begin{pmatrix}\mathbb{O}_{l\times k} & \mathbb{I}_l\end{pmatrix}=\begin{pmatrix}\mathbb{O}_{k\times k} & -\left(\sigma_s\cdot\sigma_s^T\right)^{-1}\cdot\sigma_s\cdot c^T\cdot(\nabla_A V)^T\cdot\left(\frac{1}{V}\mathbb{I}_l\right)\\ \mathbb{O}_{l\times k} & \mathbb{I}_l\end{pmatrix} \tag{4.8.10}$$

and

$$\begin{aligned}\mathbb{I}_{k+l}-\left(\sigma_a\cdot\sigma_a^T\right)^{-1}\cdot\mu^T\cdot\left(\mu\cdot\left(\sigma_a\cdot\sigma_a^T\right)^{-1}\cdot\mu^T\right)^{-1}\cdot\mu&=\begin{pmatrix}\mathbb{I}_k & \mathbb{O}_{k\times l}\\ \mathbb{O}_{l\times k} & \mathbb{I}_l\end{pmatrix}\\ &-\begin{pmatrix}\mathbb{O}_{k\times k} & -\left(\sigma_s\cdot\sigma_s^T\right)^{-1}\cdot\sigma_s\cdot c^T\cdot(\nabla_A V)^T\cdot\left(\frac{1}{V}\mathbb{I}_l\right)\\ \mathbb{O}_{l\times k} & \mathbb{I}_l\end{pmatrix}\\ &=\begin{pmatrix}\mathbb{I}_k & \left(\sigma_s\cdot\sigma_s^T\right)^{-1}\cdot\sigma_s\cdot c^T\cdot(\nabla_A V)^T\cdot\left(\frac{1}{V}\mathbb{I}_l\right)\\ \mathbb{O}_{l\times k} & \mathbb{O}_l\end{pmatrix}\end{aligned} \tag{4.8.11}$$

and finally,

$$\begin{aligned}&\left(\mathbb{I}_{k+l}-\left(\sigma_a\cdot\sigma_a^T\right)^{-1}\cdot\mu^T\cdot\left(\mu\cdot\left(\sigma_a\cdot\sigma_a^T\right)^{-1}\cdot\mu^T\right)^{-1}\cdot\mu\right)\cdot\left(\sigma_a\cdot\sigma_a^T\right)^{-1}\\ &=\begin{pmatrix}\mathbb{I}_k & \left(\sigma_s\cdot\sigma_s^T\right)^{-1}\cdot\sigma_s\cdot c^T\cdot(\nabla_A V)^T\cdot\left(\frac{1}{V}\mathbb{I}_l\right)\\ \mathbb{O}_{l\times k} & \mathbb{O}_l\end{pmatrix}\cdot\begin{pmatrix}\mathfrak{S}_{1,1} & \mathfrak{S}_{1,2}\\ \mathfrak{S}_{2,1} & \mathfrak{S}_{2,2}\end{pmatrix}\\ &=\left(\begin{matrix}\mathfrak{S}_{1,1}+\left(\sigma_s\cdot\sigma_s^T\right)^{-1}\cdot\sigma_s\cdot c^T\cdot(\nabla_A V)^T\cdot\left(\frac{1}{V}\mathbb{I}_l\right)\cdot\mathfrak{S}_{2,1}\mathfrak{S}_{1,2}\\ \mathbb{O}_{l\times k}\end{matrix}\right.\\ &\qquad\left.\begin{matrix}+\left(\sigma_s\cdot\sigma_s^T\right)^{-1}\cdot\sigma_s\cdot c^T\cdot(\nabla_A V)^T\cdot\left(\frac{1}{V}\mathbb{I}_l\right)\cdot\mathfrak{S}_{2,2}\\ \mathbb{O}_l\end{matrix}\right)\\ &=\begin{pmatrix}\left(\sigma_s\cdot\sigma_s^T\right)^{-1} & \mathbb{O}_{k\times l}\\ \mathbb{O}_{l\times k} & \mathbb{O}_l\end{pmatrix}.\end{aligned} \tag{4.8.12}$$

So we transform (3.5.1) into

$$\begin{aligned}0=&\frac{\partial\varphi_a}{\partial t}\frac{\partial^2\varphi_a}{\partial X^2}+\frac{1}{2}\frac{\partial^2\varphi_a}{\partial X^2}\mathrm{Tr}\left(c\cdot c^T\cdot\nabla_A\nabla_A\varphi_a\right)-\frac{1}{2}\left(\frac{\partial\varphi_a}{\partial X}\right)^2(a_a-r)\\ &\cdot\left(\mathbb{I}_{k+l}-\left(\sigma_a\cdot\sigma_a^T\right)^{-1}\cdot\mu^T\cdot\left(\mu\cdot\left(\sigma_a\cdot\sigma_a^T\right)^{-1}\cdot\mu^T\right)^{-1}\cdot\mu\right)\cdot\left(\sigma_a\cdot\sigma_a^T\right)^{-1}\end{aligned}$$

$$
\begin{aligned}
&\cdot (a_{\mathrm{a}}-r) - \frac{\partial \varphi_{\mathrm{a}}}{\partial X}\left(\nabla_A \frac{\partial \varphi_{\mathrm{a}}}{\partial X}\right)\cdot c\cdot \sigma_{\mathrm{a}}^{\mathrm{T}}\cdot\left(\mathbb{I}_{k+l} - \left(\sigma_{\mathrm{a}}\cdot\sigma_{\mathrm{a}}^{\mathrm{T}}\right)^{-1}\cdot\mu^{\mathrm{T}}\right.\\
&\left.\cdot\left(\mu\cdot\left(\sigma_{\mathrm{a}}\cdot\sigma_{\mathrm{a}}^{\mathrm{T}}\right)^{-1}\cdot\mu^{\mathrm{T}}\right)^{-1}\cdot\mu\right)\cdot\left(\sigma_{\mathrm{a}}\cdot\sigma_{\mathrm{a}}^{\mathrm{T}}\right)^{-1}\cdot(a_{\mathrm{a}}-r) + \frac{\partial \varphi_{\mathrm{a}}}{\partial X}\frac{\partial^2 \varphi_{\mathrm{a}}}{\partial X^2}Xr\\
&+\frac{\partial \varphi_{\mathrm{a}}}{\partial X}\frac{\partial^2 \varphi_{\mathrm{a}}}{\partial X^2}\xi.\left(\mu\cdot\left(\sigma_{\mathrm{a}}\cdot\sigma_{\mathrm{a}}^{\mathrm{T}}\right)^{-1}\cdot\mu^{\mathrm{T}}\right)^{-1}\cdot\mu\cdot\left(\sigma_{\mathrm{a}}\cdot\sigma_{\mathrm{a}}^{\mathrm{T}}\right)^{-1}\cdot(a_{\mathrm{a}}-r)\\
&+\frac{\partial^2 \varphi_{\mathrm{a}}}{\partial X^2}b\cdot\nabla_A\varphi - \frac{1}{2}\left(\nabla_A\frac{\partial \varphi_{\mathrm{a}}}{\partial X}\right)\cdot c\cdot\sigma_{\mathrm{a}}^{\mathrm{T}}\cdot\left(\sigma_{\mathrm{a}}\cdot\sigma_{\mathrm{a}}^{\mathrm{T}}\right)^{-1}\\
&\cdot\left(\mathbb{I}_{k+l} - \mu^{\mathrm{T}}\cdot\left(\mu\cdot\left(\sigma_{\mathrm{a}}\cdot\sigma_{\mathrm{a}}^{\mathrm{T}}\right)^{-1}\cdot\mu^{\mathrm{T}}\right)^{-1}\cdot\mu.\left(\sigma_{\mathrm{a}}\cdot\sigma_{\mathrm{a}}^{\mathrm{T}}\right)^{-1}\right)\cdot\sigma_{\mathrm{a}}\cdot c^{\mathrm{T}}\cdot\left(\nabla_A\frac{\partial \varphi_{\mathrm{a}}}{\partial X}\right)\\
&+\frac{1}{2}\left(\frac{\partial^2 \varphi_{\mathrm{a}}}{\partial X^2}\right)^2\xi\cdot\left(\mu\cdot\left(\sigma_{\mathrm{a}}\cdot\sigma_{\mathrm{a}}^{\mathrm{T}}\right)^{-1}\cdot\mu^{\mathrm{T}}\right)^{-1}\cdot\xi + \frac{\partial^2 \varphi_{\mathrm{a}}}{\partial X^2}\xi\cdot\left(\mu\cdot\left(\sigma_{\mathrm{a}}\cdot\sigma_{\mathrm{a}}^{\mathrm{T}}\right)^{-1}\cdot\mu^{\mathrm{T}}\right)^{-1}\\
&\cdot\mu\cdot\left(\sigma_{\mathrm{a}}\cdot\sigma_{\mathrm{a}}^{\mathrm{T}}\right)^{-1}\cdot\sigma_{\mathrm{a}}\cdot c^{\mathrm{T}}\cdot\left(\nabla_A\frac{\partial \varphi_{\mathrm{a}}}{\partial X}\right)\\
=\;&\frac{\partial \varphi_{\mathrm{a}}}{\partial t}\frac{\partial^2 \varphi_{\mathrm{a}}}{\partial X^2} + \frac{1}{2}\frac{\partial^2 \varphi_{\mathrm{a}}}{\partial X^2}\mathrm{Tr}\left(c\cdot c^{\mathrm{T}}\cdot\nabla_A\nabla_A\varphi_{\mathrm{a}}\right) - \frac{1}{2}\left(\frac{\partial \varphi_{\mathrm{a}}}{\partial X}\right)^2(a_s-r)\cdot\left(\sigma_s\cdot\sigma_s^{\mathrm{T}}\right)^{-1}\\
&\cdot(a_s-r) - \frac{\partial \varphi_{\mathrm{a}}}{\partial X}\left(\nabla_A\frac{\partial \varphi_{\mathrm{a}}}{\partial X}\right)\cdot c\cdot\sigma_s^{\mathrm{T}}\cdot\left(\sigma_s\cdot\sigma_s^{\mathrm{T}}\right)^{-1}\cdot(a_s-r) + \frac{\partial \varphi_{\mathrm{a}}}{\partial X}\frac{\partial^2 \varphi_{\mathrm{a}}}{\partial X^2}Xr\\
&+\frac{\partial \varphi_{\mathrm{a}}}{\partial X}\frac{\partial^2 \varphi_{\mathrm{a}}}{\partial X^2}\xi.\left(\left(\frac{-1}{V}\mathbb{I}_l\right)\cdot\nabla_A V\cdot c\cdot\sigma_s^{\mathrm{T}}\cdot\left(\sigma_s\cdot\sigma_s^{\mathrm{T}}\right)^{-1}\cdot(a_s-r)\right.\\
&\left.+\frac{1}{V}\left(\frac{\partial V}{\partial t} + \frac{1}{2}\mathrm{tr}\left(\nabla_A\nabla_A V\cdot c\cdot c^{\mathrm{T}}\right) + \nabla_A V.b + \mathcal{D} - rV\right)\right) + \frac{\partial^2 \varphi_{\mathrm{a}}}{\partial X^2}b\cdot\nabla_A\varphi_{\mathrm{a}}\\
&-\frac{1}{2}\left(\nabla_A\frac{\partial \varphi_{\mathrm{a}}}{\partial X}\right)\cdot c\cdot\sigma_s^{\mathrm{T}}\cdot\left(\sigma_s\cdot\sigma_s^{\mathrm{T}}\right)^{-1}\cdot\sigma_s\cdot c^{\mathrm{T}}\cdot\left(\nabla_A\frac{\partial \varphi_{\mathrm{a}}}{\partial X}\right) + \frac{1}{2}\left(\frac{\partial^2 \varphi_{\mathrm{a}}}{\partial X^2}\right)^2\xi\\
&\cdot\left(\frac{1}{V}\mathbb{I}_l\right)\cdot\mathfrak{X}\cdot\left(\frac{1}{V}\mathbb{I}_l\right)\cdot\xi + \frac{\partial^2 \varphi_{\mathrm{a}}}{\partial X^2}\xi.\left(\left(\frac{1}{V}\mathbb{I}_l\right)\cdot\mathfrak{X}\cdot\left(\frac{1}{V}\mathbb{I}_l\right)\cdot\left(-(V\mathbb{I}_l)\right.\right.\\
&\left.\cdot\mathfrak{X}^{-1}\cdot\nabla_A V\cdot c\cdot\sigma_s^{\mathrm{T}}\cdot\left(\sigma_s\cdot\sigma_s^{\mathrm{T}}\right)^{-1}\right)\cdot\sigma_s + \left(\frac{1}{V}\mathbb{I}_l\right)\cdot\mathfrak{X}\cdot\left(\frac{1}{V}\mathbb{I}_l\right)\cdot(V\mathbb{I}_l)\\
&\left.\cdot\mathfrak{X}^{-1}\cdot(V\mathbb{I}_l)\cdot\frac{\nabla_A V}{V}\cdot c\right)\cdot c^{\mathrm{T}}\cdot\left(\nabla_A\frac{\partial \varphi_{\mathrm{a}}}{\partial X}\right),
\end{aligned}
$$

and therefore

$$
\begin{aligned}
&\mathcal{M}(\varphi_{\mathrm{a}})+\frac{\partial^2\varphi_{\mathrm{a}}}{\partial X^2}\left(\frac{\partial\varphi_{\mathrm{a}}}{\partial X}\xi\cdot\left(-\left(\frac{1}{V}\mathbb{I}_l\right)\cdot\nabla_A V\cdot c\cdot\sigma_s^{\mathrm{T}}\cdot\left(\sigma_s\cdot\sigma_s^{\mathrm{T}}\right)^{-1}\cdot(a_s-r)\right.\right.\\
&\left.\left.+\frac{1}{V}\left(\frac{\partial V}{\partial t}+\frac{1}{2}\mathrm{tr}\left(\nabla_A\nabla_A V\cdot c\cdot c^{\mathrm{T}}\right)+\nabla_A V.b-rV+\mathcal{D}\right)\right)\right)+\frac{1}{2}\frac{\partial^2\varphi_{\mathrm{a}}}{\partial X^2}\xi\\
&\cdot\left(\frac{1}{V}\mathbb{I}_l\right)\cdot\nabla_A V\cdot c\cdot\left(\mathbb{I}_n-\sigma_s^{\mathrm{T}}\cdot\left(\sigma_s\cdot\sigma_s^{\mathrm{T}}\right)^{-1}\cdot\sigma_s\right)\cdot c^{\mathrm{T}}\cdot\left(\nabla_A V\right)^{\mathrm{T}}\cdot\left(\frac{1}{V}\mathbb{I}_l\right)\cdot\xi+\xi\\
&\cdot\left(-\left(\frac{1}{V}\mathbb{I}_l\right)\cdot\nabla_A V\cdot c\cdot\sigma_s^{\mathrm{T}}\cdot\left(\sigma_s\cdot\sigma_s^{\mathrm{T}}\right)^{-1}\cdot\sigma_s+\frac{\nabla_A V}{V}\cdot c\right)\cdot c^{\mathrm{T}}\cdot\left(\nabla_A\frac{\partial\varphi_{\mathrm{a}}}{\partial X}\right)\right)=0.
\end{aligned}
\tag{4.8.13}
$$

From Definition 4.3.6 and (3.4.2), it follows that $\mathcal{M}(\varphi_a)=0$, and then from (4.8.13), we conclude that

$$
\begin{aligned}
\xi\cdot&\left(\frac{\partial\varphi_{\mathrm{a}}}{\partial X}\left(-\left(\frac{1}{V}\mathbb{I}_l\right)\cdot\nabla_A V\cdot c\cdot\sigma_s^{\mathrm{T}}\cdot\left(\sigma_s\cdot\sigma_s^{\mathrm{T}}\right)^{-1}\cdot(a_s-r)\right.\right.\\
&\left.\left.+\frac{1}{V}\left(\frac{\partial V}{\partial t}+\frac{1}{2}\mathrm{tr}\left(\nabla_A\nabla_A V\cdot c\cdot c^{\mathrm{T}}\right)+\nabla_A V.b-rV+\mathcal{D}\right)\right)\right)\\
&+\frac{1}{2}\frac{\partial^2\varphi_{\mathrm{a}}}{\partial X^2}\left(\frac{1}{V}\mathbb{I}_l\right)\cdot\nabla_A V\cdot c\cdot\left(\mathbb{I}_n-\sigma_s^{\mathrm{T}}\cdot\left(\sigma_s.\sigma_s^{\mathrm{T}}\right)^{-1}\cdot\sigma_s\right)\cdot c^{\mathrm{T}}\cdot\left(\nabla_A V\right)^{\mathrm{T}}\cdot\left(\frac{1}{V}\mathbb{I}_l\right)\cdot\xi\\
&+\left(-\left(\frac{1}{V}\mathbb{I}_l\right)\cdot\nabla_A V\cdot c\cdot\sigma_s^{\mathrm{T}}\cdot\left(\sigma_s\cdot\sigma_s^{\mathrm{T}}\right)^{-1}\cdot\sigma_s+\frac{\nabla_A V}{V}\cdot c\right)\cdot c^{\mathrm{T}}\cdot\left(\nabla_A\frac{\partial\varphi_{\mathrm{a}}}{\partial X}\right)\right)=0,
\end{aligned}
$$

and since $\xi=\kappa V$,

$$
\begin{aligned}
\xi\cdot&\left(-\nabla_A V\cdot c\cdot\sigma_s^{\mathrm{T}}\cdot\left(\sigma_s\cdot\sigma_s^{\mathrm{T}}\right)^{-1}\cdot(a_s-r)+\frac{\partial V}{\partial t}+\frac{1}{2}\mathrm{tr}\left(\nabla_A\nabla_A V\cdot c\cdot c^{\mathrm{T}}\right)\right.\\
&+\nabla_A V.b-rV+\mathcal{D}+\frac{1}{2}\left(\frac{\partial^2\varphi_{\mathrm{a}}}{\partial X^2}\Big/\frac{\partial\varphi_{\mathrm{a}}}{\partial X}\right)\nabla_A V\cdot c\cdot\left(\mathbb{I}_n-\sigma_s^{\mathrm{T}}\cdot\left(\sigma_s\cdot\sigma_s^{\mathrm{T}}\right)^{-1}\cdot\sigma_s\right)\cdot c^{\mathrm{T}}\\
&\cdot\left(\nabla_A V\right)^{\mathrm{T}}\cdot\kappa+\left(-\nabla_A V\cdot c\cdot\sigma_s^{\mathrm{T}}\cdot\left(\sigma_s\cdot\sigma_s^{\mathrm{T}}\right)^{-1}\cdot\sigma_s+\nabla_A V\cdot c\cdot c^{\mathrm{T}}\right)\\
&\left.\cdot\left(\nabla_A\frac{\partial\varphi_{\mathrm{a}}}{\partial X}\Big/\frac{\partial\varphi_{\mathrm{a}}}{\partial X}\right)\right)=0.
\end{aligned}
\tag{4.8.14}
$$

Equation (4.8.2) follows (in the case $l=1$ and *conjecturally*in the case $l\geq 2$; this is the only place where the difficulty arises for the case $l\geq 2$). □

Corollary 4.8.2. *If a market is complete, then for any position $\kappa=\kappa(t,X,A)$ and any utility of wealth $\psi=\psi(X)$, the ψ-κ-indifference price (per single contract)*

$V = V_{\psi,\kappa}(t,A) = \{V_{\psi,\kappa,1}(t,A),\ldots,V_{\psi,\kappa,l}(t,A)\}$ *is equal to the* ψ-κ-*neutral price, and it is the solution of*

$$\frac{\partial V}{\partial t} + \frac{1}{2}\mathrm{tr}\left(\nabla_A \nabla_A V \cdot c \cdot c^{\mathrm{T}}\right) + \nabla_A V. \left(b - (a_s - r)\cdot\left(\sigma_s \cdot \sigma_s^{\mathrm{T}}\right)^{-1}\cdot \sigma_s \cdot c^{\mathrm{T}}\right) - rV = -\mathcal{D} \tag{4.8.15}$$

for $t < T$, together with the terminal condition $V(T,A) = \upsilon(A)$.

Proof. The proof is the same as before: If the market is complete, then $c \cdot \left(\mathbb{I}_n - \sigma_s^{\mathrm{T}}\cdot\left(\sigma_s\cdot\sigma_s^{\mathrm{T}}\right)^{-1}\cdot\sigma_s\right)\cdot c^{\mathrm{T}} = \mathbb{O}_m$. □

Remark 4.8.2. Comparing Theorems 4.9.1 and 4.8.1, we can see that Theorem 4.8.1 is easier to implement, since (3.4.2) is uncoupled from the system (4.8.2), so that it can be solved first, and then one can solve system (4.8.2). On the other hand, (4.7.1) is coupled with (4.7.2), so that they need to be solved simultaneously.

Remark 4.8.3. (The relationship between neutral and indifference prices.) Comparing Theorems 4.9.1 and 4.8.1, and, for example, assuming that the φ 's characterized by (3.4.2) and (4.7.1) respectively both have the form $\varphi(t,X,A) = (X^{1-\gamma}e^{g(t)} - 1)/(1-\gamma)$, with possibly different $g(t)$'s (it *is* possible to have an incomplete market and yet to have such a simple form of φ), so that $\frac{\partial^2\varphi}{\partial X^2}/\frac{\partial\varphi}{\partial X} = -\frac{\gamma}{X}$ and $\nabla_A\frac{\partial\varphi}{\partial X}/\frac{\partial\varphi}{\partial X} = 0$, then equations (4.7.2) and (4.8.2) simplify to

$$\begin{aligned}&\frac{\partial V_{\psi,\kappa}}{\partial t} + \frac{1}{2}\mathrm{tr}\left(\nabla_A\nabla_A V_{\psi,\kappa}\cdot c\cdot c^{\mathrm{T}}\right) + \nabla_A V_{\psi,\kappa}\cdot\left(b - (a_s - r)\cdot\left(\sigma_s\cdot\sigma_s^{\mathrm{T}}\right)^{-1}\cdot\sigma_s\cdot c^{\mathrm{T}}\right)\\&-\frac{\gamma}{X}\nabla_A V_{\psi,\kappa}\cdot c\cdot\left(\mathbb{I}_n - \sigma_s^{\mathrm{T}}\cdot\left(\sigma_s\cdot\sigma_s^{\mathrm{T}}\right)^{-1}\cdot\sigma_s\right)\cdot c^{\mathrm{T}}\cdot\left(\nabla_A V_{\psi,\kappa}\right)^{\mathrm{T}}\\&\cdot\kappa - rV_{\psi,\kappa} = -\mathcal{D}\end{aligned} \tag{4.8.16}$$

for $t < T$, and

$$\begin{aligned}&\frac{\partial V_{\psi,\iota,\kappa}}{\partial t} + \frac{1}{2}\mathrm{tr}\left(\nabla_A\nabla_A V_{\psi,\iota,\kappa}\cdot c\cdot c^{\mathrm{T}}\right) + \nabla_A V_{\psi,\iota,\kappa}\cdot\left(b - (a_s - r)\cdot\left(\sigma_s\cdot\sigma_s^{\mathrm{T}}\right)^{-1}\cdot\sigma_s\cdot c^{\mathrm{T}}\right)\\&-\frac{1}{2}\frac{\gamma}{X}\nabla_A V_{\psi,\iota,\kappa}\cdot c\cdot\left(\mathbb{I}_n - \sigma_s^{\mathrm{T}}\cdot\left(\sigma_s\cdot\sigma_s^{\mathrm{T}}\right)^{-1}\cdot\sigma_s\right)\cdot c^{\mathrm{T}}\cdot\left(\nabla_A V_{\psi,\iota,\kappa}\right)^{\mathrm{T}}\\&\cdot\kappa - rV_{\psi,\iota,\kappa} = -\mathcal{D}\end{aligned} \tag{4.8.17}$$

for $t < T$, respectively, so that the corresponding prices, neutral and indifferent, are related by $V_{\psi,\kappa/2} = V_{\psi,\iota,\kappa}$. So under such simplifying circumstances, *if an investor is "indifferent" about holding a position* κ, *then it is optimal to hold a position* $\kappa/2$.

Remark 4.8.5. If $\kappa = 0$, then (4.7.2) and (4.8.2), and therefore also the corresponding prices, are equal. This consistency also, in a way, further confirms the conjecture part of Theorem 4.8.1.

4.9 CRRA Neutral Pricing

We now specialize to the most convenient utility function of all, the CRRA utility, the constant relative risk-aversion utility, given in (3.2.7). In the case of CRRA utility and κ-neutral pricing, in order to achieve simplification of the above pricing system, i.e., to have Hypothesis 1 fulfilled, we shall have to assume that the position κ has the special form

$$\kappa(t,X,A) = X\kappa_0(t,A). \tag{4.9.1}$$

Remark 4.9.1. Since κ is the position, i.e., the number of contracts held, it follows that $\kappa_0 = \kappa/X$ is the number of contracts held per currency unit of wealth, say per each dollar of the available wealth, which is a unit that takes some getting used to (see Examples 6.4.1 and 6.4.2 below).

Remark 4.9.2. The assumption (4.9.1), of course, includes a very important special case, namely $\kappa = \kappa_0 = 0$, which was solved by the author in [40], by deriving system (4.9.6)–(4.9.7) below.

Theorem 4.9.1. *In the above framework, for a position $\kappa(t,X,A) = X\kappa_0(t,A)$ and CRRA utility of wealth, the γ-κ-neutral price (per single contract) is a function of t and A only, i.e., Hypothesis 1 is satisfied, and $V = V_{\gamma,\kappa}(t,A) = \{V_{\gamma,\kappa,1}(t,A),\ldots, V_{\gamma,\kappa,l}(t,A)\}$ is the solution of the following pricing system of PDEs:*

$$\begin{aligned}
&\frac{\partial g_\gamma}{\partial t} + \frac{1}{2}\mathrm{Tr}\left(c\cdot c^{\mathrm{T}}\cdot\nabla_A\nabla_A g_\gamma\right) + \left(b - \frac{\gamma-1}{\gamma}(a_s - r)\cdot\left(\sigma_s\cdot\sigma_s^{\mathrm{T}}\right)^{-1}\cdot\sigma_s\cdot c^{\mathrm{T}}\right)\cdot\nabla_A g_\gamma \\
&\quad + \frac{1}{2}\nabla_A g_\gamma\cdot c\cdot\left(\mathbb{I}_n - \frac{\gamma-1}{\gamma}\sigma_s^{\mathrm{T}}\cdot\left(\sigma_s\cdot\sigma_s^{\mathrm{T}}\right)^{-1}\cdot\sigma_s\right)\cdot c^{\mathrm{T}}\cdot\nabla_A g_\gamma \\
&\quad - \frac{1}{2}\gamma(\gamma-1)\kappa_0(t,A)\cdot\nabla_A V\cdot c\cdot\left(\mathbb{I}_n - \sigma_s^{\mathrm{T}}\cdot\left(\sigma_s\cdot\sigma_s^{\mathrm{T}}\right)^{-1}\cdot\sigma_s\right) \\
&\quad \cdot c^{\mathrm{T}}\cdot\left(\nabla_A V\right)^{\mathrm{T}}\cdot\kappa_0(t,A) \\
&= \frac{1}{2}\frac{\gamma-1}{\gamma}\left((a_s - r)\cdot\left(\sigma_s\cdot\sigma_s^{\mathrm{T}}\right)^{-1}\cdot(a_s - r) + 2r\gamma\right)
\end{aligned} \tag{4.9.2}$$

for $t < T < \infty$, together with the terminal condition $g_\gamma(T,A) = 0$, and

$$\begin{aligned}
&\frac{\partial V}{\partial t} + \frac{1}{2}\mathrm{tr}\left(\nabla_A\nabla_A V\cdot c\cdot c^{\mathrm{T}}\right) + \nabla_A V.\left(b - (a_s - r)\cdot\left(\sigma_s\cdot\sigma_s^{\mathrm{T}}\right)^{-1}\cdot\sigma_s\cdot c^{\mathrm{T}}\right) \\
&\quad + \nabla_A g_\gamma\cdot c\cdot\left(\mathbb{I}_n - \sigma_s^{\mathrm{T}}\cdot\left(\sigma_s\cdot\sigma_s^{\mathrm{T}}\right)^{-1}\cdot\sigma_s\right)\cdot c^{\mathrm{T}}\cdot\left(\nabla_A V\right)^{\mathrm{T}} \\
&\quad - \gamma\kappa_0(t,A)\cdot\nabla_A V\cdot c\cdot\left(\mathbb{I}_n - \sigma_s^{\mathrm{T}}\cdot\left(\sigma_s\cdot\sigma_s^{\mathrm{T}}\right)^{-1}\cdot\sigma_s\right)\cdot c^{\mathrm{T}}\cdot\left(\nabla_A V\right)^{\mathrm{T}} - rV \\
&= -\mathcal{D}
\end{aligned} \tag{4.9.3}$$

for $t < T \le \infty$, together with the terminal condition $V(T,A) = \upsilon(A)$.

Proof. As before, setting $\varphi(t,X,A) = \left(X^{1-\gamma}e^{g_\gamma(t,A)} - 1\right) \Big/ (1-\gamma)$ in (4.7.1), we get

$$
\begin{aligned}
&-\frac{X^{-2\gamma}\gamma}{1-\gamma}\frac{\partial g_\gamma}{\partial t}e^{2g_\gamma} - \frac{1}{2}\frac{e^{2g_\gamma}X^{-2\gamma}\gamma}{1-\gamma}\left(\nabla_A g_\gamma \cdot c \cdot c^{\mathrm{T}} \cdot \nabla_A g_\gamma + \mathrm{Tr}\left(c \cdot c^{\mathrm{T}} \cdot \nabla_A \nabla_A g_\gamma\right)\right) \\
&-\frac{1}{2}e^{2g_\gamma}X^{-2\gamma}(a_s - r)\cdot\left(\sigma_s \cdot \sigma_s^{\mathrm{T}}\right)^{-1}\cdot(a_s - r) - \frac{e^{2g_\gamma}X^{-2\gamma}\gamma b \cdot \nabla_A g_\gamma}{1-\gamma} \\
&-X^{-2\gamma}e^{2g_\gamma}(a_s - r)\cdot\left(\sigma_s \cdot \sigma_s^{\mathrm{T}}\right)^{-1}\cdot\sigma_s\cdot c^{\mathrm{T}}\cdot\nabla_A g_\gamma \\
&-\frac{1}{2}e^{2g_\gamma}X^{-2\gamma}\nabla_A g_\gamma \cdot c \cdot \sigma_s^{\mathrm{T}} \cdot \left(\sigma_s \cdot \sigma_s^{\mathrm{T}}\right)^{-1}\cdot\sigma_s . c^{\mathrm{T}}\cdot\nabla_A g_\gamma - e^{2g_\gamma}rX^{-2\gamma}\gamma \\
&-\frac{1}{2}e^{2g_\gamma}X^{-2-2\gamma}\gamma^2\left(X\kappa_0(t,A)\right)\cdot\nabla_A V \cdot c \cdot \left(\mathbb{I}_n - \sigma_s^{\mathrm{T}}\cdot\left(\sigma_s\cdot\sigma_s^{\mathrm{T}}\right)^{-1}\cdot\sigma_s\right) \\
&\cdot c^{\mathrm{T}}\cdot\left(\nabla_A V\right)^{\mathrm{T}}\cdot\left(X\kappa_0(t,A)\right) = 0. \qquad (4.9.4)
\end{aligned}
$$

On the other hand, (4.7.2) becomes

$$
\begin{aligned}
&\frac{\partial V}{\partial t} + \frac{1}{2}\mathrm{tr}\left(\nabla_A\nabla_A V\cdot c\cdot c^{\mathrm{T}}\right) + \nabla_A V . \left(b - (a_s - r)\cdot\left(\sigma_s\cdot\sigma_s^{\mathrm{T}}\right)^{-1}\cdot\sigma_s\cdot c^{\mathrm{T}}\right) \\
&+\nabla_A g_\gamma \cdot c\cdot\left(\mathbb{I}_n - \sigma_s^{\mathrm{T}}\cdot\left(\sigma_s\cdot\sigma_s^{\mathrm{T}}\right)^{-1}\cdot\sigma_s\right)\cdot c^{\mathrm{T}}\cdot\left(\nabla_A V\right)^{\mathrm{T}} \\
&-\frac{\gamma}{X}\nabla_A V\cdot c\cdot\left(\mathbb{I}_n - \sigma_s^{\mathrm{T}}\cdot\left(\sigma_s\cdot\sigma_s^{\mathrm{T}}\right)^{-1}\cdot\sigma_s\right)\cdot c^{\mathrm{T}}\cdot\left(\nabla_A V\right)^{\mathrm{T}}\cdot\kappa - rV = -\mathcal{D}, \qquad (4.9.5)
\end{aligned}
$$

from which (4.9.3) follows. □

Remark 4.9.3. If $\kappa_0(t,A) \neq 0$, then (4.9.2)–(4.9.3) is a coupled system. If $\kappa_0(t,A) = 0 = \kappa$, then (4.9.2) becomes (3.6.17), and can be solved first, after which (4.9.3) is solved.

To emphasize this special, yet important, case, we state the result as a theorem.

Theorem 4.9.2 ([40]). *In the above framework, if $\kappa = 0$, then the γ-neutral price (per single contract) $V = V_{\gamma,\kappa}(t,A) = \left\{V_{\gamma,\kappa,1}(t,A), \ldots, V_{\gamma,\kappa,l}(t,A)\right\}$ is the solution of the following uncoupled pricing system of PDEs:*

$$
\begin{aligned}
&\frac{\partial g_\gamma}{\partial t} + \frac{1}{2}\mathrm{Tr}\left(c\cdot c^{\mathrm{T}}\cdot\nabla_A\nabla_A g_\gamma\right) + \left(b - \frac{\gamma-1}{\gamma}(a_s - r)\cdot\left(\sigma_s\cdot\sigma_s^{\mathrm{T}}\right)^{-1}\cdot\sigma_s\cdot c^{\mathrm{T}}\right)\cdot\nabla_A g_\gamma \\
&+\frac{1}{2}\nabla_A g_\gamma\cdot c\cdot\left(\mathbb{I}_n - \frac{\gamma-1}{\gamma}\sigma_s^{\mathrm{T}}\cdot\left(\sigma_s\cdot\sigma_s^{\mathrm{T}}\right)^{-1}\cdot\sigma_s\right)\cdot c^{\mathrm{T}}\cdot\nabla_A g_\gamma \\
&= \frac{1}{2}\frac{\gamma-1}{\gamma}\left((a_s - r)\cdot\left(\sigma_s\cdot\sigma_s^{\mathrm{T}}\right)^{-1}\cdot(a_s - r) + 2r\gamma\right) \qquad (4.9.6)
\end{aligned}
$$

for $t < T$, together with the terminal condition $g_\gamma(T,A) = 0$, and

$$\frac{\partial V}{\partial t} + \frac{1}{2}\mathrm{tr}\left(\nabla_A \nabla_A V \cdot c \cdot c^{\mathrm{T}}\right) + \nabla_A V.\left(b - (a_s - r)\cdot\left(\sigma_s \cdot \sigma_s^{\mathrm{T}}\right)^{-1}\cdot \sigma_s \cdot c^{\mathrm{T}}\right)$$

$$+\nabla_A g_\gamma \cdot c \cdot \left(\mathbb{I}_n - \sigma_s^{\mathrm{T}}\cdot\left(\sigma_s\cdot\sigma_s^{\mathrm{T}}\right)^{-1}\cdot\sigma_s\right)\cdot c^{\mathrm{T}}\cdot\left(\nabla_A V\right)^{\mathrm{T}} - rV = -\mathcal{D} \quad (4.9.7)$$

for $t < T$, together with the terminal condition $V(T,A) = \upsilon(A)$.

If $\kappa_0(t,A) = \text{const}$, then system (4.9.2)– can be decoupled as follows. Set

$$\mathcal{V}(t,A) = \kappa_0.V(t,A). \quad (4.9.8)$$

Then we have the following.

Theorem 4.9.3. *In the above framework, for a position $\kappa(t,X,A) = X\kappa_0$ and CRRA utility of wealth, the γ-κ-neutral price (per single contract) $V = V_{\gamma,\kappa}(t,A) = \{V_{\gamma,\kappa,1}(t,A),\ldots,V_{\gamma,\kappa,l}(t,X,A)\}$ can be computed in the following way. First solve the following system of two scalar PDEs:*

$$\frac{\partial g_\gamma}{\partial t} + \frac{1}{2}\mathrm{Tr}\left(c\cdot c^{\mathrm{T}}\cdot\nabla_A\nabla_A g_\gamma\right) + \left(b - \frac{\gamma-1}{\gamma}(a_s - r)\cdot\left(\sigma_s\cdot\sigma_s^{\mathrm{T}}\right)^{-1}\cdot\sigma_s\cdot c^{\mathrm{T}}\right)\cdot\nabla_A g_\gamma$$

$$+\frac{1}{2}\nabla_A g_\gamma\cdot c\cdot\left(\mathbb{I}_n - \frac{\gamma-1}{\gamma}\sigma_s^{\mathrm{T}}\cdot\left(\sigma_s\cdot\sigma_s^{\mathrm{T}}\right)^{-1}\cdot\sigma_s\right)\cdot c^{\mathrm{T}}\cdot\nabla_A g_\gamma$$

$$-\frac{1}{2}\gamma(\gamma-1)\nabla_A\mathcal{V}\cdot c\cdot\left(\mathbb{I}_n - \sigma_s^{\mathrm{T}}\cdot\left(\sigma_s\cdot\sigma_s^{\mathrm{T}}\right)^{-1}\cdot\sigma_s\right)\cdot c^{\mathrm{T}}\cdot\nabla_A\mathcal{V}$$

$$= \frac{1}{2}\frac{\gamma-1}{\gamma}\left((a_s - r)\cdot\left(\sigma_s\cdot\sigma_s^{\mathrm{T}}\right)^{-1}\cdot(a_s - r) + 2r\gamma\right) \quad (4.9.9)$$

for $t < T$, together with the terminal condition $g_\gamma(T,A) = 0$, and

$$\frac{\partial\mathcal{V}}{\partial t} + \frac{1}{2}\mathrm{tr}\left(\nabla_A\nabla_A\mathcal{V}\cdot c\cdot c^{\mathrm{T}}\right) + \nabla_A\mathcal{V}\cdot\left(b - (a_s - r)\cdot\left(\sigma_s\cdot\sigma_s^{\mathrm{T}}\right)^{-1}\cdot\sigma_s\cdot c^{\mathrm{T}}\right)$$

$$+\nabla_A g_\gamma\cdot c\cdot\left(\mathbb{I}_n - \sigma_s^{\mathrm{T}}\cdot\left(\sigma_s\cdot\sigma_s^{\mathrm{T}}\right)^{-1}\cdot\sigma_s\right)\cdot c^{\mathrm{T}}\cdot\mathcal{V} \quad (4.9.10)$$

$$-\gamma\nabla_A\mathcal{V}\cdot c\cdot\left(\mathbb{I}_n - \sigma_s^{\mathrm{T}}\cdot\left(\sigma_s\cdot\sigma_s^{\mathrm{T}}\right)^{-1}\cdot\sigma_s\right)\cdot c^{\mathrm{T}}\cdot\nabla_A\mathcal{V} - r\mathcal{V}$$

$$= -\kappa_0\cdot\mathcal{D} \quad (4.9.11)$$

for $t < T$, together with the terminal condition $\mathcal{V}(T,A) = \kappa_0.\upsilon(A)$. Once system (4.9.9)–(4.9.11) has been solved, the (vector) price is computed as a solution of the uncoupled linear system

$$\frac{\partial V}{\partial t} + \frac{1}{2}\mathrm{tr}\left(\nabla_A\nabla_A V\cdot c\cdot c^{\mathrm{T}}\right) + \nabla_A V.\left(b - (a_s - r)\cdot\left(\sigma_s\cdot\sigma_s^{\mathrm{T}}\right)^{-1}\cdot\sigma_s\cdot c^{\mathrm{T}}\right) \quad (4.9.12)$$

$$+\left(\nabla_A g_\gamma - \gamma \nabla_A \mathcal{V}\right) \cdot c \cdot \left(\mathbb{I}_n - \sigma_s^{\mathrm{T}} \cdot \left(\sigma_s \cdot \sigma_s^{\mathrm{T}}\right)^{-1} \cdot \sigma_s\right) \cdot c^{\mathrm{T}} \cdot \left(\nabla_A V\right)^{\mathrm{T}} - rV$$
$$= -\mathcal{D}. \tag{4.9.13}$$

Proof. By inspection. □

4.10 CRRA Indifference Pricing

Of course, the reader should keep in mind that in the case $l \neq 1$, the indifference pricing theorem (Theorem 4.8.1) was only a conjecture, and therefore what follows in this section in that case is only a (very plausible) conjecture.

Theorem 4.10.1. *In the above framework, for a position $\kappa(t,X,A) = X\kappa_0(t,A)$ and CRRA utility of wealth, the γ-κ-indifference price (per single contract) $V = V_{\gamma,\kappa}(t,A) = \{V_{\gamma,\kappa,1}(t,A), \ldots, V_{\gamma,\kappa,l}(t,A)\}$ is the solution of the following pricing system of PDEs:*

$$\frac{\partial g_\gamma}{\partial t} + \frac{1}{2} Tr\left(c \cdot c^{\mathrm{T}} \cdot \nabla_A \nabla_A g_\gamma\right) + \left(b - \frac{\gamma - 1}{\gamma}(a_s - r) \cdot \left(\sigma_s \cdot \sigma_s^{\mathrm{T}}\right)^{-1} \cdot \sigma_s \cdot c^{\mathrm{T}}\right) \cdot \nabla_A g_\gamma$$
$$+\frac{1}{2} \nabla_A g_\gamma \cdot c \cdot \left(\mathbb{I}_n - \frac{\gamma - 1}{\gamma} \sigma_s^{\mathrm{T}} \cdot \left(\sigma_s \cdot \sigma_s^{\mathrm{T}}\right)^{-1} \cdot \sigma_s\right) \cdot c^{\mathrm{T}} \cdot \nabla_A g_\gamma$$
$$= \frac{1}{2} \frac{\gamma - 1}{\gamma} \left((a_s - r) \cdot \left(\sigma_s \cdot \sigma_s^{\mathrm{T}}\right)^{-1} \cdot (a_s - r) + 2r\gamma\right) \tag{4.10.1}$$

for $t < T$, together with the terminal condition $g_\gamma(T,A) = 0$ (which is (3.6.17)), and

$$\frac{\partial V}{\partial t} + \frac{1}{2} tr\left(\nabla_A \nabla_A V \cdot c \cdot c^{\mathrm{T}}\right) + \nabla_A V . \left(b - (a_s - r) \cdot \left(\sigma_s \cdot \sigma_s^{\mathrm{T}}\right)^{-1} \cdot \sigma_s \cdot c^{\mathrm{T}}\right)$$
$$+\nabla_A g_\gamma \cdot c \cdot \left(\mathbb{I}_n - \sigma_s^{\mathrm{T}} \cdot \left(\sigma_s \cdot \sigma_s^{\mathrm{T}}\right)^{-1} \cdot \sigma_s\right) \cdot c^{\mathrm{T}} \cdot \left(\nabla_A V\right)^{\mathrm{T}}$$
$$-\frac{1}{2} \gamma \kappa_0(t,A) \cdot \nabla_A V \cdot c \cdot \left(\mathbb{I}_n - \sigma_s^{\mathrm{T}} \cdot \left(\sigma_s \cdot \sigma_s^{\mathrm{T}}\right)^{-1} \cdot \sigma_s\right) \cdot c^{\mathrm{T}} \cdot \left(\nabla_A V\right)^{\mathrm{T}} - rV$$
$$= -\mathcal{D} \tag{4.10.2}$$

for $t < T$, together with the terminal condition $V(T,A) = \upsilon(A)$.

Proof. Analogously as in the previous section, set $\varphi(t,X,A) = (X^{1-\gamma} e^{g_\gamma(t,A)} - 1)/(1-\gamma)$ in (4.8.2) and (3.4.2). □

Remark 4.10.1. Equation (4.10.1) is solved first (it is uncoupled), and then system (4.10.2) is solved. Thus indifference pricing is easier than κ-neutral pricing (when $\kappa \neq 0$).

Remark 4.10.2. If $\kappa = 0$, then the CRRA neutral price is equal to the CRRA indifference price.

4.11 CARA Neutral and Indifference Pricing

In the case of CARA utility (3.2.12), to have Hypothesis 1 satisfied, it is assumed that the position κ has the form $\kappa(t,X,A) = \kappa(t,A)$. Additionally, as before, when we were studying portfolio optimization, we shall have to assume that the interest rates are deterministic. We have the following theorem.

Theorem 4.11.1. *In the above framework, assuming that the interest rates are deterministic, for the position $\kappa(t,X,A) = \kappa(t,A)$, and for the CARA utility of wealth $\psi(X) = \left(1 - \mathrm{e}^{(1-X)\omega}\right) / \omega$, Hypothesis 1is satisfied, and the ψ-κ-neutral price (per single contract) $V = V_{\omega,\kappa}(t,A) = \{V_{\omega,\kappa,1}(t,A), \ldots, \ \{.V_{\omega,\kappa,l}(t,A)\}$ is the solution of the following pricing system of PDEs:*

$$\begin{aligned}
&\frac{\partial g_\omega(t,A)}{\partial t} + \frac{1}{2} Tr\left(c \cdot c^{\mathrm{T}} \cdot \nabla_A \nabla_A g_\omega(t,A)\right) + \left(b - (a_s - r) \cdot \left(\sigma_s \cdot \sigma_s^{\mathrm{T}}\right)^{-1} \cdot \sigma_s \cdot c^{\mathrm{T}}\right) \\
&\quad \cdot \nabla_A g_\omega(t,A) + \frac{1}{2} \nabla_A g_\omega(t,A) \cdot c \cdot \left(\mathbb{I}_n - \sigma_s^{\mathrm{T}} \cdot \left(\sigma_s \cdot \sigma_s^{\mathrm{T}}\right)^{-1} \cdot \sigma_s\right) \cdot c^{\mathrm{T}} \cdot \nabla_A g_\omega(t,A) \\
&\quad - \frac{1}{2} \mathrm{e}^{2\int_t^{\mathrm{T}} r(\tau)\,\mathrm{d}\tau} \omega^2 \kappa . \nabla_A V \cdot c \cdot \left(\mathbb{I}_n - \sigma_s^{\mathrm{T}} \cdot \left(\sigma_s \cdot \sigma_s^{\mathrm{T}}\right)^{-1} \cdot \sigma_s\right) \cdot c^{\mathrm{T}} \cdot (\nabla_A V)^{\mathrm{T}} \cdot \kappa \\
&= \frac{1}{2} (a_s - r) \cdot \left(\sigma_s \cdot \sigma_s^{\mathrm{T}}\right)^{-1} \cdot (a_s - r) \qquad (4.11.1)
\end{aligned}$$

for $t < T$, together with the terminal condition $g_\omega(T,A) = 0$, and

$$\begin{aligned}
&\frac{\partial V}{\partial t} + \frac{1}{2} tr\left(\nabla_A \nabla_A V \cdot c \cdot c^{\mathrm{T}}\right) + \nabla_A V . \left(b - (a_s - r) \cdot \left(\sigma_s \cdot \sigma_s^{\mathrm{T}}\right)^{-1} \cdot \sigma_s \cdot c^{\mathrm{T}}\right) \\
&\quad + \nabla_A g_\omega \cdot c \cdot \left(\mathbb{I}_n - \sigma_s^{\mathrm{T}} \cdot \left(\sigma_s \cdot \sigma_s^{\mathrm{T}}\right)^{-1} \cdot \sigma_s\right) \cdot c^{\mathrm{T}} \cdot (\nabla_A V)^{\mathrm{T}} \\
&\quad - \mathrm{e}^{\int_t^{\mathrm{T}} r(\tau)\,\mathrm{d}\tau} \omega \nabla_A V \cdot c \cdot \left(\mathbb{I}_n - \sigma_s^{\mathrm{T}} \cdot \left(\sigma_s \cdot \sigma_s^{\mathrm{T}}\right)^{-1} \cdot \sigma_s\right) \cdot c^{\mathrm{T}} \cdot (\nabla_A V)^{\mathrm{T}} \cdot \kappa - rV \\
&= -\mathcal{D} \qquad (4.11.2)
\end{aligned}$$

for $t < T$, together with the terminal condition $V(T,A) = \upsilon(A)$.

Proof. As before, we set $\varphi(t,X,A) = (1 - \mathrm{e}^{-\mathrm{e}^{\int_t^{\mathrm{T}} r(\tau)\,\mathrm{d}\tau} X\omega + \omega + g_\omega(t,A)})/\omega$ in (4.7.1). There is only one new term (in comparison to (3.4.2) and (3.8.17)):

$$-\frac{1}{2}\left(\frac{\partial^2\varphi}{\partial X^2}\right)^2 \kappa.\nabla_A V\cdot c\cdot\left(\mathbb{I}_n-\sigma_s^{\mathrm{T}}\cdot\left(\sigma_s\cdot\sigma_s^{\mathrm{T}}\right)^{-1}\cdot\sigma_s\right)\cdot c^{\mathrm{T}}\cdot\left(\nabla_A V\right)^{\mathrm{T}}\cdot\kappa$$
$$= \mathrm{e}^{2\left(-\mathrm{e}^{\int_t^{\mathrm{T}} r(\tau)\,\mathrm{d}\tau}X\omega+\omega+\int_t^{\mathrm{T}} r(\tau)\,\mathrm{d}\tau+g_\omega(t,A)\right)}$$
$$\left(-\frac{1}{2}\mathrm{e}^{2\int_t^{\mathrm{T}} r(\tau)\,\mathrm{d}\tau}\omega^2\right)\kappa.\nabla_A V\cdot c\cdot\left(\mathbb{I}_n-\sigma_s^{\mathrm{T}}\cdot\left(\sigma_s\cdot\sigma_s^{\mathrm{T}}\right)^{-1}\cdot\sigma_s\right)$$
$$\cdot c^{\mathrm{T}}\cdot\left(\nabla_A V\right)^{\mathrm{T}}\cdot\kappa. \tag{4.11.3}$$

□

Therefore, after dividing by $\mathrm{e}^{2\left(-\mathrm{e}^{\int_t^{\mathrm{T}} r(\tau)\,\mathrm{d}\tau}X\omega+\omega+\int_t^{\mathrm{T}} r(\tau)\,\mathrm{d}\tau+g_\omega(t,A)\right)}$, (4.7.1) becomes (4.11.1). Also

$$\frac{\partial^2\varphi}{\partial X^2}\Big/\frac{\partial\varphi}{\partial X}$$
$$= -\mathrm{e}^{-\mathrm{e}^{\int_t^{\mathrm{T}} r(\tau)\,\mathrm{d}\tau}X\omega+\omega+2\int_t^{\mathrm{T}} r(\tau)\,\mathrm{d}\tau+g_\omega(t,A)}\omega/\mathrm{e}^{-\mathrm{e}^{\int_t^{\mathrm{T}} r(\tau)\,\mathrm{d}\tau}X\omega+\omega+\int_t^{\mathrm{T}} r(\tau)\,\mathrm{d}\tau+g_\omega(t,A)}$$
$$= -\mathrm{e}^{\int_t^{\mathrm{T}} r(\tau)\,\mathrm{d}\tau}\omega \tag{4.11.4}$$

and

$$\nabla_A\frac{\partial\varphi}{\partial X}\Big/\frac{\partial\varphi}{\partial X}$$
$$= \mathrm{e}^{-\mathrm{e}^{\int_t^{\mathrm{T}} r(\tau)\,\mathrm{d}\tau}X\omega+\omega+\int_t^{\mathrm{T}} r(\tau)\,\mathrm{d}\tau+g_\omega(t,A)}\nabla_A g_\omega/\mathrm{e}^{-\mathrm{e}^{\int_t^{\mathrm{T}} r(\tau)\,\mathrm{d}\tau}X\omega+\omega+\int_t^{\mathrm{T}} r(\tau)\,\mathrm{d}\tau+g_\omega(t,A)}$$
$$= \nabla_A g_\omega, \tag{4.11.5}$$

and therefore (4.11.2) follows.

Remark 4.11.1. In the above framework, for a position $\kappa = 0$ and CARA utility of wealth $\psi(X) = (1-\mathrm{e}^{(1-X)\omega}/\omega$, the ψ-ω-neutral price V does not depend on ω, and it is the solution of the following *uncoupled pricing system of PDEs*:

$$\frac{\partial g(t,A)}{\partial t}+\frac{1}{2}\mathrm{Tr}\left(c\cdot c^{\mathrm{T}}\cdot\nabla_A\nabla_A g(t,A)\right)$$
$$+\left(b-(a_s-r)\cdot\left(\sigma_s\cdot\sigma_s^{\mathrm{T}}\right)^{-1}\cdot\sigma_s\cdot c^{\mathrm{T}}\right)\cdot\nabla_A g(t,A)$$
$$+\frac{1}{2}\nabla_A g(t,A)\cdot c\cdot\left(\mathbb{I}_n-\sigma_s^{\mathrm{T}}\cdot\left(\sigma_s\cdot\sigma_s^{\mathrm{T}}\right)^{-1}\cdot\sigma_s\right)\cdot c^{\mathrm{T}}\cdot\nabla_A g(t,A)$$
$$= \frac{1}{2}(a_s-r)\cdot\left(\sigma_s\cdot\sigma_s^{\mathrm{T}}\right)^{-1}\cdot(a_s-r) \tag{4.11.6}$$

for $t < T$, together with the terminal condition $g(T,A) = 0$, and

$$\begin{aligned}
&\frac{\partial V}{\partial t} + \frac{1}{2}\mathrm{tr}\left(\nabla_A \nabla_A V \cdot c \cdot c^{\mathrm{T}}\right) + \nabla_A V.\left(b - (a_s - r)\cdot\left(\sigma_s \cdot \sigma_s^{\mathrm{T}}\right)^{-1}\cdot \sigma_s \cdot c^{\mathrm{T}}\right) \\
&\quad + \nabla_A g \cdot c \cdot \left(\mathbb{I}_n - \sigma_s^{\mathrm{T}}\cdot\left(\sigma_s\cdot\sigma_s^{\mathrm{T}}\right)^{-1}\cdot\sigma_s\right)\cdot c^{\mathrm{T}}\cdot\left(\nabla_A V\right)^{\mathrm{T}} - rV \\
&= -\mathcal{D}
\end{aligned} \tag{4.11.7}$$

for $t < T$, together with the terminal condition $V(T,X,A) = \upsilon(A)$. Similarly, we can prove the following result.

Theorem 4.11.2 ([17]). *In the above framework, assuming that the interest rates are deterministic, for the position $\kappa(t,X,A) = \kappa(t,A)$ and CARA utility of wealth $\psi(X) = \left(1 - e^{(1-X)\omega}\right)/\omega$, Hypothesis 1 is satisfied, and the ψ-κ-indifference price (per single contract) V is the solution of the following pricing system of PDEs:*

$$\begin{aligned}
&\frac{\partial g(t,A)}{\partial t} + \frac{1}{2}Tr\left(c\cdot c^{\mathrm{T}}\cdot\nabla_A\nabla_A g(t,A)\right) + \left(b - (a_s - r)\cdot\left(\sigma_s\cdot\sigma_s^{\mathrm{T}}\right)^{-1}\cdot\sigma_s\cdot c^{\mathrm{T}}\right) \\
&\quad \cdot\nabla_A g(t,A) + \frac{1}{2}\nabla_A g(t,A)\cdot c\cdot\left(\mathbb{I}_n - \sigma_s^{\mathrm{T}}\cdot\left(\sigma_s\cdot\sigma_s^{\mathrm{T}}\right)^{-1}\cdot\sigma_s\right)\cdot c^{\mathrm{T}}\cdot\nabla_A g(t,A) \\
&= \frac{1}{2}(a_s - r)\cdot\left(\sigma_s\cdot\sigma_s^{\mathrm{T}}\right)^{-1}\cdot(a_s - r)
\end{aligned} \tag{4.11.8}$$

for $t < T$, together with the terminal condition $g(T,A) = 0$, and

$$\begin{aligned}
&\frac{\partial V}{\partial t} + \frac{1}{2}tr\left(\nabla_A\nabla_A V\cdot c\cdot c^{\mathrm{T}}\right) + \nabla_A V.\left(b - (a_s - r)\cdot\left(\sigma_s\cdot\sigma_s^{\mathrm{T}}\right)^{-1}\cdot\sigma_s\cdot c^{\mathrm{T}}\right) \\
&\quad + \nabla_A g\cdot c\cdot\left(\mathbb{I}_n - \sigma_s^{\mathrm{T}}\cdot\left(\sigma_s\cdot\sigma_s^{\mathrm{T}}\right)^{-1}\cdot\sigma_s\right)\cdot c^{\mathrm{T}}\cdot\left(\nabla_A V\right)^{\mathrm{T}} \\
&\quad - \frac{1}{2}e^{\int_t^T r(\tau)\,d\tau}\omega\nabla_A V\cdot c\cdot\left(\mathbb{I}_n - \sigma_s^{\mathrm{T}}\cdot\left(\sigma_s\cdot\sigma_s^{\mathrm{T}}\right)^{-1}\cdot\sigma_s\right)\cdot c^{\mathrm{T}}\cdot\left(\nabla_A V\right)^{\mathrm{T}}\cdot\kappa - rV \\
&= -\mathcal{D}
\end{aligned} \tag{4.11.9}$$

for $t < T$, together with the terminal condition $V(T,A) = \upsilon(A)$.

4.12 Pricing: Consistency with the Usual Theory

4.12.1 Black–Scholes Model

Continuing Sect. 2.4.1, recall that $S = A = \{S_1\}$ and that the state space is equal to $\mathcal{A} = (0, \infty)$. Also, as concluded there, the market is complete, and all of the pricing PDEs, neutral and indifference, and under any utility of wealth, reduce to (4.7.18).

Further deriving the pricing PDE, we calculate (set $V = \{V_1\}$)

$$\mathrm{Tr}\left(\nabla_A \nabla_A V_1 \cdot c \cdot c^{\mathrm{T}}\right) = S_1^2 \mathbf{p}^2 V_1^{(0,2)}(t, S_1) \tag{4.12.1}$$

and

$$b - (a_s - r) \cdot \left(\sigma_s \cdot \sigma_s^{\mathrm{T}}\right)^{-1} \cdot \sigma_s \cdot c^{\mathrm{T}} = \{S_1(\mathbf{a}_1 - \mathbb{D}_1) - S_1(\mathbf{a}_1 - r)\} = \{S_1(r - \mathbb{D}_1)\}, \tag{4.12.2}$$

and therefore (let $\mathcal{D} = 0$)

$$V_1^{(1,0)}(t, S_1) + \frac{1}{2} S_1^2 V_1^{(0,2)}(t, S_1)\, \mathbf{p}^2 + S_1(r - \mathbb{D}_1) V_1^{(0,1)}(t, S_1) - r V_1(t, S_1) = 0 \tag{4.12.3}$$

for $t < T, S_1 > 0$, with the terminal condition $V_1(T, S_1) = \upsilon(S_1)$, where $\upsilon(Y)$ is the contract payoff, which is the classical Black–Scholes PDE.

4.12.2 Black–Scholes Model 2 (Without Hedging)

We are interested here in the CRRA neutral price under zero position ($\kappa = 0$). So we shall apply Theorem 4.9.

What if no hedging in the underlying is possible? This time, $S(t) = \emptyset, A(t) = \{S_1(t)\}$, which, however, falls outside the simple-economy framework, because of the nonredundancy condition ($\sigma_s \cdot \sigma_s^{\mathrm{T}} > 0$). Therefore, we extend the Black–Scholes model, by considering also an auxiliary tradable $S_2(t)$:

$$\begin{aligned} \mathrm{d}\, S_1(t) &= S_1(t)((\mathbf{a}_1 - \mathbb{D}_1)\,\mathrm{d}t + \mathbf{p}_1 \mathrm{d}\, B_1(t)), \\ \mathrm{d}\, S_2(t) &= S_2(t)((\mathbf{a}_2 - \mathbb{D}_2)\,\mathrm{d}t + \mathbf{p}_2 \mathrm{d}\, \mathbb{B}_2(t)), \end{aligned} \tag{4.12.4}$$

with $\mathrm{d}\, B_1(t) \mathrm{d}\, \mathbb{B}_2(t) = 0$ (since S_2 is only an auxiliary tradable, to ensure that no hedging interference occurs, we assume zero correlation). So, $S = \{S_2\}$, $A = \{S_1, S_2\}$, and

$$\begin{aligned} a_s &= \{\mathbf{a}_2\}, l\{S_1(\mathbf{a}_1 - \mathbb{D}_1), S_2(\mathbf{a}_2 - \mathbb{D}_2)\}, \sigma_s = \left(0 \;\; \mathbf{p}_2\right), \\ c &= \begin{pmatrix} S_1 \mathbf{p}_1 & 0 \\ 0 & S_2 \mathbf{p}_2 \end{pmatrix} \end{aligned} \tag{4.12.5}$$

Again, we first decide on completeness/incompleteness:

$$c \cdot \left(\mathbb{I}_2 - \sigma_s^{\mathrm{T}} \cdot \left(\sigma_s \cdot \sigma_s^{\mathrm{T}}\right)^{-1} \cdot \sigma_s\right) \cdot c^{\mathrm{T}} = \begin{pmatrix} S_1^2 \mathbf{p}_1^2 & 0 \\ 0 & 0 \end{pmatrix}, \tag{4.12.6}$$

so unless $\mathbf{p}_1 = 0$, this market is incomplete.

Equation (4.9.6) can be solved in the form $g(t, Y_1, Y_2) = g(t)$, since it reduces to

$$\begin{aligned} g^{(1)}(t) &= \frac{\gamma - 1}{\gamma} \left(\frac{(\mathbf{a}_2 - r)^2}{2\mathbf{p}_2^2} + r\gamma \right), \\ g(T) &= 0. \end{aligned} \tag{4.12.7}$$

Consequently, $\nabla_A g_\gamma = 0$. The pricing PDE (4.9.7) becomes ($\mathcal{D} = 0$)

$$\begin{aligned} &V_1^{(1,0,0)}(t, S_1, S_2) + \frac{1}{2}\left(S_1^2 V_1^{(0,2,0)}(t, S_1, S_2)\, \mathbf{p}_1^2 + S_2^2 \mathbf{p}_2^2 V_1^{(0,0,2)}(t, S_1, S_2)\right) + S_2 \\ &\quad (r - \mathbb{D}_2)\, V_1^{(0,0,1)}(t, S_1, S_2) + S_1 (\mathbf{a}_1 - \mathbb{D}_1)\, V_1^{(0,1,0)}(t, S_1, S_2) - r V_1(t, S_1, S_2) \\ &= 0 \end{aligned} \tag{4.12.8}$$

for $t < T, S_1 > 0, S_2 > 0$, together with the terminal condition $V_1(T, S_1, S_2) = \upsilon(S_1, S_2)$.

Finally, if $\upsilon(S_1, S_2) = \upsilon(S_1)$, we look for the solution in the form $V_1(t, S_1)$, obtaining

$$V_1^{(1,0)}(t, S_1) + \frac{1}{2} S_1^2 V_1^{(0,2)}(t, S_1)\, \mathbf{p}_1^2 + S_1 (\mathbf{a}_1 - \mathbb{D}_1)\, V_1^{(0,1)}(t, S_1) - r V_1(t, S_1) = 0 \tag{4.12.9}$$

for $t < T, S_1 > 0$, together with the terminal condition $V_1(T, S_1) = \upsilon(S_1)$.

The only difference between the classical Black–Scholes PDE and (4.12.9) is in their first-order coefficients, and the only reason for that difference is the hedging vs. no-hedging.

4.12.3 Pricing Equations for the Musiela and Zariphopoulou Model

In [30] Musiela and Zariphopoulou studied indifference pricing under CARA utility for the market model

$$\begin{aligned} \mathrm{d}S_1(t) &= S_1(t)(\mathbf{a} - \mathbb{D}_1)\,\mathrm{d}t + S_1(t)\mathbf{p}_1 \mathrm{d}B_1(t), \\ \mathrm{d}Y(t) &= \mathbf{b}(t, Y(t))\mathrm{d}t + w(t, Y)\mathrm{d}\mathbb{B}_2(t), \end{aligned} \tag{4.12.10}$$

where $\mathrm{d}B_1(t)\mathrm{d}\mathbb{B}_2(t) = \rho_{2,1}\mathrm{d}t$, and where only S_1 is tradable. Here, and in some later sections as well, we compare our results with theirs for the sake of checking consistency, and we provide some additional results. Their model can be put in our context using $A = \{S_1, Y\}, S = \{S_1\}$, with state space being equal to $\mathcal{A} = (0,\infty) \times \mathcal{Y}$, where $\mathcal{Y}$ is the set of all Y values that can be reached by the process $Y(t)$ (due to the unspecified nature of the coefficients $\mathbf{b}(t,Y)$ and $w(t,Y)$, the precise nature of $\mathcal{Y}$ is left unspecified), and

$$a_s = \{\mathbf{a}\}, \sigma_s = \begin{pmatrix} \mathbf{p}_1 & 0 \end{pmatrix}, b = \{S_1(\mathbf{a} - \mathbb{D}_1), \mathbf{b}(t,Y)\},$$

$$c = \begin{pmatrix} S_1\mathbf{p}_1 & 0 \\ w(t,Y)\rho_{2,1} & w(t,Y)\sqrt{1-\rho_{2,1}^2} \end{pmatrix}, \tag{4.12.11}$$

as the market coefficients. Then

$$c \cdot \left(\mathbb{I}_2 - \sigma_s^{\mathrm{T}} \cdot \left(\sigma_s \cdot \sigma_s^{\mathrm{T}}\right)^{-1} \cdot \sigma_s\right) \cdot c^{\mathrm{T}} = \begin{pmatrix} 0 & 0 \\ 0 & \left(1-\rho_{2,1}^2\right) w(t,Y)^2 \end{pmatrix},$$

and the market is complete if and only if $\left(1-\rho_{2,1}^2\right) w(t,Y)^2 = 0$.

We shall first check for the consistency between our results and those of Musiela and Zariphopoulou in [30], i.e., we first solve the problem of the CARA indifference pricing for the above model, i.e., we first implement (apply) Theorem 4.11.2 Equation (4.11.8) reads

$$\begin{aligned} &\frac{1}{2}\left(1-\rho_{2,1}^2\right) w(t,Y)^2 g^{(0,0,1)}(t,S_1,Y)^2 \\ &\quad + \left(\mathbf{b}(t,Y) - \frac{(\mathbf{a}-r)\rho_{2,1} w(t,Y)}{\mathbf{p}_1}\right) g^{(0,0,1)}(t,S_1,Y) + S_1(r-\mathbb{D}_1) g^{(0,1,0)}(t,S_1,Y) \\ &\quad + \frac{1}{2}\left(S_1^2 g^{(0,2,0)}(t,S_1,Y)\mathbf{p}_1^2 + 2S_1\rho_{2,1} w(t,Y) g^{(0,1,1)}(t,S_1,Y)\mathbf{p}_1\right. \\ &\qquad \left. + w(t,Y)^2 g^{(0,0,2)}(t,S_1,Y)\right) + g^{(1,0,0)}(t,S_1,Y) = \frac{(\mathbf{a}-r)^2}{2\mathbf{p}_1^2} \end{aligned} \tag{4.12.12}$$

for $t < T, S_1 > 0, Y \in \mathcal{Y}$, with terminal condition $g(T,S_1,Y) = 0$, and since $\mathbf{a}, \mathbf{p}_1$, and r are assumed to be constants, by looking for a solution in the form $g(t,S_1,Y) = g(t)$, (4.12.12) simplifies into the ODE

$$g'(t) = \frac{(\mathbf{a}-r)^2}{2\mathbf{p}_1^2} \tag{4.12.13}$$

for $t < T$, with terminal condition $g(T) = 0$. Consequently, $\nabla_A g = 0$, and therefore, for a position $\kappa = \{k\}$, the pricing PDE (4.11.9) reads

$$-\frac{1}{2}e^{r(T-t)}k\omega\left(1-\rho_{2,1}^2\right)V^{(0,0,1)}(t,S_1,Y)^2 w(t,Y)^2 - rV(t,S_1,Y) + \left(\mathbf{b}(t,Y) - \frac{(\mathbf{a}-r)\rho_{2,1}w(t,Y)}{\mathbf{p}_1}\right)V^{(0,0,1)}(t,S_1,Y) + S_1(r-\mathbb{D}_1)V^{(0,1,0)}(t,S_1,Y) + \frac{1}{2}\left(S_1^2 V^{(0,2,0)}(t,S_1,Y)\mathbf{p}_1^2 + 2S_1\rho_{2,1}w(t,Y)V^{(0,1,1)}(t,S_1,Y)\mathbf{p}_1 + w(t,Y)^2 V^{(0,0,2)}(t,S_1,Y)\right) + V^{(1,0,0)}(t,S_1,Y) = 0 \tag{4.12.14}$$

for $t < T, S_1 > 0, Y \in \mathcal{Y}$, together with the terminal condition $V(t,S_1,Y) = \upsilon(S_1,Y)$. If, furthermore, $\upsilon(S_1,Y) = \upsilon(Y)$, $k = -1$ (the "seller's price"), and $r = 0$ (as assumed in [30]), then the above pricing equation simplifies into the CARA indifference pricing PDE

$$V^{(1,0)}(t,Y) + \frac{1}{2}V^{(0,2)}(t,Y)w(t,Y)^2 + \left(\mathbf{b}(t,Y) - \frac{\mathbf{a}\rho_{2,1}w(t,Y)}{\mathbf{p}_1}\right)V^{(0,1)}(t,Y) + \frac{1}{2}\omega\left(1-\rho_{2,1}^2\right)V^{(0,1)}(t,Y)^2 w(t,Y)^2 = 0 \tag{4.12.15}$$

for $t < T, Y \in \mathcal{Y}$, together with the terminal condition $V(t,Y) = \upsilon(Y)$, which is exactly equation (21) in [30], and hence we have established the consistency of our results with the results of [30] (the consistency in regard to hedging will be established in the next chapter).

For some other pricing results for the Musiela and Zariphopoulou model, see Example 4.16.1. For the hedging result in the case of the Musiela and Zariphopoulou model, see Example 5.2.4.

4.13 Case Study: Stochastic Volatility (Continuation: CRRA Neutral Pricing)

We continue Sect. 3.10. As in Sect. 2.4.3, we assume the stochastic volatility model (2.4.9), with $S = \{S_1\}, A = \{S_1, \mathbf{v}\}$, and market coefficients (2.4.13). For a position $\kappa(t,X,A) = X\kappa_0(t,A) = X\{k_0(t,A)\}$, the neutral pricing system (4.9.2)–(4.9.3) becomes

$$g_\gamma^{(1,0,0)}(t,S_1,\mathbf{v}) + \frac{1}{2}\mathbf{v}\left(g_\gamma^{(0,2,0)}(t,S_1,\mathbf{v})S_1^2 + 2\sigma_v\rho_{2,1}g_\gamma^{(0,1,1)}(t,S_1,\mathbf{v})S_1 + \sigma_v^2 g_\gamma^{(0,0,2)}(t,S_1,\mathbf{v})\right) + \left(k_v - \mathbf{v}K_v - \frac{1}{\gamma}(\gamma-1)(\mathbf{a}_0 + \mathbf{v}\lambda_s - r)\sigma_v\rho_{2,1}\right)$$

$$g_\gamma^{(0,0,1)}(t,S_1,\mathbf{v})+\left(S_1\left(\mathbf{a}_0+\mathbf{v}\lambda_s-\mathbb{D}\right)-\frac{(\gamma-1)S_1\left(\mathbf{a}_0+\mathbf{v}\lambda_s-r\right)}{\gamma}\right)g_\gamma^{(0,1,0)}$$
$$(t,S_1,\mathbf{v})-\frac{1}{2}\mathbf{v}(\gamma-1)\gamma k_0^2\sigma_v^2\left(1-\rho_{2,1}^2\right)V_1^{(0,0,1)}(t,S_1,\mathbf{v})^2$$
$$+\frac{1}{2\gamma}\mathbf{v}\left(\sigma_v^2\left(\gamma-(\gamma-1)\rho_{2,1}^2\right)(t,S_1,\mathbf{v})^2+2S_1\sigma_v\rho_{2,1}g_\gamma^{(0,1,0)}(t,S_1,\mathbf{v})g_\gamma^{(0,0,1)}\right.$$
$$\left.(t,S_1,\mathbf{v})+S_1^2g_\gamma^{(0,1,0)}(t,S_1,\mathbf{v})^2\right)g_\gamma^{(0,0,1)}=\frac{\gamma-1}{\gamma}\left(\frac{\left(\mathbf{a}_0+\mathbf{v}\lambda_s-r\right)^2}{2\mathbf{v}}+r\gamma\right) \tag{4.13.1}$$

for $t<T$, together with the terminal condition $g_\gamma(T,S_1,\mathbf{v})=0$, coupled with the equation

$$V_1^{(1,0,0)}(t,S_1,\mathbf{v})+\frac{1}{2}\mathbf{v}\left(V_1^{(0,2,0)}(t,S_1,\mathbf{v})S_1^2+2\sigma_v\rho_{2,1}V_1^{(0,1,1)}(t,S_1,\mathbf{v})S_1\right.$$
$$\left.+\sigma_v^2V_1^{(0,0,2)}(t,S_1,\mathbf{v})\right)+\left(k_v-\mathbf{v}K_v-\left(\mathbf{a}_0+\mathbf{v}\lambda_s-r\right)\sigma_v\rho_{2,1}\right)V_1^{(0,0,1)}(t,S_1,\mathbf{v})$$
$$+(r-\mathbb{D})S_1V_1^{(0,1,0)}(t,S_1,\mathbf{v})-\mathbf{v}\gamma k_0\left(1-\rho_{2,1}^2\right)V_1^{(0,0,1)}(t,S_1,\mathbf{v})^2\sigma_v^2$$
$$+\mathbf{v}\left(1-\rho_{2,1}^2\right)g_\gamma^{(0,0,1)}(t,S_1,\mathbf{v})V_1^{(0,0,1)}(t,S_1,\mathbf{v})\sigma_v^2-rV_1(t,S_1,\mathbf{v})=0 \tag{4.13.2}$$

for $t<T$, together with the terminal condition $V_1(T,S_1,\mathbf{v})=\upsilon(S_1,\mathbf{v})$, where υ is the contract terminal payoff.

We consider first what turns out to be a very simple case: $\upsilon=\upsilon(S_1)=S_1-\mathfrak{k}$, the payoff of a *forward* contract with strike $\mathfrak{k}$. It is easy to see that in spite of the apparent complexity of the pricing system (4.13.1)–(4.14.2), and irrespective of the position k_0, in such a case $V_1=V_1(t,S_1)$, and moreover,

$$V_1(t,S_1)=S_1\mathrm{e}^{(t-T)\mathbb{D}}-\mathfrak{k}\mathrm{e}^{r(t-T)}, \tag{4.13.3}$$

and by solving equation $V_1(t,S_1)=0$ for $\mathfrak{k}$, we get the forward price, and since the interest rate is deterministic (constant), the forward price is equal to the futures price, and they are, as in (1.6.39), both equal to

$$F_1(t,S_1)=\mathrm{e}^{(T-t)(r-\mathbb{D})}S_1. \tag{4.13.4}$$

Now consider a payoff $\upsilon=\upsilon(\mathbf{v})=\mathbf{v}-\mathfrak{k}$, a forward contract for instantaneous variance $\mathbf{v}$. To obtain a very convenient solution, we shall need to assume that

$\mathbf{a}_0 = r$. Under such an assumption, we can look for the solution of the pricing system (4.13.1)–(4.13.2) in the form $\{g_\gamma(t,\mathbf{v}), V_1(t,\mathbf{v})\}$, thereby solving

$$g_\gamma^{(1,0)}(t,\mathbf{v}) + \frac{1}{2}\mathbf{v}\sigma_v^2 g_\gamma^{(0,2)}(t,\mathbf{v}) - \frac{1}{2}\mathbf{v}(\gamma-1)\gamma k_0^2\sigma_v^2\left(1-\rho_{2,1}^2\right)V_1^{(0,1)}(t,\mathbf{v})^2$$
$$+\left(k_v - \mathbf{v}K_v - \frac{\mathbf{v}(\gamma-1)\lambda_s\sigma_v\rho_{2,1}}{\gamma}\right)g_\gamma^{(0,1)}(t,\mathbf{v}) + \frac{1}{2}\mathbf{v}\sigma_v^2\left(1-\frac{\gamma-1}{\gamma}\rho_{2,1}^2\right)$$
$$g_\gamma^{(0,1)}(t,\mathbf{v})^2 = \frac{\gamma-1}{\gamma}\left(\frac{\mathbf{v}\lambda_s^2}{2}+r\gamma\right) \tag{4.13.5}$$

for $t < T$, together with the terminal condition $g_\gamma(T,\mathbf{v}) = 0$, coupled with the equation

$$V_1^{(1,0)}(t,\mathbf{v}) + \frac{1}{2}\mathbf{v}\sigma_v^2 V_1^{(0,2)}(t,\mathbf{v}) + (k_v - \mathbf{v}(K_v + \lambda_s\sigma_v\rho_{2,1}))V_1^{(0,1)}(t,\mathbf{v}) - \mathbf{v}\gamma k_0$$
$$\left(1-\rho_{2,1}^2\right)V_1^{(0,1)}(t,\mathbf{v})^2\sigma_v^2 + \mathbf{v}\left(1-\rho_{2,1}^2\right)g_\gamma^{(0,1)}(t,\mathbf{v})V_1^{(0,1)}(t,\mathbf{v})\sigma_v^2 - rV_1(t,\mathbf{v}) = 0 \tag{4.13.6}$$

for $t < T$, together with the terminal condition $V_1(T,\mathbf{v}) = \upsilon(\mathbf{v}) = \mathbf{v} - \mathfrak{k}$. The linearity of the payoff is quite important here; a nonlinear PDE system will admit a linear (in $\mathbf{v}$) solution. Indeed, we look for the solution in the form

$$\{g_\gamma(t,\mathbf{v}), V_1(t,\mathbf{v})\} = \{\mathbb{G}_0(t) + \mathbf{v}\mathbb{G}_1(t), \mathbb{V}_0(t) + \mathbf{v}\mathbb{V}_1(t)\}. \tag{4.13.7}$$

Consequently, plugging (4.13.7) into the system (4.13.5)–(4.13.6), we arrive at

$$-\frac{1}{2}\mathbf{v}(\gamma-1)\gamma k_0^2\sigma_v^2\left(1-\rho_{2,1}^2\right)\mathbb{V}_1(t)^2 + \left(k_v - \mathbf{v}K_v - \frac{\mathbf{v}(\gamma-1)\lambda_s\sigma_v\rho_{2,1}}{\gamma}\right)\mathbb{G}_1(t)$$
$$+\frac{1}{2}\mathbf{v}\sigma_v^2\left(1-\frac{\gamma-1}{\gamma}\rho_{2,1}^2\right)\mathbb{G}_1(t)^2 + \mathbb{G}_0'(t) + \mathbf{v}\mathbb{G}_1'(t) = \frac{\gamma-1}{\gamma}\left(\frac{\mathbf{v}\lambda_s^2}{2}+r\gamma\right) \tag{4.13.8}$$

for $t < T$, for any $\mathbf{v} > 0$, together with the terminal condition $\mathbb{G}_0(T) = \mathbb{G}_1(T) = 0$, coupled with

$$-\mathbf{v}\gamma k_0\left(1-\rho_{2,1}^2\right)\mathbb{V}_1(t)^2\sigma_v^2 + \mathbf{v}\left(1-\rho_{2,1}^2\right)\mathbb{G}_1(t)\mathbb{V}_1(t)\sigma_v^2 + (k_v - \mathbf{v}K_v - \mathbf{v}\lambda_s\sigma_v\rho_{2,1})$$
$$\mathbb{V}_1(t) - r(\mathbb{V}_0(t) + \mathbf{v}\mathbb{V}_1(t)) + \mathbb{V}_0'(t) + \mathbf{v}\mathbb{V}_1'(t) = 0 \tag{4.13.9}$$

for $t < T$, for any $\mathbf{v} > 0$, together with the terminal condition $\mathbb{V}_0(T) = -\mathfrak{k}, \mathbb{V}_1(T) = 1$. This implies a system of four ODEs characterizing $\{\mathbb{G}_0(t), \mathbb{G}_1(t), \mathbb{V}_0(t), \mathbb{V}_1(t)\}$:

$$\mathbb{G}_0'(t) + r(1-\gamma) + k_v \mathbb{G}_1(t) = 0, \tag{4.13.10}$$

$$\mathbb{G}_1'(t) + \left(\frac{1-\gamma}{\gamma}\sigma_v \rho_{2,1}\lambda_s - K_v\right)\mathbb{G}_1(t) + \frac{1}{2}\left(1 - \frac{\gamma-1}{\gamma}\right)\sigma_v^2\rho_{2,1}^2\mathbb{G}_1(t)^2 + \frac{1}{2}\sigma_v^2 \tag{4.13.11}$$

$$\left(1-\rho_{2,1}^2\right)\mathbb{G}_1(t)^2 + \frac{1}{2}(1-\gamma)\gamma k_0^2\sigma_v^2\left(1-\rho_{2,1}^2\right)\mathbb{V}_1(t)^2 + \frac{(1-\gamma)\lambda_s^2}{2\gamma} = 0,$$

$$\mathbb{V}_0'(t) - r\mathbb{V}_0(t) + k_v\mathbb{V}_1(t) = 0, \tag{4.13.12}$$

$$\begin{aligned}&\mathbb{V}_1'(t) - (r + K_v + \lambda_s\sigma_v\rho_{2,1})\mathbb{V}_1(t) - \gamma k_0\left(1-\rho_{2,1}^2\right)\mathbb{V}_1(t)^2\sigma_v^2\\ &\quad + \left(1-\rho_{2,1}^2\right)\mathbb{G}_1(t)\mathbb{V}_1(t)\sigma_v^2 = 0,\end{aligned} \tag{4.13.13}$$

for $t < T$, together with the terminal condition $\mathbb{G}_0(T) = \mathbb{G}_1(T) = 0, \mathbb{V}_0(T) = -\mathfrak{k}, \mathbb{V}_1(T) = 1$. Summarizing, we have the following result.

Proposition 4.13.1. *In the above context, if $\mathbf{a}_0 = r$, then for the CRRA utility of wealth and the position $\kappa(t,X,A) = X\kappa_0 = X\{k_0\}$, the κ-neutral value of the forward contract $V_1(t,\mathbf{v})$ with payoff $\upsilon(\mathbf{v}) = \mathbf{v} - \mathfrak{k}$ is given by*

$$V_1(t,\mathbf{v}) = \mathbb{V}_0(t) + \mathbf{v}\mathbb{V}_1(t) = -\mathrm{e}^{-r(T-t)}\mathfrak{k} + k_v\int_t^T \mathrm{e}^{-r(\tau-t)}\mathbb{V}_1(\tau)\mathrm{d}\tau + \mathbf{v}\mathbb{V}_1(t) \tag{4.13.14}$$

for $t < T$, where $\mathbb{V}_1(t)$ is the second component of the solution $\{\mathbb{G}_1(t), \mathbb{V}_1(t)\}$ of the coupled ODE system (4.13.11), (4.13.13). The corresponding forward price, i.e., the "forward variance," is equal to

$$F_1(t,\mathbf{v}) = k_v\int_t^T \mathrm{e}^{r(T-\tau)}\mathbb{V}_1(\tau)\mathrm{d}\tau + \mathrm{e}^{r(T-t)}\mathbf{v}\mathbb{V}_1(t) \tag{4.13.15}$$

for $t < T$, where $\mathbb{V}_1(t)$ is the second component of the solution of the coupled ODE system (4.13.11), (4.13.13).

Proof. Formula (4.13.14) follows from integrating (4.13.12). Formula (4.13.15) follows from solving the equation $V_1(t,\mathbf{v}) = 0$ for $\mathfrak{k}$, which completes the proof of the proposition. □

Exercise 4.13.1. Consider Heston's stochastic volatility model

$$\begin{aligned}\mathrm{d}S_1(t) &= S_1(t)\left(r + \lambda_s\mathbf{v}(t) - \mathbb{D}_1\right)\mathrm{d}t + S_1(t)\sqrt{\mathbf{v}(t)}\mathrm{d}B_1(t),\\ \mathrm{d}\mathbf{v}(t) &= \left(k_v - K_v\mathbf{v}(t)\right)\mathrm{d}t + \sigma_v\sqrt{\mathbf{v}(t)}\mathrm{d}\mathbb{B}_2(t),\end{aligned}$$

where $dB_1(t)d\mathbb{B}_2(t) = \rho_{2,1}dt$, and r is the constant interest rate. For the position $\kappa = \{\kappa_b, \kappa_f\}$ in bonds and forward contracts on variance $\mathbf{v}$, what is the CARA neutral price (up to solving ODEs) of a single bond and single forward contract with terminal payoffs equal to $\upsilon(\mathbf{v}) = \{1, \mathbf{v} - \mathfrak{k}\}$?

4.14 Case Study: Stochastic Interest Rates (Continuation: CRRA Neutral Pricing)

4.14.1 Bonds

We continue Sect. 3.11. As in Sect. 2.4.5, we assume the interest rate model (2.4.23), with $S = \{S_1\}$, $A = \{S_1, r\}$, and market coefficients (2.4.27).

We shall price a bond using neutral (or indifference) pricing for the bond position $\kappa = 0$ (if $\kappa = 0$, then neutral and indifference pricing are equivalent). We shall use the CRRA utility of wealth (as in Sect. 3.11), and the "perpetual risk premium" $\lim_{t\to-\infty} g_\gamma^{(0,1)}(t,r) = \mathbb{H}_0 + \mathbb{H}_1 r$, with $\mathbb{H}_0$ and $\mathbb{H}_1$ characterized in (3.11.13) and (3.11.14). Consequently, (4.9.7) for $V = \{V_1(t,r)\}$ (assuming, since it is a bond that we are pricing, that the price of the contract does not depend on S_1) becomes

$$(\mathbb{H}_0 + r\mathbb{H}_1)\left(1-\rho_{2,1}^2\right)V_1^{(0,1)}(t,r)w^2 - rV_1(t,r) + \left(q_0 + rq_1 - \frac{w(\beta r - r + \mathbf{a}_0)\rho_{2,1}}{\mathbf{p}_1}\right)$$
$$V_1^{(0,1)}(t,r) + \frac{1}{2}w^2 V_1^{(0,2)}(t,r) + V_1^{(1,0)}(t,r) = 0 \tag{4.14.1}$$

for $t < T$, with the terminal condition (bond payoff) $V_1(T,r) = 1$. Looking for the solution of (4.14.1) in the form

$$V_1(t,r) := e^{\mathbb{A}_0(t) + r\mathbb{A}_1(t)}, \tag{4.14.2}$$

we see that (4.14.1) becomes

$$e^{\mathbb{A}_0(t)+r\mathbb{A}_1(t)}(\mathbb{H}_0 + r\mathbb{H}_1)\left(1-\rho_{2,1}^2\right)\mathbb{A}_1(t)w^2 - e^{\mathbb{A}_0(t)+r\mathbb{A}_1(t)}r + e^{\mathbb{A}_0(t)+r\mathbb{A}_1(t)}$$
$$\left(q_0 + rq_1 - \frac{w(\beta r - r + \mathbf{a}_0)\rho_{2,1}}{\mathbf{p}_1}\right)\mathbb{A}_1(t) + \frac{1}{2}\left(e^{\mathbb{A}_0(t)+r\mathbb{A}_1(t)}w^2\rho_{2,1}^2\mathbb{A}_1(t)^2\right.$$
$$\left. + e^{\mathbb{A}_0(t)+r\mathbb{A}_1(t)}w^2\left(1-\rho_{2,1}^2\right)\mathbb{A}_1(t)^2\right) + e^{\mathbb{A}_0(t)+r\mathbb{A}_1(t)}\left(\mathbb{A}_0'(t) + r\mathbb{A}_1'(t)\right) = 0$$
$$\tag{4.14.3}$$

for $t < T$, with terminal conditions $\mathbb{A}_0(T) = \mathbb{A}_1(T) = 0$ and for $-\infty < r < \infty$. Consequently, we arrive at two ODEs characterizing $\mathbb{A}_0(t)$ and $\mathbb{A}_1(t)$:

$$\frac{1}{2}\mathbb{A}_1(t)^2 w^2 + \left(q_0 - w\left(\frac{\mathbf{a}_0}{\mathbf{p}_1}\rho_{2,1} + w\mathbb{H}_0\left(\rho_{2,1}^2 - 1\right)\right)\right)\mathbb{A}_1(t) + \mathbb{A}_0'(t) = 0 \quad (4.14.4)$$

and

$$\mathbb{A}_1'(t) + \left(q_1 + w\left((1-\beta)\frac{\rho_{2,1}}{\mathbf{p}_1} + w\mathbb{H}_1\left(1 - \rho_{2,1}^2\right)\right)\right)\mathbb{A}_1(t) - 1 = 0 \quad (4.14.5)$$

for $t < T$, with terminal conditions $\mathbb{A}_0(T) = \mathbb{A}_1(T) = 0$. Equation (4.14.5) is solved, and we obtain

$$\mathbb{A}_1(t) = \frac{1 - e^{(T-t)\alpha}}{\alpha} \quad (4.14.6)$$

for

$$\alpha = q_1 + w\left((1-\beta)\frac{\rho_{2,1}}{\mathbf{p}_1} + w\mathbb{H}_1\left(1 - \rho_{2,1}^2\right)\right), \quad (4.14.7)$$

and then (4.14.4) is solved, and we obtain

$$\begin{aligned}\mathbb{A}_0(t) = \frac{1}{\alpha^3}\Big(&w\alpha\left(-(T-t)\alpha - 1 + e^{(T-t)\alpha}\right)\frac{\mathbf{a}_0}{\mathbf{p}_1}\rho_{2,1} - \frac{w^2}{4}\big(-2(T-t)\alpha - 3\\ &-e^{2(T-t)\alpha} + 4e^{(T-t)\alpha}\big) - \alpha\left(-(T-t)\alpha - 1 + e^{(T-t)\alpha}\right)\big(q_0 - w^2\mathbb{H}_0\\ &(\rho_{2,1}^2 - 1)\big)\Big). \end{aligned} \quad (4.14.8)$$

In the simpler case $\beta = 1$, using (3.11.15) and (3.11.16), we conclude from (4.14.7) that $\alpha = q_1$ and that therefore the price of the bond in such a case is given by (4.14.2), where $\mathbb{A}_0(t)$ and $\mathbb{A}_1(t)$ are explicitly given as

$$\begin{aligned}\mathbb{A}_0(t) = \frac{1}{q_1^3}\Big(&wq_1\left(-(T-t)q_1 - 1 + e^{(T-t)q_1}\right)\frac{\mathbf{a}_0}{\mathbf{p}_1}\rho_{2,1} - \frac{w^2}{4}\big(-2(T-t)q_1\\ &-3 - e^{2(T-t)q_1} + 4e^{(T-t)q_1}\big) - q_1\left(-(T-t)q_1 - 1 + e^{(T-t)q_1}\right)\big(q_0 - w^2\\ &\frac{\gamma-1}{q_1}(\rho_{2,1}^2 - 1)\big)\Big) \end{aligned} \quad (4.14.9)$$

and

$$\mathbb{A}_1(t) = \frac{1 - e^{(T-t)q_1}}{q_1}. \quad (4.14.10)$$

We therefore have the following result.

Proposition 4.14.1. *In the above context, for the CRRA utility of wealth and position $\kappa = 0$, the neutral/indifference price of a bond with unit payoff is given by (4.14.2), where $\mathbb{A}_0(t)$ is given in (4.14.8), $\mathbb{A}_1(t)$ is given in (4.14.6), and α is given in (4.14.7). If $\beta = 1$, the price of the bond is given by (4.14.2), where $\mathbb{A}_0(t)$ and $\mathbb{A}_1(t)$ are given by (4.14.9) and (4.14.10), respectively.*

4.14.2 Forward Contract Prices, Forwards, and Futures Under Stochastic Interest Rates, the Comparison of Forwards and Futures

We now calculate the forward contract prices, forward prices, and futures prices under the stochastic interest rate model (2.4.23), with $S = \{S_1\}$, $A = \{S_1, r\}$ and market coefficients (2.4.27) for the position $\kappa = 0$, CRRA utility of wealth, and $\beta = 1$.

We shall first calculate the price of the forward contract with the delivery price $\mathfrak{k}$. The pricing PDE (4.9.7) becomes

$$V_1^{(1,0,0)}(t,S_1,r) + \frac{1}{2}\Big(V_1^{(0,0,2)}(t,S_1,r)\,w^2 + 2S_1\mathbf{p}_1\rho_{2,1}V_1^{(0,1,1)}(t,S_1,r)\,w$$
$$+S_1^2\mathbf{p}_1^2V_1^{(0,2,0)}(t,S_1,r)\Big) + (r-\mathbb{D})S_1V_1^{(0,1,0)}(t,S_1,r) + \Big(q_0 + rq_1$$
$$-\frac{\mathbf{w}\mathbf{a}_0\rho_{2,1}}{\mathbf{p}_1}\Big)V_1^{(0,0,1)}(t,S_1,r) + \left(1-\rho_{2,1}^2\right)g^{(0,0,1)}(t,S_1,r)$$
$$V_1^{(0,0,1)}(t,S_1,r)\,w^2 - rV_1(t,S_1,r) = 0 \tag{4.14.11}$$

for $t < T$, with the terminal condition $V_1(T,S_1,r) = S_1 - \mathfrak{k}$. Recall that under the above assumption, using (3.11.9), we have

$$g^{(0,0,1)}(t,S_1,r) = \mathbb{G}_1(t) = (\gamma-1)\frac{1-\mathrm{e}^{(T-t)q_1}}{q_1}, \tag{4.14.12}$$

so that the pricing equation (4.14.11) is equal to

$$V_1^{(1,0,0)}(t,S_1,r) + \frac{1}{2}\Big(V_1^{(0,0,2)}(t,S_1,r)\,w^2 + 2S_1\mathbf{p}_1\rho_{2,1}V_1^{(0,1,1)}(t,S_1,r)\,w + S_1^2$$
$$\mathbf{p}_1^2V_1^{(0,2,0)}(t,S_1,r)\Big) + (r-\mathbb{D})S_1V_1^{(0,1,0)}(t,S_1,r) + \Big(q_0 + rq_1$$
$$-\frac{\mathbf{w}\mathbf{a}_0\rho_{2,1}}{\mathbf{p}_1}\Big)V_1^{(0,0,1)}(t,S_1,r) + \left(1-\rho_{2,1}^2\right)(\gamma-1)\frac{1-\mathrm{e}^{(T-t)q_1}}{q_1}$$
$$V_1^{(0,0,1)}(t,S_1,r)\,w^2 - rV_1(t,S_1,r) = 0 \tag{4.14.13}$$

for $t < T$, with the terminal condition $V_1(T, S_1, r) = S_1 - \text{ŧ}$. We look for the solution of the pricing equation (4.14.13) in the form

$$V_1(t, S_1, r) = f(t)S_1 - \text{ŧ}e^{\mathbb{A}_0(t)+\mathbb{A}_1(t)r}, \tag{4.14.14}$$

obtaining

$$\begin{aligned}
&-\frac{1}{2}e^{\mathbb{A}_0(t)+r\mathbb{A}_1(t)}\text{ŧ}\mathbb{A}_1(t)^2 w^2 - \frac{1}{q_1}e^{\mathbb{A}_0(t)+r\mathbb{A}_1(t)}\left(1 - e^{(T-t)q_1}\right)\text{ŧ}(\gamma - 1)\left(1 - \rho_{2,1}^2\right)\\
&\mathbb{A}_1(t)w^2 + (r - \mathbb{D})f(t)S_1 - r\left(f(t)S_1 - e^{\mathbb{A}_0(t)+r\mathbb{A}_1(t)}\text{ŧ}\right) - e^{\mathbb{A}_0(t)+r\mathbb{A}_1(t)}\text{ŧ}\\
&\left(q_0 + rq_1 - \frac{w\mathbf{a}_0\rho_{2,1}}{\mathbf{p}_1}\right)\mathbb{A}_1(t) + S_1 f'(t) - e^{\mathbb{A}_0(t)+r\mathbb{A}_1(t)}\text{ŧ}\left(\mathbb{A}_0'(t) + r\mathbb{A}_1'(t)\right) = 0
\end{aligned} \tag{4.14.15}$$

for $t < T$, $S_1 > 0$, and $-\infty < r < \infty$, with the terminal conditions $f(T) = 1$, $\mathbb{A}_0(T) = \mathbb{A}_1(T) = 0$, and therefore we arrive at three ODEs:

$$\begin{aligned}
&f'(t) - \mathbb{D}f(t) = 0,\\
&-\frac{1}{2q_1\mathbf{p}_1}e^{\mathbb{A}_0(t)+r\mathbb{A}_1(t)}\text{ŧ}\Big(\mathbf{p}_1\Big(w^2 q_1\mathbb{A}_1(t)^2 + 2\Big((\gamma - 1)\left(\rho_{2,1}^2 - 1\right)\\
&\left(-1 + e^{(T-t)q_1}\right)w^2 + q_0 q_1\Big)\mathbb{A}_1(t) + 2q_1\mathbb{A}_0'(t)\Big) - 2wq_1\mathbf{a}_0\rho_{2,1}\mathbb{A}_1(t)\Big) = 0,\\
&-e^{\mathbb{A}_0(t)+r\mathbb{A}_1(t)}\text{ŧ}\left(q_1\mathbb{A}_1(t) + \mathbb{A}_1'(t) - 1\right) = 0,
\end{aligned} \tag{4.14.16}$$

for $t < T$, with the terminal conditions $f(T) = 1$, $\mathbb{A}_0(T) = \mathbb{A}_1(T) = 0$. The last two ODEs can be simplified, obviously, into

$$\begin{aligned}
&\mathbf{p}_1\Big(w^2 q_1\mathbb{A}_1(t)^2 + 2\Big((\gamma - 1)\left(\rho_{2,1}^2 - 1\right)\left(-1 + e^{(T-t)q_1}\right)w^2 + q_0 q_1\Big)\mathbb{A}_1(t)\\
&\quad + 2q_1\mathbb{A}_0'(t)\Big) - 2wq_1\mathbf{a}_0\rho_{2,1}\mathbb{A}_1(t) = 0,
\end{aligned} \tag{4.14.17}$$

$$q_1\mathbb{A}_1(t) + \mathbb{A}_1'(t) - 1 = 0. \tag{4.14.18}$$

Solving the system (4.14.16)–(4.14.17), we get

$$f(t) = e^{-(T-t)\mathbb{D}}, \tag{4.14.19}$$

$$\mathbb{A}_1(t) = \frac{1 - e^{(T-t)q_1}}{q_1}, \tag{4.14.20}$$

and then

$$\begin{aligned}
\mathbb{A}_0(t) = &-\frac{1}{4q_1^3\mathbf{p}_1}e^{-2tq_1}\Big(\mathbf{p}_1\Big(\left(3e^{2tq_1} + e^{2Tq_1} - 4e^{(t+T)q_1}\right)\Big(2(\gamma - 1)\rho_{2,1}^2\\
&-2\gamma + 1\Big)w^2 + 4e^{2tq_1}(t - T)q_0 q_1^2 - 2e^{tq_1}q_1\left(e^{tq_1}(t - T)\right.
\end{aligned}$$

$$\left(2(\gamma-1)\rho_{2,1}^2-2\gamma+1\right)w^2+2\left(e^{tq_1}-e^{Tq_1}\right)q_0\Big)\Big)$$

$$-4wq_1\left(e^{2tq_1}(t-T)q_1-e^{2tq_1}+e^{(t+T)q_1}\right)\mathbf{a}_0\rho_{2,1}\Big), \tag{4.14.21}$$

and consequently

$$V_1(t,S_1,r)=e^{-(T-t)\mathbb{D}}S_1-\mathfrak{k}\mathrm{Exp}\left[\frac{1}{4q_1^3\mathbf{p}_1}\left(4wq_1\left((t-T)q_1-1+e^{(T-t)q_1}\right)\mathbf{a}_0\rho_{2,1}\right.\right.$$
$$+\mathbf{p}_1\left(-\left(3+e^{2(T-t)q_1}-4e^{(T-t)q_1}\right)\left(2(\gamma-1)\rho_{2,1}^2-2\gamma+1\right)w^2\right.$$
$$-4(t-T)q_0q_1^2+2q_1\left((t-T)\left(2(\gamma-1)\rho_{2,1}^2\right.\right.$$
$$\left.\left.\left.\left.-2\gamma+1\right)\ w^2+2\left(1-e^{(T-t)q_1}\right)q_0\right)\right)\right)+\frac{1-e^{(T-t)q_1}}{q_1}r\Bigg], \tag{4.14.22}$$

the price of the forward contract on the underlying with spot price S_1 and delivery price $\mathfrak{k}$. Solving the equation $V_1(t,S_1,r)=0$ for $\mathfrak{k}$, we get the forward price of the underlying with the spot price S_1 to be equal to

$$F_1(t,S_1,r)=S_1e^{(t-T)\mathbb{D}}\mathrm{Exp}\left[-\frac{\left(1-e^{(T-t)q_1}\right)q_0}{q_1^2}-\frac{\left(1-e^{(T-t)q_1}\right)r}{q_1}+\frac{(t-T)q_0}{q_1}\right.$$
$$-\frac{1}{q_1^2\mathbf{p}_1}w\left(-1+e^{(T-t)q_1}+(t-T)q_1\right)\mathbf{a}_0\rho_{2,1}-\frac{1}{4q_1^3}\left(-3+4e^{(T-t)q_1}\right.$$
$$\left.-e^{2(T-t)q_1}\right)w^2\left(1-2\gamma+2(-1+\gamma)\rho_{2,1}^2\right)-\frac{1}{2q_1^2}(t-T)w^2$$
$$\left.\left(1-2\gamma+2(-1+\gamma)\rho_{2,1}^2\right)\right] \tag{4.14.23}$$

for $t<T$.

Remark 4.14.1. Observe that

$$F_1(t,S_1,r)=\frac{S_1e^{(t-T)\mathbb{D}}}{P(t,r)}, \tag{4.14.24}$$

where $P(t,r)$ is the price of the bond. We note that $P(t,r)$ is somewhat different from the price derived in the previous section, since here we have used the exact form of the risk premium, and not its perpetual version, as was done in the previous section.

On the other hand, rewriting (4.14.13) to account for the contract dividend in the amount of $\mathcal{D} = rV_1$, i.e., to obtain the futures pricing equation, we get (we shall denote by V_2 the futures price)

$$V_2^{(1,0,0)}(t,S_1,r) + \frac{1}{2}\left(V_2^{(0,0,2)}(t,S_1,r)w^2 + 2S_1\mathbf{p}_1\rho_{2,1}V_2^{(0,1,1)}(t,S_1,r)w\right.$$
$$\left.+S_1^2\mathbf{p}_1^2V_2^{(0,2,0)}(t,S_1,r)\right) + (r-\mathbb{D})S_1V_2^{(0,1,0)}(t,S_1,r) + \left(q_0 + rq_1\right.$$
$$\left.-\frac{w\mathbf{a}_0\rho_{2,1}}{\mathbf{p}_1}\right)V_2^{(0,0,1)}(t,S_1,r) + \left(1-\rho_{2,1}^2\right)(\gamma-1)\frac{1-e^{(T-t)q_1}}{q_1}V_2^{(0,0,1)}$$
$$(t,S_1,r)w^2 = 0 \tag{4.14.25}$$

for $t < T$, with the terminal condition $V_2(T,S_1,r) = S_1$. We look for the solution of the futures pricing equation (4.14.25) in the form

$$V_2(t,S_1,r) = S_1 e^{\mathbb{B}_0(t)+\mathbb{B}_1(t)r}, \tag{4.14.26}$$

obtaining

$$\frac{1}{q_1}e^{\mathbb{B}_0(t)+r\mathbb{B}_1(t)}\left(1-e^{(T-t)q_1}\right)(\gamma-1)S_1\left(1-\rho_{2,1}^2\right)\mathbb{B}_1(t)w^2 + e^{\mathbb{B}_0(t)+r\mathbb{B}_1(t)}$$
$$(r-\mathbb{D})S_1 + e^{\mathbb{B}_0(t)+r\mathbb{B}_1(t)}S_1\left(q_0 + rq_1 - \frac{w\mathbf{a}_0\rho_{2,1}}{\mathbf{p}_1}\right)\mathbb{B}_1(t)$$
$$+\frac{1}{2}\left(e^{\mathbb{B}_0(t)+r\mathbb{B}_1(t)}w^2S_1\mathbb{B}_1(t)^2 + 2e^{\mathbb{B}_0(t)+r\mathbb{B}_1(t)}wS_1\mathbf{p}_1\rho_{2,1}\mathbb{B}_1(t)\right)$$
$$+e^{\mathbb{B}_0(t)+r\mathbb{B}_1(t)}S_1\left(\mathbb{B}_0'(t) + r\mathbb{B}_1'(t)\right) = 0 \tag{4.14.27}$$

for $t < T, S_1 > 0$, and $-\infty < r < \infty$, with the terminal conditions $\mathbb{B}_0(T) = \mathbb{B}_1(T) = 0$, and therefore we arrive at two ODEs:

$$\frac{1}{2q_1\mathbf{p}_1}e^{\mathbb{B}_0(t)+r\mathbb{B}_1(t)}\left(2\left(-1+e^{(T-t)q_1}\right)(\gamma-1)\mathbf{p}_1\left(\rho_{2,1}^2-1\right)\mathbb{B}_1(t)w^2 + q_1\right.$$
$$\left(2w\rho_{2,1}\mathbb{B}_1(t)\mathbf{p}_1^2 + \left(w^2\mathbb{B}_1(t)^2 + 2q_0\mathbb{B}_1(t) - 2\mathbb{D} + 2\mathbb{B}_0'(t)\right)\mathbf{p}_1 - 2\right.$$
$$\left.\left.w\mathbf{a}_0\rho_{2,1}\mathbb{B}_1(t)\right)\right) = 0 \tag{4.14.28}$$

and

$$e^{\mathbb{B}_0(t)+r\mathbb{B}_1(t)}\left(q_1\mathbb{B}_1(t) + \mathbb{B}_1'(t) + 1\right) = 0, \tag{4.14.29}$$

which simplify to

$$2\left(-1+e^{(T-t)q_1}\right)(\gamma-1)\mathbf{p}_1\left(\rho_{2,1}^2-1\right)\mathbb{B}_1(t)w^2+q_1\left(2w\rho_{2,1}\mathbb{B}_1(t)\mathbf{p}_1^2+\left(w^2\mathbb{B}_1(t)^2\right.\right.$$
$$\left.\left.+2q_0\mathbb{B}_1(t)-2\mathbb{D}+2\mathbb{B}_0'(t)\right)\mathbf{p}_1-2w\mathbf{a}_0\rho_{2,1}\mathbb{B}_1(t)\right)=0 \tag{4.14.30}$$

and

$$q_1\mathbb{B}_1(t)+\mathbb{B}_1'(t)+1=0 \tag{4.14.31}$$

for $t<T$, with the terminal conditions $\mathbb{B}_0(T)=\mathbb{B}_1(T)=0$. Solving (4.14.31), we get

$$\mathbb{B}_1(t)=\frac{e^{(T-t)q_1}-1}{q_1} \tag{4.14.32}$$

and then

$$\begin{aligned}\mathbb{B}_0(t)=&-\frac{1}{2q_1^2\mathbf{p}_1}\left(-2w\left((t-T)q_1+e^{(T-t)q_1}-1\right)\rho_{2,1}\mathbf{p}_1^2\right.\\&+\left(-\frac{1}{2q_1}\left(-2(t-T)\,q_1-4e^{(T-t)q_1}+e^{2(T-t)q_1}+3\right)\right.\\&\left(2(\gamma-1)\rho_{2,1}^2-2\gamma+3\right)w^2-2t\mathbb{D}q_1^2\\&\left.+2T\mathbb{D}q_1^2-2q_0\left((t-T)q_1+e^{(T-t)q_1}-1\right)\right)\mathbf{p}_1\\&\left.+2w\left((t-T)q_1+e^{(T-t)q_1}-1\right)\mathbf{a}_0\rho_{2,1}\right).\end{aligned} \tag{4.14.33}$$

Summarizing, the futures pricing equation (4.14.25) admits the solutions

$$\begin{aligned}V_2(t,S_1,r)=S_1\mathrm{Exp}\Bigg[&-\frac{1}{2q_1^2\mathbf{p}_1}\left(-2w\left((t-T)q_1+e^{(T-t)q_1}-1\right)\rho_{2,1}\mathbf{p}_1^2\right.\\&+\left(-\frac{1}{2q_1}\left(-2(t-T)q_1-4e^{(T-t)q_1}+e^{2(T-t)q_1}+3\right)\right.\\&\left(2(\gamma-1)\rho_{2,1}^2-2\gamma+3\right)w^2-2t\mathbb{D}q_1^2+2T\mathbb{D}q_1^2\\&\left.-2\,q_0\left((t-T)q_1+e^{(T-t)q_1}-1\right)\right)\mathbf{p}_1+2w((t-T)q_1\\&\left.+e^{(T-t)q_1}-1\right)\mathbf{a}_0\rho_{2,1}\Big)+\frac{e^{(T-t)q_1}-1}{q_1}r\Bigg].\end{aligned} \tag{4.14.34}$$

It is anticipated, since this is a stochastic interest rate model, that the forward price will be different from the futures price, in contrast to the situation in the deterministic interest rate case.

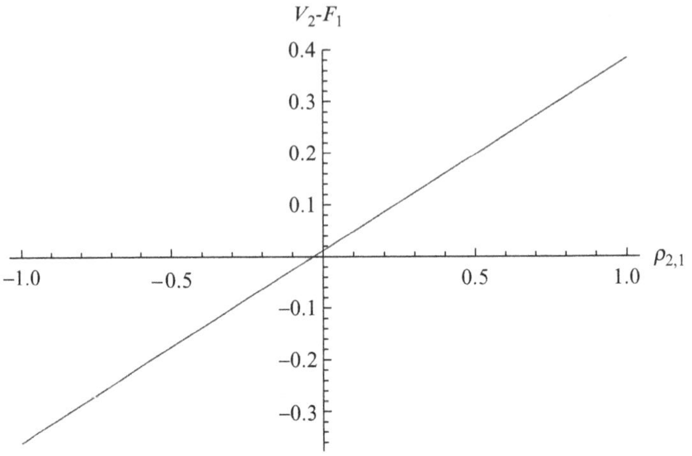

Fig. 4.1 Difference between futures and forwards

The stochasticity of the interest rate can be eliminated by setting $w = 0$ (see (2.4.23). So by setting $w = 0$ in (4.14.23) and (4.14.34), we get

$$F_1(t, S_1, r) = \mathrm{e}^{-\frac{\left(1-\mathrm{e}^{(T-t)q_1}\right)r}{q_1}+(t-T)\left(\mathbb{D}+\frac{q_0}{q_1}\right)-\frac{\left(1-\mathrm{e}^{(T-t)q_1}\right)q_0}{q_1^2}} S_1 = V_2(t, S_1, r) \quad (4.14.35)$$

as anticipated. On the other hand, if $w \neq 0$, then $F_1(t, S_1, r) \neq V_2(t, S_1, r)$. For example, for

$$\begin{aligned} S_1 &= 900, T = 1, t = 0, w = 0.00638909, q_0 = 0.00224622, \\ q_1 &= -0.056621, \mathbf{a}_0 = 0.0432591, \gamma = 10, \mathbf{p}_1 = 0.130523, \\ \mathbb{D} &= 0.01, r = 0.02, \end{aligned} \quad (4.14.36)$$

in Fig. 4.1 we plot the difference between futures prices and forward prices $(V_2 - F_1)$ as a function of the instantaneous correlation between the underlying tradable and the short interest rate.

Remark 4.14.3. If, for example, $\rho_{2,1}$ is close to 1, $\mathrm{d}S_1$ and $\mathrm{d}r$ are highly correlated, so that $\mathrm{d}S_1 > 0$, which causes $\mathrm{d}V_2 > 0$, and which also causes a positive cash flow at the end of the trading day, and also causes $\mathrm{d}r > 0$, so that such a cash flow is compensated at a higher rate. Conversely, if $\mathrm{d}S_1 < 0$, then $\mathrm{d}V_2 < 0$ is penalized at a lower rate, since by correlation, $\mathrm{d}r < 0$. Therefore, if $\rho_{2,1}$ is close to 1, then $V_2(t, S_1, r) - F_1(t, S_1, r) > 0$, as Fig. 4.1 shows more transparently.

4.15 Stochastic Interest, Dividend, and Appreciation Rate Model: Part 3 (CRRA Neutral Pricing of Futures)

Continuing Sect. 2.4.8 and Sect. 3.12, following [21], we solve the problem of zero-position ($\kappa = 0$) CRRA pricing of a futures contract on the underlying with price S_1 for the market model (2.4.39)–(2.4.42). Denote the price of such a contract by V_1.

Equation (4.9.7), after dropping the zero-order term for the sake of pricing a futures contract, with the terminal condition $V_1(T, S_1, \mathbf{v}, r, \mathbb{D}) = S_1$, and using (3.12.2), admits a solution of the form

$$V_1(t, S_1, \mathbf{v}, r, \mathbb{D}) = e^{g_1(t)+rg_2(t)+\mathbf{v}g_3(t)+\mathbb{D}g_4(t)} S_1. \tag{4.15.1}$$

Moreover, this yields four ODEs characterizing functions $g_1(t), g_2(t), g_3(t)$, and $g_4(t)$:

$$\begin{aligned}
&(\sigma_{\mathbf{r}}\rho_{3,1}g_2(t) + \sigma_v\rho_{2,1}g_3(t) + \sigma_d\rho_{4,1}g_4(t))(\sigma_v\rho_{2,1}\mathbb{H}_1(t) + \sigma_{\mathbf{r}}\rho_{3,1}\mathbb{H}_4(t)) \\
&\quad = \frac{1}{2}g_3(t)^2\sigma_v^2 + g_3(t)\mathbb{H}_1(t)\sigma_v^2 + \sigma_{\mathbf{r}}\rho_{3,2}g_2(t)g_3(t)\sigma_v + \sigma_d\rho_{4,2}g_3(t)g_4(t)\sigma_v \\
&\quad\quad +\sigma_{\mathbf{r}}\rho_{3,2}g_2(t)\mathbb{H}_1(t)\sigma_v + \sigma_d\rho_{4,2}g_4(t)\mathbb{H}_1(t)\sigma_v + \sigma_{\mathbf{r}}\rho_{3,2}g_3(t)\mathbb{H}_4(t)\sigma_v \\
&\quad\quad +\frac{1}{2}\sigma_{\mathbf{r}}^2 g_2(t)^2 + \frac{1}{2}\sigma_d^2 g_4(t)^2 + \alpha_{\mathbf{r}}\theta_{\mathbf{r}}g_2(t) + \alpha_v\theta_v g_3(t) + \alpha_d\theta_d g_4(t) \\
&\quad\quad +\sigma_d\sigma_{\mathbf{r}}\rho_{4,3}g_2(t)g_4(t) + \sigma_S(\sigma_{\mathbf{r}}\rho_{3,1}g_2(t) + \sigma_v\rho_{2,1}g_3(t) + \sigma_d\rho_{4,1}g_4(t)) \\
&\quad\quad +\sigma_{\mathbf{r}}^2 g_2(t)\mathbb{H}_4(t) + \sigma_d\sigma_{\mathbf{r}}\rho_{4,3}g_4(t)\mathbb{H}_4(t) + g_1'(t),
\end{aligned} \tag{4.15.2}$$

$$\begin{aligned}
&\frac{1}{\sigma_S}((\beta - 1)(\sigma_{\mathbf{r}}\rho_{3,1}g_2(t) + \sigma_v\rho_{2,1}g_3(t) + \sigma_d\rho_{4,1}g_4(t)) \\
&\quad +\sigma_S((\rho_{2,1}^2 - 1)g_3(t)\mathbb{H}_3(t)\,\sigma_v^2 + (\sigma_d(\rho_{2,1}\rho_{4,1} - \rho_{4,2})g_4(t)\mathbb{H}_3(t) \\
&\quad +\sigma_{\mathbf{r}}(\rho_{2,1}\rho_{3,1} - \rho_{3,2})(g_2(t)\mathbb{H}_3(t) + 2g_3(t)\mathbb{H}_5(t)))\sigma_v + \alpha_{\mathbf{r}}g_2(t) \\
&\quad -2\sigma_{\mathbf{r}}^2 g_2(t)\mathbb{H}_5(t) + 2\sigma_{\mathbf{r}}^2\rho_{3,1}^2 g_2(t)\mathbb{H}_5(t) + 2\sigma_d\sigma_{\mathbf{r}}\rho_{3,1}\rho_{4,1}g_4(t)\mathbb{H}_5(t) \\
&\quad -2\sigma_d\sigma_{\mathbf{r}}\rho_{4,3}g_4(t)\mathbb{H}_5(t) - g_2'(t) - 1)) = 0,
\end{aligned} \tag{4.15.3}$$

$$\begin{aligned}
&\frac{1}{\sigma_S}(2\sigma_S\rho_{2,1}^2 g_3(t)\mathbb{H}_2(t)\sigma_v^2 - 2\sigma_S g_3(t)\mathbb{H}_2(t)\sigma_v^2 + \rho_{2,1}g_3(t)\sigma_v \\
&\quad +2\sigma_S\sigma_d\rho_{2,1}\rho_{4,1}g_4(t)\mathbb{H}_2(t)\sigma_v - 2\sigma_S\sigma_d\rho_{4,2}g_4(t)\mathbb{H}_2(t)\sigma_v + \alpha_v\sigma_S g_3(t) \\
&\quad +\sigma_d\rho_{4,1}g_4(t) + \sigma_S\sigma_{\mathbf{r}}^2(\rho_{3,1}^2 - 1)g_2(t)\mathbb{H}_3(t) + \sigma_{\mathbf{r}}(\rho_{3,1}(g_2(t)(2\sigma_S\sigma_v\rho_{2,1}\mathbb{H}_2(t) \\
&\quad +1) + \sigma_S(\sigma_v\rho_{2,1}g_3(t) + \sigma_d\rho_{4,1}g_4(t))\mathbb{H}_3(t)) - \sigma_S(\sigma_d\rho_{4,3}g_4(t)\mathbb{H}_3(t) \\
&\quad +\sigma_v\rho_{3,2}(2g_2(t)\mathbb{H}_2(t) + g_3(t)\mathbb{H}_3(t)))) - \sigma_S g_3'(t)) = 0,
\end{aligned} \tag{4.15.4}$$

$$\alpha_d g_4(t) + 1 = g_4'(t), \tag{4.15.5}$$

for $t < T$, with the terminal condition $g_i(T) = 0$, for $i = 1,\ldots,4$. Equation (4.15.5) is easily solved by

$$g_4(t) = -\frac{1 - \mathrm{e}^{-(T-t)\alpha_d}}{\alpha_d}, \tag{4.15.6}$$

while the rest of the ODEs above are again solved numerically. We summarize all of this in the following proposition.

Proposition 4.15.1. *The zero-position ($\kappa = 0$) CRRA price of a futures contract on the underlying with price S_1 for the market model (2.4.39)–(2.4.42) is given by (4.15.1), where $g_4(t)$ is given by (4.15.6, and $g_1(t), g_2(t)$, and $g_3(t)$ are found by solving the system of ODEs (4.15.2)–(4.15.4), where the functions $\mathbb{H}_i(t)$, for $i = 1,\ldots,5$, are found by solving the system of ODEs (3.12.4)–(3.12.8).*

4.16 Circumventing Hypothesis 4.3.1

Here we shall consider only the ($\kappa = 0$) pricing.

The idea is to use the results already derived, such as the fundamental matrix, since much effort was invested in doing so.

From (3.3.2) and (3.4.3), we know that for any wealth utility $\psi(X)$, the optimal portfolio strategy $\Pi^\star$ in the original economy $\mathfrak{E}$ implies the wealth evolution

$$\begin{aligned} \mathrm{d}X^{\Pi^\star}(t) = \mathrm{d}X(t) = &\left(-\left(1 \Big/ \frac{\partial^2\varphi}{\partial X^2}\right)\left(\frac{\partial\varphi}{\partial X}(a_s - r) + \left(\nabla_A \frac{\partial\varphi}{\partial X}\right)\cdot c\cdot\sigma_s^{\mathrm{T}}\right)\right. \\ &\left.\cdot\left(\sigma_s\cdot\sigma_s^{\mathrm{T}}\right)^{-1}\cdot(a_s - r) + rX(t)\right)\mathrm{d}t - \left(1 \Big/ \frac{\partial^2\varphi}{\partial X^2}\right)\left(\frac{\partial\varphi}{\partial X}(a_s - r)\right. \\ &\left. + \left(\nabla_A \frac{\partial\varphi}{\partial X}\right)\cdot c\cdot\sigma_s^{\mathrm{T}}\right)\cdot\left(\sigma_s\cdot\sigma_s^{\mathrm{T}}\right)^{-1}\cdot\sigma_s.\mathrm{d}B(t), \end{aligned} \tag{4.16.1}$$

where $\varphi = \varphi(t,X,A)$ is the solution of $\mathcal{M}(\varphi) = 0$, with $\mathcal{M}$ defined in (3.4.1). If Hypothesis 1 is violated, i.e., if $V = V(t,X,A)$, we declare X to be an additional factor, and to fix ideas, we define an extended, or *auxiliary,* factor vector

$$A_a := \{A, X\} = \{A_1,\ldots,A_m,X\}. \tag{4.16.2}$$

Extending (2.2.2), the auxiliary factor A_a evolves according to

$$\mathrm{d}A_a(t) = b_a(t,A_a(t))\mathrm{d}t + c_a(t,A_a(t))\cdot\mathrm{d}B(t), \tag{4.16.3}$$

with, due to (4.16.1),

$$b_a(t,A_a) = \left\{ b(t,A), -\left(1\Big/\frac{\partial^2\varphi}{\partial X^2}\right)\left(\frac{\partial\varphi}{\partial X}(a_s-r)+\left(\nabla_A\frac{\partial\varphi}{\partial X}\right)\cdot c\cdot\sigma_s^{\mathrm{T}}\right) \cdot\left(\sigma_s\cdot\sigma_s^{\mathrm{T}}\right)^{-1}\cdot(a_s-r)+rX \right\} \tag{4.16.4}$$

and

$$c_a(t,A_a) = \begin{pmatrix} c(t,A) \\ -\left(1\Big/\frac{\partial^2\varphi}{\partial X^2}\right)\left(\frac{\partial\varphi}{\partial X}(a_s-r)+\left(\nabla_A\frac{\partial\varphi}{\partial X}\right)\cdot c\cdot\sigma_s^{\mathrm{T}}\right)\cdot\left(\sigma_s\cdot\sigma_s^{\mathrm{T}}\right)^{-1}\cdot\sigma_s \end{pmatrix}. \tag{4.16.5}$$

Observe that coefficients b_a and c_a are written explicitly in terms of the market coefficients of the original economy (and also in terms of φ); they do not involve the price V. We have the following theorem.

Theorem 4.16.1. *In the above framework, for the zero position $\kappa = 0$, and for the utility of wealth $\psi = \psi(X)$, the ψ-0- neutral/indifference price (per single contract) $V = V_{\psi,\iota,0} = \{V_{\psi,\iota,\kappa,1}(t,X,A),\ldots,\; V_{\psi,\iota,\kappa,l}(t,X,A)\}$ is the solution of the following pricing system of PDEs:*

$$\mathcal{M}(\varphi) = 0 \tag{4.16.6}$$

for $t < T$, together with the terminal condition $\varphi(T,X,A) = \psi(X)$, where the operator $\mathcal{M}$ is defined in (3.4.1), and

$$\begin{aligned}\frac{\partial V}{\partial t} &+ \frac{1}{2}\mathrm{tr}\left(\nabla_{A_a}\nabla_{A_a}V\cdot c_a\cdot c_a^{\mathrm{T}}\right)+\nabla_{A_a}V.\left(b_a-(a_s-r)\cdot\left(\sigma_s\cdot\sigma_s^{\mathrm{T}}\right)^{-1}\cdot\sigma_s\cdot c_a^{\mathrm{T}}\right)\\ &+\left(\nabla_{A_a}\frac{\partial\varphi}{\partial X}\Big/\frac{\partial\varphi}{\partial X}\right)\cdot c_a\cdot\left(\mathbb{I}_n-\sigma_s^{\mathrm{T}}\cdot\left(\sigma_s\cdot\sigma_s^{\mathrm{T}}\right)^{-1}\cdot\sigma_s\right)\cdot c_a^{\mathrm{T}}\cdot\left(\nabla_{A_a}V\right)^{\mathrm{T}}-rV\\ &= -\mathcal{D}\end{aligned} \tag{4.16.7}$$

for $t < T$, together with the terminal condition $V(T,X,A) = \upsilon(A)$.

Remark 4.16.1. We remark that the simplicity of (4.16.1) implies the feasibility of Theorem 4.16.1.

Example 4.16.1. (Indifference pricing for the model (4.12.10).) Set $\kappa = \{0\}$. We now apply Theorem 4.16.1. Equation (4.16.6) reads

$$-\frac{(\mathbf{a}-r)^2\varphi^{(0,1,0,0)}(t,X,S_1,Y)^2}{2\mathbf{p}_1^2}-\left(\frac{1}{\mathbf{p}_1}(\mathbf{a}-r)\rho_{2,1}w(t,Y)\varphi^{(0,1,0,1)}(t,X,S_1,Y)\right.$$
$$\left.+(\mathbf{a}-r)S_1\varphi^{(0,1,1,0)}(t,X,S_1,Y)\right)\varphi^{(0,1,0,0)}(t,X,S_1,Y)$$

$$+rX\varphi^{(0,2,0,0)}(t,X,S_1,Y)\,\varphi^{(0,1,0,0)}(t,X,S_1,Y)+\frac{1}{2}\Big(-\rho_{2,1}w(t,Y)\varphi^{(0,1,0,1)}$$
$$(t,X,S_1,Y)\Big(\rho_{2,1}w(t,Y)\varphi^{(0,1,0,1)}(t,X,S_1,Y)+S_1\mathbf{p}_1\varphi^{(0,1,1,0)}(t,X,S_1,Y)\Big)$$
$$-S_1\mathbf{p}_1\varphi^{(0,1,1,0)}(t,X,S_1,Y)\Big(\rho_{2,1}w(t,Y)\varphi^{(0,1,0,1)}(t,X,S_1,Y)$$
$$+S_1\mathbf{p}_1\varphi^{(0,1,1,0)}(t,X,S_1,Y)\Big)\Big)+\Big(\mathbf{b}(t,Y)\varphi^{(0,0,0,1)}(t,X,S_1,Y)$$
$$+S_1(\mathbf{a}-\mathbb{D}_1)\varphi^{(0,0,1,0)}(t,X,S_1,Y)\Big)\varphi^{(0,2,0,0)}(t,X,S_1,Y)$$
$$+\frac{1}{2}\Big(S_1^2\varphi^{(0,0,2,0)}(t,X,S_1,Y)\mathbf{p}_1^2+2S_1\rho_{2,1}w(t,Y)\varphi^{(0,0,1,1)}(t,X,S_1,Y)\mathbf{p}_1$$
$$+\left(\rho_{2,1}^2w(t,Y)^2+\left(1-\rho_{2,1}^2\right)\;w(t,Y)^2\right)\varphi^{(0,0,0,2)}(t,X,S_1,Y)\Big)\varphi^{(0,2,0,0)}$$
$$(t,X,S_1,Y)+\varphi^{(0,2,0,0)}(t,X,S_1,Y)\,\varphi^{(1,0,0,0)}(t,X,S_1,Y)=0 \tag{4.16.8}$$

for $t<T$, with terminal condition $\varphi(T,X,S_1,Y)=\psi(X)$. Equation (4.16.8) looks complicated, yet it admits a solution in the form $\varphi=\varphi(t,X)$, therefore solving just

$$-\frac{(\mathbf{a}-r)^2\varphi^{(0,1)}(t,X)^2}{2\mathbf{p}_1^2}+rX\varphi^{(0,2)}(t,X)\varphi^{(0,1)}(t,X)+\varphi^{(0,2)}(t,X)\varphi^{(1,0)}(t,X)=0 \tag{4.16.9}$$

for $t<T$, with terminal condition $\varphi(T,X)=\psi(X)$, which can be solved first. This also implies that the pricing equation (4.16.7) reads

$$\frac{1}{2}S_1^2V_1^{(0,0,2,0)}(t,X,S_1,Y)\mathbf{p}_1^2+S_1\rho_{2,1}w(t,Y)V_1^{(0,0,1,1)}(t,X,S_1,Y)\mathbf{p}_1$$
$$-rV_1(t,X,S_1,Y)+\left(\frac{(r-\mathbf{a})\rho_{2,1}w(t,Y)}{\mathbf{p}_1}+\mathbf{b}(t,Y)\right)V_1^{(0,0,0,1)}(t,X,S_1,Y)$$
$$+\frac{1}{2}w(t,Y)^2V_1^{(0,0,0,2)}(t,X,S_1,Y)+S_1(r-\mathbb{D}_1)V_1^{(0,0,1,0)}(t,X,S_1,Y)$$
$$+rXV_1^{(0,1,0,0)}(t,X,S_1,Y)+\Big((r-\mathbf{a})\rho_{2,1}w(t,Y)\varphi^{(0,1)}(t,X)V_1^{(0,1,0,1)}$$
$$(t,X,S_1,Y)\Big)\Big/\Big(\mathbf{p}_1\varphi^{(0,2)}(t,X)\Big)+\Big((r-\mathbf{a})S_1\varphi^{(0,1)}(t,X)V_1^{(0,1,1,0)}(t,X,S_1,Y)\Big)\Big/$$
$$\varphi^{(0,2)}(t,X)+\Big((r-\mathbf{a})^2\varphi^{(0,1)}(t,X)^2V_1^{(0,2,0,0)}(t,X,S_1,Y)\Big)\Big/\Big(2\mathbf{p}_1^2\varphi^{(0,2)}(t,X)^2\Big)$$
$$+V_1^{(1,0,0,0)}(t,X,S_1,Y)=0 \tag{4.16.10}$$

for $t < T$, with terminal condition $V(T,X,S_1,Y) = \upsilon(S_1,Y)$. Assuming $\upsilon(S_1,Y) = \upsilon(Y)$, we get

$$V_1^{(1,0,0)}(t,X,Y) + \frac{1}{2}V_1^{(0,0,2)}(t,X,Y)w(t,Y)^2 + \Big((r-\mathbf{a})\rho_{2,1}\varphi^{(0,1)}(t,X)V_1^{(0,1,1)}$$
$$(t,X,Y)w(t,Y)\Big)/\left(\mathbf{p}_1\varphi^{(0,2)}(t,X)\right) + \left(\frac{(r-\mathbf{a})\rho_{2,1}w(t,Y)}{\mathbf{p}_1} + \mathbf{b}(t,Y)\right)$$
$$V_1^{(0,0,1)}(t,X,Y) + rXV_1^{(0,1,0)}(t,X,Y) + \Big((r-\mathbf{a})^2\varphi^{(0,1)}(t,X)^2V_1^{(0,2,0)}$$
$$(t,X,Y)\Big)\Big/\left(2\mathbf{p}_1^2\varphi^{(0,2)}(t,X)^2\right) - rV_1(t,X,Y) = 0 \tag{4.16.11}$$

for $t < T$, with terminal condition $V(T,X,Y) = \upsilon(Y)$. Assuming furthermore, as in [30], that $r = 0$, (4.16.9) becomes

$$\varphi^{(0,2)}(t,X)\varphi^{(1,0)}(t,X) - \frac{\mathbf{a}^2\varphi^{(0,1)}(t,X)^2}{2\mathbf{p}_1^2} = 0 \tag{4.16.12}$$

for $t < T$, with the terminal condition $\varphi(T,X) = \psi(X)$, while (4.16.11) becomes

$$V_1^{(1,0,0)}(t,X,Y) + \frac{\mathbf{a}^2\varphi^{(0,1)}(t,X)^2V_1^{(0,2,0)}(t,X,Y)}{2\mathbf{p}_1^2\varphi^{(0,2)}(t,X)^2} - \Big(\mathbf{a}\rho_{2,1}\varphi^{(0,1)}(t,X)V_1^{(0,1,1)}$$
$$(t,X,Y)w(t,Y)\Big)\Big/\left(\mathbf{p}_1\varphi^{(0,2)}(t,X)\right) + \frac{1}{2}V_1^{(0,0,2)}(t,X,Y)w(t,Y)^2$$
$$+\left(\mathbf{b}(t,Y) - \frac{\mathbf{a}\rho_{2,1}w(t,Y)}{\mathbf{p}_1}\right)V_1^{(0,0,1)}(t,X,Y) = 0 \tag{4.16.13}$$

for $t < T$, with terminal condition $V(T,X,Y) = \upsilon(Y)$. System (4.16.12)–(4.16.13) characterizes solution of the indifference pricing problem for any utility of wealth $\psi(X)$.

It is interesting to study the degeneracy property of the differential operator in (4.16.13) with respect to the X variable, since we anticipate that (4.16.13) will have an entire solution in the domain of the utility of the wealth function $\psi(X)$, i.e., in $\mathcal{X} = \text{dom}(\psi)$. To do so, we specialize the type of considered wealth utility.

Consider HARA utility (3.2.4), and let

$$\psi(X) = \alpha(\eta + \omega X)^{1-\gamma} \tag{4.16.14}$$

and look for the solution of (4.16.12) in the form $\varphi(t,X) := f(t)\psi(X) = f(t)\alpha(\eta + \omega X)^{1-\gamma}$, which yields the ODE characterizing $f(t)$:

$$\mathbf{a}^2(\gamma-1)f(t) - 2\gamma\mathbf{p}_1^2 f'(t) = 0, \tag{4.16.15}$$

with the terminal condition $f(T) = 1$ (solving for $f(t)$ is not needed). Consequently, (4.16.13) reduces to

$$V_1^{(1,0,0)}(t,X,Y) + \frac{\mathbf{a}^2 V_1^{(0,2,0)}(t,X,Y)(\eta + X\omega)^2}{2\gamma^2\omega^2\mathbf{p}_1^2} + \frac{1}{\gamma\omega\mathbf{p}_1}\mathbf{a}\rho_{2,1}w(t,Y)V_1^{(0,1,1)}$$
$$(t,X,Y)(\eta + X\omega) + \frac{1}{2}w(t,Y)^2 V_1^{(0,0,2)}(t,X,Y) + \left(\mathbf{b}(t,Y) - \frac{\mathbf{a}\rho_{2,1}w(t,Y)}{\mathbf{p}_1}\right)$$
$$V_1^{(0,0,1)}(t,X,Y) = 0 \tag{4.16.16}$$

for $t < T$, for $X \in \mathcal{X} = \mathrm{dom}(\psi)$ ($\mathcal{X}$ is a bit complicated to describe; see (3.2.15), and for appropriate values of Y (due to the unspecified nature of the coefficients in the second SDE in (4.12.10) we do not specify what appropriate values of Y are) and terminal condition $V(T,X,Y) = \upsilon(Y)$.

First, set $\eta = 0$ (see (3.2.17)). Then (4.16.16) becomes

$$V_1^{(1,0,0)}(t,X,Y) + \frac{X^2 V_1^{(0,2,0)}(t,X,Y)\mathbf{a}^2}{2\gamma^2\mathbf{p}_1^2} + \frac{X\rho_{2,1}w(t,Y)V_1^{(0,1,1)}(t,X,Y)\mathbf{a}}{\gamma\mathbf{p}_1}$$
$$+\left(\mathbf{b}(t,Y) - \frac{\mathbf{a}\rho_{2,1}w(t,Y)}{\mathbf{p}_1}\right)V_1^{(0,0,1)}(t,X,Y) + \frac{1}{2}w(t,Y)^2 V_1^{(0,0,2)}(t,X,Y) = 0$$
$$\tag{4.16.17}$$

for $t < T$, $X \in \mathcal{X} = \{X > 0\}$ (indeed, in this case we are back to the CRRA utility (see (3.2.17))), and with the terminal condition $V(T,X,Y) = \upsilon(Y)$. We observe that the nature of the degeneracy of the differential operators

$$V_1| \to \frac{X^2 V_1^{(0,2,0)}(t,X,Y)\mathbf{a}^2}{2\gamma^2\mathbf{p}_1^2} \tag{4.16.18}$$

and

$$V_1| \to \frac{X\rho_{2,1}w(t,Y)V_1^{(0,1,1)}(t,X,Y)\mathbf{a}}{\gamma\mathbf{p}_1} \tag{4.16.19}$$

in the X variable at $X = 0$ ensures that no boundary condition is needed (nor is it possible to impose one). More explicitly, (4.16.17) admits a solution in the form $V_1(t,X,Y) = V_1(t,Y)$, with

$$V_1^{(1,0)}(t,Y) + \frac{1}{2}V_1^{(0,2)}(t,Y)w(t,Y)^2 + \left(\mathbf{b}(t,Y) - \frac{\mathbf{a}\rho_{2,1}w(t,Y)}{\mathbf{p}_1}\right)V_1^{(0,1)}(t,Y) = 0$$
$$\tag{4.16.20}$$

for $t < T$, with the terminal condition $V(T,Y) = \upsilon(Y)$, and Hypothesis 1 is satisfied.

On the other hand, if $\eta = -\gamma, \gamma \to -\infty, -\omega > 0$ (see (3.2.16)), then (4.16.16) becomes

$$V_1^{(1,0,0)}(t,X,Y) + \frac{\mathbf{a}^2 V_1^{(0,2,0)}(t,X,Y)}{2\omega^2 \mathbf{p}_1^2} - \frac{\mathbf{a}\rho_{2,1} w(t,Y) V_1^{(0,1,1)}(t,X,Y)}{\omega \mathbf{p}_1}$$
$$+\frac{1}{2} w(t,Y)^2 V_1^{(0,0,2)}(t,X,Y) + \left(\mathbf{b}(t,Y) - \frac{\mathbf{a}\rho_{2,1} w(t,Y)}{\mathbf{p}_1}\right) V_1^{(0,0,1)}(t,X,Y) = 0 \tag{4.16.21}$$

for $t < T$, $X \in \mathcal{X} = (-\infty, \infty)$ (indeed, in this case we are back to the CARA utility (see (3.2.16))), and with the terminal condition $V(T,X,Y) = \upsilon(Y)$. Notice that above, $-\omega$ corresponds to the absolute risk-aversion parameter ω. We remark that now, as compared with (4.16.18) and (4.16.19), the differential operators

$$V_1| \rightarrow \frac{\mathbf{a}^2 V_1^{(0,2,0)}(t,X,Y)}{2\omega^2 \mathbf{p}_1^2} \tag{4.16.22}$$

and

$$V_1| \rightarrow -\frac{\mathbf{a}\rho_{2,1} w(t,Y) V_1^{(0,1,1)}(t,X,Y)}{\omega \mathbf{p}_1} \tag{4.16.23}$$

are not degenerate in the X variable, which is consistent with the above observation that $X \in \mathcal{X} = (-\infty, \infty)$. Also, as above, (4.16.21) simplifies to

$$V_1^{(1,0)}(t,Y) + \frac{1}{2} V_1^{(0,2)}(t,Y) w(t,Y)^2 + \left(\mathbf{b}(t,Y) - \frac{\mathbf{a}\rho_{2,1} w(t,Y)}{\mathbf{p}_1}\right) V_1^{(0,1)}(t,Y) = 0 \tag{4.16.24}$$

for $t < T$, with the terminal condition $V(T,Y) = \upsilon(Y)$, and Hypothesis 1 is satisfied.

Question 4.16.1. What if Hypothesis 1 is violated in the case of $(\kappa \neq 0)$ pricing?

Chapter 5
Hedging

Portfolio Optimization Approach to Hedging in Incomplete Markets

5.1 Introduction

In the case of complete markets, such as in the classical Black–Scholes case, perfect hedging, which eliminates all randomness in the evolution of the portfolio, is (theoretically) possible. This, of course, is not (even theoretically) possible in an incomplete market. Nevertheless, it is possible to hedge the risk that is correlated with the tradables – partial hedging is possible. So the goal is to have a formula that will coincide with the Black–Scholes formula when the market is complete, while hedging all the risk that is correlated with the available tradables when the market is incomplete. Moreover, the goal is to have a formula applicable to any financial engineering model, i.e., applicable in the context of simple economies, and to be applicable for hedging of any portfolio of contracts.

In this chapter we present such a formula. It is called the *most conservative hedging formula*, and it was introduced by the author in [40]. It is completely consistent with the classical Black–Scholes hedging formula, as well as with well-known hedging formulas in some particular incomplete market models. As a matter of fact, after having developed the machinery in Chap. 4 (and in particular, the fundamental matrix of derivative pricing and hedging (Theorem 4.6.1)), and also in Chap. 3, it is now a simple observation to derive the most conservative hedging formula.

It was more complicated to derive a general, relaxed hedging formula, applicable if an investor wishes to hedge less for the sake of the possibility of greater profit. In this chapter we shall present such results as well.

As a matter of fact, we could have done all that in Chap. 4, in parallel with deriving the pricing results, but since the calculations were already quite complicated, and since the calculations to be done for (relaxed) hedging are also somewhat complicated, those two efforts, pricing and hedging derivations, were separated mostly for pedagogical reasons.

S. Stojanovic, *Neutral and Indifference Portfolio Pricing, Hedging and Investing: With Applications in Equity and FX*, DOI 10.1007/978-0-387-71418-9_5,

Remark 5.1.1. Throughout this chapter we shall assume Hypothesis 4.3.1, and we refer the reader to Sect. 4.16 on how to circumvent that hypothesis.

5.2 Derivation of the Most Conservative Hedging Formula

The classical Black–Scholes formula can be derived also as the wealth volatility, or equivalently, wealth-variance minimization portfolio strategy. Indeed, since Black–Scholes hedging yields the elimination of randomness, the resulting wealth volatility is zero, i.e., the volatility minimum is achieved. The advantage of wealth-volatility minimization over the classical Black–Scholes randomness elimination is that it is applicable also in the full generality of incomplete markets, and, as we shall describe here, in the generality of simple economies and arbitrary portfolios of hedged financial contracts.

The wealth-volatility minimization problem was discussed and solved in Remark 3.7.2. Indeed, consider a simple economy $\mathfrak{E}$, as in Definition 2.2.1, with k tradables $S(t) = \{S_1(t),\ldots,S_k(t)\}$ and m factors $A(t) = \{A_1(t),\ldots,A_m(t)\}$, assume the market dynamics (2.2.1)–(2.2.2), and consider the portfolio constraint (3.3.4). Then according to Remark 3.7.2, the wealth-volatility-minimization portfolio strategy is given by formula (3.7.18). We shall use that result to derive the most conservative hedging formula (recall (3.7.12) – "most conservative" refers to the fact that $\gamma \to \infty$).

Consider a self-financing portfolio strategy $\Pi(t,X,A) = \{\Pi_1,\ldots,\Pi_k,\Pi_{k+1},\ldots,\Pi_{k+l}\}$ in the auxiliary economy $\mathfrak{E}_a$, where $\{\Pi_1,\ldots,\Pi_k\}$ is the cash value of the investments into the market S, and $\{\Pi_{k+1},\ldots,\Pi_{k+l}\}$ is the cash value of the investments into the considered contracts, with previously determined (vector) price $V = V(t,A)$.

Consider constraint (4.3.5) with (4.3.6) and (4.3.7), and recall formula (4.8.9). We observe that they were all established in [40], and that they were established there in the current (hedging) context.

Applying (4.8.9), we conclude (recall that $V\kappa = \{V_1,\ldots,V_l\}\{\kappa_1,\ldots,\kappa_l\} := \{V_1\kappa_1,\ldots,V_l\kappa_l\}$) that

$$
\begin{aligned}
&\left(\sigma_{\mathrm{a}}\cdot\sigma_{\mathrm{a}}^{\mathrm{T}}\right)^{-1}\cdot\mu^{\mathrm{T}}\cdot\left(\mu\cdot\left(\sigma_{\mathrm{a}}\cdot\sigma_{\mathrm{a}}^{\mathrm{T}}\right)^{-1}\cdot\mu^{\mathrm{T}}\right)^{-1}\cdot\xi \\
&= \begin{pmatrix} -\left(\sigma_s\cdot\sigma_s^{\mathrm{T}}\right)^{-1}\cdot\sigma_s\cdot c^{\mathrm{T}}\cdot(\nabla_A V)^{\mathrm{T}}\cdot\left(\frac{1}{V}\mathbb{I}_l\right) \\ \mathbb{I}_l \end{pmatrix}\cdot(V\kappa) \\
&= \left\{-\left(\sigma_s\cdot\sigma_s^{\mathrm{T}}\right)^{-1}\cdot\sigma_s\cdot c^{\mathrm{T}}\cdot(\nabla_A V)^{\mathrm{T}}\cdot\left(\frac{1}{V}\mathbb{I}_l\right)\cdot(V\kappa),\mathbb{I}_l\cdot(V\kappa)\right\} \\
&= \left\{-\left(\sigma_s\cdot\sigma_s^{\mathrm{T}}\right)^{-1}\cdot\sigma_s\cdot c^{\mathrm{T}}\cdot(\nabla_A V)^{\mathrm{T}}\cdot\kappa, V\kappa\right\},
\end{aligned}
\tag{5.2.1}
$$

and therefore we may state the following theorem.

Theorem 5.2.1 (Stojanovic [40]). *The most conservative hedging of* $\kappa = \{\kappa_1, \ldots, \kappa_l\}$ *contracts with prices* $V(t,A) = \{V_1(t,A), \ldots, V_l(t,A)\}$*, in any simple economy* $\mathfrak{E}$*, is given by the formula*

$$\Pi^{a,\star}(t,X,A) = \left\{-\left(\sigma_s \cdot \sigma_s^{\mathrm{T}}\right)^{-1} \cdot \sigma_s \cdot c^{\mathrm{T}} \cdot (\nabla_A V)^{\mathrm{T}} \cdot \kappa, V\kappa\right\}, \tag{5.2.2}$$

i.e., by

$$\Pi^{\mathbf{h}}(t,X,A) = -\left(\sigma_s \cdot \sigma_s^{\mathrm{T}}\right)^{-1} \cdot \sigma_s \cdot c^{\mathrm{T}} \cdot (\nabla_A V)^{\mathrm{T}} \cdot \kappa = -\kappa \cdot \nabla_A V \cdot c \cdot \sigma_s^{\mathrm{T}} \cdot \left(\sigma_s \cdot \sigma_s^{\mathrm{T}}\right)^{-1} \tag{5.2.3}$$

for $t < T$.

Remark 5.2.1. The market risk-aversion, i.e., the nature of the wealth utility and the intensity of the risk-aversion, affects the (most conservative) hedging formula (5.2.3) only through the price V.

Example 5.2.1. (Log-normal market – the Black–Scholes hedging formula.) In the log-normal (Black–Scholes) market, as discussed in Sect. 2.4.1, we have $S = A = \{S_1\}$, and the (relevant to hedging) market coefficients are given by

$$\sigma_s = (\mathbf{p}_1), c = (S_1 \mathbf{p}_1), \tag{5.2.4}$$

yielding

$$-\left(\sigma_s \cdot \sigma_s^{\mathrm{T}}\right)^{-1} \cdot \sigma_s \cdot c^{\mathrm{T}} = (-S_1), \tag{5.2.5}$$

and then if we are hedging a single contract, by setting $V = \{V_1(t,S_1)\}, \kappa = \{k_1\}$ in (5.2.3), we conclude that

$$\begin{aligned}\Pi^{\mathbf{h}}(t,X,A) &= \Pi^{\mathbf{h}}(t,A) = \Pi^{\mathbf{h}}(t,Y)\\ &= -\left(\sigma_s \cdot \sigma_s^{\mathrm{T}}\right)^{-1} \cdot \sigma_s \cdot c^{\mathrm{T}} \cdot (\nabla_A V)^{\mathrm{T}} \cdot \kappa = (-S_1) \cdot \left\{k_1 V_1^{(0,1)}(t,S_1)\right\}\\ &= \left\{-k_1 S_1 V_1^{(0,1)}(t,S_1)\right\},\end{aligned} \tag{5.2.6}$$

which is Black–Scholes hedging. Indeed, setting $k_1 = 1$, a single contract is hedged by $-V_1^{(0,1)}(t,S_1)$ units of the underlying with spot price S_1, i.e., the hedging position has the value $-S_1 V_1^{(0,1)}(t,S_1)$, which is the meaning of (5.2.6).

Example 5.2.2 (Case study: stochastic volatility – hedging.). We continue Sects. 3.10 and 4.13. As in Sect. 2.4.3, we assume the stochastic volatility model (2.4.9), with $S = \{S_1\}$, $A = \{S_1, \mathbf{v}\}$, and market coefficients (2.4.13). Consequently,

$$\begin{aligned}\left(\sigma_s \cdot \sigma_s^{\mathrm{T}}\right)^{-1} &= \left(\frac{1}{\mathbf{v}}\right),\\ \sigma_s \cdot c^{\mathrm{T}} &= \left(\mathbf{v}S_1 \quad \mathbf{v}\sigma_v\rho_{2,1}\right),\\ -\left(\sigma_s \cdot \sigma_s^{\mathrm{T}}\right)^{-1} \cdot \sigma_s \cdot c^{\mathrm{T}} &= \left(-S_1 \quad -\sigma_v\rho_{2,1}\right),\end{aligned} \tag{5.2.7}$$

and if we are hedging a portfolio of two kinds of contracts, or more precisely a position $\kappa = \{k_1, k_2\}$, we can set $V = \{V_1(t, S_1, \mathbf{v}), V_2(t, S_1, \mathbf{v})\}$ in (5.2.2) to conclude that

$$\begin{aligned}\Pi^{a,\star}(t,A) &= \left\{-\left(\sigma_{s}.\sigma_s^{\mathrm{T}}\right)^{-1}\cdot\sigma_{s}.c^{\mathrm{T}}\cdot(\nabla_A V)^{\mathrm{T}}\cdot\kappa, V\kappa\right\} \\ &= \left\{\left\{-k_1\left(\sigma_v\rho_{2,1}V_1^{(0,0,1)}(t,S_1,\mathbf{v})+S_1V_1^{(0,1,0)}(t,S_1,\mathbf{v})\right)\right.\right. \\ &\qquad \left. -k_2\left(\sigma_v\rho_{2,1}V_2^{(0,0,1)}(t,S_1,\mathbf{v})+S_1V_2^{(0,1,0)}(t,S_1,\mathbf{v})\right)\right\}, \\ &\qquad \left.\{k_1V_1(t,S_1,\mathbf{v}),k_2V_2(t,S_1,\mathbf{v})\}\right\}. \end{aligned} \tag{5.2.8}$$

Example 5.2.3 (Stochastic volatility: the alternative model.). In the stochastic volatility model, as discussed in Sect. 2.4.4, we have $S = \{S_1\}, A = \{S_1, \mathbf{v}\}$, and the market coefficients are given in (2.4.20). Consequently,

$$\begin{aligned}\left(\sigma_s\cdot\sigma_s^{\mathrm{T}}\right)^{-1} &= \left(\frac{1}{\mathbf{v}^2}\right) \\ \sigma_s\cdot c^{\mathrm{T}} &= \left(\mathbf{v}^2 S_1 \quad \mathbf{v}^2\sigma_v\rho_{2,1}\right) \\ -\left(\sigma_s\cdot\sigma_s^{\mathrm{T}}\right)^{-1}\cdot\sigma_s\cdot c^{\mathrm{T}} &= \left(-S_1 \quad -\sigma_v\rho_{2,1}\right), \end{aligned} \tag{5.2.9}$$

and if we are hedging a position $\kappa = \{k_1, k_2\}$, we again get formula (5.2.8).

Example 5.2.4. (Musiela and Zariphopoulou model; see [30].) As discussed in Sect. 4.12.3, we have $S = \{S_1\}, A = \{S_1, Y\}$, and the market coefficients are given in (4.12.11). Consequently,

$$\begin{aligned}\left(\sigma_s\cdot\sigma_s^{\mathrm{T}}\right)^{-1} &= \left(\frac{1}{\mathbf{p}_1^2}\right), \\ \sigma_s\cdot c^{\mathrm{T}} &= \left(S_1\mathbf{p}_1^2 \quad \mathbf{p}_1\rho_{2,1}w(t,Y)\right), \\ -\left(\sigma_s\cdot\sigma_s^{\mathrm{T}}\right)^{-1}\cdot\sigma_s\cdot c^{\mathrm{T}} &= \left(-S_1 \quad -\frac{\rho_{2,1}w(t,Y)}{\mathbf{p}_1}\right), \end{aligned} \tag{5.2.10}$$

and if we are hedging a position $\kappa = \{k_1\}$, we conclude, assuming that price does not depend on X, that

$$\begin{aligned}\Pi^{a,\star}(t,A) &= \left\{-\left(\sigma_s\cdot\sigma_s^{\mathrm{T}}\right)^{-1}\cdot\sigma_s\cdot c^{\mathrm{T}}\cdot(\nabla_A V)^{\mathrm{T}}\cdot\kappa, V\kappa\right\} \\ &= \left\{\left\{-k_1\left(\frac{1}{\mathbf{p}_1}\rho_{2,1}w(t,Y)V_1^{(0,0,1)}(t,S_1,Y)\right.\right.\right. \\ &\qquad \left.\left. +S_1V_1^{(0,1,0)}(t,S_1,Y)\right)\right\},\left.\{k_1V_1(t,S_1,Y)\}\right\}. \end{aligned} \tag{5.2.11}$$

Finally, assuming, as in [30], that the price does not depend on S_1, the hedging formula becomes

$$\Pi^{a,\star}(t,A) = \left\{-\left(\sigma_s \cdot \sigma_s^{\mathrm{T}}\right)^{-1} \cdot \sigma_s \cdot c^{\mathrm{T}} \cdot (\nabla_A V)^{\mathrm{T}} \cdot \kappa, V\kappa\right\}$$

$$= \left\{\left\{-\frac{k_1 \rho_{2,1} w(t,Y) V_1^{(0,1)}(t,Y)}{\mathbf{p}_1}\right\}, \{k_1 V_1(t,Y)\}\right\}, \quad (5.2.12)$$

which, after setting $k_1 = -1$, is equivalent to formula (32) in [30].

Remark 5.2.2. By means of the hedging result above and pricing results from Sect. 4.12.3, the full consistency between our general results, applicable in any simple economy, and well-known results obtained by other means and for a particular financial engineering model has now been fully established.

Example 5.2.5. (The stochastic interest, dividend, and appreciation rate model: part 4.) Continuing Sects. 2.4.8, 3.12, and 4.15, following [21], we derive the most conservative hedging formula for k_1 futures contracts, each with price V_1. We have

$$\left(\sigma_s \cdot \sigma_s^{\mathrm{T}}\right)^{-1} = \left(\frac{1}{\sigma_S^2}\right), \quad (5.2.13)$$

$$\sigma_s \cdot c^{\mathrm{T}} = \left(S_1 \sigma_S^2 \quad \sigma_S \sigma_v \rho_{2,1} \quad \sigma_S \sigma_{\mathbf{r}} \rho_{3,1} \quad \sigma_S \sigma_d \rho_{4,1}\right), \quad (5.2.14)$$

$$-\left(\sigma_s \cdot \sigma_s^{\mathrm{T}}\right)^{-1} \cdot \sigma_s \cdot c^{\mathrm{T}} = \left(-S_1 \quad -\frac{\sigma_v \rho_{2,1}}{\sigma_S} \quad -\frac{\sigma_{\mathbf{r}} \rho_{3,1}}{\sigma_S} \quad -\frac{\sigma_d \rho_{4,1}}{\sigma_S}\right), \quad (5.2.15)$$

and therefore

$$\Pi^{a,\star}(t,A) = \Bigg\{\Bigg\{-k_1 \Bigg(\frac{1}{\sigma_S} \sigma_d \rho_{4,1} V_1^{(0,0,0,0,1)}(t,S_1,\mathbf{v},r,\mathbb{D}) + \frac{1}{\sigma_S} \sigma_{\mathbf{r}} \rho_{3,1} V_1^{(0,0,0,1,0)}$$
$$(t,S_1,\mathbf{v},r,\mathbb{D}) + \frac{1}{\sigma_S} \sigma_v \rho_{2,1} V_1^{(0,0,1,0,0)}(t,S_1,\mathbf{v},r,\mathbb{D})$$
$$+ S_1 V_1^{(0,1,0,0,0)}(t,S_1,\mathbf{v},r,\mathbb{D})\Bigg)\Bigg\}, \{k_1 V_1(t,S_1,\mathbf{v},r,\mathbb{D})\}\Bigg\}$$
$$= \Bigg\{\Bigg\{-k_1 \Bigg(e^{g_1(t)+r g_2(t)+\mathbf{v} g_3(t)+\mathbb{D} g_4(t)} S_1 + \frac{1}{\sigma_S} e^{g_1(t)+r g_2(t)+\mathbf{v} g_3(t)+\mathbb{D} g_4(t)}$$
$$\sigma_{\mathbf{r}} \rho_{3,1} g_2(t) S_1 + \frac{1}{\sigma_S} e^{g_1(t)+r g_2(t)+\mathbf{v} g_3(t)+\mathbb{D} g_4(t)}$$
$$\sigma_v \rho_{2,1} g_3(t) S_1 + \frac{1}{\sigma_S} e^{g_1(t)+r g_2(t)+\mathbf{v} g_3(t)+\mathbb{D} g_4(t)}$$
$$\sigma_d \rho_{4,1} g_4(t) S_1\Bigg)\Bigg\}, \left\{e^{g_1(t)+r g_2(t)+\mathbf{v} g_3(t)+\mathbb{D} g_4(t)} k_1 S_1\right\}\Bigg\}.$$

5.3 Relaxed Hedging via Neutral Pricing: Relaxed Hedging for Neutral Prices

When we were deriving the neutral pricing PDE in Sect. 4.7, the key was the fundamental matrix of derivative pricing and hedging (4.6.3)–(4.6.4). The full matrix was used when Lemmas 4.7.1, 4.7.2, and 4.7.3 were proved. Yet for the rest of the results (see, e.g., (4.7.12) and (4.7.13)), we didn't use the first block column of the fundamental matrix, since our goal was to draw conclusions in regard to pricing. Here, on the other hand, since instead of pricing, we are studying hedging, the situation is reversed.

Continuing from (4.7.12), we calculate

$$
\begin{aligned}
c\cdot\sigma_{\mathrm{a}}^{\mathrm{T}}\cdot\left(\sigma_{\mathrm{a}}\cdot\sigma_{\mathrm{a}}^{\mathrm{T}}\right)^{-1} &= \left(c\cdot\sigma_s^{\mathrm{T}} \quad c\cdot c^{\mathrm{T}}\cdot\left(\frac{\nabla_A V}{V}\right)^{\mathrm{T}}\right)\cdot\left(\sigma_{\mathrm{a}}\cdot\sigma_{\mathrm{a}}^{\mathrm{T}}\right)^{-1} = \left(c\cdot\sigma_s^{\mathrm{T}} \quad c\cdot c^{\mathrm{T}}\cdot\left(\frac{\nabla_A V}{V}\right)^{\mathrm{T}}\right)\cdot \\
&\quad \begin{pmatrix} \left(\sigma_s\cdot\sigma_s^{\mathrm{T}}\right)^{-1}+\left(\sigma_s\cdot\sigma_s^{\mathrm{T}}\right)^{-1}\cdot\sigma_s\cdot c^{\mathrm{T}}\cdot\left(\nabla_A V\right)^{\mathrm{T}}\cdot\mathfrak{X}^{-1}\cdot\nabla_A V\cdot c\cdot\sigma_s^{\mathrm{T}}\cdot\left(\sigma_s\cdot\sigma_s^{\mathrm{T}}\right)^{-1} & * \\ -\left(V\mathbb{I}_l\right)\cdot\mathfrak{X}^{-1}\cdot\nabla_A V\cdot c\cdot\sigma_s^{\mathrm{T}}\cdot\left(\sigma_s\cdot\sigma_s^{\mathrm{T}}\right)^{-1} & * \end{pmatrix} \\
&= \left(c\cdot\sigma_s^{\mathrm{T}}\cdot\left(\sigma_s\cdot\sigma_s^{\mathrm{T}}\right)^{-1} - c\cdot\left(\mathbb{I}_n-\sigma_s^{\mathrm{T}}\cdot\left(\sigma_s\cdot\sigma_s^{\mathrm{T}}\right)^{-1}\cdot\sigma_s\right)\cdot c^{\mathrm{T}}\cdot\left(\nabla_A V\right)^{\mathrm{T}}\cdot\mathfrak{X}^{-1}\cdot\nabla_A V\right. \\
&\quad \left.\cdot c\cdot\sigma_s^{\mathrm{T}}\cdot\left(\sigma_s\cdot\sigma_s^{\mathrm{T}}\right)^{-1} \quad *\right)
\end{aligned}
\tag{5.3.1}
$$

and also

$$
\begin{aligned}
\left(a_{\mathrm{a}}-r\right)\cdot\left(\sigma_{\mathrm{a}}\cdot\sigma_{\mathrm{a}}^{\mathrm{T}}\right)^{-1} &= \left\{a_s-r, \frac{1}{V}\left(\frac{\partial V}{\partial t}+\frac{1}{2}\mathrm{tr}\left(\nabla_A\nabla_A V\cdot c\cdot c^{\mathrm{T}}\right)+\nabla_A V\cdot b+\mathcal{D}\right)-r\right\} \\
&\quad \cdot\begin{pmatrix} \left(\sigma_s\cdot\sigma_s^{\mathrm{T}}\right)^{-1}+\left(\sigma_s\cdot\sigma_s^{\mathrm{T}}\right)^{-1}\cdot\sigma_s\cdot c^{\mathrm{T}}\cdot\left(\nabla_A V\right)^{\mathrm{T}}\cdot\mathfrak{X}^{-1}\cdot\nabla_A V\cdot c\cdot\sigma_s^{\mathrm{T}}\cdot\left(\sigma_s\cdot\sigma_s^{\mathrm{T}}\right)^{-1} & * \\ -\left(V\mathbb{I}_l\right)\cdot\mathfrak{X}^{-1}\cdot\nabla_A V\cdot c\cdot\sigma_s^{\mathrm{T}}\cdot\left(\sigma_s\cdot\sigma_s^{\mathrm{T}}\right)^{-1} & * \end{pmatrix} \\
&= \left\{\left(a_s-r\right)\cdot\left(\sigma_s\cdot\sigma_s^{\mathrm{T}}\right)^{-1}-\left(\frac{\partial V}{\partial t}+\frac{1}{2}\mathrm{tr}\left(\nabla_A\nabla_A V\cdot c\cdot c^{\mathrm{T}}\right)+\nabla_A V\cdot b+\mathcal{D}-rV-\left(a_s-r\right)\right.\right. \\
&\quad \left.\left.\cdot\left(\sigma_s\cdot\sigma_s^{\mathrm{T}}\right)^{-1}\cdot\sigma_s\cdot c^{\mathrm{T}}\cdot\left(\nabla_A V\right)^{\mathrm{T}}\right)\cdot\mathfrak{X}^{-1}\cdot\nabla_A V\cdot c\cdot\sigma_s^{\mathrm{T}}\cdot\left(\sigma_s\cdot\sigma_s^{\mathrm{T}}\right)^{-1}, *\right\}.
\end{aligned}
$$

Therefore, assuming that $\{\varphi(t,X,A),V(t,A)\}$ is the solution of (4.7.1)–(4.7.2), from the above, we conclude that (4.7.14) is also equal to

$$
\begin{aligned}
\Pi_{\psi}^{\star}(t,X,A) &= -\left(1\Big/\frac{\partial^2\varphi}{\partial X^2}\right)\left(\frac{\partial\varphi}{\partial X}\left(a_{\mathrm{a}}-r\right)+\left(\nabla_A\frac{\partial\varphi}{\partial X}\right)\cdot c\cdot\sigma_{\mathrm{a}}^{\mathrm{T}}\right)\cdot\left(\sigma_{\mathrm{a}}\cdot\sigma_{\mathrm{a}}^{\mathrm{T}}\right)^{-1} \\
&= -\left(1\Big/\frac{\partial^2\varphi}{\partial X^2}\right)\left(\frac{\partial\varphi}{\partial X}\left\{\left(a_s-r\right)\cdot\left(\sigma_s\cdot\sigma_s^{\mathrm{T}}\right)^{-1}\right.\right. \\
&\quad -\left(\frac{\partial V}{\partial t}+\frac{1}{2}\mathrm{tr}\left(\nabla_A\nabla_A V\cdot c\cdot c^{\mathrm{T}}\right)+\nabla_A V\cdot b+\mathcal{D}-rV\right. \\
&\quad \left.-\left(a_s-r\right)\cdot\left(\sigma_s\cdot\sigma_s^{\mathrm{T}}\right)^{-1}\cdot\sigma_s\cdot c^{\mathrm{T}}\cdot\left(\nabla_A V\right)^{\mathrm{T}}\right)\cdot\mathfrak{X}^{-1}\cdot\nabla_A V
\end{aligned}
$$

$$\cdot c\cdot\sigma_s^{\mathrm{T}}\cdot\left(\sigma_s\cdot\sigma_s^{\mathrm{T}}\right)^{-1},*\Big\}+\left(\nabla_A\frac{\partial\varphi}{\partial X}\right)\cdot\Big(c\cdot\sigma_s^{\mathrm{T}}\cdot\left(\sigma_s\cdot\sigma_s^{\mathrm{T}}\right)^{-1}$$

$$-c\cdot\left(\mathbb{I}_n-\sigma_s^{\mathrm{T}}\cdot\left(\sigma_s\cdot\sigma_s^{\mathrm{T}}\right)^{-1}\cdot\sigma_s\right)\cdot c^{\mathrm{T}}\cdot(\nabla_A V)^{\mathrm{T}}\cdot\mathfrak{X}^{-1}\cdot\nabla_A V\cdot c\cdot\sigma_s^{\mathrm{T}}$$

$$\cdot\left(\sigma_s\cdot\sigma_s^{\mathrm{T}}\right)^{-1}\quad *\Big)\Big)$$

$$=\Big\{-\left(1\Big/\frac{\partial^2\varphi}{\partial X^2}\right)\left(\frac{\partial\varphi}{\partial X}(a_s-r)+\left(\nabla_A\frac{\partial\varphi}{\partial X}\right)\cdot c\cdot\sigma_s^{\mathrm{T}}\right)$$

$$\cdot\left(\sigma_s\cdot\sigma_s^{\mathrm{T}}\right)^{-1},*\Big\}+\left(\frac{\partial\varphi}{\partial X}\Big/\frac{\partial^2\varphi}{\partial X^2}\right)\Big\{\Big(\frac{\partial V}{\partial t}+\frac{1}{2}\mathrm{tr}\left(\nabla_A\nabla_A V\cdot c\cdot c^{\mathrm{T}}\right)$$

$$+\nabla_A V\cdot b+\mathcal{D}-rV-(a_s-r)\cdot\left(\sigma_s\cdot\sigma_s^{\mathrm{T}}\right)^{-1}\cdot\sigma_s\cdot c^{\mathrm{T}}\cdot(\nabla_A V)^{\mathrm{T}}\Big)$$

$$\cdot\mathfrak{X}^{-1}\cdot\nabla_A V\cdot c\cdot\sigma_s^{\mathrm{T}}\cdot\left(\sigma_s\cdot\sigma_s^{\mathrm{T}}\right)^{-1},*\Big\}+\left(\left(\nabla_A\frac{\partial\varphi}{\partial X}\right)\Big/\frac{\partial^2\varphi}{\partial X^2}\right)$$

$$\cdot\Big(c\cdot\left(\mathbb{I}_n-\sigma_s^{\mathrm{T}}\cdot\left(\sigma_s\cdot\sigma_s^{\mathrm{T}}\right)^{-1}\cdot\sigma_s\right)\cdot c^{\mathrm{T}}\cdot(\nabla_A V)^{\mathrm{T}}\cdot\mathfrak{X}^{-1}\cdot\nabla_A V$$

$$\cdot c\cdot\sigma_s^{\mathrm{T}}\cdot\left(\sigma_s\cdot\sigma_s^{\mathrm{T}}\right)^{-1}\quad *\Big)$$

$$=\Big\{-\left(1\Big/\frac{\partial^2\varphi}{\partial X^2}\right)\left(\frac{\partial\varphi}{\partial X}(a_s-r)+\left(\nabla_A\frac{\partial\varphi}{\partial X}\right)\cdot c\cdot\sigma_s^{\mathrm{T}}\right)$$

$$\cdot\left(\sigma_s\cdot\sigma_s^{\mathrm{T}}\right)^{-1},*\Big\}+\Big\{\left(\frac{\partial\varphi}{\partial X}\Big/\frac{\partial^2\varphi}{\partial X^2}\right)\Big\{\Big(\frac{\partial V}{\partial t}+\frac{1}{2}\mathrm{tr}\left(\nabla_A\nabla_A V\cdot c\cdot c^{\mathrm{T}}\right)$$

$$+\nabla_A V\cdot b+\mathcal{D}-rV-(a_s-r)\cdot\left(\sigma_s\cdot\sigma_s^{\mathrm{T}}\right)^{-1}\cdot\sigma_s\cdot c^{\mathrm{T}}\cdot(\nabla_A V)^{\mathrm{T}}\Big)$$

$$\cdot\mathfrak{X}^{-1}\cdot\nabla_A V\cdot c\cdot\sigma_s^{\mathrm{T}}\cdot\left(\sigma_s\cdot\sigma_s^{\mathrm{T}}\right)^{-1},*\Big\}+\left(\left(\nabla_A\frac{\partial\varphi}{\partial X}\right)\Big/\frac{\partial^2\varphi}{\partial X^2}\right)$$

$$\cdot\Big(c\cdot\left(\mathbb{I}_n-\sigma_s^{\mathrm{T}}\cdot\left(\sigma_s\cdot\sigma_s^{\mathrm{T}}\right)^{-1}\cdot\sigma_s\right)\cdot c^{\mathrm{T}}\cdot(\nabla_A V)^{\mathrm{T}}\cdot\mathfrak{X}^{-1}\cdot\nabla_A V\cdot c$$

$$\cdot\sigma_s^{\mathrm{T}}\cdot\left(\sigma_s\cdot\sigma_s^{\mathrm{T}}\right)^{-1}\quad *\Big),$$

and then, using (4.7.15),

$$\Pi_\psi^\star(t,X,A)=\Big\{-\left(1\Big/\frac{\partial^2\varphi}{\partial X^2}\right)\Big(\frac{\partial\varphi}{\partial X}(a_s-r)$$

$$+\left(\nabla_A\frac{\partial\varphi}{\partial X}\right)\cdot c\cdot\sigma_s^{\mathrm{T}}\Big)\cdot\left(\sigma_s\cdot\sigma_s^{\mathrm{T}}\right)^{-1},*\Big\}$$

$$+\Big\{-\kappa\cdot\nabla_A V\cdot c\cdot\sigma_s^{\mathrm{T}}\cdot\left(\sigma_s\cdot\sigma_s^{\mathrm{T}}\right)^{-1},*\Big\}$$

$$= \left\{ -\left(1 \Big/ \frac{\partial^2 \varphi}{\partial X^2}\right) \left(\frac{\partial \varphi}{\partial X}(a_s - r) + \left(\nabla_A \frac{\partial \varphi}{\partial X}\right) \cdot c \cdot \sigma_s^{\mathrm{T}} \right) \cdot \left(\sigma_s \cdot \sigma_s^{\mathrm{T}}\right)^{-1} \right.$$
$$\left. - \kappa \cdot \nabla_A V \cdot c \cdot \sigma_s^{\mathrm{T}} \cdot \left(\sigma_s \cdot \sigma_s^{\mathrm{T}}\right)^{-1}, \kappa V \right\}.$$

We have proved the following theorem.

Theorem 5.3.1. *Assuming Hypothesis 4.3.1, then in the framework of Chap. 4, for the utility of wealth $\psi = \psi(X)$ and the ψ-κ-neutral price (per single contract) V, with $\{\varphi, V\}$ the solution of the pricing system (4.7.1)–(4.7.2), a relaxed hedging of the position $\kappa = \kappa(t,X,A)$ is given by the formula*

$$\Pi^{a,\star}(t,X,A) = \left\{ -\left(1 \Big/ \frac{\partial^2 \varphi}{\partial X^2}\right) \left(\frac{\partial \varphi}{\partial X}(a_s - r) + \left(\nabla_A \frac{\partial \varphi}{\partial X}\right) \cdot c \cdot \sigma_s^{\mathrm{T}} \right) \cdot \left(\sigma_s \cdot \sigma_s^{\mathrm{T}}\right)^{-1} \right.$$
$$\left. - \left(\sigma_s \cdot \sigma_s^{\mathrm{T}}\right)^{-1} \cdot \sigma_s \cdot c^{\mathrm{T}} \cdot (\nabla_A V)^{\mathrm{T}} \cdot \kappa, \kappa V \right\}, \tag{5.3.2}$$

i.e., by

$$\Pi^{\mathbf{h}}(t,X,A) = -\left(1 \Big/ \frac{\partial^2 \varphi}{\partial X^2}\right) \left(\frac{\partial \varphi}{\partial X}(a_s - r) + \left(\nabla_A \frac{\partial \varphi}{\partial X}\right) \cdot c \cdot \sigma_s^{\mathrm{T}} \right) \cdot \left(\sigma_s \cdot \sigma_s^{\mathrm{T}}\right)^{-1}$$
$$- \left(\sigma_s \cdot \sigma_s^{\mathrm{T}}\right)^{-1} \cdot \sigma_s \cdot c^{\mathrm{T}} \cdot (\nabla_A V)^{\mathrm{T}} \cdot \kappa \tag{5.3.3}$$

for $t < T$.

Corollary 5.3.1. *In the framework of Chap. 4 and for the CRRA neutral price (per single contract) V, with $\{g_\gamma(t,A), V(t,A)\}$ computed as a solution of the pricing system (4.9.2)–(4.9.3), a relaxed hedging of the position $\kappa = \kappa(t,X,A) = X\kappa_0(t,A)$ is given by the formula*

$$\Pi^{a,\star}(t,X,A) = \left\{ \frac{X}{\gamma}\left(a_s - r + \nabla_A g_\gamma \cdot c \cdot \sigma_s^{\mathrm{T}}\right) \cdot \left(\sigma_s \cdot \sigma_s^{\mathrm{T}}\right)^{-1} \right.$$
$$\left. - \kappa \cdot \nabla_A V \cdot c \cdot \sigma_s^{\mathrm{T}} \cdot \left(\sigma_s \cdot \sigma_s^{\mathrm{T}}\right)^{-1}, \kappa V \right\}, \tag{5.3.4}$$

i.e., by

$$\Pi^{\mathbf{h}}(t,X,A) = \frac{X}{\gamma}\left(a_s - r + \nabla_A g_\gamma \cdot c \cdot \sigma_s^{\mathrm{T}}\right) \cdot \left(\sigma_s \cdot \sigma_s^{\mathrm{T}}\right)^{-1} - \kappa \cdot \nabla_A V \cdot c \cdot \sigma_s^{\mathrm{T}} \cdot \left(\sigma_s \cdot \sigma_s^{\mathrm{T}}\right)^{-1} \tag{5.3.5}$$

for $t < T$.

Proof. Set $\varphi(t,X,A) := (X^{1-\gamma} e^{g(t,A)} - 1)/(1-\gamma)$ in (5.3.2), etc. □

Corollary 5.3.2. *For the deterministic interest rate $r(t)$, in the framework of Chap. 4 and for the CARA neutral price (per single contract) V, with $\{g_\omega(t,A), V(t,A)\}$ computed as a solution of the pricing system (4.11.1)–(4.11.2), a relaxed hedging of the position $\kappa(t,X,A) = \kappa(t,A)$ is given by the formula*

$$\Pi^{a,\star}(t,A) = \left\{ \frac{e^{-\int_t^T r(\tau)\,d\tau}}{\omega} \left(a_s - r + \nabla_A g_\omega \cdot c \cdot \sigma_s^T\right) \cdot \left(\sigma_s \cdot \sigma_s^T\right)^{-1} \right.$$
$$\left. - \kappa \cdot \nabla_A V \cdot c \cdot \sigma_s^T \cdot \left(\sigma_s \cdot \sigma_s^T\right)^{-1}, \kappa V \right\}, \tag{5.3.6}$$

i.e., by

$$\Pi^{\mathbf{h}}(t,A) = \frac{e^{-\int_t^T r(\tau)\,d\tau}}{\omega} \left(a_s - r + \nabla_A g_\omega \cdot c \cdot \sigma_s^T\right) \cdot \left(\sigma_s \cdot \sigma_s^T\right)^{-1}$$
$$- \kappa \cdot \nabla_A V \cdot c \cdot \sigma_s^T \cdot \left(\sigma_s \cdot \sigma_s^T\right)^{-1} \tag{5.3.7}$$

for $t < T$.

Proof. Set $\varphi(t,X,A) = (1 - e^{-e^{\int_t^T r(\tau)\,d\tau} X\omega + \omega + g_\omega(t,A)})/\omega$ in (5.3.2), etc. □

Remark 5.3.1. Theorem 5.3.1 confirms the importance of the most conservative hedging formula (5.2.6), since the general relaxed hedging (5.3.2) is just a combination of the most conservative hedging (5.2.6) and the portfolio optimization formula (3.4.3).

Remark 5.3.2. Sending $\gamma \to \infty$ in (5.3.4), or sending $\omega \to \infty$ in (5.3.6), we *formally* conclude (5.2.6), and thereby we have yet another justification for the name "the most conservative" hedging.

Example 5.3.1. (Relaxed Black–Scholes hedging.) We continue Sect. 2.4.1, Exercise 3.4.1, and Example 5.2.1. We have $S = A = \{S_1\}$, while the market coefficients are given in (2.4.2). In the case of general wealth utility $\psi(X)$, continuing Exercise 3.4.1 and applying (5.3.2), we get

$$\Pi^{a,\star}(t,X,A) = \left\{ -\frac{(\mathbf{a}_1 - r)\,\varphi^{(0,1)}(t,X)}{\mathbf{p}_1^2 \varphi^{(0,2)}(t,X)} - k_1 S_1 V_1^{(0,1)}(t,S_1), k_1 V_1(t,S_1) \right\}, \tag{5.3.8}$$

where φ is the solution of (3.4.6) (cf. (3.4.17) and also (5.2.6)). Furthermore, in the case of CRRA utility of wealth, applying (5.3.4), we get

$$\Pi^{a,\star}(t,X,A) = \left\{ \frac{X(\mathbf{a}_1 - r)}{\gamma \mathbf{p}_1^2} - k_1 S_1 V_1^{(0,1)}(t,S_1), k_1 V_1(t,S_1) \right\}, \tag{5.3.9}$$

while in the case of deterministic interest rates and CARA utility of wealth, applying (5.3.6), we get

$$\Pi^{a,\star}(t,A)=\left\{\mathrm{e}^{-\int_t^{\mathrm{T}} r\,\mathrm{d}\tau}\frac{\mathbf{a}_1-r}{\omega\mathbf{p}_1^2}-k_1S_1V_1^{(0,1)}(t,S_1),k_1V_1(t,S_1)\right\}. \tag{5.3.10}$$

In all three cases, the Black–Scholes hedging $\left\{-k_1S_1V_1^{(0,1)}(t,S_1),k_1V_1(t,S_1)\right\}$ is combined with an optimal investment in the underlying (see Remark 5.2.2).

5.4 Relaxed Hedging via Constrained Portfolio Optimization: Relaxed Hedging for Arbitrary and Indifference Prices

Consider now the auxiliary economy version of the affine constraint optimal portfolio formula (3.5.2):

$$\begin{aligned}\Pi^{a,\star}(t,X,A) = &-\left(\frac{\partial\varphi}{\partial X}\Big/\frac{\partial^2\varphi}{\partial X^2}\right)(a_{\mathrm{a}}-r)\cdot\left(\mathbb{I}_{k+l}-\left(\sigma_{\mathrm{a}}\cdot\sigma_{\mathrm{a}}^{\mathrm{T}}\right)^{-1}\cdot\mu^{\mathrm{T}}\right.\\ &\left.\cdot\left(\mu\cdot\left(\sigma_{\mathrm{a}}\cdot\sigma_{\mathrm{a}}^{\mathrm{T}}\right)^{-1}\cdot\mu^{\mathrm{T}}\right)^{-1}\cdot\mu\right)\cdot\left(\sigma_{\mathrm{a}}\cdot\sigma_{\mathrm{a}}^{\mathrm{T}}\right)^{-1}-\left(\nabla_A\frac{\partial\varphi}{\partial X}\Big/\frac{\partial^2\varphi}{\partial X^2}\right)\\ &\cdot c\cdot\sigma_{\mathrm{a}}^{\mathrm{T}}\cdot\left(\mathbb{I}_{k+l}-\left(\sigma_{\mathrm{a}}\cdot\sigma_{\mathrm{a}}^{\mathrm{T}}\right)^{-1}\cdot\mu^{\mathrm{T}}\cdot\left(\mu\cdot\left(\sigma_{\mathrm{a}}\cdot\sigma_{\mathrm{a}}^{\mathrm{T}}\right)^{-1}\cdot\mu^{\mathrm{T}}\right)^{-1}\cdot\mu\right)\\ &\cdot\left(\sigma_{\mathrm{a}}\cdot\sigma_{\mathrm{a}}^{\mathrm{T}}\right)^{-1}+\xi\cdot\left(\mu\cdot\left(\sigma_{\mathrm{a}}\cdot\sigma_{\mathrm{a}}^{\mathrm{T}}\right)^{-1}\cdot\mu^{\mathrm{T}}\right)^{-1}\cdot\mu\cdot\left(\sigma_{\mathrm{a}}\cdot\sigma_{\mathrm{a}}^{\mathrm{T}}\right)^{-1}.\end{aligned} \tag{5.4.1}$$

Using (4.3.6)–(4.3.7), we have

$$\begin{aligned}\Pi^{a,\star}(t,X,A) = &-\left(\frac{\partial\varphi}{\partial X}\Big/\frac{\partial^2\varphi}{\partial X^2}\right)(a_{\mathrm{a}}-r)\cdot\begin{pmatrix}\left(\sigma_s\cdot\sigma_s^{\mathrm{T}}\right)^{-1} & \mathbb{O}_{k\times l}\\ \mathbb{O}_{l\times k} & \mathbb{O}_l\end{pmatrix}\\ &-\left(\nabla_A\frac{\partial\varphi}{\partial X}\Big/\frac{\partial^2\varphi}{\partial X^2}\right)\cdot c\cdot\sigma_{\mathrm{a}}^{\mathrm{T}}\cdot\begin{pmatrix}\left(\sigma_s\cdot\sigma_s^{\mathrm{T}}\right)^{-1} & \mathbb{O}_{k\times l}\\ \mathbb{O}_{l\times k} & \mathbb{O}_l\end{pmatrix}\\ &+\begin{pmatrix}-\left(\sigma_s\cdot\sigma_s^{\mathrm{T}}\right)^{-1}\cdot\sigma_s\cdot c^{\mathrm{T}}\cdot\left(\nabla_AV\right)^{\mathrm{T}}\cdot\left(\frac{1}{V}\mathbb{I}_l\right)\\ \mathbb{I}_l\end{pmatrix}\cdot\xi\\ = &-\left(\frac{\partial\varphi}{\partial X}\Big/\frac{\partial^2\varphi}{\partial X^2}\right)\left\{a_s-r,\frac{1}{V}\left(\frac{\partial V}{\partial t}+\frac{1}{2}\mathrm{tr}\left(\nabla_A\nabla_AV\cdot c\cdot c^{\mathrm{T}}\right)\right.\right.\\ &\left.\left.+\nabla_AV\cdot b+\mathcal{D}\right)-r\right\}\end{aligned}$$

$$
\cdot \begin{pmatrix} (\sigma_s \cdot \sigma_s^{\mathrm{T}})^{-1} & \mathbb{O}_{k\times l} \\ \mathbb{O}_{l\times k} & \mathbb{O}_l \end{pmatrix} - \left(\nabla_A \frac{\partial \varphi}{\partial X} \Big/ \frac{\partial^2 \varphi}{\partial X^2}\right) \cdot c
$$

$$
\cdot \sigma_{\mathrm{a}}^{\mathrm{T}} \cdot \begin{pmatrix} (\sigma_s \cdot \sigma_s^{\mathrm{T}})^{-1} & \mathbb{O}_{k\times l} \\ \mathbb{O}_{l\times k} & \mathbb{O}_l \end{pmatrix}
$$

$$
+ \begin{pmatrix} -(\sigma_s \cdot \sigma_s^{\mathrm{T}})^{-1} \cdot \sigma_s \cdot c^{\mathrm{T}} \cdot (\nabla_A V)^{\mathrm{T}} \cdot (\frac{1}{V}\mathbb{I}_l) \\ \mathbb{I}_l \end{pmatrix} \cdot \xi
$$

$$
= -\left(\frac{\partial \varphi}{\partial X} \Big/ \frac{\partial^2 \varphi}{\partial X^2}\right) \left\{ (a_s - r) \cdot (\sigma_s \cdot \sigma_s^{\mathrm{T}})^{-1}, 0 \right\} - \left(\nabla_A \frac{\partial \varphi}{\partial X} \Big/ \frac{\partial^2 \varphi}{\partial X^2}\right) \cdot c
$$

$$
\cdot \left(\sigma_s^{\mathrm{T}} \; c^{\mathrm{T}} \cdot \left(\tfrac{\nabla_A V}{V}\right)^{\mathrm{T}}\right) \cdot \begin{pmatrix} (\sigma_s \cdot \sigma_s^{\mathrm{T}})^{-1} & \mathbb{O}_{k\times l} \\ \mathbb{O}_{l\times k} & \mathbb{O}_l \end{pmatrix}
$$

$$
+ \begin{pmatrix} -(\sigma_s \cdot \sigma_s^{\mathrm{T}})^{-1} \cdot \sigma_s \cdot c^{\mathrm{T}} \cdot (\nabla_A V)^{\mathrm{T}} \cdot (\frac{1}{V}\mathbb{I}_l) \\ \mathbb{I}_l \end{pmatrix} \cdot \xi
$$

$$
= -\left(\frac{\partial \varphi}{\partial X} \Big/ \frac{\partial^2 \varphi}{\partial X^2}\right) \left\{ (a_s - r) \cdot (\sigma_s \cdot \sigma_s^{\mathrm{T}})^{-1}, 0 \right\} - \left(\nabla_A \frac{\partial \varphi}{\partial X} \Big/ \frac{\partial^2 \varphi}{\partial X^2}\right) \cdot c
$$

$$
\cdot \left(\sigma_s^{\mathrm{T}} \cdot (\sigma_s \cdot \sigma_s^{\mathrm{T}})^{-1} \; \mathbb{O}_{n\times l}\right)
$$

$$
+ \begin{pmatrix} -(\sigma_s \cdot \sigma_s^{\mathrm{T}})^{-1} \cdot \sigma_s \cdot c^{\mathrm{T}} \cdot (\nabla_A V)^{\mathrm{T}} \cdot (\frac{1}{V}\mathbb{I}_l) \\ \mathbb{I}_l \end{pmatrix} \cdot \xi
$$

$$
= -\left(\frac{\partial \varphi}{\partial X} \Big/ \frac{\partial^2 \varphi}{\partial X^2}\right) \left\{ (a_s - r) \cdot (\sigma_s \cdot \sigma_s^{\mathrm{T}})^{-1}, 0 \right\}
$$

$$
- \left\{ \left(\nabla_A \frac{\partial \varphi}{\partial X} \Big/ \frac{\partial^2 \varphi}{\partial X^2}\right) \cdot c \cdot \sigma_s^{\mathrm{T}} \cdot (\sigma_s \cdot \sigma_s^{\mathrm{T}})^{-1}, 0 \right\}
$$

$$
+ \left\{ -(\sigma_s \cdot \sigma_s^{\mathrm{T}})^{-1} \cdot \sigma_s \cdot c^{\mathrm{T}} \cdot (\nabla_A V)^{\mathrm{T}} \cdot \left(\frac{1}{V}\mathbb{I}_l\right) \cdot \xi, \xi \right\}
$$

$$
= -\left(\frac{\partial \varphi}{\partial X} \Big/ \frac{\partial^2 \varphi}{\partial X^2}\right) \left\{ (a_s - r) \cdot (\sigma_s \cdot \sigma_s^{\mathrm{T}})^{-1}, 0 \right\}
$$

$$
- \left\{ \left(\nabla_A \frac{\partial \varphi}{\partial X} \Big/ \frac{\partial^2 \varphi}{\partial X^2}\right) \cdot c \cdot \sigma_s^{\mathrm{T}} \cdot (\sigma_s \cdot \sigma_s^{\mathrm{T}})^{-1}, 0 \right\}
$$

$$
+ \left\{ -(\sigma_s \cdot \sigma_s^{\mathrm{T}})^{-1} \cdot \sigma_s \cdot c^{\mathrm{T}} \cdot (\nabla_A V)^{\mathrm{T}} \cdot \kappa, V\kappa \right\}, \tag{5.4.2}
$$

where φ is the solution of (we rewrite (3.4.1) with $\xi = \kappa V$)

$$\mathcal{M}(\varphi_{\mathrm{a}}) + \frac{\partial^2 \varphi_{\mathrm{a}}}{\partial X^2}\left(\frac{\partial \varphi_{\mathrm{a}}}{\partial X}\kappa \cdot \left(-\nabla_A V \cdot c \cdot \sigma_s^{\mathrm{T}} \cdot \left(\sigma_s \cdot \sigma_s^{\mathrm{T}}\right)^{-1} \cdot (a_s - r) + \frac{\partial V}{\partial t} \right.\right.$$
$$\left. + \frac{1}{2}\mathrm{tr}\left(\nabla_A \nabla_A V \cdot c \cdot c^{\mathrm{T}}\right) + \nabla_A V \cdot b - rV + \mathcal{D}\right)$$
$$+ \frac{1}{2}\frac{\partial^2 \varphi_{\mathrm{a}}}{\partial X^2}\kappa \cdot \nabla_A V \cdot c \cdot \left(\mathbb{I}_n - \sigma_s^{\mathrm{T}} \cdot \left(\sigma_s \cdot \sigma_s^{\mathrm{T}}\right)^{-1} \cdot \sigma_s\right) \cdot c^{\mathrm{T}} \cdot \left(\nabla_A V\right)^{\mathrm{T}} \cdot \kappa$$
$$\left. + \kappa \cdot \left(-\nabla_A V \cdot c \cdot \sigma_s^{\mathrm{T}} \cdot \left(\sigma_s \cdot \sigma_s^{\mathrm{T}}\right)^{-1} \cdot \sigma_s + \nabla_A V \cdot c\right) \cdot c^{\mathrm{T}} \cdot \left(\nabla_A \frac{\partial \varphi_{\mathrm{a}}}{\partial X}\right)\right) = 0 \tag{5.4.3}$$

for $t < T$, where $\mathcal{M}$ is defined in (3.4.1), with the terminal condition $\varphi_{\mathrm{a}}(t,X,A) = \psi(X)$. We may therefore state the following theorem.

Theorem 5.4.1. *In the framework of Chap. 4, for the utility of wealth $\psi = \psi(X)$ and for any given price (per single contract) $V = V(t,A)$, a relaxed hedging of the position $\kappa = \kappa(t,X,A)$ is given by the formula*

$$\Pi^{a,\star}(t,X,A) = \left\{ -\left(1 \Big/ \frac{\partial^2 \varphi_{\mathrm{a}}}{\partial X^2}\right)\left(\frac{\partial \varphi_{\mathrm{a}}}{\partial X}(a_s - r) + \left(\nabla_A \frac{\partial \varphi_{\mathrm{a}}}{\partial X}\right) \cdot c \cdot \sigma_s^{\mathrm{T}}\right) \cdot \left(\sigma_s \cdot \sigma_s^{\mathrm{T}}\right)^{-1} \right.$$
$$\left. - \left(\sigma_s \cdot \sigma_s^{\mathrm{T}}\right)^{-1} \cdot \sigma_s \cdot c^{\mathrm{T}} \cdot \left(\nabla_A V\right)^{\mathrm{T}} \cdot \kappa, \kappa V \right\} \tag{5.4.4}$$

for $t < T$, where φ_{a} is the solution of (5.4.3).

Corollary 5.4.1. *In the framework of Chap. 4, for the utility of wealth $\psi = \psi(X)$, and for the ψ-κ-indifference price (per single contract) V, with $\{\varphi(t,X,A), V(t,A)\}$ the solution of the pricing system (4.8.2)–(4.8.1), a relaxed hedging of the position $\kappa = \kappa(t,X,A)$ is given by the formula*

$$\Pi^{a,\star}(t,X,A) = \left\{ -\left(1 \Big/ \frac{\partial^2 \varphi}{\partial X^2}\right)\left(\frac{\partial \varphi}{\partial X}(a_s - r) + \left(\nabla_A \frac{\partial \varphi}{\partial X}\right) \cdot c \cdot \sigma_s^{\mathrm{T}}\right) \cdot \left(\sigma_s \cdot \sigma_s^{\mathrm{T}}\right)^{-1} \right.$$
$$\left. - \left(\sigma_s \cdot \sigma_s^{\mathrm{T}}\right)^{-1} \cdot \sigma_s \cdot c^{\mathrm{T}} \cdot \left(\nabla_A V\right)^{\mathrm{T}} \cdot \kappa, \kappa V \right\}, \tag{5.4.5}$$

i.e., by

$$\Pi^{\mathbf{h}}(t,X,A) = -\left(1 \Big/ \frac{\partial^2 \varphi}{\partial X^2}\right)\left(\frac{\partial \varphi}{\partial X}(a_s - r) + \left(\nabla_A \frac{\partial \varphi}{\partial X}\right) \cdot c \cdot \sigma_s^{\mathrm{T}}\right) \cdot \left(\sigma_s \cdot \sigma_s^{\mathrm{T}}\right)^{-1}$$
$$- \left(\sigma_s \cdot \sigma_s^{\mathrm{T}}\right)^{-1} \cdot \sigma_s \cdot c^{\mathrm{T}} \cdot \left(\nabla_A V\right)^{\mathrm{T}} \cdot \kappa \tag{5.4.6}$$

for $t < T$.

Remark 5.4.1. Are Theorems 5.3.1–5.4.1 consistent? The hedging formulas (5.3.2)–(5.4.4) are identical, so the question is whether in the case of neutral pricing, φ in Theorem 5.3.1 is equal to φ_a in Theorem 5.4.1. The answer is affirmative. We shall prove that if (4.7.1) and (4.7.2) hold, then (5.4.3) holds as well. So, first using (4.7.2), we have

$$
\begin{aligned}
&\mathcal{M}(\varphi_a) + \frac{\partial^2 \varphi_a}{\partial X^2}\left(\frac{\partial \varphi_a}{\partial X}\kappa \cdot \left(-\nabla_A V \cdot c \cdot \sigma_s^{\mathrm{T}} \cdot \left(\sigma_s \cdot \sigma_s^{\mathrm{T}}\right)^{-1} \cdot (a_s - r) + \frac{\partial V}{\partial t}\right.\right.\\
&\quad \left. + \frac{1}{2}\operatorname{tr}\left(\nabla_A \nabla_A V \cdot c \cdot c^{\mathrm{T}}\right) + \nabla_A V \cdot b - rV + \mathcal{D}\right) + \frac{1}{2}\frac{\partial^2 \varphi_a}{\partial X^2}\kappa \cdot \nabla_A V \cdot c\\
&\quad \cdot \left(\mathbb{I}_n - \sigma_s^{\mathrm{T}} \cdot \left(\sigma_s \cdot \sigma_s^{\mathrm{T}}\right)^{-1} \cdot \sigma_s\right) \cdot c^{\mathrm{T}} \cdot \left(\nabla_A V\right)^{\mathrm{T}} \cdot \kappa\\
&\quad \left. + \kappa \cdot \left(-\nabla_A V \cdot c \cdot \sigma_s^{\mathrm{T}} \cdot \left(\sigma_s \cdot \sigma_s^{\mathrm{T}}\right)^{-1} \cdot \sigma_s + \nabla_A V \cdot c\right) \cdot c^{\mathrm{T}} \cdot \left(\nabla_A \frac{\partial \varphi_a}{\partial X}\right)\right)\\
&= \mathcal{M}(\varphi_a) + \frac{\partial^2 \varphi_a}{\partial X^2}\frac{\partial \varphi_a}{\partial X}\kappa \cdot \left(-\nabla_A V \cdot c \cdot \sigma_s^{\mathrm{T}} \cdot \left(\sigma_s \cdot \sigma_s^{\mathrm{T}}\right)^{-1} \cdot (a_s - r) + \frac{\partial V}{\partial t}\right.\\
&\quad + \frac{1}{2}\operatorname{tr}\left(\nabla_A \nabla_A V \cdot c \cdot c^{\mathrm{T}}\right) + \nabla_A V \cdot b - rV + \mathcal{D} + \frac{1}{2}\left(\frac{\partial^2 \varphi_a}{\partial X^2} \bigg/ \frac{\partial \varphi_a}{\partial X}\right)\nabla_A V \cdot c\\
&\quad \cdot \left(\mathbb{I}_n - \sigma_s^{\mathrm{T}} \cdot \left(\sigma_s \cdot \sigma_s^{\mathrm{T}}\right)^{-1} \cdot \sigma_s\right) \cdot c^{\mathrm{T}} \cdot \left(\nabla_A V\right)^{\mathrm{T}} \cdot \kappa - \nabla_A V \cdot c\\
&\quad \left. \cdot \left(\sigma_s^{\mathrm{T}} \cdot \left(\sigma_s \cdot \sigma_s^{\mathrm{T}}\right)^{-1} \cdot \sigma_s - \mathbb{I}_n\right) \cdot c^{\mathrm{T}} \cdot \left(\nabla_A \frac{\partial \varphi}{\partial X} \bigg/ \frac{\partial^2 \varphi}{\partial X^2}\right)\right)\\
&= \mathcal{M}(\varphi_a) + \frac{\partial^2 \varphi_a}{\partial X^2}\frac{\partial \varphi_a}{\partial X}\kappa \cdot \left(-\frac{1}{2}\left(\frac{\partial^2 \varphi_a}{\partial X^2} \bigg/ \frac{\partial \varphi_a}{\partial X}\right)\nabla_A V \cdot c\right.\\
&\quad \left. \cdot \left(\mathbb{I}_n - \sigma_s^{\mathrm{T}} \cdot \left(\sigma_s \cdot \sigma_s^{\mathrm{T}}\right)^{-1} \cdot \sigma_s\right) \cdot c^{\mathrm{T}} \cdot \left(\nabla_A V\right)^{\mathrm{T}} \cdot \kappa\right)\\
&= \mathcal{M}(\varphi_a) - \frac{1}{2}\left(\frac{\partial^2 \varphi_a}{\partial X^2}\right)^2 \kappa \cdot \nabla_A V \cdot c \cdot \left(\mathbb{I}_n - \sigma_s^{\mathrm{T}} \cdot \left(\sigma_s \cdot \sigma_s^{\mathrm{T}}\right)^{-1} \cdot \sigma_s\right) \cdot c^{\mathrm{T}} \cdot \left(\nabla_A V\right)^{\mathrm{T}} \cdot \kappa,
\end{aligned}
$$

which is the left-hand side of (4.7.2), and hence equal to zero, and therefore (5.4.3) holds.

Chapter 6
Equity Valuation and Investing: Continuous-time Accounting

Equity Portfolios: Market Sentiment and Dividend Policy Effect on Equity Value; Prices vs. Investing; Stock Buyback and Dilution Effect on Equity Value

Wall Street Journal (WSJ.com) U.S. Edition
MAY 21, 2010, 8:55 A.M. ET
US Stocks Futures Slip; Sentiment Low Despite German Votes

Apple CEO Jobs Favors Flexibility of Cash Hoard
By THE ASSOCIATED PRESS
Published: February 25, 2010
CUPERTINO, Calif. (AP) – Apple Inc., buoyed by the success of the iPhones, iPods and computers it churns out to breathless buyers, could give some of its $25 billion in cash back to shareholders. But CEO Steve Jobs said Thursday that he thinks Apple is better off keeping that money stockpiled.

Resona Holdings Tumbles Most in 7 Years in Tokyo on Share-Issue Plan
Blumberg, by Shigeru Sato and Takako Taniguchi – Nov 4, 2010
Resona Holdings Inc. fell the most in more than 7 years in Tokyo trading after reports said the Japanese bank plans to sell shares to help repay government bailout funds.... The dilution effect may be limited if Resona decides to buy back the shares and retire them."

6.1 Introduction

Equity valuation has a long history, longer than that of "financial engineering." Yet as became apparent not too long ago, and to be elaborated in this chapter, the modern financial mathematics of incomplete markets can bring some new, more complete, and deeper insights into the theory of equity valuation.

While since the 1970's there were tremendous developments in mathematical finance (as also shown in this book already), and framed under the term

S. Stojanovic, *Neutral and Indifference Portfolio Pricing, Hedging and Investing: With Applications in Equity and FX*, DOI 10.1007/978-0-387-71418-9_6,

Continuous-time Finance (see [28]), there were no comparable parallel developments in the area of (corporate) Accounting. In this chapter we employ the above developed theory of neutral and indifference pricing and hedging of portfolios of financial contracts in incomplete markets to solve a few problems in fundamental valuation. So this chapter can be read also as an invitation into the Continuous-time Accounting.

If a market is complete, then prices, even equity prices, and even equity prices calculated via "financial engineering" methods, will be unique, i.e., independent of the market sentiment, i.e., independent of the investor's risk-aversion. On the other hand, as is well known in practice, the market valuation of stocks is very much influenced by market sentiment. More precisely, an increase in market risk-aversion causes equity prices to decline, and vice versa. So here we shall introduce a simple incomplete market model, referred to as the "basic equity model," of neutral and indifference equity pricing that will allow, under nonzero market exposure, for equity prices to depend on risk-aversion in such a way.

One of the most celebrated results in equity valuation is the paper of Miller and Modigliani [29]. One of the consequences of their results has become the famous "dividend puzzle," as discussed by Black in [3]. Indeed, it was asserted by Miller and Modigliani's theorem, and disputed by Black and others, that the dividend policy of a firm does not affect the value of investment into that firm: whatever is gained through the dividend is lost through equity depreciation, and vice versa. This, as was observed previously, is true only if the firm's expenditures (to its employees and management, for example) are not affected by the amount of the company's available cash, in other words, only if there is no "competition" between the owners (the shareholders) and employees/management for the disbursement of the available cash. Real-world observation suggests that such competition does exist, and therefore, more-comprehensive equity valuation models are necessary to capture such a phenomenon. We provide such a model, mentioned above, called the "basic equity model."

The neutral pricing of equities can be used to find a fundamental, and also invertible, relationship between prices and optimal investing decisions, invertible in the sense that we can compute equity prices if an equity portfolio position is given, or we can compute the optimal equity portfolio position if equity (market) prices are given.

It can also be used for understanding and quantifying the effects of stock buyback and/or dilution, taking into account the price at which the buyback and/or dilution is implemented, as well as many other parameters that come into play.

There is no limit on how complicated/comprehensive equity valuation modeling can be developed in the general framework to be presented in this chapter – the price to pay is the complexity of the corresponding pricing equations. We therefore limit our discussion here to the models that allow for explicit solutions.

6.2 Definition of the Basic Equity Model

Definition 6.2.1. A *stock* of a company is a *perpetual* financial contract between the company and the owner of the stock that entitles the owner of the stock a dividend payment from the company in the (possibly random) amount of d units of currency per year.

Company 1, whose stock price $V_1(t)$ is to be determined, has cash holdings (*per unit stock*) amounting to $\mathbb{C}_1(t)$ at time t. The cash holding $\mathbb{C}_1(t)$ depends on various cash flows, such as income, cost, interest on the company's cash holdings, and dividends paid.

Following [41], our basic assumption is that the company's cash holdings (*or book value; per unit stock*) evolve according to (*the cash or book value evolution equation*)

$$\mathrm{d}\mathbb{C}_1(t) = (i_1(t) - q_1(t))\,\mathrm{d}t + (r(t) - \mathbf{q}_{1,1}(t))\,\mathbb{C}_1(t)\mathrm{d}t - d_1(\ldots)\mathrm{d}t, \tag{6.2.1}$$

where $i_1(t)$ is the *income rate*, $r(t) > 0$ is the *interest rate*, $d_1(\ldots)$ is the *dividend rate*, and $\mathbf{p}(t) = q_1(t) + \mathbf{q}_{1,1}(t)\mathbb{C}_1(t)$ is the *cost rate*, including payroll, pensions, corporate bonds issues, etc., i.e., the conglomerate of all cash liabilities rate; $q_1(t)$ will be referred to as the *basic cost rate*, while $\mathbf{q}_{1,1}(t)$ will be referred to as the *surcharge (cost) rate*.

To embed the equity valuation into the general concept of derivative pricing in incomplete markets elaborated in previous chapters, consider a simple economy $\mathfrak{E}$ with a single tradable security (used for hedging the equity position) $S = \{S_0\}$, and a vector of factors $A = \{S_0, \mathbb{C}_1, i_1, q_1\}$, where $\mathbb{C}_1, i_1, q_1$ are as introduced above. Aiming for explicit valuation solutions, suppose that the company's dividend policy has the following simple form:

$$d_1 = d_1(\mathbb{C}_1, i_1, q_1) := \alpha_{1,0} + \mathbb{C}_1\alpha_{1,1} + i_1\alpha_{1,2} + q_1\alpha_{1,3} \tag{6.2.2}$$

units of currency, per year, per stock outstanding. We summarize the above in the following definition.

Definition 6.2.2. A *basic model for equity (or stock) valuation/pricing* will refer to the following "market" model:

$$\mathrm{d}S_0(t) = S_0(t)(\mathbf{a}_0 - \mathbb{D}_0)\,\mathrm{d}t + S_0(t)\mathbf{p}_0\mathrm{d}\mathbb{B}_1(t), \tag{6.2.3}$$

$$\begin{aligned}\mathrm{d}\mathbb{C}_1(t) &= (i_1(t) - q_1(t) + \mathbb{C}_1(t)(r - \mathbf{q}_{1,1}) - d_1)\,\mathrm{d}t \\ &= (i_1(t) - q_1(t) + \mathbb{C}_1(t)(r - \mathbf{q}_{1,1}) - (\alpha_{1,0} + \mathbb{C}_1(t)\alpha_{1,1} \\ &\quad + i_1(t)\alpha_{1,2} + q_1(t)\alpha_{1,3}))\,\mathrm{d}t,\end{aligned} \tag{6.2.4}$$

$$\mathrm{d}i_1(t) = \left(\mu_{1,\mathbf{i},0} + \mu_{1,\mathbf{i},1}i_1(t)\right)\cdot\mathrm{d}t + \left(w_{1,\mathbf{i},0} + w_{1,\mathbf{i},1}i_1(t)\right)\mathrm{d}\mathbb{B}_2(t), \tag{6.2.5}$$

$$\mathrm{d}q_1(t) = \left(\mu_{1,\mathbf{q}_0,0} + \mu_{1,\mathbf{q}_0,1}q_1(t)\right)\mathrm{d}t + \left(w_{1,\mathbf{q}_0,0} + w_{1,\mathbf{q}_0,1}q_1(t)\right)\mathrm{d}\mathbb{B}_3(t), \tag{6.2.6}$$

where $d\mathbb{B}_1(t)d\mathbb{B}_2(t) = \rho_{2,1}dt$, $d\mathbb{B}_1(t)d\mathbb{B}_3(t) = \rho_{3,1}dt$, and $d\mathbb{B}_2(t)d\mathbb{B}_3(t) = \rho_{3,2}dt$. The above dynamics (6.2.3)–(6.2.6) are described via the following market coefficients for the simple economy $\mathfrak{E}$:

$$a_s = \{\mathbf{a}_0\}, \sigma_s = \left(\mathbf{p}_0\ 0\ 0\right), \tag{6.2.7}$$

$$b = \{S_0(\mathbf{a}_0 - \mathbb{D}_0), -\alpha_{1,2}i_1 + i_1 - q_1 + \mathbb{C}_1(r - \mathbf{q}_{1,1}) - \alpha_{1,0} - \mathbb{C}_1\alpha_{1,1} - q_1\alpha_{1,3},$$
$$\mu_{1,\mathbf{i},0} + i_1\mu_{1,\mathbf{i},1}, \mu_{1,\mathbf{q}_0,0} + q_1\mu_{1,\mathbf{q}_0,1}\}, \tag{6.2.8}$$

$$c = \begin{pmatrix} S_0\mathbf{p}_0 & 0 & 0 \\ 0 & 0 & 0 \\ \rho_{2,1}\left(w_{1,\mathbf{i},0} + i_1 w_{1,\mathbf{i},1}\right) & \sqrt{1-\rho_{2,1}^2}\left(w_{1,\mathbf{i},0} + i_1 w_{1,\mathbf{i},1}\right) & 0 \\ \rho_{3,1}\left(w_{1,\mathbf{q}_0,0} + q_1 w_{1,\mathbf{q}_0,1}\right) & \frac{\left(\rho_{3,2}-\rho_{2,1}\rho_{3,1}\right)\left(w_{1,\mathbf{q}_0,0}+q_1 w_{1,\mathbf{q}_0,1}\right)}{\sqrt{1-\rho_{2,1}^2}} & \left(w_{1,\mathbf{q}_0,0} + q_1 w_{1,\mathbf{q}_0,1}\right)\mathbf{s}_{3,3} \end{pmatrix}, \tag{6.2.9}$$

where $\mathbf{s}_{3,3}$ is defined in the appendix. We remark that this market is complete if and only if

$$c.\left(\mathbb{I}_n - \sigma_s{}^{\mathrm{T}} \cdot \left(\sigma_s.\sigma_s{}^{\mathrm{T}}\right)^{-1} \cdot \sigma_s\right) \cdot c^{\mathrm{T}} = \mathbb{O}_{4\times 4}, \tag{6.2.10}$$

or, after calculation, if and only if

$$0 = \left(1-\rho_{2,1}^2\right)\left(w_{1,\mathbf{i},0} + i_1 w_{1,\mathbf{i},1}\right)^2 \wedge 0 = \left(\rho_{3,2} - \rho_{2,1}\rho_{3,1}\right)\left(w_{1,\mathbf{i},0} + i_1 w_{1,\mathbf{i},1}\right)$$
$$\left(w_{1,\mathbf{q}_0,0} + q_1 w_{1,\mathbf{q}_0,1}\right) \wedge 0 = \left(1-\rho_{3,1}^2\right)\left(w_{1,\mathbf{q}_0,0} + q_1 w_{1,\mathbf{q}_0,1}\right)^2.$$

6.3 Basic Equity Model for a Portfolio of Stocks

The basic equity model above can be extended to consider, instead of a single stock, a portfolio of stocks. For example, for simplicity of presentation, consider a portfolio of length 2. Then the tradable and factors will be denoted by $S = \{S_0\}$, $A = \{S_0, \mathbb{C}_1, i_1, q_1, \mathbb{C}_2, i_2, q_2\}$, the dividend policies will be defined by

$$d_1 = \alpha_{1,0} + \mathbb{C}_1\alpha_{1,1} + i_1\alpha_{1,2} + q_1\alpha_{1,3}, \tag{6.3.1}$$

$$d_2 = \alpha_{2,0} + \mathbb{C}_2\alpha_{2,1} + i_2\alpha_{2,2} + q_2\alpha_{2,3}, \tag{6.3.2}$$

while the market coefficients (6.2.7)–(6.2.9) extend into

$$a_s = \{\mathbf{a}_0\}, \sigma_s = \left(\mathbf{p}_0\ 0\ 0\ 0\ 0\right), \tag{6.3.3}$$

$$b = \{S_0(\mathbf{a}_0 - \mathbb{D}_0),\ -\alpha_{1,2}i_1 + i_1 - q_1 + \mathbb{C}_1(r - \mathbf{q}_{1,1}) - \alpha_{1,0} - \mathbb{C}_1\alpha_{1,1} - q_1\alpha_{1,3},$$
$$\mu_{1,\mathbf{i},0} + i_1\mu_{1,\mathbf{i},1}, \mu_{1,\mathbf{q}_0,0} + q_1\mu_{1,\mathbf{q}_0,1},$$
$$-\alpha_{2,2}i_2 + i_2 - q_2 + \mathbb{C}_2(r - \mathbf{q}_{2,1}) - \alpha_{2,0} - \mathbb{C}_2\alpha_{2,1} - q_2\alpha_{2,3}, \mu_{2,\mathbf{i},0} + i_2\mu_{2,\mathbf{i},1},$$
$$\mu_{2,\mathbf{q}_0,0} + q_2\mu_{2,\mathbf{q}_0,1}\}, \tag{6.3.4}$$

and the (due to its size, only partially displayed) matrix $c(7 \times 5\text{matrix})$:

$$c=\begin{pmatrix} S_0\mathbf{p}_0 & 0 & 0 & 0 & 0 \\ 0 & 0 & 0 & 0 & 0 \\ \rho_{2,1}\left(w_{1,\mathbf{i},0}+i_1 w_{1,\mathbf{i},1}\right) & \sqrt{1-\rho_{2,1}^2}\left(w_{1,\mathbf{i},0}+i_1 w_{1,\mathbf{i},1}\right) & 0 & 0 & 0 \\ \rho_{3,1}\left(w_{1,\mathbf{q}_0,0}+q_1 w_{1,\mathbf{q}_0,1}\right) & \frac{\left(\rho_{3,2}-\rho_{2,1}\rho_{3,1}\right)\left(w_{1,\mathbf{q}_0,0}+q_1 w_{1,\mathbf{q}_0,1}\right)}{\sqrt{1-\rho_{2,1}^2}} & c_{4,3} & 0 & 0 \\ 0 & 0 & 0 & 0 & 0 \\ \rho_{4,1}\left(w_{2,\mathbf{i},0}+i_2 w_{2,\mathbf{i},1}\right) & \frac{\left(\rho_{4,2}-\rho_{2,1}\rho_{4,1}\right)\left(w_{2,\mathbf{i},0}+i_2 w_{2,\mathbf{i},1}\right)}{\sqrt{1-\rho_{2,1}^2}} & c_{6,3} & c_{6,4} & 0 \\ \rho_{5,1}\left(w_{2,\mathbf{q}_0,0}+q_2 w_{2,\mathbf{q}_0,1}\right) & \frac{\left(\rho_{5,2}-\rho_{2,1}\rho_{5,1}\right)\left(w_{2,\mathbf{q}_0,0}+q_2 w_{2,\mathbf{q}_0,1}\right)}{\sqrt{1-\rho_{2,1}^2}} & c_{7,3} & c_{7,4} & c_{7,5} \end{pmatrix} \tag{6.3.5}$$

(the precise formulation of the matrix c can be deduced using the appendix). We note that this market is complete if and only if

$$c.\left(\mathbb{I}_n-\sigma_s{}^{\mathrm{T}}\cdot\left(\sigma_s.\sigma_s{}^{\mathrm{T}}\right)^{-1}\cdot\sigma_s\right)\cdot c^{\mathrm{T}}=\mathbb{O}_{7\times 7}, \tag{6.3.6}$$

or, after a calculation, if and only if

$$\begin{aligned} 0= & \left(1-\rho_{2,1}^2\right)\left(w_{1,\mathbf{i},0}+i_1 w_{1,\mathbf{i},1}\right)^2 \wedge 0=\left(\rho_{3,2}-\rho_{2,1}\rho_{3,1}\right)\left(w_{1,\mathbf{i},0}+i_1 w_{1,\mathbf{i},1}\right)\left(w_{1,\mathbf{q}_0,0}\right. \\ & \left.+q_1 w_{1,\mathbf{q}_0,1}\right) \wedge 0=\left(1-\rho_{3,1}^2\right)\left(w_{1,\mathbf{q}_0,0}+q_1 w_{1,\mathbf{q}_0,1}\right)^2 \wedge 0=\left(\rho_{4,2}-\rho_{2,1}\rho_{4,1}\right) \\ & \left(w_{1,\mathbf{i},0}+i_1 w_{1,\mathbf{i},1}\right)\left(w_{2,\mathbf{i},0}+i_2 w_{2,\mathbf{i},1}\right) \wedge 0=\left(\rho_{4,3}-\rho_{3,1}\rho_{4,1}\right) \\ & \left(w_{1,\mathbf{q}_0,0}+q_1 w_{1,\mathbf{q}_0,1}\right)\left(w_{2,\mathbf{i},0}+i_2 w_{2,\mathbf{i},1}\right) \wedge 0=\left(1-\rho_{4,1}^2\right)\left(w_{2,\mathbf{i},0}+i_2 w_{2,\mathbf{i},1}\right)^2 \wedge \\ 0= & \left(\rho_{5,2}-\rho_{2,1}\rho_{5,1}\right)\left(w_{1,\mathbf{i},0}+i_1 w_{1,\mathbf{i},1}\right)\left(w_{2,\mathbf{q}_0,0}+q_2 w_{2,\mathbf{q}_0,1}\right) \wedge 0=\left(\rho_{5,3}-\rho_{3,1}\rho_{5,1}\right) \\ & \left(w_{1,\mathbf{q}_0,0}+q_1 w_{1,\mathbf{q}_0,1}\right)\left(w_{2,\mathbf{q}_0,0}+q_2 w_{2,\mathbf{q}_0,1}\right) \wedge 0=\left(\rho_{5,4}-\rho_{4,1}\rho_{5,1}\right) \\ & \left(w_{2,\mathbf{i},0}+i_2 w_{2,\mathbf{i},1}\right)\left(w_{2,\mathbf{q}_0,0}+q_2 w_{2,\mathbf{q}_0,1}\right) \wedge 0=\left(1-\rho_{5,1}^2\right)\left(w_{2,\mathbf{q}_0,0}+q_2 w_{2,\mathbf{q}_0,1}\right)^2. \end{aligned} \tag{6.3.7}$$

6.4 Basic Equity Model: CRRA Neutral Pricing of a Single (Type of) Stock

6.4.1 Valuation Formula Derivation

We now solve the problem of CRRA neutral pricing for the model (6.2.3)–(6.2.6), i.e., market coefficients (6.2.7)–(6.2.9), and for the market position $\kappa = X\kappa_0$, with $\kappa_0 = \{k_0\}$, with k_0 a constant. Since stocks are perpetual contracts, an entire solution (i.e., for $T = \infty$) of the pricing PDE (4.9.3) needs to be found. On the other hand, the corresponding $\nabla_A g_\gamma$ is found by solving the PDE (4.9.2) for $T < \infty$, and then by passing to the limit $T \to \infty$ in $\nabla_A g_\gamma = \lim_{T\to\infty} \nabla_A g_{T,\gamma}$.

We shall consider two cases:

$$\kappa = 0 \tag{6.4.1}$$

and

$$\kappa = X\kappa_0 = X\{k_0\} \wedge \left(w_{1,\mathbf{q}_0,1} = 0 \wedge w_{1,\mathbf{i},1} = 0\right), \tag{6.4.2}$$

with k_0 a constant. In the former case, (6.4.1), assuming $g_\gamma = g(t)$, (4.9.2) simplifies to

$$g'(t) = \frac{\gamma-1}{\gamma}\left(\frac{(r-\mathbf{a}_0)^2}{2\mathbf{p}_0^2} + r\gamma\right) \tag{6.4.3}$$

for $t < T < \infty$, with the terminal condition $g(T) = 0$, confirming that $g_\gamma = g(t)$ and $\nabla_A g_\gamma = 0$.

In the latter case, (6.4.2), assuming $g_\gamma = g(t)$ and assuming that the solution of the pricing PDE (4.9.3) has the form $V = \{V_1\}$ with

$$V_1(t, S_0, \mathbb{C}_1, i_1, q_1) := \mathbb{A}_{1,1} + \mathbb{C}_1\mathbb{A}_{1,3} + i_1\mathbb{A}_{1,4} + q_1\mathbb{A}_{1,5}, \tag{6.4.4}$$

with $\mathbb{A}_{1,1}$, $\mathbb{A}_{1,3}$, $\mathbb{A}_{1,4}$, and $\mathbb{A}_{1,5}$ constants, (4.9.2) simplifies to

$$\begin{aligned}&\frac{1}{2}(\gamma-1)\gamma\big(\mathbb{A}_{1,4}^2\left(\rho_{2,1}^2-1\right)w_{1,\mathbf{i},0}^2 + 2\mathbb{A}_{1,4}\mathbb{A}_{1,5}\left(\rho_{2,1}\rho_{3,1}-\rho_{3,2}\right)w_{1,\mathbf{q}_0,0}w_{1,\mathbf{i},0}\\ &\quad +\mathbb{A}_{1,5}^2\left(\rho_{3,1}^2-1\right)w_{1,\mathbf{q}_0,0}^2\big)k_1^2 + g'(t) = \frac{\gamma-1}{\gamma}\left(\frac{(\mathbf{a}_0-r)^2}{2\mathbf{p}_0^2} + r\gamma\right)\end{aligned} \tag{6.4.5}$$

for $t < T$, with the terminal condition $g(T) = 0$, confirming again, conditionally on the assumption (6.4.4), that $g_\gamma = g(t)$, and therefore that $\nabla_A g_\gamma = 0$. In the former case, (6.4.1), assuming (6.4.4), the pricing PDE (4.9.3) becomes

$$\begin{aligned}&-r\left(\mathbb{A}_{1,1} + \mathbb{C}_1\mathbb{A}_{1,3} + i_1\mathbb{A}_{1,4} + q_1\mathbb{A}_{1,5}\right) + \mathbb{A}_{1,3}\big(-\alpha_{1,2}i_1 + i_1 - q_1 + \mathbb{C}_1\left(r - \mathbf{q}_{1,1}\right)\\ &\quad -\alpha_{1,0} - \mathbb{C}_1\alpha_{1,1} - q_1\alpha_{1,3}\big) + \mathbb{A}_{1,4}\left(-\frac{1}{\mathbf{p}_0}(\mathbf{a}_0 - r)\rho_{2,1}\left(w_{1,\mathbf{i},0} + i_1 w_{1,\mathbf{i},1}\right)\right.\\ &\quad \left. + \mu_{1,\mathbf{i},0} + i_1\mu_{1,\mathbf{i},1}\right) + \mathbb{A}_{1,5}\left(-\frac{1}{\mathbf{p}_0}(\mathbf{a}_0 - r)\rho_{3,1}\left(w_{1,\mathbf{q}_0,0} + q_1 w_{1,\mathbf{q}_0,1}\right) + \mu_{1,\mathbf{q}_0,0}\right.\\ &\quad \left. + q_1\mu_{1,\mathbf{q}_0,1}\right) = -\alpha_{1,0} - \mathbb{C}_1\alpha_{1,1} - i_1\alpha_{1,2} - q_1\alpha_{1,3},\end{aligned} \tag{6.4.6}$$

confirming assumption (6.4.4), while in the latter case, (6.4.2), assuming (6.4.4), the pricing PDE (4.9.3) becomes

$$
\begin{aligned}
&-r\left(\mathbb{A}_{1,1}+\mathbb{C}_1\mathbb{A}_{1,3}+i_1\mathbb{A}_{1,4}+q_1\mathbb{A}_{1,5}\right)+\mathbb{A}_{1,3}\left(-\alpha_{1,2}i_1+i_1-q_1+\mathbb{C}_1\left(r-\mathbf{q}_{1,1}\right)\right.\\
&\left.-\alpha_{1,0}-\mathbb{C}_1\alpha_{1,1}-q_1\alpha_{1,3}\right)+\gamma k_0\left(\mathbb{A}_{1,4}^2\left(-1+\rho_{2,1}^2\right)w_{1,\mathbf{i},0}^2+2\mathbb{A}_{1,4}\mathbb{A}_{1,5}\left(\rho_{2,1}\rho_{3,1}\right.\right.\\
&\left.\left.-\rho_{3,2}\right)w_{1,\mathbf{i},0}w_{1,\mathbf{q}_0,0}+\mathbb{A}_{1,5}^2\left(-1+\rho_{3,1}^2\right)w_{1,\mathbf{q}_0,0}^2\right)+\mathbb{A}_{1,4}\\
&\left(-\frac{(\mathbf{a}_0-r)\rho_{2,1}w_{1,\mathbf{i},0}}{\mathbf{p}_0}+\mu_{1,\mathbf{i},0}+i_1\mu_{1,\mathbf{i},1}\right)+\mathbb{A}_{1,5}\left(-\frac{(\mathbf{a}_0-r)\rho_{3,1}w_{1,\mathbf{q}_0,0}}{\mathbf{p}_0}\right.\\
&\left.+\mu_{1,\mathbf{q}_0,0}+q_1\mu_{1,\mathbf{q}_0,1}\right)=-\alpha_{1,0}-\mathbb{C}_1\alpha_{1,1}-i_1\alpha_{1,2}-q_1\alpha_{1,3},
\end{aligned}
\tag{6.4.7}
$$

confirming again assumption (6.4.4).

In the former case, (6.4.1), collecting coefficients for $\mathbb{C}_1, i_1, q_1$, and setting them equal to zero, we get

$$
\begin{aligned}
&-r\mathbb{A}_{1,1}-\mathbb{A}_{1,3}\alpha_{1,0}+\alpha_{1,0}+\frac{(r-\mathbf{a}_0)\mathbb{A}_{1,4}\rho_{2,1}w_{1,\mathbf{i},0}}{\mathbf{p}_0}+\frac{(r-\mathbf{a}_0)\mathbb{A}_{1,5}\rho_{3,1}w_{1,\mathbf{q}_0,0}}{\mathbf{p}_0}\\
&+\mathbb{A}_{1,4}\mu_{1,\mathbf{i},0}+\mathbb{A}_{1,5}\mu_{1,\mathbf{q}_0,0}=0,
\end{aligned}
\tag{6.4.8}
$$

$$
-\alpha_{1,3}\mathbb{A}_{1,3}-\mathbb{A}_{1,3}-r\mathbb{A}_{1,5}+\alpha_{1,3}+\frac{(r-\mathbf{a}_0)\mathbb{A}_{1,5}\rho_{3,1}w_{1,\mathbf{q}_0,1}}{\mathbf{p}_0}+\mathbb{A}_{1,5}\mu_{1,\mathbf{q}_0,1}=0, \tag{6.4.9}
$$

$$
-\alpha_{1,2}\mathbb{A}_{1,3}+\mathbb{A}_{1,3}-r\mathbb{A}_{1,4}+\alpha_{1,2}+\frac{(r-\mathbf{a}_0)\mathbb{A}_{1,4}\rho_{2,1}w_{1,\mathbf{i},1}}{\mathbf{p}_0}+\mathbb{A}_{1,4}\mu_{1,\mathbf{i},1}=0, \tag{6.4.10}
$$

$$
-r\mathbb{A}_{1,3}+\left(r-\mathbf{q}_{1,1}\right)\mathbb{A}_{1,3}-\alpha_{1,1}\mathbb{A}_{1,3}+\alpha_{1,1}=0. \tag{6.4.11}
$$

Equation (6.4.11) is solved first, yielding

$$
\mathbb{A}_{1,3}=\frac{\alpha_{1,1}}{\mathbf{q}_{1,1}+\alpha_{1,1}}. \tag{6.4.12}
$$

Then equations (6.4.9) and (6.4.10) are solved, and we obtain

$$
\mathbb{A}_{1,5}=\frac{\alpha_{1,3}-\frac{\alpha_{1,1}(1+\alpha_{1,3})}{\mathbf{q}_{1,1}+\alpha_{1,1}}}{r-\mu_{1,\mathbf{q}_0,1}+\frac{\mathbf{a}_0-r}{\mathbf{p}_0}\rho_{3,1}w_{1,\mathbf{q}_0,1}} \tag{6.4.13}
$$

and

$$
\mathbb{A}_{1,4}=\frac{\alpha_{1,2}-\frac{\alpha_{1,1}(-1+\alpha_{1,2})}{\mathbf{q}_{1,1}+\alpha_{1,1}}}{r-\mu_{1,\mathbf{i},1}+\frac{\mathbf{a}_0-r}{\mathbf{p}_0}\rho_{2,1}w_{1,\mathbf{i},1}}. \tag{6.4.14}
$$

Finally, we solve equation (6.4.8), getting

$$\mathbb{A}_{1,1} = \frac{1}{r}\left(\frac{\mathbf{a}_0 - r}{\mathbf{p}_0}\left(\left(\left(\frac{\alpha_{1,1}(\alpha_{1,2}-1)}{\mathbf{q}_{1,1}+\alpha_{1,1}} - \alpha_{1,2}\right)\rho_{2,1}w_{1,\mathbf{i},0}\right)\Big/\left(r-\mu_{1,\mathbf{i},1}\right.\right.\right.$$
$$\left.+\frac{\mathbf{a}_0 - r}{\mathbf{p}_0}\rho_{2,1}w_{1,\mathbf{i},1}\right) + \left(\left(\frac{\alpha_{1,1}(\alpha_{1,3}+1)}{\mathbf{q}_{1,1}+\alpha_{1,1}} - \alpha_{1,3}\right)\rho_{3,1}w_{1,\mathbf{q}_0,0}\right)\Big/\left(r-\mu_{1,\mathbf{q}_0,1}\right.$$
$$\left.\left.\left.+\frac{\mathbf{a}_0 - r}{\mathbf{p}_0}\rho_{3,1}w_{1,\mathbf{q}_0,1}\right)\right) + \alpha_{1,0}\left(1-\frac{\alpha_{1,1}}{\mathbf{q}_{1,1}+\alpha_{1,1}}\right) + \frac{\left(\alpha_{1,2} - \frac{\alpha_{1,1}(\alpha_{1,2}-1)}{\mathbf{q}_{1,1}+\alpha_{1,1}}\right)\mu_{1,\mathbf{i},0}}{r-\mu_{1,\mathbf{i},1}+\frac{\mathbf{a}_0-r}{\mathbf{p}_0}\rho_{2,1}w_{1,\mathbf{i},1}}$$
$$\left.+\left(\left(\alpha_{1,3} - \frac{\alpha_{1,1}(\alpha_{1,3}+1)}{\mathbf{q}_{1,1}+\alpha_{1,1}}\right)\mu_{1,\mathbf{q}_0,0}\right)\Big/\left(r-\mu_{1,\mathbf{q}_0,1}+\frac{\mathbf{a}_0-r}{\mathbf{p}_0}\rho_{3,1}w_{1,\mathbf{q}_0,1}\right)\right). \tag{6.4.15}$$

In the latter case, (6.4.2), collecting coefficients for $\mathbb{C}_1, i_1, q_1$, and setting them equal to zero, we get

$$-r\mathbb{A}_{1,1} - \mathbb{A}_{1,3}\alpha_{1,0} + \alpha_{1,0} + \gamma k_0\left(\mathbb{A}_{1,4}^2\left(\rho_{2,1}^2-1\right)w_{1,\mathbf{i},0}^2 + 2\mathbb{A}_{1,4}\mathbb{A}_{1,5}\left(\rho_{2,1}\rho_{3,1}\right.\right.$$
$$\left.\left. - \rho_{3,2}\right)w_{1,\mathbf{q}_0,0}w_{1,\mathbf{i},0} + \mathbb{A}_{1,5}^2\left(\rho_{3,1}^2-1\right)w_{1,\mathbf{q}_0,0}^2\right) + \mathbb{A}_{1,4}\mu_{1,\mathbf{i},0} + \mathbb{A}_{1,5}$$
$$\mu_{1,\mathbf{q}_0,0} - \frac{(\mathbf{a}_0-r)\mathbb{A}_{1,4}\rho_{2,1}w_{1,\mathbf{i},0}}{\mathbf{p}_0} - \frac{(\mathbf{a}_0-r)\mathbb{A}_{1,5}\rho_{3,1}w_{1,\mathbf{q}_0,0}}{\mathbf{p}_0} = 0, \tag{6.4.16}$$

$$-\alpha_{1,3}\mathbb{A}_{1,3} - \mathbb{A}_{1,3} - r\mathbb{A}_{1,5} + \alpha_{1,3} + \mathbb{A}_{1,5}\mu_{1,\mathbf{q}_0,1} = 0, \tag{6.4.17}$$
$$-\alpha_{1,2}\mathbb{A}_{1,3} + \mathbb{A}_{1,3} - r\mathbb{A}_{1,4} + \alpha_{1,2} + \mathbb{A}_{1,4}\mu_{1,\mathbf{i},1} = 0, \tag{6.4.18}$$
$$-r\mathbb{A}_{1,3} + (r-\mathbf{q}_{1,1})\mathbb{A}_{1,3} - \alpha_{1,1}\mathbb{A}_{1,3} + \alpha_{1,1} = 0, \tag{6.4.19}$$

yielding (6.4.12), and also

$$\mathbb{A}_{1,5} = \frac{\alpha_{1,3} - \frac{\alpha_{1,1}(\alpha_{1,3}+1)}{\mathbf{q}_{1,1}+\alpha_{1,1}}}{r-\mu_{1,\mathbf{q}_0,1}}, \tag{6.4.20}$$

$$\mathbb{A}_{1,4} = \frac{\alpha_{1,2} - \frac{\alpha_{1,1}(\alpha_{1,2}-1)}{\mathbf{q}_{1,1}+\alpha_{1,1}}}{r-\mu_{1,\mathbf{i},1}}, \tag{6.4.21}$$

and

$$\mathbb{A}_{1,1} = \frac{1}{r}\left(\left(\frac{r-\mathbf{a}_0}{\mathbf{p}_0}\right)\left(\left(\left(\alpha_{1,2} - \frac{\alpha_{1,1}(\alpha_{1,2}-1)}{\mathbf{q}_{1,1}+\alpha_{1,1}}\right)\rho_{2,1}w_{1,\mathbf{i},0}\right)\Big/\left(r-\mu_{1,\mathbf{i},1}\right)\right.\right.$$
$$\left.+\left(\left(\alpha_{1,3} - \frac{\alpha_{1,1}(\alpha_{1,3}+1)}{\mathbf{q}_{1,1}+\alpha_{1,1}}\right)\rho_{3,1}w_{1,\mathbf{q}_0,0}\right)\Big/\left(r-\mu_{1,\mathbf{q}_0,1}\right)\right)$$

$$+\left(\alpha_{1,0}\left(1-\frac{\alpha_{1,1}}{\mathbf{q}_{1,1}+\alpha_{1,1}}\right)+\gamma k_0\left(\left(\left(\alpha_{1,2}-\frac{\alpha_{1,1}(\alpha_{1,2}-1)}{\mathbf{q}_{1,1}+\alpha_{1,1}}\right)^2\right.\right.\right.$$

$$\left.\left(\rho_{2,1}^2-1\right)w_{1,\mathbf{i},0}^2\right)\Big/\left(r-\mu_{1,\mathbf{i},1}\right)^2+\left(2\left(\alpha_{1,2}-\frac{\alpha_{1,1}(\alpha_{1,2}-1)}{\mathbf{q}_{1,1}+\alpha_{1,1}}\right)\right.$$

$$\left.\left(\alpha_{1,3}-\frac{\alpha_{1,1}(\alpha_{1,3}+1)}{\mathbf{q}_{1,1}+\alpha_{1,1}}\right)(\rho_{2,1}\rho_{3,1}-\rho_{3,2})w_{1,\mathbf{i},0}w_{1,\mathbf{q}_0,0}\right)\Big/$$

$$\left(\left(r-\mu_{1,\mathbf{i},1}\right)\left(r-\mu_{1,\mathbf{q}_0,1}\right)\right)+\left(\left(\alpha_{1,3}-\frac{\alpha_{1,1}(\alpha_{1,3}+1)}{\mathbf{q}_{1,1}+\alpha_{1,1}}\right)^2\right.$$

$$\left.\left.\left(\rho_{3,1}^2-1\right)w_{1,\mathbf{q}_0,0}^2\right)\Big/\left(r-\mu_{1,\mathbf{q}_0,1}\right)^2\right)+\left(\left(\alpha_{1,2}-\frac{\alpha_{1,1}(\alpha_{1,2}-1)}{\mathbf{q}_{1,1}+\alpha_{1,1}}\right)\right.$$

$$\left.\mu_{1,\mathbf{i},0}\right)\Big/\left(r-\mu_{1,\mathbf{i},1}\right)+\left(\left(\alpha_{1,3}-\frac{\alpha_{1,1}(\alpha_{1,3}+1)}{\mathbf{q}_{1,1}+\alpha_{1,1}}\right)\mu_{1,\mathbf{q}_0,0}\right)\Big/$$

$$\left.\left.\left(r-\mu_{1,\mathbf{q}_0,1}\right)\right)\right). \tag{6.4.22}$$

We have therefore the following theorem.

Theorem 6.4.1. *In the above basic equity framework, assuming (6.4.1), the CRRA $(\kappa=0)$–equity value/price is independent of the relative risk-aversion parameter γ, and it is equal to*

$$V_1(t,S_0,\mathbb{C}_1,i_1,q_1)=V_1(\mathbb{C}_1,i_1,q_1)$$

$$=\frac{1}{r}\left(\frac{\mathbf{a}_0-r}{\mathbf{p}_0}\left(\left(\left(\frac{\alpha_{1,1}(\alpha_{1,2}-1)}{\mathbf{q}_{1,1}+\alpha_{1,1}}-\alpha_{1,2}\right)\rho_{2,1}\;w_{1,\mathbf{i},0}\right)\Big/\right.\right.$$

$$\left(r-\mu_{1,\mathbf{i},1}+\frac{\mathbf{a}_0-r}{\mathbf{p}_0}\rho_{2,1}w_{1,\mathbf{i},1}\right)+\left(\left(\frac{\alpha_{1,1}(\alpha_{1,3}+1)}{\mathbf{q}_{1,1}+\alpha_{1,1}}-\alpha_{1,3}\right)\rho_{3,1}w_{1,\mathbf{q}_0,0}\right)\Big/$$

$$\left.\left.\left(r-\mu_{1,\mathbf{q}_0,1}+\frac{\mathbf{a}_0-r}{\mathbf{p}_0}\rho_{3,1}w_{1,\mathbf{q}_0,1}\right)\right)\right)+\alpha_{1,0}\left(1-\frac{\alpha_{1,1}}{\mathbf{q}_{1,1}+\alpha_{1,1}}\right)$$

$$+\frac{\left(\alpha_{1,2}-\frac{\alpha_{1,1}(\alpha_{1,2}-1)}{\mathbf{q}_{1,1}+\alpha_{1,1}}\right)\mu_{1,\mathbf{i},0}}{r-\mu_{1,\mathbf{i},1}+\frac{\mathbf{a}_0-r}{\mathbf{p}_0}\rho_{2,1}w_{1,\mathbf{i},1}}+\left(\left(\alpha_{1,3}-\frac{\alpha_{1,1}(\alpha_{1,3}+1)}{\mathbf{q}_{1,1}+\alpha_{1,1}}\right)\mu_{1,\mathbf{q}_0,0}\right)\Big/$$

$$\left(r-\mu_{1,\mathbf{q}_0,1}+\frac{\mathbf{a}_0-r}{\mathbf{p}_0}\rho_{3,1}w_{1,\mathbf{q}_0,1}\right)\bigg)\bigg)+\mathbb{C}_1\frac{\alpha_{1,1}}{\mathbf{q}_{1,1}+\alpha_{1,1}}$$

$$+i_1\frac{\alpha_{1,2}-\frac{\alpha_{1,1}(-1+\alpha_{1,2})}{\mathbf{q}_{1,1}+\alpha_{1,1}}}{r-\mu_{1,\mathbf{i},1}+\frac{\mathbf{a}_0-r}{\mathbf{p}_0}\rho_{2,1}w_{1,\mathbf{i},1}}+q_1\frac{\alpha_{1,3}-\frac{\alpha_{1,1}(1+\alpha_{1,3})}{\mathbf{q}_{1,1}+\alpha_{1,1}}}{r-\mu_{1,\mathbf{q}_0,1}+\frac{\mathbf{a}_0-r}{\mathbf{p}_0}\rho_{3,1}w_{1,\mathbf{q}_0,1}} \tag{6.4.23}$$

for $-\infty<t<\infty$. In the above basic equity framework, assuming (6.4.2), the CRRA $\kappa=X\{k_0\}$-neutral equity value/price is dependent on the relative risk-aversion parameter γ, and it is equal to

$$
\begin{aligned}
&V_1(t,S_0,\mathbb{C}_1,i_1,q_1)=V_1(\mathbb{C}_1,i_1,q_1)\\
&=\frac{1}{r}\left(\left(\frac{r-\mathbf{a}_0}{\mathbf{p}_0}\right)\left(\left(\left(\alpha_{1,2}-\frac{\alpha_{1,1}(\alpha_{1,2}-1)}{\mathbf{q}_{1,1}+\alpha_{1,1}}\right)\rho_{2,1}w_{1,\mathbf{i},0}\right)\Big/\left(r-\mu_{1,\mathbf{i},1}\right)\right.\right.\\
&\quad\left.+\left(\left(\alpha_{1,3}-\frac{\alpha_{1,1}(\alpha_{1,3}+1)}{\mathbf{q}_{1,1}+\alpha_{1,1}}\right)\rho_{3,1}\ w_{1,\mathbf{q}_0,0}\right)\Big/\left(r-\mu_{1,\mathbf{q}_0,1}\right)\right)\\
&\quad+\left(\alpha_{1,0}\left(1-\frac{\alpha_{1,1}}{\mathbf{q}_{1,1}+\alpha_{1,1}}\right)+\gamma k_0\left(\left(\left(\alpha_{1,2}-\frac{\alpha_{1,1}(\alpha_{1,2}-1)}{\mathbf{q}_{1,1}+\alpha_{1,1}}\right)^2\right.\right.\right.\\
&\quad\left.\left(\rho_{2,1}^2-1\right)w_{1,\mathbf{i},0}^2\right)\Big/\left(r-\mu_{1,\mathbf{i},1}\right)^2+\left(2\left(\alpha_{1,2}-\frac{\alpha_{1,1}(\alpha_{1,2}-1)}{\mathbf{q}_{1,1}+\alpha_{1,1}}\right)\right.\\
&\quad\left.\left(\alpha_{1,3}-\frac{\alpha_{1,1}(\alpha_{1,3}+1)}{\mathbf{q}_{1,1}+\alpha_{1,1}}\right)\left(\rho_{2,1}\ \rho_{3,1}-\rho_{3,2}\right)w_{1,\mathbf{i},0}w_{1,\mathbf{q}_0,0}\right)\Big/\\
&\quad\left(\left(r-\mu_{1,\mathbf{i},1}\right)\left(r-\mu_{1,\mathbf{q}_0,1}\right)\right)+\left(\left(\alpha_{1,3}-\frac{\alpha_{1,1}(\alpha_{1,3}+1)}{\mathbf{q}_{1,1}+\alpha_{1,1}}\right)^2\right.\\
&\quad\left.\left.\left(\rho_{3,1}^2-1\right)w_{1,\mathbf{q}_0,0}^2\right)\Big/\left(r-\mu_{1,\mathbf{q}_0,1}\right)^2\right)+\left(\left(\alpha_{1,2}-\frac{\alpha_{1,1}(\alpha_{1,2}-1)}{\mathbf{q}_{1,1}+\alpha_{1,1}}\right)\mu_{1,\mathbf{i},0}\right)\Big/\\
&\quad\left.\left.\left(r-\mu_{1,\mathbf{i},1}\right)+\left(\left(\alpha_{1,3}-\frac{\alpha_{1,1}(\alpha_{1,3}+1)}{\mathbf{q}_{1,1}+\alpha_{1,1}}\right)\mu_{1,\mathbf{q}_0,0}\right)\Big/\left(r-\mu_{1,\mathbf{q}_0,1}\right)\right)\right)\\
&\quad+\mathbb{C}_1\frac{\alpha_{1,1}}{\mathbf{q}_{1,1}+\alpha_{1,1}}+i_1\frac{\alpha_{1,2}-\frac{\alpha_{1,1}(\alpha_{1,2}-1)}{\mathbf{q}_{1,1}+\alpha_{1,1}}}{r-\mu_{1,\mathbf{i},1}}+q_1\frac{\alpha_{1,3}-\frac{\alpha_{1,1}(\alpha_{1,3}+1)}{\mathbf{q}_{1,1}+\alpha_{1,1}}}{r-\mu_{1,\mathbf{q}_0,1}}
\end{aligned}
\tag{6.4.24}
$$

for $-\infty<t<\infty$.

Remark 6.4.1. There are conditions, such as $r>0, r-\mu_{1,\mathbf{q}_0,1}>0, r-\mu_{1,\mathbf{i},1}>0$, on the coefficients in the basic equity model under which the above formulas represent the value of the equity.

6.4.2 The Role of Dividends in Equity Valuation

Here we reconcile the celebrated result of Miller and Modigliani [29], famously called into question by Black [3], known as the "dividend puzzle." We have the following corollary.

Corollary 1. *(The "Miller–Modigliani case.") If the surcharge cost rate is equal to zero, i.e., if* $\mathbf{q}_{1,1} = 0$*, then, assuming (6.4.1) the equity value/price is equal to*

$$
\begin{aligned}
V_1(\mathbb{C}_1, i_1, q_1) = {} & \mathbb{C}_1 + \frac{i_1}{r-} \mu_{1,\mathbf{i},1} + \frac{(\mathbf{a}_0 - r)\rho_{2,1} w_{1,\mathbf{i},1}}{\mathbf{p}_0} - \frac{q_1}{r - \mu_{1,\mathbf{q}_0,1} + \frac{(\mathbf{a}_0 - r)\rho_{3,1} w_{1,\mathbf{q}_0,1}}{\mathbf{p}_0}} \\
& + \frac{1}{r} \left(\frac{\mu_1, \mathbf{i}, 0}{-} \frac{\mathbf{a}_0 - r}{\mathbf{p}_0} \rho_{2,1} w_{1,\mathbf{i},} \right) 0r - \mu_1, \mathbf{i}, 1 + \frac{(\mathbf{a}_0 - r)\rho_{2,1} w_{1,\mathbf{i},1}}{\mathbf{p}_0} \\
& - \left(\mu_{1,\mathbf{q}_0,0} - \frac{\mathbf{a}_0 - r}{\mathbf{p}_0} \rho_{3,1} w_{1,\mathbf{q}_0,0} \right) \Big/ \left(r - \mu_{1,\mathbf{q}_0,1} + \frac{(\mathbf{a}_0 - r)\rho_{3,1} w_{1,\mathbf{q}_0,1}}{\mathbf{p}_0} \right) \Bigg) , \qquad (6.4.25)
\end{aligned}
$$

while assuming (6.4.2), it is equal to

$$
\begin{aligned}
V_1(\mathbb{C}_1, i_1, q_1) = {} & \mathbb{C}_1 + \frac{i_1}{r - \mu_1, \mathbf{i}, 1} - \frac{q_1}{r - \mu_{1,\mathbf{q}_0,1}} + \frac{1}{r} \left(\frac{\mu_{1,\mathbf{i},0} - \frac{\mathbf{a}_0 - r}{\mathbf{p}_0} \rho_{2,1} w_{1,\mathbf{i},0}}{r} \right. \\
& - \mu_{1,\mathbf{i},1} - \frac{\mu_1, \mathbf{q}_0, 0}{-} \frac{\mathbf{a}_0 - r}{\mathbf{p}_0} \rho_{3,1} w_1, \mathbf{q}_0, 0r - \mu_1, \mathbf{q}_0, 1 - \gamma k_0 \left(\frac{\left(1 - \rho_{2,1}^2\right) w_{1,\mathbf{i},0}^2}{\left(r - \mu_{1,\mathbf{i},1}\right)^2} \right. \\
& - \left(2\left(\rho_{3,2} - \rho_{2,1}\rho_{3,1}\right) w_{1,\mathbf{q}_0,0} w_{1,\mathbf{i},0}\right) / \left(\left(r - \mu_{1,\mathbf{i},1}\right)\left(r - \mu_{1,\mathbf{q}_0,1}\right)\right) \\
& \left. \left. + \frac{\left(1 - \rho_{3,1}^2\right) w_{1,\mathbf{q}_0,0}^2}{\left(r - \mu_{1,\mathbf{q}_0,1}\right)^2} \right) \right) \qquad (6.4.26)
\end{aligned}
$$

Remark 6.4.2. (The dividend puzzle unpuzzled [41].) If the surcharge cost rate is equal to zero, i.e., if $\mathbf{q}_{1,1} = 0$, then the equity value *function* is independent of the coefficients $\alpha_{1,i} (i = 0, 1, 2, 3)$, i.e., the equity value function is independent of the dividend policy (6.2.2). So in the case $\mathbf{q}_{1,1} = 0$, the famous, yet often disputed (see [3]), assertion of Miller and Modigliani (see [29]) is correct, while in the case $\mathbf{q}_{1,1} \neq 0$, it is not: if the cost to the firm depends on the available cash (which seems to be the case in reality), then the Miller–Modigliani assertion about dividend policy irrelevancy is incorrect, and more-elaborate models for equity valuation, such as those presented here, seem to be more appropriate/accurate.

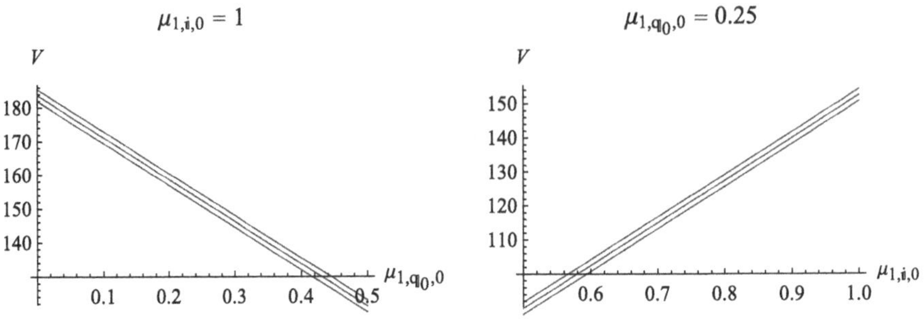

Fig. 6.1 The equity (dollar) value under three different market positions as a function of the cost/income growth rate

6.4.3 *The Role of Market Sentiment in Equity Valuation*

How does market risk-aversion, such as the CRRA relative risk-aversion parameter γ, affect the value of equity? Empirically, the higher the risk-aversion, the lower the equity market prices. Is this phenomenon captured by the above simple model?

Example 6.4.1. For the position $\kappa = X\kappa_0$, with $\kappa_0 = \{k_0\}$, with

$$k_0 = -0.01, 0, 0.01, \tag{6.4.27}$$

short position, no position, and long position (it is important to recall the meaning of κ_0, or equivalently, of k_0; it is the number of *contracts*, in this example the number of *stocks, per unit of wealth*; so, if the wealth is counted in USD, and let's say $X = 10000$, then $k_0 = 0.01$ means 100 stocks; if the wealth is counted in cents (i.e., in 1/100 USD), then $X = 1000000$, and 100 stocks would mean $k_0 = 0.0001$; for consistency, see Example (6.4.2)), and if the rest of the data (expressed in USD) are

$$\begin{aligned}
&r = 0.05, \mathbf{a}_0 = 0.07, \mathbf{p}_0 = 0.2, \mu_{1,\mathbf{i},1} = -0.03, \mu_{1,\mathbf{q}_0,1} = -0.03, w_{1,\mathbf{i},0} = 0.1,\\
&w_{1,\mathbf{q}_0,0} = 0.1, \rho_{i,j} = 0.85, i_1 = 6, q_1 = 1, \mathbb{C}_1 = 15, \gamma = 50, \mathbf{q}_{1,1} = 0.1,\\
&\alpha_{1,0} = 2, \alpha_{1,1} = 0.1, \alpha_{1,2} = 0, \alpha_{1,3} = 0,
\end{aligned} \tag{6.4.28}$$

then the corresponding equity value/price, for fixed income rate dynamics and for fixed basic cost dynamics, respectively, as a function of the basic cost rate dynamics, and as a function of the basic income rate dynamics, respectively, are plotted in Fig. 6.1. As expected, equity value declines as the prospects of the basic cost rate increase, while its value increases as the prospects of the income rate increase. We note that the price is higher for the short position than for no position, which is higher than the price for the long position.

The risk-aversion γ affects the width of the band around the mid-price, obtained for $\gamma = 0$, or also for $\kappa = 0$. The larger the value of γ, the wider the band, since

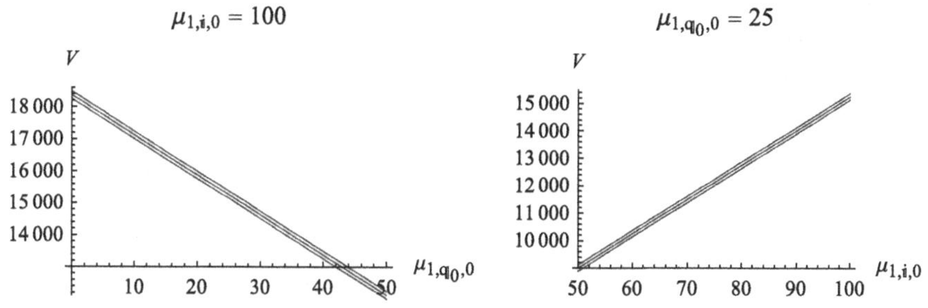

Fig. 6.2 The equity (cents) value under three different market positions as a function of the cost/income growth rate

smaller prices are required by a risk-averse investor to enter a long position, and conversely, higher prices are required for such an investor to enter a short position.

Example 6.4.2. To check for consistency, we rewrite the above example using cents (1/100 USD) instead of USD. The data become

$$r = 0.05, \mathbf{a}_0 = 0.07, \mathbf{p}_0 = 0.2, \mu_{1,\mathbf{i},1} = -0.03, \mu_{1,\mathbf{q}_0,1} = -0.03, w_{1,\mathbf{i},0} = 10,$$
$$w_{1,\mathbf{q}_0,0} = 10, \rho_{i,j} = 0.85, i_1 = 600, q_1 = 100, \mathbb{C}_1 = 1500, \gamma = 50, \mathbf{q}_{1,1} = 0.1,$$
$$\alpha_{1,0} = 200, \alpha_{1,1} = 0.1, \alpha_{1,2} = 0, \alpha_{1,3} = 0, \tag{6.4.29}$$

and instead of (6.4.27), we have

$$k_0 = -0.0001, 0, 0.0001. \tag{6.4.30}$$

The corresponding equity value/price (expressed in cents, i.e., in 1/100 USD) is plotted in Fig. 6.2, which, as can be confirmed by direct calculation, is identical to Fig. 6.1.

Example 6.4.3. We continue Example 6.4.1 by appending the data (6.4.28) with

$$\mu_{1,\mathbf{i},0} = 1, \mu_{1,\mathbf{q}_0,0} = 0.25 \tag{6.4.31}$$

and by deleting the specification for γ. Figure 6.3 shows the dependence of the equity prices on the (market) CRRA risk-aversion parameter γ. We note that the price is higher for the short position than for no position, which is higher than the price for the long position. In the case of a long position, the higher the (market) risk-aversion, the lower the equity (market) price, and conversely for the short position. One could argue that *since the aggregate market is long overall (the total number of stocks outstanding is positive), the higher mar ket risk-aversion yie lds lower prices.*

Remark 6.4.3. The basic equity model in the case of time-dependent coefficients (and $\kappa = 0$) was solved in [41].

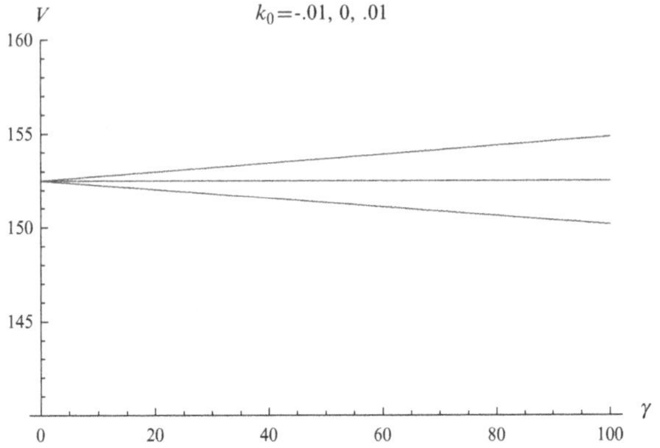

Fig. 6.3 The equity (dollar) value under three different market positions as a function of the relative risk aversion

Remark 6.4.4. An extension of the basic equity model in the case that interest rates are stochastic (and $\kappa = 0$) was studied in [16].

6.5 Basic Equity Model: CRRA Indifference Pricing a Single (Type of) Stock

In the case $\kappa = 0$, CRRA neutral and CRRA indifference prices coincide (Remark 4.10.2). So we consider only the case (6.4.2). Equation (4.10.1) reduces to (6.4.3), and therefore, again, $\nabla_A g_\gamma = 0$. Assuming (cf. (6.4.4))

$$V_1(t, X, S_0, \mathbb{C}_1, i_1, q_1) := \mathbb{A}_{1,1} + \mathbb{C}_1\mathbb{A}_{1,4} + i_1\mathbb{A}_{1,5} + q_1\mathbb{A}_{1,6} \tag{6.5.1}$$

and using (4.10.2), to find the indifference price, we just need to solve (cf. (6.4.7))

$$\begin{aligned}
&-r(\mathbb{A}_{1,1} + \mathbb{C}_1\mathbb{A}_{1,4} + i_1\mathbb{A}_{1,5} + q_1\mathbb{A}_{1,6}) + \mathbb{A}_{1,4}(-\alpha_{1,2}i_1 + i_1 - q_1 + \mathbb{C}_1(r - \mathbf{q}_{1,1})\\
&\quad - \alpha_{1,0} - \mathbb{C}_1\alpha_{1,1} - q_1\alpha_{1,3}) - \frac{1}{2}\gamma k_0\left(-\mathbb{A}_{1,5}^2\left(\rho_{2,1}^2 - 1\right)w_{1,\mathbf{i},0}^2\right.\\
&\quad \left. + 2\mathbb{A}_{1,5}\mathbb{A}_{1,6}\left(\rho_{3,2} - \rho_{2,1}\rho_{3,1}\right)w_{1,\mathbf{q}_0,0}w_{1,\mathbf{i},0} - \mathbb{A}_{1,6}^2\left(\rho_{3,1}^2 - 1\right)w_{1,\mathbf{q}_0,0}^2\right)\\
&\quad + \mathbb{A}_{1,5}\left(-\frac{(\mathbf{a}_0 - r)\rho_{2,1}w_{1,\mathbf{i},0}}{\mathbf{p}_0} + \mu_{1,\mathbf{i},0} + i_1\mu_{1,\mathbf{i},1}\right) + \mathbb{A}_{1,6}\left(-\frac{(\mathbf{a}_0 - r)\rho_{3,1}w_{1,\mathbf{q}_0,0}}{\mathbf{p}_0}\right.\\
&\quad \left. + \mu_{1,\mathbf{q}_0,0} + q_1\mu_{1,\mathbf{q}_0,1}\right) = -\alpha_{1,0} - \mathbb{C}_1\alpha_{1,1} - i_1\alpha_{1,2} - q_1\alpha_{1,3}.
\end{aligned} \tag{6.5.2}$$

Collecting coefficients for $\mathbb{C}_1, i_1$, and q_1, setting them equal to zero, and solving the corresponding equations, we obtain the following theorem.

Theorem 6.5.1. *In the above basic equity framework, assuming (6.4.2), the equity κ -indifference value/price is dependent on the relative risk-aversion parameter γ, and it is equal to*

$$
\begin{aligned}
&V_1(t,S_0,\mathbb{C}_1,i_1,q_1) = V_{1,indifference}(\mathbb{C}_1,i_1,q_1)\\
&= \frac{\mathbb{C}_1\alpha_{1,1}}{\mathbf{q}_{1,1}+\alpha_{1,1}} + \frac{1}{r}\left(-\frac{\alpha_{1,1}\alpha_{1,0}}{\mathbf{q}_{1,1}+\alpha_{1,1}} + \alpha_{1,0} - \frac{1}{2}\gamma(-((-\alpha_{1,1}-\mathbf{q}_{1,1}\alpha_{1,2})\right.\\
&\qquad \sqrt{1-\rho_{2,1}^2}w_{1,\mathbf{i},0}\left(-(k_0(-\alpha_{1,1}-\mathbf{q}_{1,1}\,\alpha_{1,2})\sqrt{1-\rho_{2,1}^2}w_{1,\mathbf{i},0}\right)\Big/\\
&\qquad ((\mathbf{q}_{1,1}+\alpha_{1,1})(r-\mu_{1,\mathbf{i},1})) - (k_0(\alpha_{1,1}-\mathbf{q}_{1,1}\alpha_{1,3})(\rho_{3,2}-\rho_{2,1}\rho_{3,1})w_{1,\mathbf{q}_0,0})\,/\\
&\qquad \left((\mathbf{q}_{1,1}+\alpha_{1,1})\sqrt{1-\rho_{2,1}^2}\;\; r-\mu_{1,\mathbf{q}_0,1})))\right)/((\mathbf{q}_{1,1}+\alpha_{1,1})(r-\mu_{1,\mathbf{i},1}))-1/\\
&\qquad ((\mathbf{q}_{1,1}+\alpha_{1,1})(r-\mu_{1,\mathbf{q}_0,1}))(\alpha_{1,1}-\mathbf{q}_{1,1}\alpha_{1,3})\left(\frac{1}{\sqrt{1-\rho_{2,1}^2}}\right.\\
&\qquad (\rho_{3,2}-\rho_{2,1}\rho_{3,1})w_{1,\mathbf{q}_0,0}\left(-\left(k_0(-\alpha_{1,1}-\mathbf{q}_{1,1}\alpha_{1,2})\sqrt{1-\rho_{2,1}^2}\,w_{1,\mathbf{i},0}\right)\Big/\right.\\
&\qquad ((\mathbf{q}_{1,1}+\alpha_{1,1})(r-\mu_{1,\mathbf{i},1})) - (k_0(\alpha_{1,1}-\mathbf{q}_{1,1}\alpha_{1,3})(\rho_{3,2}-\rho_{2,1}\rho_{3,1})w_{1,\mathbf{q}_0,0})\,/\\
&\qquad \left(\left(\mathbf{q}_{1,1}+\alpha_{1,1})\sqrt{1-\rho_{2,1}^2}(r-\mu_{1,\mathbf{q}_0,1})\right)\right) - (k_0(\alpha_{1,1}-\mathbf{q}_{1,1}\alpha_{1,3})\\
&\qquad \left.\left(-\rho_{3,1}^2+\frac{(\rho_{3,2}-\rho_{2,1}\rho_{3,1})^2}{\rho_{2,1}^2-1}+1\right)w_{1,\mathbf{q}_0,0}^2\right)\Big/\\
&\qquad ((\mathbf{q}_{1,1}+\alpha_{1,1})(r-\mu_{1,\mathbf{q}_0,1})))) - \frac{(-\alpha_{1,1}-\mathbf{q}_{1,1}\alpha_{1,2})\mu_{1,\mathbf{i},0}}{(\mathbf{q}_{1,1}+\alpha_{1,1})(r-\mu_{1,\mathbf{i},1})} +\\
&\qquad ((\mathbf{a}_0-r)(-\alpha_{1,1}-\mathbf{q}_{1,1}\alpha_{1,2})\rho_{2,1}w_{1,\mathbf{i},0})\,/\,(\mathbf{p}_0(\mathbf{q}_{1,1}+\alpha_{1,1})\\
&\qquad (r-\mu_{1,\mathbf{i},1})) - \frac{(\alpha_{1,1}-\mathbf{q}_{1,1}\alpha_{1,3})\mu_{1,\mathbf{q}_0,0}}{(\mathbf{q}_{1,1}+\alpha_{1,1})(r-\mu_{1,\mathbf{q}_0,1})} + ((\mathbf{a}_0-r)(\alpha_{1,1}-\mathbf{q}_{1,1}\alpha_{1,3})\\
&\qquad \rho_{3,1}w_{1,\mathbf{q}_0,0})\,/\,(\mathbf{p}_0(\mathbf{q}_{1,1}+\alpha_{1,1})(r-\mu_{1,\mathbf{q}_0,1})))\\
&\qquad -\frac{i_1(-\alpha_{1,1}-\mathbf{q}_{1,1}\alpha_{1,2})}{(\mathbf{q}_{1,1}+\alpha_{1,1})(r-\mu_{1,\mathbf{i},1})} - \frac{q_1(\alpha_{1,1}-\mathbf{q}_{1,1}\alpha_{1,3})}{(\mathbf{q}_{1,1}+\alpha_{1,1})(r-\mu_{1,\mathbf{q}_0,1})}
\end{aligned}
$$

for $-\infty < t < \infty$.

Corollary 6.5.1 (The Miller–Modigliani case). *If the surcharge cost rate is equal to zero, i.e., if $\mathbf{q}_{1,1} = 0$, then, assuming (6.4.2), the κ-indifference equity price is equal to*

$$V_{1,indifference}(\mathbb{C}_1, i_1, q_1) = \mathbb{C}_1 + \frac{i_1}{r-\mu_{1,\mathbf{i},1}} + \frac{1}{r}\left(\frac{(r-\mathbf{a}_0)\rho_{2,1}w_{1,\mathbf{i},0}}{\mathbf{p}_0(r-\mu_{1,\mathbf{i},1})} + \frac{\mu_{1,\mathbf{i},0}}{r-\mu_{1,\mathbf{i},1}}\right.$$
$$-\frac{\mu_{1,\mathbf{q}_0,0}}{r-\mu_{1,\mathbf{q}_0,1}} + \frac{(\mathbf{a}_0-r)\rho_{3,1}w_{1,\mathbf{q}_0,0}}{\mathbf{p}_0(r-\mu_{1,\mathbf{q}_0,1})} + \left(\gamma k_0\left(\left(\rho_{3,1}^2-1\right)w_{1,\mathbf{q}_0,0}^2(r-\mu_{1,\mathbf{i},1})^2\right.\right.$$
$$-2(\rho_{2,1}\rho_{3,1}-\rho_{3,2})w_{1,\mathbf{i},0}w_{1,\mathbf{q}_0,0}(r-\mu_{1,\mathbf{q}_0,1})(r-\mu_{1,\mathbf{i},1}) + (\rho_{2,1}^2-)$$
$$\left.\left.\left.w_{1,\mathbf{i},0}^2(r-\mu_{1,\mathbf{q}_0,1})^2\right)\right)\Big/\left(2(r-\mu_{1,\mathbf{i},1})^2(r-\mu_{1,\mathbf{q}_0,1})^2\right)\right) - \frac{q_1}{r-\mu_{1,\mathbf{q}_0,1}}.$$

6.6 Basic Equity Model: Relationship Between CRRA Neutral and Indifference Pricing, Investing

As observed already, if $k_0 = 0$, CRRA neutral and CRRA indifference prices are same. Here we consider the case $k_0 \neq 0$.

Remark 6.6.1. The difference between equity indifference and equity neutral price is then equal to

$$V_{1,\text{indifference}}(\mathbb{C}_1, i_1, q_1) - V_1(\mathbb{C}_1, i_1, q_1) = \frac{1}{2r(\mathbf{q}_{1,1}+\alpha_{1,1})^2}$$
$$\gamma k_0\left(-\left((\alpha_{1,1}+\mathbf{q}_{1,1}\,\alpha_{1,2})^2(\rho_{2,1}^2-1)w_{1,\mathbf{i},0}^2\right)/(r-\mu_{1,\mathbf{i},1})^2 + (2(\alpha_{1,1}+\mathbf{q}_{1,1}\alpha_{1,2})\right.$$
$$(\alpha_{1,1}-\mathbf{q}_{1,1}\alpha_{1,3})(\rho_{2,1}\rho_{3,1}-\rho_{3,2})w_{1,\mathbf{q}_0,0}w_{1,\mathbf{i},0})/((r-\mu_{1,\mathbf{i},1})(r-\mu_{1,\mathbf{q}_0,1}))$$
$$\left.-\left((\alpha_{1,1}-\mathbf{q}_{1,1}\alpha_{1,3})^2(\rho_{3,1}^2-1)w_{1,\mathbf{q}_0,0}^2\right)/(r-\mu_{1,\mathbf{q}_0,1})^2\right), \tag{6.6.1}$$

while if $\mathbf{q}_{1,1} = 0$ (the Miller–Modigliani case), it is equal to

$$V_{1,\text{indifference}}(\mathbb{C}_1, i_1, q_1) - V_1(\mathbb{C}_1, i_1, q_1)$$
$$= \frac{\gamma k_0}{2r}\left(\frac{\left(1-\rho_{2,1}^2\right)w_{1,\mathbf{i},0}^2}{(r-\mu_{1,\mathbf{i},1})^2}\left(2(\rho_{2,1}\rho_{3,1}-\rho_{3,2})w_{1,\mathbf{q}_0,0}w_{1,\mathbf{i},0}\right)\Big/\right.$$
$$\left.+((r-\mu_{1,\mathbf{i},1})(r-\mu_{1,\mathbf{q}_0,1})) + \frac{\left(1-\rho_{3,1}^2\right)w_{1,\mathbf{q}_0,0}^2}{(r-\mu_{1,\mathbf{q}_0,1})^2}\right). \tag{6.6.2}$$

So since $r > 0$, for a long position ($k_0 > 0$), $V_{1,\text{indifference}}(\mathbb{C}_1, i_1, q_1) - V_1(\mathbb{C}_1, i_1, q_1) > 0$. Also, if we denote by $V_{1,k_0=0}(\mathbb{C}_1, i_1, q_1)$ the CRRA equity price corresponding to the zero position (in which case neutral is equal to indifference), then we also have

$$\begin{aligned} &V_{1,k_0=0}\left(\mathbb{C}_1,i_1,q_1\right)-V_{1,\text{indifference}}\left(\mathbb{C}_1,i_1,q_1\right)\\ &\quad=V_{1,\text{indifference}}\left(\mathbb{C}_1,i_1,q_1\right)-V_1\left(\mathbb{C}_1,i_1,q_1\right) \end{aligned} \tag{6.6.3}$$

(whether or not $\mathbf{q}_{1,1}=0$). So the equity $(X\{k_0\})$-indifference price is "halfway between" the equity (zero position) neutral (or equivalently, indifference) price and the $(X\{k_0\})$-neutral price.

Remark 6.6.2. What is achieved here is an explicit relationship between equity prices and equity positions, including the equity optimal position (the κ-neutral price), which is the fundamental relationship for *investing*. For example, solving (6.4.24) for k_0, we get the optimal position $\kappa_{\text{opt}}=X\left\{k_{0,\text{opt}}\right\}$, long or short, as a function of the stock price V_1:

$$\begin{aligned} k_{0,\text{opt}}=&\Bigg(-V_1+\frac{\alpha_{1,0}}{r}+\frac{\mathbb{C}_1\alpha_{1,1}}{\mathbf{q}_{1,1}+\alpha_{1,1}}-\frac{\alpha_{1,0}\alpha_{1,1}}{r\left(\mathbf{q}_{1,1}+\alpha_{1,1}\right)}-\frac{i_1\left(-\alpha_{1,1}-\mathbf{q}_{1,1}\alpha_{1,2}\right)}{\left(\mathbf{q}_{1,1}+\alpha_{1,1}\right)\left(r-\mu_{1,\mathbf{i},1}\right)}\\ &-\frac{\left(-\alpha_{1,1}-\mathbf{q}_{1,1}\alpha_{1,2}\right)\mu_{1,\mathbf{i},0}}{r\left(\mathbf{q}_{1,1}+\alpha_{1,1}\right)\left(r-\mu_{1,\mathbf{i},1}\right)}+\left(\left(\mathbf{a}_0-r\right)\left(-\alpha_{1,1}-\mathbf{q}_{1,1}\alpha_{1,2}\right)\rho_{2,1}w_{1,\mathbf{i},0}\right)/\\ &\left(r\mathbf{p}_0\left(\mathbf{q}_{1,1}+\alpha_{1,1}\right)\left(r-\mu_{1,\mathbf{i},1}\right)\right)-\frac{q_1\left(\alpha_{1,1}-\mathbf{q}_{1,1}\alpha_{1,3}\right)}{\left(\mathbf{q}_{1,1}+\alpha_{1,1}\right)\left(r-\mu_{1,\mathbf{q}_0,1}\right)}-\\ &\frac{\left(\alpha_{1,1}-\mathbf{q}_{1,1}\alpha_{1,3}\right)\mu_{1,\mathbf{q}_0,0}}{r\left(\mathbf{q}_{1,1}+\alpha_{1,1}\right)\left(r-\mu_{1,\mathbf{q}_0,1}\right)}+\left(\left(\mathbf{a}_0-r\right)\left(\alpha_{1,1}-\mathbf{q}_{1,1}\alpha_{1,3}\right)\rho_{3,1}w_{1,\mathbf{q}_0,0}\right)/\\ &\left(r\mathbf{p}_0\ \mathbf{q}_{1,1}+\alpha_{1,1}\right)\left(r-\mu_{1,\mathbf{q}_0,1}\right)\Big)\Big)/\left(\left(\gamma\left(-\alpha_{1,1}-\mathbf{q}_{1,1}\alpha_{1,2}\right)^2\right.\right.\\ &\left(1-\rho_{2,1}^2\right)w_{1,\mathbf{i},0}^2\Big)/\left(r\left(\mathbf{q}_{1,1}+\alpha_{1,1}\right)^2\left(r-\mu_{1,\mathbf{i},1}\right)^2\right)+\left(2\gamma\left(-\alpha_{1,1}-\mathbf{q}_{1,1}\alpha_{1,2}\right)\right.\\ &\left(\alpha_{1,1}-\mathbf{q}_{1,1}\alpha_{1,3}\right)\left(\rho_{3,2}-\rho_{2,1}\rho_{3,1}\right)w_{1,\mathbf{q}_0,0}w_{1,\mathbf{i},0}\Big)/\left(r\left(\mathbf{q}_{1,1}+\alpha_{1,1}\right)^2\right.\\ &\left(r-\mu_{1,\mathbf{i},1}\right)\left(r-\mu_{1,\mathbf{q}_0,1}\right)\Big)+\Bigg(\gamma\left(\alpha_{1,1}-\mathbf{q}_{1,1}\alpha_{1,3}\right)^2\\ &\left(-\rho_{3,1}^2+\frac{\left(\rho_{3,2}-\rho_{2,1}\rho_{3,1}\right)^2}{\rho_{2,1}^2-1}+1\right)w_{1,\mathbf{q}_0,0}^2\Bigg)\Bigg/\left(r\left(\mathbf{q}_{1,1}+\alpha_{1,1}\right)^2\right.\\ &\left(r-\mu_{1,\mathbf{q}_0,1}\right)^2\Big)+\left(\gamma\left(\alpha_{1,1}-\mathbf{q}_{1,1}\alpha_{1,3}\right)^2\left(\rho_{3,2}-\rho_{2,1}\rho_{3,1}\right)^2\right.\\ &w_{1,\mathbf{q}_0,0}^2\Big)/\left(r\left(\mathbf{q}_{1,1}+\alpha_{1,1}\right)^2\left(1-\rho_{2,1}^2\right)\left(r-\mu_{1,\mathbf{q}_0,1}\right)^2\right)\Big), \end{aligned} \tag{6.6.4}$$

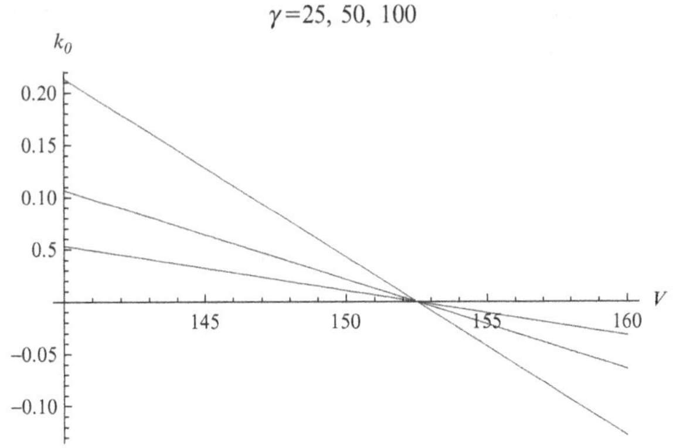

Fig. 6.4 Optimal (market) position as a function of (market) prices (for three different values of the relative risk aversion)

or in the Miller–Modigliani case ($\mathbf{q}_{1,1} = 0$),

$$
\begin{aligned}
k_{0,\text{opt}} = \Bigg(& -V_1 + \frac{i_1}{r-\mu_{1,\mathbf{i},1}} + \mathbb{C}_1 + \frac{\mu_{1,\mathbf{i},0}}{r\left(r-\mu_{1,\mathbf{i},1}\right)} - \frac{\left(\mathbf{a}_0 - r\right)\rho_{2,1} w_{1,\mathbf{i},0}}{r\mathbf{p}_0\left(r-\mu_{1,\mathbf{i},1}\right)} - \frac{q_1}{r-\mu_{1,\mathbf{q}_0,1}} \\
& - \frac{\mu_{1,\mathbf{q}_0,0}}{r\left(r-\mu_{1,\mathbf{q}_0,1}\right)} + \frac{\left(\mathbf{a}_0 - r\right)\rho_{3,1} w_{1,\mathbf{q}_0,0}}{r\mathbf{p}_0\left(r-\mu_{1,\mathbf{q}_0,1}\right)} \Bigg) \Bigg/ \left(\frac{\gamma\left(1-\rho_{2,1}^2\right) w_{1,\mathbf{i},0}^2}{r\left(r-\mu_{1,\mathbf{i},1}\right)^2} \right. \\
& - \left(2\gamma\left(\rho_{3,2} - \rho_{2,1}\rho_{3,1}\right) w_{1,\mathbf{q}_0,0} w_{1,\mathbf{i},0}\right) \Big/ \left(r\left(r-\mu_{1,\mathbf{i},1}\right)\left(r-\mu_{1,\mathbf{q}_0,1}\right)\right) \\
& + \left(\gamma\left(-\rho_{3,1}^2 + \frac{\left(\rho_{3,2} - \rho_{2,1}\rho_{3,1}\right)^2}{\rho_{2,1}^2 - 1} + 1\right) w_{1,\mathbf{q}_0,0}^2\right) \Big/ \left(r\left(r-\mu_{1,\mathbf{q}_0,1}\right)^2\right) \\
& \left. + \frac{\gamma\left(\rho_{3,2} - \rho_{2,1}\rho_{3,1}\right)^2 w_{1,\mathbf{q}_0,0}^2}{r\left(1-\rho_{2,1}^2\right)\left(r-\mu_{1,\mathbf{q}_0,1}\right)^2} \right).
\end{aligned}
$$

Remark 6.6.3. Much more generally, these explicit results are examples of a general framework of connecting prices and optimal positions, provided by the concept of κ-neutral pricing.

Example 6.6.1. We continue Examples 6.4.1 and 6.4.3, using the same data. Figure 6.4 shows the dependence of the optimal position k_0 as a function of the equity price V, for different values of the CRRA risk-aversion parameter γ. Higher γ implies a less-pronounced position k_0, long or short, and conversely. At the price $V = 152.5$, all three optimal positions, regardless of the risk-aversion, are equal to $k_0 = 0$.

6.7 Pricing of and Investing in Equity Portfolios

Above, we were able to find equity price in a very simple form, as a linear function of factors. We found a linear solution of a nonlinear pricing PDE. When pricing portfolios, we have to solve a nonlinear system of pricing PDEs. Will the same strategy still work? Can we still get explicit formulas in the case of equity portfolios? Moreover, if we succeed in finding pricing formulas, can we invert them and find optimal position (investment) formulas, analogous to (6.6.4)? If so, such formulas could be quite useful, since equity investing is usually done for portfolios of stocks, not just for a single one. Also, if so, will there be some substantive difference in the conclusions, for example in regard to Miller–Modigliani's dividend policy irrelevancy context (in the case $\mathbf{q}_{j,1} = 0$)?

As in Sect. 6.3, for simplicity of presentation, consider a portfolio of length 2, i.e., assume that $\kappa = \{\kappa_1, \kappa_2\}$, with a tradable $S = \{S_0\}$ and factors $A = \{S_0, \mathbb{C}_1, i_1, q_1, \mathbb{C}_2, i_2, q_2\}$, and with the dividend policies and market coefficients as given in Sect. 6.3.

For example, we shall consider CRRA κ-neutral pricing.

If $\kappa = 0$, considering a portfolio reduces to considering individual components – nothing new is achieved, so we skip that case, and concentrate on the analogue of the case (6.4.2), i.e., we shall assume that

$$\kappa = X\kappa_0 = X\{k_1, k_2\} \wedge \left(w_{1,\mathbf{q}_0,1} = 0 \wedge w_{1,\mathbf{i},1} = 0 \wedge w_{2,\mathbf{q}_0,1} = 0 \wedge w_{2,\mathbf{i},1} = 0\right) \quad (6.7.1)$$

with k_1, k_2 constants. Assuming $g_\gamma = g(t)$ and assuming that the solution of the pricing PDE (4.9.3) has the form $V = \{V_1, V_2\}$ with

$$\begin{aligned} &V_j(t, S_0, \mathbb{C}_1, i_1, q_1, \mathbb{C}_2, i_2, q_2) \\ &\quad := \mathbb{A}_{j,1} + \mathbb{C}_1\mathbb{A}_{j,3} + i_1\mathbb{A}_{j,4} + q_1\mathbb{A}_{j,5} + \mathbb{C}_2\mathbb{A}_{j,6} + i_2\mathbb{A}_{j,7} + q_2\mathbb{A}_{j,8} \quad (6.7.2) \end{aligned}$$

for $j = 1, 2$, with $\mathbb{A}_{1,1}, \mathbb{A}_{1,3}, \mathbb{A}_{1,4}, \mathbb{A}_{1,5}, \mathbb{A}_{1,6}, \mathbb{A}_{1,7}, \mathbb{A}_{1,8}, \mathbb{A}_{2,1}, \mathbb{A}_{2,3}, \mathbb{A}_{2,4}, \mathbb{A}_{2,5}, \mathbb{A}_{2,6}$, $\mathbb{A}_{2,7}$, and $\mathbb{A}_{2,8}$ constants, (4.9.2) simplifies (instead of to (6.4.5)) to

$$\begin{aligned} &\frac{1}{2}(\gamma-1)\gamma\Big(\Big(\mathbb{A}_{1,4}^2\left(\rho_{2,1}^2-1\right)w_{1,\mathbf{i},0}^2 + 2\mathbb{A}_{1,4}\Big(\mathbb{A}_{1,5}\left(\rho_{2,1}\rho_{3,1}-\rho_{3,2}\right)w_{1,\mathbf{q}_0,0} \\ &\quad + \mathbb{A}_{1,7}\left(\rho_{2,1}\rho_{4,1}-\rho_{4,2}\right)w_{2,\mathbf{i},0} + \mathbb{A}_{1,8}\left(\rho_{2,1}\rho_{5,1}-\rho_{5,2}\right)w_{2,\mathbf{q}_0,0}\Big)w_{1,\mathbf{i},0} \\ &\quad + \mathbb{A}_{1,5}^2\left(\rho_{3,1}^2-1\right)w_{1,\mathbf{q}_0,0}^2 - \mathbb{A}_{1,7}^2 w_{2,\mathbf{i},0}^2 + \mathbb{A}_{1,7}^2\rho_{4,1}^2 w_{2,\mathbf{i},0}^2 \\ &\quad - \mathbb{A}_{1,8}^2 w_{2,\mathbf{q}_0,0}^2 + \mathbb{A}_{1,8}^2\rho_{5,1}^2 w_{2,\mathbf{q}_0,0}^2 + 2\mathbb{A}_{1,7}\mathbb{A}_{1,8}\rho_{4,1}\rho_{5,1}w_{2,\mathbf{i},0}w_{2,\mathbf{q}_0,0} \\ &\quad - 2\mathbb{A}_{1,7}\mathbb{A}_{1,8}\rho_{5,4}w_{2,\mathbf{i},0}w_{2,\mathbf{q}_0,0} + 2\mathbb{A}_{1,5}w_{1,\mathbf{q}_0,0}\left(\mathbb{A}_{1,7}\left(\rho_{3,1}\rho_{4,1}-\rho_{4,3}\right)w_{2,\mathbf{i},0}\right. \\ &\quad \left. + \mathbb{A}_{1,8}\left(\rho_{3,1}\rho_{5,1}-\rho_{5,3}\right)w_{2,\mathbf{q}_0,0}\right)\Big)k_1^2 + 2k_2\Big(\mathbb{A}_{1,7}\mathbb{A}_{2,7}\rho_{4,1}^2 w_{2,\mathbf{i},0}^2 - \mathbb{A}_{1,7}\mathbb{A}_{2,7}w_{2,\mathbf{i},0}^2 \\ &\quad + \mathbb{A}_{1,7}\mathbb{A}_{2,4}\rho_{2,1}\rho_{4,1}w_{1,\mathbf{i},0}w_{2,\mathbf{i},0} - \mathbb{A}_{1,7}\mathbb{A}_{2,4}\rho_{4,2}w_{1,\mathbf{i},0}w_{2,\mathbf{i},0} \end{aligned}$$

$$
\begin{aligned}
&+\mathbb{A}_{1,7}\mathbb{A}_{2,5}\rho_{3,1}\rho_{4,1}w_{1,\mathbf{q}_0,0}w_{2,\mathbf{i},0}-\mathbb{A}_{1,7}\mathbb{A}_{2,5}\rho_{4,3}w_{1,\mathbf{q}_0,0}w_{2,\mathbf{i},0}\\
&+\mathbb{A}_{1,8}\mathbb{A}_{2,7}\rho_{4,1}\rho_{5,1}w_{2,\mathbf{q}_0,0}w_{2,\mathbf{i},0}+\mathbb{A}_{1,7}\mathbb{A}_{2,8}\rho_{4,1}\rho_{5,1}w_{2,\mathbf{q}_0,0}w_{2,\mathbf{i},0}\\
&-\mathbb{A}_{1,8}\mathbb{A}_{2,7}\rho_{5,4}w_{2,\mathbf{q}_0,0}w_{2,\mathbf{i},0}-\mathbb{A}_{1,7}\mathbb{A}_{2,8}\rho_{5,4}w_{2,\mathbf{q}_0,0}w_{2,\mathbf{i},0}+\mathbb{A}_{1,8}\mathbb{A}_{2,8}\rho_{5,1}^2w_{2,\mathbf{q}_0,0}^2\\
&-\mathbb{A}_{1,8}\mathbb{A}_{2,8}w_{2,\mathbf{q}_0,0}^2+\mathbb{A}_{1,8}\mathbb{A}_{2,4}\rho_{2,1}\rho_{5,1}w_{1,\mathbf{i},0}w_{2,\mathbf{q}_0,0}-\mathbb{A}_{1,8}\mathbb{A}_{2,4}\rho_{5,2}w_{1,\mathbf{i},0}w_{2,\mathbf{q}_0,0}\\
&+\mathbb{A}_{1,8}\mathbb{A}_{2,5}\rho_{3,1}\rho_{5,1}w_{1,\mathbf{q}_0,0}w_{2,\mathbf{q}_0,0}-\mathbb{A}_{1,8}\mathbb{A}_{2,5}\rho_{5,3}w_{1,\mathbf{q}_0,0}w_{2,\mathbf{q}_0,0}\\
&+\mathbb{A}_{1,4}w_{1,\mathbf{i},0}\left(\mathbb{A}_{2,4}\left(\rho_{2,1}^2-1\right)w_{1,\mathbf{i},0}+\mathbb{A}_{2,5}\left(\rho_{2,1}\rho_{3,1}-\rho_{3,2}\right)w_{1,\mathbf{q}_0,0}\right.\\
&\left.+\mathbb{A}_{2,7}\rho_{2,1}\rho_{4,1}w_{2,\mathbf{i},0}-\mathbb{A}_{2,7}\rho_{4,2}w_{2,\mathbf{i},0}+\mathbb{A}_{2,8}\rho_{2,1}\rho_{5,1}w_{2,\mathbf{q}_0,0}-\mathbb{A}_{2,8}\rho_{5,2}w_{2,\mathbf{q}_0,0}\right)\\
&+\mathbb{A}_{1,5}w_{1,\mathbf{q}_0,0}\left(\mathbb{A}_{2,4}\left(\rho_{2,1}\rho_{3,1}-\rho_{3,2}\right)w_{1,\mathbf{i},0}+\mathbb{A}_{2,5}\left(\rho_{3,1}^2-1\right)w_{1,\mathbf{q}_0,0}\right.\\
&\left.\left.+\mathbb{A}_{2,7}\rho_{3,1}\rho_{4,1}w_{2,\mathbf{i},0}-\mathbb{A}_{2,7}\rho_{4,3}w_{2,\mathbf{i},0}+\mathbb{A}_{2,8}\rho_{3,1}\rho_{5,1}w_{2,\mathbf{q}_0,0}-\mathbb{A}_{2,8}\rho_{5,3}w_{2,\mathbf{q}_0,0}\right)\right)k_1\\
&+k_2^2\left(\mathbb{A}_{2,4}^2\left(\rho_{2,1}^2-1\right)w_{1,\mathbf{i},0}^2+2\mathbb{A}_{2,4}\left(\mathbb{A}_{2,5}\left(\rho_{2,1}\rho_{3,1}-\rho_{3,2}\right)w_{1,\mathbf{q}_0,0}\right.\right.\\
&\left.+\mathbb{A}_{2,7}\left(\rho_{2,1}\rho_{4,1}-\rho_{4,2}\right)w_{2,\mathbf{i},0}+\mathbb{A}_{2,8}\left(\rho_{2,1}\rho_{5,1}-\rho_{5,2}\right)w_{2,\mathbf{q}_0,0}\right)w_{1,\mathbf{i},0}\\
&+\mathbb{A}_{2,5}^2\left(\rho_{3,1}^2-1\right)w_{1,\mathbf{q}_0,0}^2-\mathbb{A}_{2,7}^2w_{2,\mathbf{i},0}^2+\mathbb{A}_{2,7}^2\rho_{4,1}^2w_{2,\mathbf{i},0}^2\\
&-\mathbb{A}_{2,8}^2w_{2,\mathbf{q}_0,0}^2+\mathbb{A}_{2,8}^2\rho_{5,1}^2w_{2,\mathbf{q}_0,0}^2+2\mathbb{A}_{2,7}\mathbb{A}_{2,8}\rho_{4,1}\rho_{5,1}w_{2,\mathbf{i},0}w_{2,\mathbf{q}_0,0}\\
&-2\mathbb{A}_{2,7}\mathbb{A}_{2,8}\rho_{5,4}w_{2,\mathbf{i},0}w_{2,\mathbf{q}_0,0}+2\mathbb{A}_{2,5}w_{1,\mathbf{q}_0,0}\left(\mathbb{A}_{2,7}\left(\rho_{3,1}\rho_{4,1}-\rho_{4,3}\right)w_{2,\mathbf{i},0}\right.\\
&\left.\left.\left.+\mathbb{A}_{2,8}\left(\rho_{3,1}\rho_{5,1}-\rho_{5,3}\right)w_{2,\mathbf{q}_0,0}\right)\right)\right)+g'(t)=\frac{\gamma-1}{\gamma}\left(\frac{(\mathbf{a}_0-r)^2}{2\mathbf{p}_0^2}+r\gamma\right) \qquad (6.7.3)
\end{aligned}
$$

for $t<T$, with the terminal condition $g(T)=0$, confirming again, conditionally on the assumption (6.7.2), that $g_\gamma = g(t)$, and therefore, that $\nabla_A g_\gamma = 0$. Assuming now (6.7.2), the pricing PDE system (4.9.3) is indeed solved in that form, and collecting coefficients for $\mathbb{C}_j$, i_j, and q_j, for $j=1,2$, setting them equal to zero, and solving the corresponding equations, we conclude that

$$
\begin{aligned}
\{V_1,V_2\} &= \{\beta_1+k_1\delta_{1,1}+k_2\delta_{1,2},\beta_2+k_1\delta_{2,1}+k_2\delta_{2,2}\}\\
&=\begin{pmatrix}\delta_{1,1} & \delta_{1,2}\\ \delta_{2,1} & \delta_{2,2}\end{pmatrix}\cdot\{k_1,k_2\}+\{\beta_1,\beta_2\} \qquad (6.7.4)
\end{aligned}
$$

with

$$
\begin{aligned}
\delta_{1,1} = &\frac{1}{r(\mathbf{q}_{1,1}+\alpha_{1,1})^2}\gamma\left(\left(\left(\alpha_{1,1}+\mathbf{q}_{1,1}\alpha_{1,2}\right)^2\left(\rho_{2,1}^2-1\right)w_{1,\mathbf{i},0}^2\right)/\left(r-\mu_{1,\mathbf{i},1}\right)^2\right.\\
&-\left(2\left(\alpha_{1,1}+\mathbf{q}_{1,1}\alpha_{1,2}\right)\left(\alpha_{1,1}-\mathbf{q}_{1,1}\alpha_{1,3}\right)\left(\rho_{2,1}\rho_{3,1}-\rho_{3,2}\right)w_{1,\mathbf{q}_0,0}w_{1,\mathbf{i},0}\right)/\\
&\left(\left(r-\mu_{1,\mathbf{i},1}\right)\left(r-\mu_{1,\mathbf{q}_0,1}\right)\right)-\left(\left(\alpha_{1,1}-\mathbf{q}_{1,1}\alpha_{1,3}\right)^2\left(-\rho_{3,1}^2+\frac{\left(\rho_{3,2}-\rho_{2,1}\rho_{3,1}\right)^2}{\rho_{2,1}^2-1}+1\right)\right.
\end{aligned}
$$

$$w_{1,\mathbf{q}_0,0}^2\Big) \Big/ (r-\mu_{1,\mathbf{q}_0,1})^2 + ((\alpha_{1,1}-\mathbf{q}_{1,1}\alpha_{1,3})^2 (\rho_{3,2}-\rho_{2,1}\rho_{3,1})^2\ w_{1,\mathbf{q}_0,0}^2) /$$

$$((\rho_{2,1}^2-1)(r-\mu_{1,\mathbf{q}_0,1})^2)), \tag{6.7.5}$$

$$\begin{aligned}
\delta_{1,2} = {} & (\gamma(((\alpha_{1,1}+\mathbf{q}_{1,1}\alpha_{1,2})(\alpha_{2,1}+\mathbf{q}_{2,1}\alpha_{2,2})(\rho_{2,1}\rho_{4,1}-\rho_{4,2})w_{1,\mathbf{i},0}w_{2,\mathbf{i},0}) / \\
& ((r-\mu_{1,\mathbf{i},1})(r-\mu_{2,\mathbf{i},1})) - ((\alpha_{1,1}-\mathbf{q}_{1,1}\alpha_{1,3})(\alpha_{2,1}+\mathbf{q}_{2,1}\ \alpha_{2,2})(\rho_{2,1}) \cdot \\
& \rho_{3,1}-\rho_{3,2})(\rho_{2,1}\rho_{4,1}-\rho_{4,2})w_{1,\mathbf{q}_0,0}w_{2,\mathbf{i},0}) / ((\rho_{2,1}^2-1)\,r-\mu_{1,\mathbf{q}_0,1}) \\
& (r-\mu_{2,\mathbf{i},1})) + ((\alpha_{1,1}-\mathbf{q}_{1,1}\alpha_{1,3})(\alpha_{2,1}+\mathbf{q}_{2,1}\alpha_{2,2})(\rho_{4,3}\rho_{2,1}^2-\rho_{2,1}\rho_{3,2}\rho_{4,1} \\
& +\rho_{3,2}\rho_{4,2}+\rho_{3,1}(\rho_{4,1}-\rho_{2,1}\rho_{4,2})-\rho_{4,3})w_{1,\mathbf{q}_0,0}w_{2,\mathbf{i},0}) / \\
& ((\rho_{2,1}^2-1)(r-\mu_{1,\mathbf{q}_0,1})(r-\mu_{2,\mathbf{i},1})) - ((\alpha_{1,1}+\mathbf{q}_{1,1}\alpha_{1,2}) \\
& (\alpha_{2,1}-\mathbf{q}_{2,1}\alpha_{2,3})(\rho_{2,1}\rho_{5,1}-\rho_{5,2})w_{1,\mathbf{i},0}w_{2,\mathbf{q}_0,0}) / \\
& ((r-\mu_{1,\mathbf{i},1})(r-\mu_{2,\mathbf{q}_0,1})) + ((\alpha_{1,1}-\mathbf{q}_{1,1}\alpha_{1,3})(\alpha_{2,1}-\mathbf{q}_{2,1}\alpha_{2,3}) \\
& (\rho_{2,1}\ \rho_{3,1}-\rho_{3,2})(\rho_{2,1}\rho_{5,1}-\rho_{5,2})w_{1,\mathbf{q}_0,0}w_{2,\mathbf{q}_0,0}) / ((\rho_{2,1}^2-1)(r-\mu_{1,\mathbf{q}_0,1}) \\
& (r-\mu_{2,\mathbf{q}_0,1})) - ((\alpha_{1,1}-\mathbf{q}_{1,1}\alpha_{1,3})(\alpha_{2,1}-\mathbf{q}_{2,1}\alpha_{2,3})(\rho_{5,3}\rho_{2,1}^2-\rho_{2,1}\rho_{3,2}\rho_{5,1} \\
& +\rho_{3,2}\rho_{5,2}+\rho_{3,1}(\rho_{5,1}-\rho_{2,1}\rho_{5,2})-\rho_{5,3})w_{1,\mathbf{q}_0,0}w_{2,\mathbf{q}_0,0}) / \\
& ((\rho_{2,1}^2-1)(r-\mu_{1,\mathbf{q}_0,1})(r-\mu_{2,\mathbf{q}_0,1})))) / \\
& (r(\mathbf{q}_{1,1}+\alpha_{1,1})(\mathbf{q}_{2,1}+\alpha_{2,1})),
\end{aligned} \tag{6.7.6}$$

$$\begin{aligned}
\beta_1 = {} & \frac{\alpha_{1,0}}{r} + \frac{\mathbb{C}_1\alpha_{1,1}}{\mathbf{q}_{1,1}+\alpha_{1,1}} - \frac{\alpha_{1,0}\alpha_{1,1}}{r(\mathbf{q}_{1,1}+\alpha_{1,1})} - \frac{i_1(-\alpha_{1,1}-\mathbf{q}_{1,1}\alpha_{1,2})}{(\mathbf{q}_{1,1}+\alpha_{1,1})(r-\mu_{1,\mathbf{i},1})} \\
& + ((-r+\mathbf{a}_0) - \alpha_{1,1}-\mathbf{q}_{1,1}\alpha_{1,2})\rho_{2,1}w_{1,\mathbf{i},0}) / (r\mathbf{p}_0(\mathbf{q}_{1,1}+\alpha_{1,1})(r-\mu_{1,\mathbf{i},1})) \\
& - \frac{(-\alpha_{1,1}-\mathbf{q}_{1,1}\alpha_{1,2})\mu_{1,\mathbf{i},0}}{r(\mathbf{q}_{1,1}+\alpha_{1,1})(r-\mu_{1,\mathbf{i},1})} - \frac{q_1(\alpha_{1,1}-\mathbf{q}_{1,1}\alpha_{1,3})}{(\mathbf{q}_{1,1}+\alpha_{1,1})(r-\mu_{1,\mathbf{q}_0,1})} + ((-r+\mathbf{a}_0) \\
& (\alpha_{1,1}-\mathbf{q}_{1,1}\ \alpha_{1,3})\rho_{3,1}w_{1,\mathbf{q}_0,0}) / (r\mathbf{p}_0(\mathbf{q}_{1,1}+\alpha_{1,1})(r-\mu_{1,\mathbf{q}_0,1})) \\
& - \frac{(\alpha_{1,1}-\mathbf{q}_{1,1}\alpha_{1,3})\mu_{1,\mathbf{q}_0,0}}{r(\mathbf{q}_{1,1}+\alpha_{1,1})(r-\mu_{1,\mathbf{q}_0,1})},
\end{aligned} \tag{6.7.7}$$

$$\begin{aligned}
\delta_{2,1} = {} & (\gamma(((\alpha_{1,1}+\mathbf{q}_{1,1}\alpha_{1,2})(\alpha_{2,1}+\mathbf{q}_{2,1}\alpha_{2,2})(\rho_{2,1}\rho_{4,1}-\rho_{4,2})w_{1,\mathbf{i},0}w_{2,\mathbf{i},0}) / \\
& ((r-\ \mu_{1,\mathbf{i},1})(r-\mu_{2,\mathbf{i},1})) - ((\alpha_{1,1}-\mathbf{q}_{1,1}\alpha_{1,3})(\alpha_{2,1}+\mathbf{q}_{2,1}\alpha_{2,2}) \\
& (\rho_{2,1}\rho_{3,1}-\rho_{3,2})(\rho_{2,1}\rho_{4,1}-\rho_{4,2})w_{1,\mathbf{q}_0,0}w_{2,\mathbf{i},0}) / ((\rho_{2,1}^2-1) \\
& (r-\mu_{1,\mathbf{q}_0,1})(r-\mu_{2,\mathbf{i},1})) + ((\alpha_{1,1}-\mathbf{q}_{1,1}\alpha_{1,3})(\alpha_{2,1}+\mathbf{q}_{2,1}\alpha_{2,2})
\end{aligned}$$

$$\left(\rho_{4,3}\rho_{2,1}^2-\rho_{2,1}\rho_{3,2}\rho_{4,1}+\rho_{3,2}\rho_{4,2}+\rho_{3,1}\left(\rho_{4,1}-\rho_{2,1}\rho_{4,2}\right)-\rho_{4,3}\right)$$
$$w_{1,\mathbf{q}_0,0}w_{2,\mathbf{i},0}\right)/\left(\left(\rho_{2,1}^2-1\right)\left(r-\mu_{1,\mathbf{q}_0,1}\right)\left(r-\mu_{2,\mathbf{i},1}\right)\right)$$
$$-\left(\left(\alpha_{1,1}+\mathbf{q}_{1,1}\alpha_{1,2}\right)\left(\alpha_{2,1}-\mathbf{q}_{2,1}\alpha_{2,3}\right)\left(\rho_{2,1}\rho_{5,1}-\rho_{5,2}\right)w_{1,\mathbf{i},0}w_{2,\mathbf{q}_0,0}\right)/$$
$$\left(\left(r-\mu_{1,\mathbf{i},1}\right)\left(r-\mu_{2,\mathbf{q}_0,1}\right)\right)+\left(\left(\alpha_{1,1}-\mathbf{q}_{1,1}\alpha_{1,3}\right)\left(\alpha_{2,1}-\mathbf{q}_{2,1}\alpha_{2,3}\right)\right.$$
$$\left(\rho_{2,1}\,\rho_{3,1}-\rho_{3,2}\right)\left(\rho_{2,1}\rho_{5,1}-\rho_{5,2}\right)w_{1,\mathbf{q}_0,0}w_{2,\mathbf{q}_0,0}\right)/\left(\left(\rho_{2,1}^2-1\right)\left(r-\mu_{1,\mathbf{q}_0,1}\right)\right.$$
$$\left(r-\mu_{2,\mathbf{q}_0,1}\right)\right)-\left(\left(\alpha_{1,1}-\mathbf{q}_{1,1}\alpha_{1,3}\right)\left(\alpha_{2,1}-\mathbf{q}_{2,1}\alpha_{2,3}\right)\left(\rho_{5,3}\rho_{2,1}^2-\rho_{2,1}\rho_{3,2}\rho_{5,1}\right.\right.$$
$$+\rho_{3,2}\rho_{5,2}+\rho_{3,1}\left(\rho_{5,1}-\rho_{2,1}\rho_{5,2}\right)-\rho_{5,3}\right)w_{1,\mathbf{q}_0,0}w_{2,\mathbf{q}_0,0}\right)/$$
$$\left(\left(\rho_{2,1}^2-1\right)\left(r-\mu_{1,\mathbf{q}_0,1}\right)\left(r-\mu_{2,\mathbf{q}_0,1}\right)\right)\right)\right)/\left(r\left(\mathbf{q}_{1,1}+\alpha_{1,1}\right)\left(\mathbf{q}_{2,1}+\alpha_{2,1}\right)\right), \tag{6.7.8}$$

$$\delta_{2,2}=\left(\gamma\left(\left(\alpha_{2,3}^2\left(\rho_{5,1}^2-1\right)w_{2,\mathbf{q}_0,0}^2\left(r-\mu_{2,\mathbf{i},1}\right)^2+2\alpha_{2,2}\alpha_{2,3}\left(\rho_{4,1}\rho_{5,1}-\rho_{5,4}\right)w_{2,\mathbf{i},0}\right.\right.\right.$$
$$w_{2,\mathbf{q}_0,0}\left(r-\mu_{2,\mathbf{q}_0,1}\right)\left(r-\mu_{2,\mathbf{i},1}\right)+\alpha_{2,2}^2\left(\rho_{4,1}^2-1\right)w_{2,\mathbf{i},0}^2\left(r-\mu_{2,\mathbf{q}_0,1}\right)^2\right)$$
$$\mathbf{q}_{2,1}^2+2\alpha_{2,1}\left(\alpha_{2,2}w_{2,\mathbf{i},0}\left(\left(\rho_{4,1}^2-1\right)w_{2,\mathbf{i},0}\left(r-\mu_{2,\mathbf{q}_0,1}\right)-\left(\rho_{4,1}\rho_{5,1}-\rho_{5,4}\right)\right.\right.$$
$$w_{2,\mathbf{q}_0,0}\left(r-\mu_{2,\mathbf{i},1}\right)\right)\left(r-\mu_{2,\mathbf{q}_0,1}\right)+\alpha_{2,3}w_{2,\mathbf{q}_0,0}$$
$$\left(r-\mu_{2,\mathbf{i},1}\right)\left(-\left(\rho_{5,1}^2-1\right)w_{2,\mathbf{q}_0,0}\left(r-\mu_{2,\mathbf{i},1}\right)+\rho_{4,1}\rho_{5,1}w_{2,\mathbf{i},0}\left(r-\mu_{2,\mathbf{q}_0,1}\right)\right.$$
$$\left.\left.+\rho_{5,4}w_{2,\mathbf{i},0}\left(\mu_{2,\mathbf{q}_0,1}-r\right)\right)\right)\mathbf{q}_{2,1}+\alpha_{2,1}^2\left(\left(\rho_{5,1}^2-1\right)w_{2,\mathbf{q}_0,0}^2\left(r-\mu_{2,\mathbf{i},1}\right)^2\right.$$
$$-2\left(\rho_{4,1}\rho_{5,1}-\rho_{5,4}\right)w_{2,\mathbf{i},0}w_{2,\mathbf{q}_0,0}\left(r-\mu_{2,\mathbf{q}_0,1}\right)\left(r-\mu_{2,\mathbf{i},1}\right)$$
$$\left.\left.\left.+\left(\rho_{4,1}^2-1\right)w_{2,\mathbf{i},0}^2\left(r-\mu_{2,\mathbf{q}_0,1}\right)^2\right)\right)\right)/\left(r\left(\mathbf{q}_{2,1}+\alpha_{2,1}\right)^2\right.$$
$$\left.\left(r-\mu_{2,\mathbf{i},1}\right)^2\left(r-\mu_{2,\mathbf{q}_0,1}\right)^2\right), \tag{6.7.9}$$

$$\beta_2=\frac{\alpha_{2,0}}{r}-\frac{\alpha_{2,1}\alpha_{2,0}}{r\left(\mathbf{q}_{2,1}+\alpha_{2,1}\right)}+\frac{\mathbb{C}_2\alpha_{2,1}}{\mathbf{q}_{2,1}+\alpha_{2,1}}+\frac{i_2\left(\alpha_{2,1}+\mathbf{q}_{2,1}\alpha_{2,2}\right)}{\left(\mathbf{q}_{2,1}+\alpha_{2,1}\right)\left(r-\mu_{2,\mathbf{i},1}\right)}$$
$$+\frac{\left(\alpha_{2,1}+\mathbf{q}_{2,1}\alpha_{2,2}\right)\mu_{2,\mathbf{i},0}}{r\left(\mathbf{q}_{2,1}+\alpha_{2,1}\right)\left(r-\mu_{2,\mathbf{i},1}\right)}+\left(\left(r-\mathbf{a}_0\right)\left(\alpha_{2,1}+\mathbf{q}_{2,1}\alpha_{2,2}\right)\rho_{4,1}w_{2,\mathbf{i},0}\right)/$$
$$\left(r\mathbf{p}_0\left(\mathbf{q}_{2,1}+\alpha_{2,1}\right)\left(r-\mu_{2,\mathbf{i},1}\right)\right)+\frac{q_2\left(\mathbf{q}_{2,1}\alpha_{2,3}-\alpha_{2,1}\right)}{\left(\mathbf{q}_{2,1}+\alpha_{2,1}\right)\left(r-\mu_{2,\mathbf{q}_0,1}\right)}$$
$$+\frac{\left(\mathbf{q}_{2,1}\alpha_{2,3}-\alpha_{2,1}\right)\mu_{2,\mathbf{q}_0,0}}{r\left(\mathbf{q}_{2,1}+\alpha_{2,1}\right)\left(r-\mu_{2,\mathbf{q}_0,1}\right)}+\left(\left(\mathbf{a}_0-r\right)\left(\alpha_{2,1}-\mathbf{q}_{2,1}\alpha_{2,3}\right)\rho_{5,1}w_{2,\mathbf{q}_0,0}\right)/$$
$$\left(r\mathbf{p}_0\left(\mathbf{q}_{2,1}+\alpha_{2,1}\right)\left(r-\mu_{2,\mathbf{q}_0,1}\right)\right), \tag{6.7.10}$$

while in the Miller–Modigliani case,

$$\delta_{1,1} = \Big(\gamma\Big(\left(\rho_{3,1}^2-1\right)w_{1,\mathbf{q}_0,0}^2\left(r-\mu_{1,\mathbf{i},1}\right)^2 - 2\left(\rho_{2,1}\rho_{3,1}-\rho_{3,2}\right)w_{1,\mathbf{i},0}w_{1,\mathbf{q}_0,0}$$
$$\left(r-\mu_{1,\mathbf{q}_0,1}\right)\left(r-\mu_{1,\mathbf{i},1}\right)+\left(\rho_{2,1}^2-1\right)w_{1,\mathbf{i},0}^2\left(r-\mu_{1,\mathbf{q}_0,1}\right)^2\Big)\Big)\,/$$
$$\left(r\left(r-\mu_{1,\mathbf{i},1}\right)^2\left(r-\mu_{1,\mathbf{q}_0,1}\right)^2\right), \tag{6.7.11}$$

$$\delta_{1,2} = \Big(\gamma\Big(w_{2,\mathbf{q}_0,0}\left(\rho_{5,3}w_{1,\mathbf{q}_0,0}\left(\mu_{1,\mathbf{i},1}-r\right)+\rho_{5,2}w_{1,\mathbf{i},0}\left(r-\mu_{1,\mathbf{q}_0,1}\right)\right)\left(r-\mu_{2,\mathbf{i},1}\right)$$
$$-\rho_{3,1}w_{1,\mathbf{q}_0,0}\left(r-\mu_{1,\mathbf{i},1}\right)\left(\rho_{5,1}w_{2,\mathbf{q}_0,0}\left(\mu_{2,\mathbf{i},1}-r\right)+\rho_{4,1}w_{2,\mathbf{i},0}\left(r-\mu_{2,\mathbf{q}_0,1}\right)\right)$$
$$+\rho_{2,1}w_{1,\mathbf{i},0}\left(r-\mu_{1,\mathbf{q}_0,1}\right)\left(\rho_{5,1}w_{2,\mathbf{q}_0,0}\left(\mu_{2,\mathbf{i},1}-r\right)+\rho_{4,1}w_{2,\mathbf{i},0}\left(r-\mu_{2,\mathbf{q}_0,1}\right)\right)$$
$$+\rho_{4,3}w_{1,\mathbf{q}_0,0}w_{2,\mathbf{i},0}\left(r-\mu_{1,\mathbf{i},1}\right)\left(r-\mu_{2,\mathbf{q}_0,1}\right)-\rho_{4,2}w_{1,\mathbf{i},0}w_{2,\mathbf{i},0}\left(r-\mu_{1,\mathbf{q}_0,1}\right)$$
$$\left(r-\mu_{2,\mathbf{q}_0,1}\right)\Big)\Big)\,/\left(r\left(r-\mu_{1,\mathbf{i},1}\right)\left(r-\mu_{1,\mathbf{q}_0,1}\right)\left(r-\mu_{2,\mathbf{i},1}\right)\left(r-\mu_{2,\mathbf{q}_0,1}\right)\right), \tag{6.7.12}$$

$$\beta_1 = \mathbb{C}_1+\frac{i_1}{r-\mu_{1,\mathbf{i},1}}+\frac{\left(r-\mathbf{a}_0\right)\rho_{2,1}w_{1,\mathbf{i},0}}{r\mathbf{p}_0\left(r-\mu_{1,\mathbf{i},1}\right)}+\frac{\mu_{1,\mathbf{i},0}}{r^2-r\mu_{1,\mathbf{i},1}}-\frac{q_1}{r-\mu_{1,\mathbf{q}_0,1}}$$
$$+\frac{\left(-r+\mathbf{a}_0\right)\rho_{3,1}w_{1,\mathbf{q}_0,0}}{r\mathbf{p}_0\left(r-\mu_{1,\mathbf{q}_0,1}\right)}-\frac{\mu_{1,\mathbf{q}_0,0}}{r^2-r\mu_{1,\mathbf{q}_0,1}}, \tag{6.7.13}$$

$$\delta_{2,1} = \frac{1}{r}\gamma\Bigg(\frac{\left(\rho_{2,1}\rho_{4,1}-\rho_{4,2}\right)w_{1,\mathbf{i},0}w_{2,\mathbf{i},0}}{\left(r-\mu_{1,\mathbf{i},1}\right)\left(r-\mu_{2,\mathbf{i},1}\right)}-\Big(\left(\rho_{2,1}\rho_{3,1}-\rho_{3,2}\right)\left(\rho_{2,1}\rho_{4,1}-\rho_{4,2}\right)$$
$$w_{1,\mathbf{q}_0,0}w_{2,\mathbf{i},0}\Big)\,/\left(\left(-1+\rho_{2,1}^2\right)\left(r-\mu_{1,\mathbf{q}_0,1}\right)\left(r-\mu_{2,\mathbf{i},1}\right)\right)+\Big(\big(-\rho_{2,1}\rho_{3,2}$$
$$\rho_{4,1}+\rho_{3,2}\rho_{4,2}+\rho_{3,1}\left(\rho_{4,1}-\rho_{2,1}\rho_{4,2}\right)-\rho_{4,3}+\rho_{2,1}^2\rho_{4,3}\big)$$
$$w_{1,\mathbf{q}_0,0}w_{2,\mathbf{i},0}\Big)\,/\left(\left(-1+\rho_{2,1}^2\right)\left(r-\mu_{1,\mathbf{q}_0,1}\right)\left(r-\mu_{2,\mathbf{i},1}\right)\right)$$
$$-\frac{\left(\rho_{2,1}\rho_{5,1}-\rho_{5,2}\right)w_{1,\mathbf{i},0}w_{2,\mathbf{q}_0,0}}{\left(r-\mu_{1,\mathbf{i},1}\right)\left(r-\mu_{2,\mathbf{q}_0,1}\right)}+\Big(\left(\rho_{2,1}\rho_{3,1}-\rho_{3,2}\right)\left(\rho_{2,1}\rho_{5,1}-\rho_{5,2}\right)w_{1,\mathbf{q}_0,0}$$
$$w_{2,\mathbf{q}_0,0}\Big)\,/\left(\left(-1+\rho_{2,1}^2\right)\left(r-\mu_{1,\mathbf{q}_0,1}\right)\left(r-\mu_{2,\mathbf{q}_0,1}\right)\right)-\Big(\big(-\rho_{2,1}\rho_{3,2}\rho_{5,1}$$
$$+\rho_{3,2}\rho_{5,2}+\rho_{3,1}\left(\rho_{5,1}-\rho_{2,1}\rho_{5,2}\right)-\rho_{5,3}+\rho_{2,1}^2\rho_{5,3}\big)w_{1,\mathbf{q}_0,0}w_{2,\mathbf{q}_0,0}\Big)\,/$$
$$\left(\left(-1+\rho_{2,1}^2\right)\left(r-\mu_{1,\mathbf{q}_0,1}\right)\left(r-\mu_{2,\mathbf{q}_0,1}\right)\right)\Bigg), \tag{6.7.14}$$

$$\delta_{2,2} = \Big(\gamma\Big(\left(-1+\rho_{5,1}^2\right)w_{2,\mathbf{q}_0,0}^2\left(r-\mu_{2,\mathbf{i},1}\right)^2-2\left(\rho_{4,1}\rho_{5,1}-\rho_{5,4}\right)w_{2,\mathbf{i},0}w_{2,\mathbf{q}_0,0}$$
$$\left(r-\mu_{2,\mathbf{i},1}\right)\left(r-\mu_{2,\mathbf{q}_0,1}\right)+\left(-1+\rho_{4,1}^2\right)w_{2,\mathbf{i},0}^2\left(r-\mu_{2,\mathbf{q}_0,1}\right)^2\Big)\Big)\,/$$
$$\left(r\left(r-\mu_{2,\mathbf{i},1}\right)^2\left(r-\mu_{2,\mathbf{q}_0,1}\right)^2\right), \tag{6.7.15}$$

$$\beta_2 = \mathbb{C}_2 + \frac{i_2}{r-\mu_{2,\mathbf{i},1}} + \frac{(r-\mathbf{a}_0)\,\rho_{4,1}w_{2,\mathbf{i},0}}{r\mathbf{p}_0\,(r-\mu_{2,\mathbf{i},1})} + \frac{\mu_{2,\mathbf{i},0}}{r^2 - r\mu_{2,\mathbf{i},1}} - \frac{q_2}{r-\mu_{2,\mathbf{q}_0,1}}$$
$$+ \frac{(-r+\mathbf{a}_0)\,\rho_{5,1}w_{2,\mathbf{q}_0,0}}{r\mathbf{p}_0\,(r-\mu_{2,\mathbf{q}_0,1})} - \frac{\mu_{2,\mathbf{q}_0,0}}{r^2 - r\mu_{2,\mathbf{q}_0,1}}, \tag{6.7.16}$$

so that in either case,

$$\begin{pmatrix} \delta_{1,1} & \delta_{1,2} \\ \delta_{2,1} & \delta_{2,2} \end{pmatrix} \cdot \{k_1, k_2\} = \{V_1, V_2\} - \{\beta_1, \beta_2\}, \tag{6.7.17}$$

and then

$$\begin{aligned}
\{k_{1,\mathrm{opt}}, k_{2,\mathrm{opt}}\} &= \begin{pmatrix} \delta_{1,1} & \delta_{1,2} \\ \delta_{2,1} & \delta_{2,2} \end{pmatrix}^{-1} \cdot (\{V_1, V_2\} - \{\beta_1, \beta_2\}) \\
&= \left\{ \frac{(V_1-\beta_1)\,\delta_{2,2}}{\delta_{1,1}\delta_{2,2} - \delta_{1,2}\delta_{2,1}} - \frac{(V_2-\beta_2)\,\delta_{1,2}}{\delta_{1,1}\delta_{2,2} - \delta_{1,2}\delta_{2,1}}, \frac{(V_2-\beta_2)\,\delta_{1,1}}{\delta_{1,1}\delta_{2,2} - \delta_{1,2}\delta_{2,1}} \right. \\
&\quad \left. - \frac{(V_1-\beta_1)\,\delta_{2,1}}{\delta_{1,1}\delta_{2,2} - \delta_{1,2}\delta_{2,1}} \right\} = \left\{ \frac{\beta_2\delta_{1,2} - \beta_1\delta_{2,2}}{-\delta_{1,2}\delta_{2,1} + \delta_{1,1}\delta_{2,2}} \right. \\
&\quad + \frac{\delta_{2,2}}{-\delta_{1,2}\delta_{2,1} + \delta_{1,1}\delta_{2,2}} V_1 + \frac{\delta_{1,2}}{\delta_{1,2}\delta_{2,1} - \delta_{1,1}\delta_{2,2}} V_2, \\
&\quad \frac{\beta_2\delta_{1,1} - \beta_1\delta_{2,1}}{\delta_{1,2}\delta_{2,1} - \delta_{1,1}\delta_{2,2}} + \frac{\delta_{2,1}}{\delta_{1,2}\delta_{2,1} - \delta_{1,1}\delta_{2,2}} V_1 \\
&\quad \left. + \frac{\delta_{1,1}}{-\delta_{1,2}\delta_{2,1} + \delta_{1,1}\delta_{2,2}} V_2 \right\}.
\end{aligned} \tag{6.7.18}$$

Summarizing, we have the following theorem.

Theorem 6.7.1. *(a) In the case of CRRA utility with risk-aversion parameter γ, and under the assumption (6.7.1), for the market position $\kappa = X\kappa_0 = X\{k_1, k_2\}$, the κ-neutral price of the portfolio is given by*

$$\{V_1, V_2\} = \{\beta_1 + k_1\delta_{1,1} + k_2\delta_{1,2}, \beta_2 + k_1\delta_{2,1} + k_2\delta_{2,2}\}, \tag{6.7.19}$$

with $\delta_{1,1}, \delta_{1,2}, \beta_1, \delta_{2,1}, \delta_{2,2}, \beta_2$ given in (6.7.5)–(6.7.10).

(b) Under the same assumptions, for the market price $V = \{V_1, V_2\}$, the V-optimal market position is equal to $\kappa_{opt} = X\kappa_{0,opt} = X\{k_{1,opt}, k_{2,opt}\}$, with

$$\begin{aligned}
\{k_{1,opt}, k_{2,opt}\} &= \left\{ \frac{(V_1-\beta_1)\,\delta_{2,2}}{\delta_{1,1}\delta_{2,2} - \delta_{1,2}\delta_{2,1}} - \frac{(V_2-\beta_2)\,\delta_{1,2}}{\delta_{1,1}\delta_{2,2} - \delta_{1,2}\delta_{2,1}}, \frac{(V_2-\beta_2)\,\delta_{1,1}}{\delta_{1,1}\delta_{2,2} - \delta_{1,2}\delta_{2,1}} \right. \\
&\quad \left. - \frac{(V_1-\beta_1)\,\delta_{2,1}}{\delta_{1,1}\delta_{2,2} - \delta_{1,2}\delta_{2,1}} \right\}
\end{aligned} \tag{6.7.20}$$

and with $\delta_{1,1}, \delta_{1,2}, \beta_1, \delta_{2,1}, \delta_{2,2}, \beta_2$ given in (6.7.5)–(6.7.10).

Remark 6.7.1. The cumulative market exposure in the considered equities is then equal to

$$X\left(k_{1,\mathrm{opt}}+k_{2,\mathrm{opt}}\right)=X\left(\left(V_1-\beta_1\right)\left(\delta_{2,2}-\delta_{2,1}\right)+\left(V_2-\beta_2\right)\left(\delta_{1,1}-\delta_{1,2}\right)\right)/ \left(\delta_{1,1}\delta_{2,2}-\delta_{1,2}\delta_{2,1}\right). \tag{6.7.21}$$

6.8 Hedging of Equity Portfolios

We apply Theorem 5.2.1, i.e., formula (5.2.2), to calculate the most conservative hedging for the $\kappa=\{\kappa_1,\kappa_2\}$ equity position, with the stock price $V=\{V_1,V_2\}$. Without any assumption on the price, and in particular for the price $\{V_1,V_2\}$ that appears in (6.7.20), the hedging formula is given by

$$\begin{aligned}
\Pi^{a,\star}(t,X,A) &= \left\{-\left(\sigma_s.\sigma_s^{\mathrm{T}}\right)^{-1}\cdot\sigma_s.c^{\mathrm{T}}\cdot(\nabla V)^{\mathrm{T}}\cdot\kappa, V\kappa\right\} \\
&= \Bigg\{\Bigg\{-\kappa_1\left(\frac{1}{\mathbf{p}_0}\rho_{5,1}\left(w_{2,\mathbf{q}_0,0}+q_2w_{2,\mathbf{q}_0,1}\right)V_1^{(0,0,0,0,0,0,0,1)}\left(t,S_0,\mathbb{C}_1,i_1,q_1,\mathbb{C}_2,i_2,q_2\right)\right. \\
&\quad+\frac{1}{\mathbf{p}_0}\rho_{4,1}\left(w_{2,\mathbf{i},0}+i_2w_{2,\mathbf{i},1}\right)V_1^{(0,0,0,0,0,0,1,0)}\left(t,S_0,\mathbb{C}_1,i_1,q_1,\mathbb{C}_2,i_2,q_2\right) \\
&\quad+\frac{1}{\mathbf{p}_0}\rho_{3,1}\left(w_{1,\mathbf{q}_0,0}+q_1w_{1,\mathbf{q}_0,1}\right)V_1^{(0,0,0,0,1,0,0,0)}\left(t,S_0,\mathbb{C}_1,i_1,q_1,\mathbb{C}_2,i_2,q_2\right) \\
&\quad+\frac{1}{\mathbf{p}_0}\rho_{2,1}\left(w_{1,\mathbf{i},0}+i_1w_{1,\mathbf{i},1}\right)V_1^{(0,0,0,1,0,0,0,0)}\left(t,S_0,\mathbb{C}_1,i_1,q_1,\mathbb{C}_2,i_2,q_2\right) \\
&\quad\left.+S_0V_1^{(0,1,0,0,0,0,0,0)}\left(t,S_0,\mathbb{C}_1,i_1,q_1,\mathbb{C}_2,i_2,q_2\right)\right) \\
&\quad-\kappa_2\left(\frac{1}{\mathbf{p}_0}\rho_{5,1}\left(w_{2,\mathbf{q}_0,0}+q_2w_{2,\mathbf{q}_0,1}\right)V_2^{(0,0,0,0,0,0,0,1)}\left(t,S_0,\mathbb{C}_1,i_1,q_1,\mathbb{C}_2,i_2,q_2\right)\right. \\
&\quad+\frac{1}{\mathbf{p}_0}\rho_{4,1}\left(w_{2,\mathbf{i},0}+i_2w_{2,\mathbf{i},1}\right)V_2^{(0,0,0,0,0,0,1,0)}\left(t,S_0,\mathbb{C}_1,i_1,q_1,\mathbb{C}_2,i_2,q_2\right) \\
&\quad+\frac{1}{\mathbf{p}_0}\rho_{3,1}\left(w_{1,\mathbf{q}_0,0}+q_1w_{1,\mathbf{q}_0,1}\right)V_2^{(0,0,0,0,1,0,0,0)}\left(t,S_0,\mathbb{C}_1,i_1,q_1,\mathbb{C}_2,i_2,q_2\right) \\
&\quad+\frac{1}{\mathbf{p}_0}\rho_{2,1}\left(w_{1,\mathbf{i},0}+i_1w_{1,\mathbf{i},1}\right)V_2^{(0,0,0,1,0,0,0,0)}\left(t,S_0,\mathbb{C}_1,i_1,q_1,\mathbb{C}_2,i_2,q_2\right)
\end{aligned}$$

$$+S_0 V_2{}^{(0,1,0,0,0,0,0,0)}(t,S_0,\mathbb{C}_1,i_1,q_1,\mathbb{C}_2,i_2,q_2)\Big)\Big\},$$

$$\{\kappa_1 V_1(t,X,S_0,\mathbb{C}_1,i_1,q_1,\mathbb{C}_2,i_2,q_2),\kappa_2 V_2(t,X,S_0,\mathbb{C}_1,i_1,q_1,\mathbb{C}_2,i_2,q_2)\}\}. \tag{6.8.1}$$

On the other hand, drawing on the conclusions of Sect. 6.7, we have

$$\{V_1,V_2\} = \{\mathbb{A}_{1,1}+\mathbb{C}_1\mathbb{A}_{1,3}+i_1\mathbb{A}_{1,4}+q_1\mathbb{A}_{1,5},\mathbb{A}_{2,1}+\mathbb{C}_2\mathbb{A}_{2,6}+i_2\mathbb{A}_{2,7}+q_2\mathbb{A}_{2,8}\} \tag{6.8.2}$$

with constants $\mathbb{A}_{1,1},\mathbb{A}_{1,3},\mathbb{A}_{1,4},\mathbb{A}_{1,5},\mathbb{A}_{2,1},\mathbb{A}_{2,6},\mathbb{A}_{2,7},\mathbb{A}_{2,8}$ characterized in Sect. 6.7. In such a case we have the following result.

Proposition 6.8.1. *In the above context, and in particular assuming (6.7.1), the most conservative hedging of the equity position $\kappa = \{\kappa_1,\kappa_2\}$ is given by the formula*

$$\Pi^{a,\star}(t,X,A) = \left\{\left\{-\kappa_2\left(\frac{\mathbb{A}_{2,6}\rho_{4,1}w_{2,\mathbf{i},0}}{\mathbf{p}_0}+\frac{\mathbb{A}_{2,7}\rho_{5,1}w_{2,\mathbf{q}_0,0}}{\mathbf{p}_0}\right)-\kappa_1\left(\frac{\mathbb{A}_{1,3}\rho_{2,1}w_{1,\mathbf{i},0}}{\mathbf{p}_0}\right.\right.\right.$$
$$\left.\left.\left.+\frac{\mathbb{A}_{1,4}\rho_{3,1}w_{1,\mathbf{q}_0,0}}{\mathbf{p}_0}\right)\right\},\{\kappa_1 V_1,\kappa_2 V_2\}\right\} \tag{6.8.3}$$

with constants $\mathbb{A}_{1,1},\mathbb{A}_{1,3},\mathbb{A}_{1,4},\mathbb{A}_{1,5},\mathbb{A}_{2,1},\mathbb{A}_{2,6},\mathbb{A}_{2,7},\mathbb{A}_{2,8}$ characterized in Sect. 6.7.

6.9 The Effect of Dilution/Buyback on the Value of Shares: CRRA Neutral Pricing

REUTERS – 12 : 12 PM ET 11/04/10
...BHP shares jump 6.1 percent on buyback expectation

CINCINNATI, Aug. 4, 1998 /PRNewswire/ –– Procter & Gamble (NYSE NYSE See: New York Stock Exchange : PG) announced today it will increase and accelerate its discretionary stock repurchase plan beyond its previously announced annual target of $1 billion. Specific repurchase amounts and timing were not disclosed. This plan is in addition to the ongoing repurchase program to offset any dilution effect from stock compensation programs.

6.9.1 Deterministic Dilution/Buyback

We end this chapter with a model for stock dilution and buyback. Since in such a situation, the number of shares outstanding is variable, it is necessary to declare it as an additional factor, and furthermore, the previous way of accounting cash (or book value), income, and cost *per share* becomes inconvenient, and needs to be substituted by accounting *in total* for the whole firm.

More precisely, following [8], we introduce the following model: let $S=\{S_1\}$ be a (hedging) tradable, and let the factors be $A=\{S_1,\mathbb{C},i,q,\mathcal{M}\}$, where in addition to S_1, $\mathbb{C}$ is the total cash available (or book value for the whole company), i is the income rate (per year, for the whole company), q is the basic cost rate (per year, for the whole company), and $\mathcal{M}$ is the number of common shares outstanding (adjusted for splits). They are assumed to obey the following SDE:

$$\begin{aligned}
dS_1(t) &= S_1(t)\left(\mathbf{a}_1-\mathbb{D}_1\right)dt+S_1(t)\mathbf{p}_1 dB_1(t),\\
d\mathbb{C}(t) &= (i(t)-q(t)+\mathbb{C}(t)(r-\mathbf{q}_1)-(\alpha_0+\mathbb{C}(t)\alpha_1+\alpha_2 i(t)+\alpha_3 q(t))\\
&\quad +\beta\left(q_{M,0}+\mathcal{M}(t)q_{M,1}\right)V_1\left(t,S_1(t),\mathbb{C}(t),i(t),q(t),\mathcal{M}(t)\right))\,dt,\\
di(t) &= \left(\mu_{\mathbf{i},0}+i(t)\mu_{\mathbf{i},1}\right)dt+w_{\mathbf{i},0}d\mathbb{B}_2(t),\\
dq(t) &= \left(\mu_{\mathbf{q}_0,0}+q(t)\mu_{\mathbf{q}_0,1}\right)dt+w_{\mathbf{q}_0,0}d\mathbb{B}_3(t),\\
d\mathcal{M}(t) &= \left(q_{M,0}+\mathcal{M}(t)q_{M,1}\right)dt,
\end{aligned}\tag{6.9.1}$$

where $dB_1(t)d\mathbb{B}_2(t)=\rho_{2,1}dt$, $d\mathbb{B}_1(t)d\mathbb{B}_3(t)=\rho_{3,1}dt$, and $d\mathbb{B}_2(t)d\mathbb{B}_3(t)=\rho_{3,2}dt$, where $\alpha_0+\mathbb{C}(t)\alpha_1+\alpha_2 i(t)+\alpha_3 q(t)$ is the total dividend payoff (per year, for the whole company), so that per share, the dividend payoff is equal to

$$d=\frac{1}{\mathcal{M}}\left(\alpha_0+\alpha_1\mathbb{C}+\alpha_2 i+\alpha_3 q\right).\tag{6.9.2}$$

As before, r is the interest rate, while $\mathbf{q}_1$ is the surcharge cost rate. The number of shares outstanding, $\mathcal{M}(t)$, is assumed here to be deterministic. We shall also consider a stochastic case a bit later. We observe that

$$\begin{aligned}
&\beta\left(q_{M,0}+\mathcal{M}(t)q_{M,1}\right)V_1\left(t,S_1(t),\mathbb{C}(t),i(t),q(t),\mathcal{M}(t)\right)dt\\
&\quad=\beta V_1\left(t,S_1(t),\mathbb{C}(t),i(t),q(t),\mathcal{M}(t)\right)d\mathcal{M}(t)
\end{aligned}\tag{6.9.3}$$

is the cash inflow (total for the whole company) due to stock dilution (if $d\mathcal{M}(t)>0$), or cash outflow due to stock buyback (if $d\mathcal{M}(t)<0$). The parameter β, when different from 1, allows for stock dilution and/or buyback to be done at prices different from the market price $V_1\left(t,S_1(t),\mathbb{C}(t),i(t),q(t),\mathcal{M}(t)\right)$.

Then the corresponding market coefficients are

$$
\begin{aligned}
&a_s = \{\mathbf{a}_1\},\\
&\sigma_s = \left(\mathbf{p}_1\ 0\ 0\right),\\
&b = \{S_1\left(\mathbf{a}_1 - \mathbb{D}_1\right), i - q + \mathbb{C}\left(r - \mathbf{q}_1\right) - \alpha_0 - \mathbb{C}\alpha_1 - \alpha_2 i - q\alpha_3 + \beta\left(q_{M,0}\right.\\
&\qquad \left. + \mathcal{M}q_{M,1}\right) V_1\left(t, S_1, \mathbb{C}, i, q, \mathcal{M}\right), \mu_{\mathbf{i},0} + i\mu_{\mathbf{i},1}, \mu_{\mathbf{q}_0,0} + q\mu_{\mathbf{q}_0,1}, q_{M,0} + \mathcal{M}q_{M,1}\},
\end{aligned}
$$

$$
c = \begin{pmatrix}
S_1\mathbf{p}_1 & 0 & 0\\
0 & 0 & 0\\
w_{\mathbf{i},0}\rho_{2,1} & w_{\mathbf{i},0}\sqrt{1-\rho_{2,1}^2} & 0\\
w_{\mathbf{q}_0,0}\rho_{3,1} & \frac{w_{\mathbf{q}_0,0}\left(\rho_{3,2}-\rho_{2,1}\rho_{3,1}\right)}{\sqrt{1-\rho_{2,1}^2}} & w_{\mathbf{q}_0,0}\sqrt{-\rho_{3,1}^2 + \frac{\left(\rho_{3,2}-\rho_{2,1}\rho_{3,1}\right)^2}{\rho_{2,1}^2-1} + 1}\\
0 & 0 & 0
\end{pmatrix}. \tag{6.9.4}
$$

We employ CRRA utility of wealth, and adopt the position in the form $\kappa = X\{k_1\}$, with k_1 a constant, and we hope to find the (single) stock price V_1 in the form

$$
V_1\left(t, S_1, \mathbb{C}, i, q, \mathcal{M}\right) = V_1(\mathbb{C}, i, q, \mathcal{M}) = \mathbb{A}_0(\mathcal{M}) + \mathbb{C}\mathbb{A}_1(\mathcal{M}) + i\mathbb{A}_2(\mathcal{M}) + q\mathbb{A}_3(\mathcal{M}). \tag{6.9.5}
$$

We need to consider a particular form of the stock price at this early stage, since the neutral pricing system (4.9.2)–(4.9.3) is coupled, so that V_1 may affect g_γ, and g_γ may affect V_1. So assuming (6.9.5), and looking for g_γ in the form $g_\gamma(t,\mathcal{M})$, the PDE (4.9.2) simplifies into

$$
\begin{aligned}
&g_\gamma{}^{(1,0)}(t,\mathcal{M}) + \left(q_{M,0} + \mathcal{M}q_{M,1}\right) g_\gamma{}^{(0,1)}(t,\mathcal{M}) - \frac{1}{2}(\gamma-1)\gamma\left(w_{\mathbf{i},0}^2\left(1-\rho_{2,1}^2\right)\mathbb{A}_2(\mathcal{M})^2\right.\\
&\qquad \left. + 2w_{\mathbf{i},0}w_{\mathbf{q}_0,0}\left(\rho_{3,2} - \rho_{2,1}\rho_{3,1}\right)\mathbb{A}_3(\mathcal{M})\mathbb{A}_2(\mathcal{M}) + w_{\mathbf{q}_0,0}^2\left(1-\rho_{3,1}^2\right)\mathbb{A}_3(\mathcal{M})^2\right) k_1^2\\
&= \frac{\gamma-1}{\gamma}\left(\frac{\left(\mathbf{a}_1 - r\right)^2}{2\mathbf{p}_1^2} + r\gamma\right)
\end{aligned} \tag{6.9.6}
$$

for $t < T < \infty$, with the terminal condition $g_\gamma(T,\mathcal{M}) = 0$. So unless $k_1 = 0$, g_γ does not have the form $g_\gamma(t)$, and therefore $\nabla_A g_\gamma \neq 0$. Nevertheless,

$$
\nabla_A g_\gamma = \left\{0,0,0,0,g^{(0,1)}(t,\mathcal{M})\right\}, \tag{6.9.7}
$$

and since the last row of c is zero, the last column of c^{T} is zero, and therefore

$$
c^{\mathrm{T}} \cdot \nabla_A g_\gamma = \{0,0,0\}. \tag{6.9.8}
$$

Consequently, $g_\gamma(t,\mathcal{M})$ will not affect PDE (4.9.3). Next, assuming (6.9.5), PDE (4.9.3) reduces to

$$
\begin{aligned}
&-\gamma k_1 \left(-w_{\mathbf{i},0}^2 \left(\rho_{2,1}^2 - 1\right) \mathbb{A}_2(\mathcal{M})^2 + 2 w_{\mathbf{i},0} w_{\mathbf{q}_0,0} \left(\rho_{3,2} - \rho_{2,1}\rho_{3,1}\right) \mathbb{A}_3(\mathcal{M}) \mathbb{A}_2(\mathcal{M})\right.\\
&\quad \left. - w_{\mathbf{q}_0,0}^2 \left(\rho_{3,1}^2 - 1\right) \mathbb{A}_3(\mathcal{M})^2\right) + \left(\mu_{\mathbf{i},0} + i\mu_{\mathbf{i},1} - \frac{(\mathbf{a}_1 - r)\, w_{\mathbf{i},0}\rho_{2,1}}{\mathbf{p}_1}\right) \mathbb{A}_2(\mathcal{M})\\
&\quad + \left(\mu_{\mathbf{q}_0,0} + q\mu_{\mathbf{q}_0,1} - \frac{(\mathbf{a}_1 - r)\, w_{\mathbf{q}_0,0}\rho_{3,1}}{\mathbf{p}_1}\right) \mathbb{A}_3(\mathcal{M}) - r\left(\mathbb{A}_0(\mathcal{M}) + \mathbb{C}\mathbb{A}_1(\mathcal{M})\right.\\
&\quad \left. + i\mathbb{A}_2(\mathcal{M}) + q\mathbb{A}_3(\mathcal{M})\right) + \mathbb{A}_1(\mathcal{M})\left(-\alpha_2 i + i - q + \mathbb{C}\left(r - \mathbf{q}_1\right) - \alpha_0\right.\\
&\quad -\mathbb{C}\alpha_1 - q\alpha_3 + \beta\left(q_{M,0} + \mathcal{M}q_{M,1}\right)\left(\mathbb{A}_0(\mathcal{M}) + \mathbb{C}\mathbb{A}_1(\mathcal{M}) + i\mathbb{A}_2(\mathcal{M})\right.\\
&\quad \left.\left. + q\mathbb{A}_3(\mathcal{M})\right)\right) + \left(q_{M,0} + \mathcal{M}q_{M,1}\right)\left(\mathbb{A}_0'(\mathcal{M}) + \mathbb{C}\mathbb{A}_1'(\mathcal{M}) + i\mathbb{A}_2'(\mathcal{M}) + q\mathbb{A}_3'(\mathcal{M})\right)\\
&= -\frac{\alpha_0 + \mathbb{C}\alpha_1 + i\alpha_2 + q\alpha_3}{\mathcal{M}}
\end{aligned}
\tag{6.9.9}
$$

for $t < T$ and any $\mathbb{C}, i, q$, and consequently reduces also to the following system of four first-order ODEs:

$$
\begin{aligned}
&\left(q_{M,0} + \mathcal{M}q_{M,1}\right)\mathbb{A}_0'(\mathcal{M}) - \mathbb{A}_1(\mathcal{M})\alpha_0 + \frac{\alpha_0}{\mathcal{M}} - \gamma k_1 \left(w_{\mathbf{i},0}^2\left(1 - \rho_{2,1}^2\right)\mathbb{A}_2(\mathcal{M})^2\right.\\
&\quad \left. + 2 w_{\mathbf{i},0} w_{\mathbf{q}_0,0}\left(\rho_{3,2} - \rho_{2,1}\rho_{3,1}\right)\mathbb{A}_3(\mathcal{M})\mathbb{A}_2(\mathcal{M}) + w_{\mathbf{q}_0,0}^2\left(1 - \rho_{3,1}^2\right)\mathbb{A}_3(\mathcal{M})^2\right)\\
&\quad - r\mathbb{A}_0(\mathcal{M}) + \beta\left(q_{M,0} + \mathcal{M}q_{M,1}\right)\mathbb{A}_0(\mathcal{M})\mathbb{A}_1(\mathcal{M}) + \mu_{\mathbf{i},0}\mathbb{A}_2(\mathcal{M}) + \mu_{\mathbf{q}_0,0}\mathbb{A}_3(\mathcal{M})\\
&\quad - \frac{(\mathbf{a}_1 - r)\, w_{\mathbf{i},0}\rho_{2,1}\mathbb{A}_2(\mathcal{M})}{\mathbf{p}_1} - \frac{(\mathbf{a}_1 - r)\, w_{\mathbf{q}_0,0}\rho_{3,1}\mathbb{A}_3(\mathcal{M})}{\mathbf{p}_1} = 0,
\end{aligned}
\tag{6.9.10}
$$

$$
\begin{aligned}
&\beta\left(q_{M,0} + \mathcal{M}q_{M,1}\right)\mathbb{A}_1(\mathcal{M})^2 - r\mathbb{A}_1(\mathcal{M}) + \left(r - \mathbf{q}_1\right)\mathbb{A}_1(\mathcal{M})\\
&\quad - \alpha_1\mathbb{A}_1(\mathcal{M}) + \frac{\alpha_1}{\mathcal{M}} + \left(q_{M,0} + \mathcal{M}q_{M,1}\right)\mathbb{A}_1'(\mathcal{M}) = 0,
\end{aligned}
\tag{6.9.11}
$$

$$
\begin{aligned}
&-\mathbb{A}_1(\mathcal{M})\alpha_2 + \frac{\alpha_2}{\mathcal{M}} + \mathbb{A}_1(\mathcal{M}) - r\mathbb{A}_2(\mathcal{M}) + \mu_{\mathbf{i},1}\mathbb{A}_2(\mathcal{M})\\
&\quad + \beta\left(q_{M,0} + \mathcal{M}q_{M,1}\right)\mathbb{A}_1(\mathcal{M})\mathbb{A}_2(\mathcal{M}) + \left(q_{M,0} + \mathcal{M}q_{M,1}\right)\mathbb{A}_2'(\mathcal{M}) = 0,
\end{aligned}
\tag{6.9.12}
$$

$$
\begin{aligned}
&-\mathbb{A}_1(\mathcal{M})\alpha_3 + \frac{\alpha_3}{\mathcal{M}} - \mathbb{A}_1(\mathcal{M}) - r\mathbb{A}_3(\mathcal{M}) + \mu_{\mathbf{q}_0,1}\mathbb{A}_3(\mathcal{M})\\
&\quad + \beta\left(q_{M,0} + \mathcal{M}q_{M,1}\right)\mathbb{A}_1(\mathcal{M})\mathbb{A}_3(\mathcal{M}) + \left(q_{M,0} + \mathcal{M}q_{M,1}\right)\mathbb{A}_3'(\mathcal{M}) = 0,
\end{aligned}
\tag{6.9.13}
$$

for $\mathcal{M} > 0$ (the entire solution). Summarizing, we have the following result.

Proposition 6.9.1. *In the above deterministic stock dilution/buyback model (6.9.1), and for (possibly nonzero) position $\kappa = X\{k_1\}$, the CRRA neutral price of a single stock V_1 is equal to (6.9.5), where $\mathbb{A}_0(\mathcal{M})$, $\mathbb{A}_0(\mathcal{M})$, $\mathbb{A}_0(\mathcal{M})$, and $\mathbb{A}_0(\mathcal{M})$ are characterized as $\{\mathcal{M} > 0\}$-entire solutions of ODEs (6.9.10)–(6.9.13).*

Remark 6.9.1. Equation (6.9.11) is quadratic and can be solved for $\mathbb{A}_1(\mathcal{M})$ first, since it does not depend on any other $\mathbb{A}_i$'s. After $\mathbb{A}_1(\mathcal{M})$ is found, then equations (6.9.12) and (6.9.13) are solved for $\mathbb{A}_2(\mathcal{M})$ and $\mathbb{A}_3(\mathcal{M})$. Finally, (6.9.10) is solved for $\mathbb{A}_0(\mathcal{M})$. In general, these ODEs must be solved numerically, as we showcase below.

There is a special case, though, when the above system has an explicit $\{\mathcal{M}>0\}$-entire solution. If $\mathbf{q}_1 = 0$ (the "Miller–Modigliani case"–no surcharge cost), $\beta = 1$ (stock dilution/buyback is done at the market price), and $k_1 = 0$ (the zero position), then we can find an explicit solution. Indeed, (6.9.11) simplifies to

$$-\mathbf{q}_1\mathbb{A}_1(\mathcal{M}) + \frac{\alpha_1\left(1-\mathcal{M}\mathbb{A}_1(\mathcal{M})\right)}{\mathcal{M}} + (q_{M,0}+\mathcal{M}q_{M,1})\left(\beta\mathbb{A}_1(\mathcal{M})^2+\mathbb{A}_1'(\mathcal{M})\right) = 0, \tag{6.9.14}$$

and its $\{\mathcal{M}>0\}$-entire solution is

$$\mathbb{A}_1(\mathcal{M}) = \frac{1}{\mathcal{M}}. \tag{6.9.15}$$

Notice that in the case $q_{M,0} = 0$, there is an additional $\{\mathcal{M}>0\}$-entire solution of (6.9.14): $\mathbb{A}_1(\mathcal{M}) = \frac{\alpha_1}{q_{M,1}}\frac{1}{\mathcal{M}}$, which is eliminated, since it blows up when $q_{M,1}\to 0$. Consequently, equations (6.9.12) and (6.9.13) simplify to

$$-r\mathbb{A}_2(\mathcal{M}) + \frac{(q_{M,0}+\mathcal{M}q_{M,1})\,\mathbb{A}_2(\mathcal{M})}{\mathcal{M}} + \mu_{\mathbf{i},1}\mathbb{A}_2(\mathcal{M}) + (q_{M,0}+\mathcal{M}q_{M,1})\,\mathbb{A}_2'(\mathcal{M}) + \frac{1}{\mathcal{M}} = 0 \tag{6.9.16}$$

and

$$-r\mathbb{A}_3(\mathcal{M}) + \frac{(q_{M,0}+\mathcal{M}q_{M,1})\,\mathbb{A}_3(\mathcal{M})}{\mathcal{M}} + \mu_{\mathbf{q}_0,1}\mathbb{A}_3(\mathcal{M}) + (q_{M,0}+\mathcal{M}\,q_{M,1})\,\mathbb{A}_3'(\mathcal{M}) - \frac{1}{\mathcal{M}} = 0, \tag{6.9.17}$$

admitting $\{\mathcal{M}>0\}$-entire solutions

$$\mathbb{A}_2(\mathcal{M}) = \frac{1}{r-\mu_{\mathbf{i},1}}\frac{1}{\mathcal{M}} \tag{6.9.18}$$

and

$$\mathbb{A}_3(\mathcal{M}) = -\frac{1}{r-\mu_{\mathbf{q}_0,1}}\frac{1}{\mathcal{M}}, \tag{6.9.19}$$

respectively. Finally, (6.9.10) is then simplified to

$$(q_{M,0}+\mathcal{M}q_{M,1})\,\mathbb{A}_0'(\mathcal{M})+\left(\frac{q_{M,0}+\mathcal{M}q_{M,1}}{\mathcal{M}}-r\right)\mathbb{A}_0(\mathcal{M})-\frac{\mathbf{a}_1-r}{\mathbf{p}_1}$$
$$\left(\frac{w_{\mathbf{i},0}\rho_{2,1}}{\mathcal{M}\left(r-\mu_{\mathbf{i},1}\right)}-\frac{w_{\mathbf{q}_0,0}\rho_{3,1}}{\mathcal{M}\left(r-\mu_{\mathbf{q}_0,1}\right)}\right)+\frac{\mu_{\mathbf{i},0}}{\mathcal{M}\left(r-\mu_{\mathbf{i},1}\right)}-\frac{\mu_{\mathbf{q}_0,0}}{\mathcal{M}\left(r-\mu_{\mathbf{q}_0,1}\right)}=0, \tag{6.9.20}$$

admitting the $\{\mathcal{M}>0\}$-entire solution

$$\mathbb{A}_0(\mathcal{M})=\left(\frac{\mu_{\mathbf{i},0}}{r\left(r-\mu_{\mathbf{i},1}\right)}-\frac{\mu_{\mathbf{q}_0,0}}{r\left(r-\mu_{\mathbf{q}_0,1}\right)}+\frac{\mathbf{a}_1-r}{\mathbf{p}_1}\left(\frac{w_{\mathbf{q}_0,0}\rho_{3,1}}{r\left(r-\mu_{\mathbf{q}_0,1}\right)}-\frac{w_{\mathbf{i},0}\rho_{2,1}}{r\left(r-\mu_{\mathbf{i},1}\right)}\right)\right)\frac{1}{\mathcal{M}}. \tag{6.9.21}$$

Summarizing, we may state the following proposition.

Proposition 6.9.2. *In the above deterministic stock dilution/buyback model (6.9.1), if* $\mathbf{q}_1=0$, $\beta=1$, *and* $k_1=0$, *then the price of a single stock* V_1 *is equal to*

$$V_1\left(t,S_1,\mathbb{C},i,q,\mathcal{M}\right)=V_1(\mathbb{C},i,q,\mathcal{M})$$
$$=\frac{1}{\mathcal{M}}\left(\mathbb{C}+\frac{i}{r-\mu_{\mathbf{i},1}}-\frac{q}{r-\mu_{\mathbf{q}_0,1}}+\frac{\mu_{\mathbf{i},0}-\frac{(\mathbf{a}_1-r)w_{\mathbf{i},0}\rho_{2,1}}{\mathbf{p}_1}}{r\left(r-\mu_{\mathbf{i},1}\right)}\right.$$
$$\left.-\frac{\mu_{\mathbf{q}_0,0}-\frac{(\mathbf{a}_1-r)w_{\mathbf{q}_0,0}\rho_{3,1}}{\mathbf{p}_1}}{r\left(r-\mu_{\mathbf{q}_0,1}\right)}\right), \tag{6.9.22}$$

while the value of the whole company $V=\mathcal{M}V_1$ *is independent of the number of shares outstanding* $\mathcal{M}$ *and is equal to*

$$V_1\left(t,S_1,\mathbb{C},i,q\right)=V(\mathbb{C},i,q)$$
$$=\mathbb{C}+\frac{i}{r-\mu_{\mathbf{i},1}}-\frac{q}{r-\mu_{\mathbf{q}_0,1}}+\frac{\mu_{\mathbf{i},0}-\frac{(\mathbf{a}_1-r)w_{\mathbf{i},0}\rho_{2,1}}{\mathbf{p}_1}}{r\left(r-\mu_{\mathbf{i},1}\right)}$$
$$-\frac{\mu_{\mathbf{q}_0,0}-\frac{(\mathbf{a}_1-r)w_{\mathbf{q}_0,0}\rho_{3,1}}{\mathbf{p}_1}}{r\left(r-\mu_{\mathbf{q}_0,1}\right)}. \tag{6.9.23}$$

Obviously, above it is also assumed that $r>0$, $r>\mu_{\mathbf{i},1}$, $r>\mu_{\mathbf{q}_0,1}$.

Remark 6.9.2. In the "Miller–Modigliani case" ($\mathbf{q}_1=0$), and also assuming $\beta=1$ and $k_1=0$, the (deterministic) stock dilution/buyback does *not* affect the value of the equity – the price of the stock is independent of $q_{M,0}+\mathcal{M}q_{M,1}=\mathrm{d}\mathcal{M}$. This analytic result contradicts the empirical evidence such as that quoted in the opening of this section.

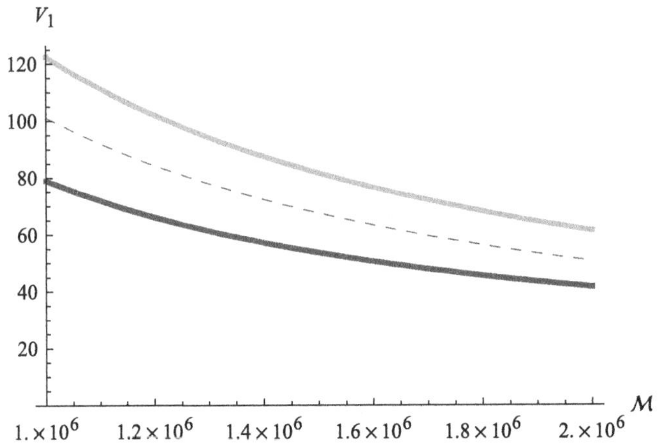

Fig. 6.5 Comparing the stock prices under buyback and dilution, as a function of the number of stocks outstanding

Remark 6.9.3. A numerical solution of the ODE system (6.9.10)–(6.9.13) is fairly simple. For example, if the data are

$$\begin{aligned} &q_{M,0}=0, q_{M,1}=\pm 0.015, \mathbf{q}_1=0.01, \beta=1, \alpha_0=0, \alpha_1=0.03, \alpha_2=0, \alpha_3=0,\\ &\mu_{\mathbf{i},0}=10^5, \mu_{\mathbf{i},1}=0.015, \mu_{\mathbf{q}_0,0}=510^4, \mu_{\mathbf{q}_0,1}=0.01, r=0.05, \mathbf{a}_1=0.06,\\ &w_{\mathbf{i},0}=10^3, w_{\mathbf{q}_0,0}=10^3, \rho_{2,1}=0.5, \rho_{3,1}=0.333333, \mathbf{p}_1=0.2, k_1=0, \gamma=50, \end{aligned} \tag{6.9.24}$$

then the buyback solution ($q_{M,1}=-0.015$) has the highest stock price, the dilution solution ($q_{M,1}=0.015$) has the lowest stock price, while the constant number of shares solution ($q_{M,1}=0$) is in the middle, as seen in Fig. 6.5. The corresponding values of the whole company (as a function of the number of shares outstanding) is shown in Fig. 6.6.

Exercise 6.9.1. Implement the numerical solution of ODEs (6.9.10)–(6.9.13) and verify the showcased solution above. In particular, since no initial condition is available, i.e., since an entire solution is sought, and since for numerical solutions an initial condition is necessary, what *artificial* initial condition is appropriate, and, importantly, depending on the different cases above, where should it be imposed?

6.9.2 *Stochastic Dilution/Buyback* ($\kappa=0$)

Can we extend the results of the previous section if we allow for stochastic dilution and/or buyback? To answer that question, we consider the following model:

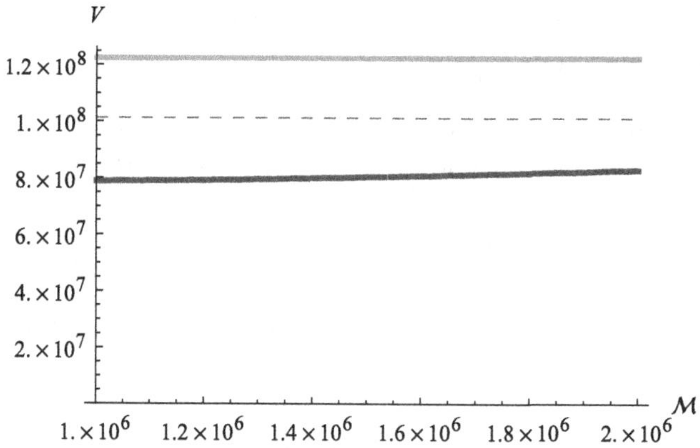

Fig. 6.6 Comparing the whole company values under buyback and dilution, as a function of the number of stocks outstanding

$$\begin{aligned}
\mathrm{d}S_1(t) &= S_1(t)\left(\mathbf{a}_1 - \mathbb{D}_1\right)\mathrm{d}t + S_1(t)\mathbf{p}_1\mathrm{d}B_1(t),\\
\mathrm{d}\mathbb{C}(t) &= \left(i(t) - q(t) + \mathbb{C}(t)\left(r - \mathbf{q}_1\right) - \left(\alpha_0 + \mathbb{C}(t)\alpha_1 + \alpha_2 i(t) + \alpha_3 q(t)\right)\right.\\
&\quad \left.+\beta\left(q_{M,0} + \mathcal{M}(t)q_{M,1}\right)V_1\left(t, S_1(t), \mathbb{C}(t), i(t), q(t), \mathcal{M}(t)\right)\right)\mathrm{d}t\\
&\quad +\beta V_1\left(t, S_1(t), \mathbb{C}(t), i(t), q(t), \mathcal{M}(t)\right)\mathcal{M}(t)w_{M,1}\mathrm{d}\mathbb{B}_4(t),\\
\mathrm{d}i(t) &= \left(\mu_{\mathbf{i},0} + i(t)\mu_{\mathbf{i},1}\right)\mathrm{d}t + w_{\mathbf{i},0}\mathrm{d}\mathbb{B}_2(t),\\
\mathrm{d}q(t) &= \left(\mu_{\mathbf{q}_0,0} + q(t)\mu_{\mathbf{q}_0,1}\right)\mathrm{d}t + w_{\mathbf{q}_0,0}\mathrm{d}\mathbb{B}_3(t),\\
\mathrm{d}\mathcal{M}(t) &= \left(q_{M,0} + \mathcal{M}(t)q_{M,1}\right)\mathrm{d}t + \mathcal{M}(t)w_{M,1}\mathrm{d}\mathbb{B}_4(t),
\end{aligned} \tag{6.9.25}$$

where additionally, $\mathrm{d}B_1(t)\mathrm{d}\mathbb{B}_4(t) = \rho_{4,1}\mathrm{d}t$, $\mathrm{d}\mathbb{B}_2(t)\mathrm{d}\mathbb{B}_4(t) = \rho_{4,2}\mathrm{d}t$, and $\mathrm{d}\mathbb{B}_3(t)$ $\mathrm{d}\mathbb{B}_3(t) = \rho_{4,3}\mathrm{d}t$. Analogous to (6.9.3), we have that

$$\begin{aligned}
&\beta\left(q_{M,0} + \mathcal{M}(t)q_{M,1}\right)V_1\left(t, S_1(t), \mathbb{C}(t), i(t), q(t), \mathcal{M}(t)\right)\mathrm{d}t\\
&\quad +\beta V_1\left(t, S_1(t), \mathbb{C}(t), i(t), q(t), \mathcal{M}(t)\right)\mathcal{M}(t)w_{M,1}\mathrm{d}\mathbb{B}_4(t)\\
&= \beta V_1\left(t, S_1(t), \mathbb{C}(t), i(t), q(t), \mathcal{M}(t)\right)\left(\left(q_{M,0} + \mathcal{M}(t)q_{M,1}\right)\mathrm{d}t\right.\\
&\quad \left.+\mathcal{M}(t)w_{M,1}\mathrm{d}\mathbb{B}_4(t)\right)\\
&= \beta V_1\left(t, S_1(t), \mathbb{C}(t), i(t), q(t), \mathcal{M}(t)\right)\mathrm{d}\mathcal{M}(t)
\end{aligned} \tag{6.9.26}$$

is the cash inflow (total for the whole company) due to stock dilution (if $\mathrm{d}\mathcal{M}(t) > 0$), or cash outflow due to stock buyback (if $\mathrm{d}\mathcal{M}(t) < 0$).

The corresponding market coefficients are therefore

$$\begin{aligned}
a_s &= \{\mathbf{a}_1\},\\
\sigma_s &= \left(\mathbf{p}_1\ 0\ 0\ 0\right),
\end{aligned}$$

$$b = \{S_1(\mathbf{a}_1 - \mathbb{D}_1), i - q + \mathbb{C}(r - \mathbf{q}_1) - \alpha_0 - \mathbb{C}\alpha_1 - \alpha_2 i - q\alpha_3 + \beta(q_{M,0} + \mathcal{M}q_{M,1}) \\ V_1(t, S_1, \mathbb{C}, i, q, \mathcal{M}), \mu_{\mathbf{i},0} + i\mu_{\mathbf{i},1}, \mu_{\mathbf{q}_0,0} + q\mu_{\mathbf{q}_0,1}, q_{M,0} + \mathcal{M}q_{M,1}\},$$

$$c = \begin{pmatrix} S_1\mathbf{p}_1 & 0 & 0 & 0 \\ \mathcal{M}\beta V_1 w_{M,1}\rho_{4,1} & \mathcal{M}\beta V_1 w_{M,1}\mathbf{s}_{4,2} & \mathcal{M}\beta V_1 w_{M,1}\mathbf{s}_{4,3} & \mathcal{M}\beta V_1 w_{M,1}\mathbf{s}_{4,4} \\ w_{\mathbf{i},0}\rho_{2,1} & w_{\mathbf{i},0}\sqrt{1-\rho_{2,1}^2} & 0 & 0 \\ w_{\mathbf{q}_0,0}\rho_{3,1} & \frac{w_{\mathbf{q}_0,0}(\rho_{3,2}-\rho_{2,1}\rho_{3,1})}{\sqrt{1-\rho_{2,1}^2}} & w_{\mathbf{q}_0,0}\mathbf{s}_{3,3} & 0 \\ \mathcal{M}w_{M,1}\rho_{4,1} & \mathcal{M}w_{M,1}\mathbf{s}_{4,2} & \mathcal{M}w_{M,1}\mathbf{s}_{4,3} & \mathcal{M}w_{M,1}\mathbf{s}_{4,4} \end{pmatrix}, \tag{6.9.27}$$

where $V_1 = V_1(t, S_1, \mathbb{C}, i, q, \mathcal{M})$, and where $\mathbf{s}_{3,3}, \mathbf{s}_{4,2}, \mathbf{s}_{4,3}, \mathbf{s}_{4,4}$ are given in the appendix. So a_s and b are unchanged, while σ_2 is a 1×4 matrix, since there are four Brownian motions now, and c is changed quite a bit to account for the diffusive part of the dilution/buyback.

Again, we employ CRRA utility of wealth, and adopt the zero position $\kappa = 0$. Consequently, we can find $g_\gamma = g(t)$ satisfying

$$g'(t) = \frac{\gamma - 1}{\gamma}\left(\frac{(\mathbf{a}_1 - r)^2}{2\mathbf{p}_1^2} + r\gamma\right), \tag{6.9.28}$$

and therefore $\nabla_A g_\gamma = 0$, and again we look for the stock price in the form

$$V_1(t, S_1, \mathbb{C}, i, q, \mathcal{M}) = V_1(\mathbb{C}, i, q, \mathcal{M}) = \mathbb{A}_0(\mathcal{M}) + \mathbb{C}\mathbb{A}_1(\mathcal{M}) + i\mathbb{A}_2(\mathcal{M}) + q\mathbb{A}_3(\mathcal{M}). \tag{6.9.29}$$

Consequently, the neutral (as well as the indifference) pricing PDE reduces to the system of four second-order ODEs:

$$\frac{1}{2}\mathcal{M}^2 w_{M,1}^2 \mathbb{A}_0''(\mathcal{M}) + \left(q_{M,0} + \mathcal{M}q_{M,1} + \frac{\mathcal{M}(r - \mathbf{a}_1) w_{M,1}\rho_{4,1}}{\mathbf{p}_1}\right)\mathbb{A}_0'(\mathcal{M}) \\ + \mathbb{A}_0(\mathcal{M})\left(\mathcal{M}^2\beta\mathbb{A}_1'(\mathcal{M})w_{M,1}^2 + \frac{1}{\mathbf{p}_1}\mathcal{M}\beta(r - \mathbf{a}_1)\rho_{4,1}\mathbb{A}_1(\mathcal{M})w_{M,1} - r \right. \\ \left. + \beta q_{M,0}\mathbb{A}_1(\mathcal{M}) + \mathcal{M}\beta q_{M,1}\mathbb{A}_1(\mathcal{M})\right) + \frac{\alpha_0(1 - \mathcal{M}\mathbb{A}_1(\mathcal{M}))}{\mathcal{M}} + \mu_{\mathbf{i},0}\mathbb{A}_2(\mathcal{M}) \\ + \frac{(r - \mathbf{a}_1) w_{\mathbf{i},0}\rho_{2,1}\mathbb{A}_2(\mathcal{M})}{\mathbf{p}_1} + \mu_{\mathbf{q}_0,0}\mathbb{A}_3(\mathcal{M}) + \frac{(r - \mathbf{a}_1) w_{\mathbf{q}_0,0}\rho_{3,1}\mathbb{A}_3(\mathcal{M})}{\mathbf{p}_1} \\ + \mathcal{M}w_{M,1}w_{\mathbf{i},0}\rho_{4,2}\mathbb{A}_2'(\mathcal{M}) + \mathcal{M}w_{M,1}w_{\mathbf{q}_0,0}\rho_{4,3}\mathbb{A}_3'(\mathcal{M}) = 0, \tag{6.9.30}$$

$$\frac{1}{2}\mathcal{M}^2 w_{M,1}^2 \mathbb{A}_1''(\mathcal{M}) + \left(q_{M,0} + \mathcal{M} q_{M,1} + \frac{\mathcal{M}(r-\mathbf{a}_1) w_{M,1}\rho_{4,1}}{\mathbf{p}_1}\right)$$

$$\mathbb{A}_1'(\mathcal{M}) - (\mathbf{q}_1 + \alpha_1)\mathbb{A}_1(\mathcal{M}) + \left(\beta q_{M,0} + \mathcal{M}\beta q_{M,1} + \frac{\mathcal{M}\beta(r-\mathbf{a}_1) w_{M,1}\rho_{4,1}}{\mathbf{p}_1}\right)$$

$$\mathbb{A}_1(\mathcal{M})^2 + \mathcal{M}^2 \beta w_{M,1}^2 \mathbb{A}_1(\mathcal{M})\mathbb{A}_1'(\mathcal{M}) + \frac{\alpha_1}{\mathcal{M}} = 0, \tag{6.9.31}$$

$$\frac{1}{2}\mathcal{M}^2 w_{M,1}^2 \mathbb{A}_2''(\mathcal{M}) + \left(q_{M,0} + \mathcal{M} q_{M,1} + \frac{\mathcal{M}(r-\mathbf{a}_1) w_{M,1}\rho_{4,1}}{\mathbf{p}_1}\right)\mathbb{A}_2'(\mathcal{M}) + \mathbb{A}_2(\mathcal{M})$$

$$\Big(\mathcal{M}^2 \beta \mathbb{A}_1'(\mathcal{M}) w_{M,1}^2 + \frac{1}{\mathbf{p}_1}\mathcal{M}\beta(r-\mathbf{a}_1)\rho_{4,1}\mathbb{A}_1(\mathcal{M}) w_{M,1} - r + \mu_{\mathbf{i},1} + \beta q_{M,0}$$

$$\mathbb{A}_1(\mathcal{M}) + \mathcal{M}\beta q_{M,1}\mathbb{A}_1(\mathcal{M})\Big) + \alpha_2\left(\frac{1}{\mathcal{M}} - \mathbb{A}_1(\mathcal{M})\right) + \mathbb{A}_1(\mathcal{M}) = 0, \tag{6.9.32}$$

$$\frac{1}{2}\mathcal{M}^2 w_{M,1}^2 \mathbb{A}_3''(\mathcal{M}) + \left(q_{M,0} + \mathcal{M} q_{M,1} + \frac{\mathcal{M}(r-\mathbf{a}_1) w_{M,1}\rho_{4,1}}{\mathbf{p}_1}\right)$$

$$\mathbb{A}_3'(\mathcal{M}) + \mathbb{A}_3(\mathcal{M})\Big(\mathcal{M}^2 \beta \mathbb{A}_1'(\mathcal{M}) w_{M,1}^2 + \frac{1}{\mathbf{p}_1}\mathcal{M}\beta(r-\mathbf{a}_1)\rho_{4,1}$$

$$\mathbb{A}_1(\mathcal{M}) w_{M,1} - r + \mu_{\mathbf{q}_0,1} + \beta q_{M,0}\mathbb{A}_1(\mathcal{M}) + \mathcal{M}\beta q_{M,1}$$

$$\mathbb{A}_1(\mathcal{M})\Big) + \alpha_3\left(\frac{1}{\mathcal{M}} - \mathbb{A}_1(\mathcal{M})\right) - \mathbb{A}_1(\mathcal{M}) = 0, \tag{6.9.33}$$

for $\mathcal{M} > 0$ (the entire solution). Summarizing, we have the following proposition.

Proposition 6.9.3. *In the above stochastic stock dilution/buyback model (6.9.3), and for the zero position $\kappa = 0$, the CRRA neutral price of a single stock V_1 is equal to (6.9.5), where $\mathbb{A}_0(\mathcal{M})$, $\mathbb{A}_0(\mathcal{M})$, $\mathbb{A}_0(\mathcal{M})$, and $\mathbb{A}_0(\mathcal{M})$ are characterized as $\{\mathcal{M} > 0\}$-entire solutions of ODEs (6.9.30)–(6.9.33).*

Remark 6.9.4. Again, (6.9.31) is quadratic and can be solved for $\mathbb{A}_1(\mathcal{M})$ first, since it does not depend on any other $\mathbb{A}_i$'s. After $\mathbb{A}_1(\mathcal{M})$ is found, then equations (6.9.32) and (6.9.33) are solved for $\mathbb{A}_2(\mathcal{M})$ and $\mathbb{A}_3(\mathcal{M})$. Finally, (6.9.30) is solved for $\mathbb{A}_0(\mathcal{M})$. In general, these ODEs must be solved numerically.

Again, there is a special case in which the above system has an explicit $\{\mathcal{M} > 0\}$-entire solution. Indeed, if $\mathbf{q}_1 = 0$ (the "Miller–Modigliani case" – no surcharge cost), and if $\beta = 1$ (the stock dilution/buyback are done at the market price), then (6.9.31) simplifies to

$$\mathcal{M}^2 \mathbb{A}_1(\mathcal{M})\mathbb{A}_1'(\mathcal{M}) w_{M,1}^2 + \frac{1}{2}\mathcal{M}^2 \mathbb{A}_1''(\mathcal{M}) w_{M,1}^2$$

$$+ \left(q_{M,0} + \mathcal{M} q_{M,1} + \frac{\mathcal{M}(r-\mathbf{a}_1) w_{M,1}\rho_{4,1}}{\mathbf{p}_1}\right)$$

$$\mathbb{A}_1(\mathcal{M})^2 + \frac{\alpha_1}{\mathcal{M}} - \alpha_1 \mathbb{A}_1(\mathcal{M}) + \left(q_{M,0} + \mathcal{M} q_{M,1} \right.$$
$$\left. + \frac{\mathcal{M}(r - \mathbf{a}_1) w_{M,1} \rho_{4,1}}{\mathbf{p}_1} \right) \mathbb{A}_1'(\mathcal{M}) = 0, \quad (6.9.34)$$

and its $\{\mathcal{M} > 0\}$-entire solution is again equal to

$$\mathbb{A}_1(\mathcal{M}) = \frac{1}{\mathcal{M}}. \quad (6.9.35)$$

Observe that in the case $q_{M,0} = 0$, an additional $\{\mathcal{M} > 0\}$-entire solution of (6.9.34) exists. It is equal to $\mathbb{A}_1(\mathcal{M}) = \frac{\mathbf{p}_1 \alpha_1}{\mathbf{p}_1\left(-w_{M,1}^2 + q_{M,1}\right) + (r - \mathbf{a}_1) w_{M,1} \rho_{4,1}} \frac{1}{\mathcal{M}}$, but it is dismissed, as before. Consequently, equations (6.9.32) and (6.9.33) simplify to

$$\frac{1}{2}\mathcal{M}^2 \mathbb{A}_2''(\mathcal{M}) w_{M,1}^2 + \left(-w_{M,1}^2 + \frac{(r - \mathbf{a}_1) \rho_{4,1} w_{M,1}}{\mathbf{p}_1} - r + \frac{q_{M,0}}{\mathcal{M}} + q_{M,1} + \mu_{\mathbf{i},1} \right)$$
$$\mathbb{A}_2(\mathcal{M}) + \left(q_{M,0} + \mathcal{M} q_{M,1} + \frac{\mathcal{M}(r - \mathbf{a}_1) w_{M,1} \rho_{4,1}}{\mathbf{p}_1} \right) \mathbb{A}_2'(\mathcal{M}) + \frac{1}{\mathcal{M}} = 0 \quad (6.9.36)$$

and

$$\frac{1}{2}\mathcal{M}^2 \mathbb{A}_3''(\mathcal{M}) w_{M,1}^2 + \left(-w_{M,1}^2 + \frac{(r - \mathbf{a}_1) \rho_{4,1} w_{M,1}}{\mathbf{p}_1} - r + \frac{q_{M,0}}{\mathcal{M}} + q_{M,1} + \mu_{\mathbf{q}_0,1} \right)$$
$$\mathbb{A}_3(\mathcal{M}) + \left(q_{M,0} + \mathcal{M} q_{M,1} + \frac{\mathcal{M}(r - \mathbf{a}_1) w_{M,1} \rho_{4,1}}{\mathbf{p}_1} \right) \mathbb{A}_3'(\mathcal{M}) - \frac{1}{\mathcal{M}} = 0, \quad (6.9.37)$$

admitting $\{\mathcal{M} > 0\}$-entire solutions

$$\mathbb{A}_2(\mathcal{M}) = \frac{1}{r - \mu_{\mathbf{i},1}} \frac{1}{\mathcal{M}} \quad (6.9.38)$$

and

$$\mathbb{A}_3(\mathcal{M}) = -\frac{1}{r - \mu_{\mathbf{q}_0,1}} \frac{1}{\mathcal{M}} \quad (6.9.39)$$

respectively. Finally, (6.9.40) is then simplified to

$$\frac{1}{2}\mathcal{M}^2 \mathbb{A}_0''(\mathcal{M}) w_{M,1}^2 + \frac{w_{\mathbf{q}_0,0} \rho_{4,3} w_{M,1}}{\mathcal{M}(r - \mu_{\mathbf{q}_0,1})} - \frac{w_{\mathbf{i},0} \rho_{4,2} w_{M,1}}{\mathcal{M}(r - \mu_{\mathbf{i},1})}$$
$$+ \frac{(r - \mathbf{a}_1) w_{\mathbf{i},0} \rho_{2,1}}{\mathcal{M} \mathbf{p}_1 (r - \mu_{\mathbf{i},1})} + \left(-w_{M,1}^2 + \frac{(r - \mathbf{a}_1) \rho_{4,1} w_{M,1}}{\mathbf{p}_1} - r + \frac{q_{M,0}}{\mathcal{M}} + q_{M,1} \right) \mathbb{A}_0(\mathcal{M})$$

$$+\left(q_{M,0}+\mathcal{M}q_{M,1}+\frac{\mathcal{M}\left(r-\mathbf{a}_1\right)w_{M,1}\rho_{4,1}}{\mathbf{p}_1}\right)\mathbb{A}_0'(\mathcal{M})+\frac{\mu_{\mathbf{i},0}}{\mathcal{M}\left(r-\mu_{\mathbf{i},1}\right)}$$
$$-\frac{\mu_{\mathbf{q}_0,0}}{\mathcal{M}\left(r-\mu_{\mathbf{q}_0,1}\right)}-\frac{\left(r-\mathbf{a}_1\right)w_{\mathbf{q}_0,0}\rho_{3,1}}{\mathcal{M}\mathbf{p}_1\left(r-\mu_{\mathbf{q}_0,1}\right)}=0, \tag{6.9.40}$$

admitting $\{\mathcal{M}>0\}$-entire solution

$$\mathbb{A}_0(\mathcal{M})=\left(\frac{\mu_{\mathbf{i},0}}{r\left(r-\mu_{\mathbf{i},1}\right)}-\frac{\mu_{\mathbf{q}_0,0}}{r\left(r-\mu_{\mathbf{q}_0,1}\right)}+\frac{\left(r-\mathbf{a}_1\right)w_{\mathbf{i},0}\rho_{2,1}}{r\left(r-\mu_{\mathbf{i},1}\right)\mathbf{p}_1}-\frac{\left(r-\mathbf{a}_1\right)w_{\mathbf{q}_0,0}\rho_{3,1}}{r\left(r-\mu_{\mathbf{q}_0,1}\right)\mathbf{p}_1}\right.$$
$$\left.-\frac{w_{M,1}w_{\mathbf{i},0}\rho_{4,2}}{r\left(r-\mu_{\mathbf{i},1}\right)}+\frac{w_{M,1}w_{\mathbf{q}_0,0}\rho_{4,3}}{r\left(r-\mu_{\mathbf{q}_0,1}\right)}\right)\frac{1}{\mathcal{M}}. \tag{6.9.41}$$

Summarizing, we have the following.

Proposition 6.9.4. *In the above stochastic stock dilution/buyback model (6.9.26), if* $\mathbf{q}_1=0,\beta=1$, *and* $k_1=0$, *then the price of a single stock* V_1 *is equal to*

$$V_1\left(t,S_1,\mathbb{C},i,q,\mathcal{M}\right)=V_1(\mathbb{C},i,q,\mathcal{M})$$
$$=\frac{1}{\mathcal{M}}\left(\mathbb{C}+\frac{i}{r-\mu_{\mathbf{i},1}}-\frac{q}{r-\mu_{\mathbf{q}_0,1}}+\frac{\mu_{\mathbf{i},0}-\frac{(\mathbf{a}_1-r)w_{\mathbf{i},0}\rho_{2,1}}{\mathbf{p}_1}}{r\left(r-\mu_{\mathbf{i},1}\right)}\right.$$
$$\left.-\frac{\mu_{\mathbf{q}_0,0}-\frac{(\mathbf{a}_1-r)w_{\mathbf{q}_0,0}\rho_{3,1}}{\mathbf{p}_1}}{r\left(r-\mu_{\mathbf{q}_0,1}\right)}-\frac{w_{M,1}w_{\mathbf{i},0}\rho_{4,2}}{r\left(r-\mu_{\mathbf{i},1}\right)}+\frac{w_{M,1}w_{\mathbf{q}_0,0}\rho_{4,3}}{r\left(r-\mu_{\mathbf{q}_0,1}\right)}\right), \tag{6.9.42}$$

while the value of the whole company, $V=\mathcal{M}V_1$, *is independent of the number of shares outstanding* $\mathcal{M}$, *and is equal to*

$$V\left(t,S_1,\mathbb{C},i,q\right)=V(\mathbb{C},i,q)$$
$$=\mathbb{C}+\frac{i}{r-\mu_{\mathbf{i},1}}-\frac{q}{r-\mu_{\mathbf{q}_0,1}}+\frac{\mu_{\mathbf{i},0}-\frac{(\mathbf{a}_1-r)w_{\mathbf{i},0}\rho_{2,1}}{\mathbf{p}_1}}{r\left(r-\mu_{\mathbf{i},1}\right)}-\frac{\mu_{\mathbf{q}_0,0}-\frac{(\mathbf{a}_1-r)w_{\mathbf{q}_0,0}\rho_{3,1}}{\mathbf{p}_1}}{r\left(r-\mu_{\mathbf{q}_0,1}\right)}$$
$$-\frac{w_{M,1}w_{\mathbf{i},0}\rho_{4,2}}{r\left(r-\mu_{\mathbf{i},1}\right)}+\frac{w_{M,1}w_{\mathbf{q}_0,0}\rho_{4,3}}{r\left(r-\mu_{\mathbf{q}_0,1}\right)}. \tag{6.9.43}$$

Remark 6.9.5. Notice the new terms

$$-\frac{w_{M,1}w_{\mathbf{i},0}\rho_{4,2}}{r\left(r-\mu_{\mathbf{i},1}\right)}+\frac{w_{M,1}w_{\mathbf{q}_0,0}\rho_{4,3}}{r\left(r-\mu_{\mathbf{q}_0,1}\right)} \tag{6.9.44}$$

in (6.9.43). They suggest that it is beneficial for the equity value that the correlation between income and stock dilution is negative, or alternatively, that it is beneficial for the equity value that the correlation between cost and stock dilution is positive. In other words, when income increases, the cash windfall should be used for the buyback, while when the cost increases, the cash shortfall should be compensated by stock dilution.

Chapter 7
FX Rates and FX Derivatives

A Framework for Modeling Dynamics of FX Rates and the Corresponding Models for FX Derivatives; FX Rates as a Statistic for Market Risk-Aversion

Dollar and yen lower amid risk-taking
By Neil Dennis
FINANCIAL TIMES, Published: January 19 2010

Foreign exchange investors were in the mood to take on risk yesterday, driving the dollar and yen lower amid speculation over strong Chinese data this week.
Both currencies have resumed their negative correlation with equity markets since the start of the year. Higher equities have indicated a stronger appetite for risk and benefited strategies such as the carry trade, which favour high-yielding currencies, including the Australian and New Zealand dollars and emerging market units, funded by borrowings in the low yielding dollar or yen.

NEW YORK TIMES, By Edward Wong:
...But some economists say China is likely to let the renminbi rise by about 5% to help stave off inflation, a growing concern among Chinese policy makers. (Published: Feb 20, 2010)

7.1 Introduction

In this final chapter we introduce an elaborate framework for modeling foreign exchange rates (FXR), that, although more empirical work in this area is necessary, appears to be capable of capturing many features of the foreign currencies markets. The framework is based on the assumption that FXR market dynamics prevent arbitrage/imbalance between the investment opportunities in two countries/economies, and then by extension, among any number of them. While this principle is not new, we take it to the next level of scope and generality.

We postulate that the balance between all of the investment opportunities in two countries is maintained by virtue of the FXR dynamics. If such a balance does not exist, the FXR dynamics (and this theory) need to be adjusted accordingly.

S. Stojanovic, *Neutral and Indifference Portfolio Pricing, Hedging and Investing: With Applications in Equity and FX*, DOI 10.1007/978-0-387-71418-9_7,

We shall comment briefly on what could be done in such a situation, but by and large we assume that there is a such balance. Furthermore, we postulate that in both countries/economies, the investors behave as an aggregate in an optimal way – they invest optimally according to a chosen type and intensity of risk-aversion, i.e., according to the adopted utility of wealth function. More explicitly, we shall derive the general FXR SDE dynamics in the context of simple economies for an arbitrary utility of wealth. Yet most of our results will be obtained after we specialize to the CRRA utility.

Once our study of FXR models is completed, we shall discuss pricing of FX derivatives using such underlying models. Usually in the FX derivative literature and practice (see [20]), following Garman and Kohlhagen [11], it is assumed that the only benefit from holding a foreign currency is investing in the foreign money market, and thus earning the interest at the foreign interest rate r_{f}. As we develop and then use models for FXR that assume, more generally, a possibility to invest in risky foreign markets as well, we shall also offer a corresponding extension of the FX derivative pricing methodology.

7.2 FX Rates

7.2.1 Derivation of the General FXR SDE

Investors trade in two economies, say *domestic* and *foreign*. In each of the two economies they have opportunities to invest in risky assets and in "risk-free" assets. The risk-free assets can be thought of as cash accounts, or money market accounts, or bonds with very short maturities (bonds with longer maturities would have to be priced and henceforth declared as tradables in the economy), with interest rates $r(t)$ (or, more precisely, $r_{\mathrm{d}}(t)$) the domestic interest rate and $r_{\mathrm{f}}(t)$ the foreign interest rate.

Following [43], consider two simple economies, a *domestic* economy $\mathfrak{E}_{\mathrm{d}}$ and a *foreign* economy $\mathfrak{E}_{\mathrm{f}}$, described quantitatively via a finite set of dynamic factors and tradables. Factors are considered across the two economies, and they are denoted, as before, by $A(t) = \{A_1(t), \ldots, A_m(t)\}$, while tradables are considered separately for the considered economies, and they are denoted, in the case of the domestic economy, by $S_{\mathrm{d}}(t) = \left\{S_{\mathrm{d},1}(t), \ldots, S_{\mathrm{d},k_{\mathrm{d}}}(t)\right\}$, and in the case of the foreign economy by $S_{\mathrm{f}}(t) = \left\{S_{\mathrm{f},1}(t), \ldots, S_{\mathrm{f},k_{\mathrm{f}}}(t)\right\}$. (Since it might be quite relevant for FXR modeling, and since consumer goods are the *ultimate tradables*, we remark that some kind of aggregate consumer prices can presumably also be modeled as some of the tradables in the considered economies, and not just as the factors, to account for inflation, which, of course, has a profound effect on the FXR.)The interest rates $r_{\mathrm{d}}(t)$ and $r_{\mathrm{f}}(t)$ can be stochastic, in which case $r_{\mathrm{d}}(t) = r_{\mathrm{d}}(t, A(t))$ and $r_{\mathrm{f}}(t) = r_{\mathrm{f}}(t, A(t))$. It is important to keep in mind that the presented theory will also include the cases in which either one or both of $S_d(t)$ and $S_f(t)$ are empty (i.e., $S_d(t) = \emptyset$ and/or $S_f(t) = \emptyset$), i.e., cases in which no risky assets are considered, i.e., cases in which only money markets are considered as investment opportunities. Factors and tradables are assumed to obey the Itô SDE dynamics

$$dA(t) = b(t,A(t))dt + c(t,A(t)) \cdot dB(t), \tag{7.2.1}$$

$$dS_d(t) = S_d(t)\left(a_{s,d}(t,A(t)) - \mathbb{D}_d(t,A(t))\right)dt + S_d(t)\sigma_{s,d}(t,A(t)) \cdot dB(t),$$

$$dS_f(t) = S_f(t)\left(a_{s,f}(t,A(t)) - \mathbb{D}_f(t,A(t))\right)dt + S_f(t)\sigma_{s,f}(t,A(t)) \cdot dB(t), \tag{7.2.2}$$

where $B(t) = \{B_1(t), \ldots, B_n(t)\}$ is a vector of n independent standard Brownian motions, the vector-valued function $b(t,A)$ is the m-vector of factor drifts, $c(t,A)$ is the $m \times n$ factor-diffusion matrix, $a_{s,d}(t,A)$ is the k_d-vector of appreciation rates for the tradables (before dividends) in the domestic economy, $a_{s,f}(t,A)$ is the k_f-vector of appreciation rates for the tradables (before dividends) in the foreign economy, $\mathbb{D}_d(t,A)$ and $\mathbb{D}_f(t,A)$ are the k_d- and k_f-vectors of dividend rates of the corresponding assets, $\sigma_{s,d}(t,A)$ and $\sigma_{s,f}(t,A)$ are the volatility $k_d \times n$ and $k_f \times n$ matrices. All of those functions are called *market coefficients*.

The basic (technical) assumption is the *nonredundancy* of the tradables in each of the economies, or explicitly, it is assumed throughout that

$$\sigma_{s,d} \cdot \sigma_{s,d}^{T} > 0,$$

$$\sigma_{s,f} \cdot \sigma_{s,f}^{T} > 0, \tag{7.2.3}$$

so that we necessarily have to consider risky assets in addition to the risk-free assets, which contradicts the above assertion that what follows will include the cases in which either one or both of $S_d(t)$ and $S_f(t)$ are empty. As a matter of fact, such cases are simpler and can be derived in an analogous fashion to what follows, with the final results being consistent with that to be derived below (see below for more details).

Definition 7.2.1. A foreign exchange rate (FXR) process is a stochastic process, to be denoted by $Y_{d,f}(t)$, representing the price of a foreign currency in units of the domestic currency (see (7.2.4) below). The process $Y_{f,d}(t)$, the reciprocal FXR, represents the price of the domestic currency in units of the foreign currency.

Consequently, any tradable in the foreign markets, with the foreign currency price of $S_f(t)$, can be thought of as having a price expressed in the domestic currency equal to

$$S_d(t) = Y_{d,f}(t)S_f(t). \tag{7.2.4}$$

Assuming (7.2.1)–(7.2.2), investors have an opportunity to invest in either the domestic or foreign markets. They do so in an optimal fashion, by choosing a utility of wealth function $\psi(X)$ and then solving equation (3.4.2), once for each of the considered economies. More precisely, as in (3.4.1), we define

$$
\begin{aligned}
\mathcal{M}_{\mathrm{d}}(\varphi) := &\frac{\partial^2\varphi}{\partial X^2}\frac{\partial\varphi}{\partial t} + \frac{1}{2}\frac{\partial^2\varphi}{\partial X^2}\mathrm{Tr}\left(c.c^{\mathrm{T}}\cdot\nabla_A\nabla_A\varphi\right) \\
&-\frac{1}{2}\left(\frac{\partial\varphi}{\partial X}\right)^2 (a_{s,\mathrm{d}}-r_{\mathrm{d}})\cdot\left(\sigma_{s,\mathrm{d}}\cdot\sigma_{s,\mathrm{d}}{}^{\mathrm{T}}\right)^{-1}\cdot(a_{s,\mathrm{d}}-r_{\mathrm{d}}) + (b.\nabla_A\varphi)\frac{\partial^2\varphi}{\partial X^2} \\
&-\frac{\partial\varphi}{\partial X}(a_{s,\mathrm{d}}-r_{\mathrm{d}})\cdot\left(\sigma_{s,\mathrm{d}}\cdot\sigma_{s,\mathrm{d}}{}^{\mathrm{T}}\right)^{-1}\cdot\sigma_{s,\mathrm{d}}\cdot c^{\mathrm{T}}\cdot\left(\nabla_A\frac{\partial\varphi}{\partial X}\right) \\
&-\frac{1}{2}\left(\nabla_A\frac{\partial\varphi}{\partial X}\right)\cdot c\cdot\sigma_{s,\mathrm{d}}{}^{\mathrm{T}}\cdot\left(\sigma_{s,\mathrm{d}}\cdot\sigma_{s,\mathrm{d}}{}^{\mathrm{T}}\right)^{-1}\cdot\sigma_{s,\mathrm{d}}\cdot c^{\mathrm{T}}\cdot\left(\nabla_A\frac{\partial\varphi}{\partial X}\right) \\
&+r_{\mathrm{d}}X\frac{\partial^2\varphi}{\partial X^2}\frac{\partial\varphi}{\partial X},
\end{aligned}
\tag{7.2.5}
$$

and similarly $\mathcal{M}_{\mathrm{f}}$, and solve

$$\mathcal{M}_{\mathrm{d}}(\varphi_{\mathrm{d}}) = 0 \tag{7.2.6}$$

for $t < T$, together with the terminal condition $\varphi_{\mathrm{d}}(T,X,A) = \psi_{\mathrm{d}}(X)$, and

$$\mathcal{M}_{\mathrm{f}}(\varphi_{\mathrm{f}}) = 0 \tag{7.2.7}$$

for $t < T$, together with the terminal condition $\varphi_{\mathrm{f}}(T,X,A) = \psi_{\mathrm{f}}(X)$. After solving these two equations, the optimal portfolio strategies (without constraints on the portfolio) in the two markets are

$$
\begin{aligned}
\Pi^{\star}_{\mathrm{d},T}(t,X,A) = &-\left(1\Big/\frac{\partial^2\varphi_{\mathrm{d}}}{\partial X^2}\right)\left(\frac{\partial\varphi_{\mathrm{d}}}{\partial X}(a_{s,\mathrm{d}}-r_{\mathrm{d}})\right. \\
&\left.+\left(\nabla_A\frac{\partial\varphi_{\mathrm{d}}}{\partial X}\right)\cdot c\cdot\sigma_{s,\mathrm{d}}{}^{\mathrm{T}}\right)\cdot\left(\sigma_{s,\mathrm{d}}\cdot\sigma_{s,\mathrm{d}}{}^{\mathrm{T}}\right)^{-1}
\end{aligned}
\tag{7.2.8}
$$

and

$$
\begin{aligned}
\Pi^{\star}_{\mathrm{f},T}(t,X,A) = &-\left(1\Big/\frac{\partial^2\varphi_{\mathrm{f}}}{\partial X^2}\right)\left(\frac{\partial\varphi_{\mathrm{f}}}{\partial X}(a_{s,\mathrm{f}}-r_{\mathrm{f}})+\left(\nabla_A\frac{\partial\varphi_{\mathrm{f}}}{\partial X}\right)\cdot c\cdot\sigma_{s,\mathrm{f}}{}^{\mathrm{T}}\right) \\
&\cdot\left(\sigma_{s,\mathrm{f}}\cdot\sigma_{s,\mathrm{f}}{}^{\mathrm{T}}\right)^{-1}
\end{aligned}
\tag{7.2.9}
$$

respectively. Often, in applications, as discussed previously, it is possible to pass to the limit $T \to \infty$ in (7.2.8) and (7.2.9) (in spite of the fact that it is not possible to pass to the limit $T \to \infty$ in (7.2.6) and (7.2.7)). This is desirable in modeling FXR, since FX markets presumably will never expire. So we shall assume that the limit exists, and define

$$\Pi^{\star}_{\mathrm{d}}(t,X,A) = \Pi^{\star}_{\mathrm{d},\infty}(t,X,A) = \lim_{T\to\infty}\Pi^{\star}_{\mathrm{d},T}(t,X,A) \tag{7.2.10}$$

and

$$\Pi_{\mathrm{f}}^{\star}(t,X,A) = \Pi_{\mathrm{f},\infty}^{\star}(t,X,A) = \lim_{T\to\infty} \Pi_{\mathrm{f},T}^{\star}(t,X,A). \tag{7.2.11}$$

Once optimal portfolio strategies have been implemented, the investor's (optimal) wealth evolutions (3.3.2) in the two considered economies obey

$$\begin{aligned} \mathrm{d}X_{\mathrm{d}}(t) = {} & (\Pi_{\mathrm{d}}^{\star}(t,X_{\mathrm{d}}(t),A(t)) \cdot (a_{s,\mathrm{d}}(t,A(t)) - r_{\mathrm{d}}(t,A(t))) + r_{\mathrm{d}}(t,A(t))X_{\mathrm{d}}(t))\,\mathrm{d}t \\ & + \Pi_{\mathrm{d}}^{\star}(t,X_{\mathrm{d}}(t),A(t)) \cdot \sigma_{s,\mathrm{d}}(t,A(t)) \cdot \mathrm{d}B(t) \end{aligned} \tag{7.2.12}$$

and

$$\begin{aligned} \mathrm{d}X_{\mathrm{f}}(t) = {} & (\Pi_{\mathrm{f}}^{\star}(t,X_{\mathrm{f}}(t),A(t)) \cdot (a_{s,\mathrm{f}}(t,A(t)) - r_{\mathrm{f}}(t,A(t))) + r_{\mathrm{f}}(t,A(t))X_{\mathrm{f}}(t))\,\mathrm{d}t \\ & + \Pi_{\mathrm{f}}^{\star}(t,X_{\mathrm{f}}(t),A(t)) \cdot \sigma_{s,\mathrm{f}}(t,A(t)) \cdot \mathrm{d}B(t). \end{aligned} \tag{7.2.13}$$

Motivated by the above discussion, we make the following definition.

Definition 7.2.2 (FXR SDE). A process $Y_{\mathrm{d,f}}(t)$, called the (bilateral) *foreign exchange rate* process, or FXR process for short, representing the price of the foreign currency in units of the domestic currency, is defined as a solution of the equation

$$\mathrm{d}X_{\mathrm{d}}(t) = d\,(Y_{\mathrm{d,f}}(t)X_{\mathrm{f}}(t)) \tag{7.2.14}$$

whenever $X_{\mathrm{d}}(t) = Y_{\mathrm{d,f}}(t)X_{\mathrm{f}}(t)$. In other words, given the equivalent endowments in two economies, the currency exchange will ensure that there is a balance between investment opportunities and risks in the two economies.

Remark 7.2.1. The *hypothesis* is that the currency exchange rates ensure the equivalence of overall investment opportunities (while allowing for profit-taking via carry trades, for example; the possibility of carry-trade profiting is not eliminated by Definition 7.2.2).

Remark 7.2.2. After substituting $X_{\mathrm{d}}(t) = Y_{\mathrm{d,f}}(t)X_{\mathrm{f}}(t)$, the left-hand side of (7.2.14) is calculated using (7.2.12), while the right-hand side of (7.2.14) is calculated using the Itô chain rule

$$d\,(X_{\mathrm{f}}(t)Y_{\mathrm{d,f}}(t)) = Y_{\mathrm{d,f}}(t)\mathrm{d}X_{\mathrm{f}}(t) + X_{\mathrm{f}}(t)\mathrm{d}Y_{\mathrm{d,f}}(t) + \mathrm{d}X_{\mathrm{f}}(t)\mathrm{d}Y_{\mathrm{d,f}}(t). \tag{7.2.15}$$

We therefore have

$$\frac{\mathrm{d}Y_{\mathrm{d,f}}(t)}{Y_{\mathrm{d,f}}(t)} = \frac{\mathrm{d}X_{\mathrm{d}}(t)}{Y_{\mathrm{d,f}}(t)X_{\mathrm{f}}(t)} - \frac{\mathrm{d}X_{\mathrm{f}}(t)}{X_{\mathrm{f}}(t)} - \frac{\mathrm{d}X_{\mathrm{f}}(t)}{X_{\mathrm{f}}(t)}\frac{\mathrm{d}Y_{\mathrm{d,f}}(t)}{Y_{\mathrm{d,f}}(t)}. \tag{7.2.16}$$

Postulating

$$\frac{\mathrm{d}Y_{\mathrm{d,f}}(t)}{Y_{\mathrm{d,f}}(t)} = m_1\,\mathrm{d}t + m_2.\mathrm{d}B(t) \tag{7.2.17}$$

and using (7.2.12) and (7.2.13), we have

$$
\begin{aligned}
&\frac{\mathrm{d}X_{\mathrm{f}}(t)}{X_{\mathrm{f}}(t)}\frac{\mathrm{d}Y_{\mathrm{d,f}}(t)}{Y_{\mathrm{d,f}}(t)} \\
&= \left(\left(\frac{\Pi_{\mathrm{f}}^{\star}(t,X_{\mathrm{f}}(t),A(t))}{X_{\mathrm{f}}(t)}\cdot(a_{s,\mathrm{f}}(t,A(t))-r_{\mathrm{f}}(t,A(t)))+r_{\mathrm{f}}(t,A(t))X_{\mathrm{f}}(t)\right)\mathrm{d}t\right. \\
&\quad\left.+\frac{\Pi_{f}^{\star}(t,X_{\mathrm{f}}(t),A(t))}{X_{\mathrm{f}}(t)}\cdot\sigma_{s,\mathrm{f}}(t,A(t))\cdot\mathrm{d}B(t)\right)(m_1\mathrm{d}t+m_2.\mathrm{d}B(t)) \\
&= \left(\frac{\Pi_{\mathrm{f}}^{\star}(t,X_{\mathrm{f}}(t),A(t))}{X_{\mathrm{f}}(t)}\cdot\sigma_{s,\mathrm{f}}(t,A(t))\cdot\mathrm{d}B(t)\right)(m_2.\mathrm{d}B(t)) \\
&= \frac{\Pi_{\mathrm{f}}^{\star}(t,X_{\mathrm{f}}(t),A(t))}{X_{\mathrm{f}}(t)}\cdot\sigma_{s,\mathrm{f}}(t,A(t))\cdot m_2\mathrm{d}t
\end{aligned}
\tag{7.2.18}
$$

as well as

$$
\begin{aligned}
\frac{\mathrm{d}X_{\mathrm{d}}(t)}{Y_{\mathrm{d,f}}(t)X_{\mathrm{f}}(t)} &= \frac{1}{Y_{\mathrm{d,f}}(t)X_{\mathrm{f}}(t)}((\Pi_{\mathrm{d}}^{\star}(t,X_{\mathrm{d}}(t),A(t))\cdot(a_{s,\mathrm{d}}(t,A(t))-r_{\mathrm{d}}(t,A(t))) \\
&\quad+r_{\mathrm{d}}(t,A(t))X_{\mathrm{d}}(t))\,\mathrm{d}t+\Pi_{\mathrm{d}}^{\star}(t,X_{\mathrm{d}}(t),A(t))\cdot\sigma_{s,\mathrm{d}}(t,A(t))\cdot\mathrm{d}B(t)) \\
&= \frac{1}{Y_{\mathrm{d,f}}(t)X_{\mathrm{f}}(t)}((\Pi_{\mathrm{d}}^{\star}(t,X_{\mathrm{d}}(t),A(t))\cdot(a_{s,\mathrm{d}}(t,A(t))-r_{\mathrm{d}}(t,A(t))) \\
&\quad+r_{\mathrm{d}}(t,A(t))Y_{\mathrm{d,f}}(t)X_{\mathrm{f}}(t))\,\mathrm{d}t+\Pi_{\mathrm{d}}^{\star}(t,X_{\mathrm{d}}(t),A(t)) \\
&\quad\cdot\sigma_{s,\mathrm{d}}(t,A(t))\cdot\mathrm{d}B(t)) \\
&= \left(\frac{\Pi_{\mathrm{d}}^{\star}(t,Y_{\mathrm{d,f}}(t)X_{\mathrm{f}}(t),A(t))}{Y_{\mathrm{d,f}}(t)X_{\mathrm{f}}(t)}\cdot(a_{s,\mathrm{d}}(t,A(t))-r_{\mathrm{d}}(t,A(t)))\right. \\
&\quad\left.+r_{\mathrm{d}}(t,A(t))\right)\mathrm{d}t+\frac{\Pi_{\mathrm{d}}^{\star}(t,Y_{\mathrm{d,f}}(t)X_{\mathrm{f}}(t),A(t))}{Y_{\mathrm{d,f}}(t)X_{\mathrm{f}}(t)}\cdot\sigma_{s,\mathrm{d}}(t,A(t))\cdot\mathrm{d}B(t)
\end{aligned}
\tag{7.2.19}
$$

and

$$
\begin{aligned}
\frac{\mathrm{d}X_{\mathrm{f}}(t)}{X_{\mathrm{f}}(t)} &= \left(\frac{\Pi_{\mathrm{f}}^{\star}(t,X_{\mathrm{f}}(t),A(t))}{X_{\mathrm{f}}(t)}\cdot(a_{s,\mathrm{f}}(t,A(t))-r_{\mathrm{f}}(t,A(t)))+r_{\mathrm{f}}(t,A(t))\right)\mathrm{d}t \\
&\quad+\frac{\Pi_{\mathrm{f}}^{\star}(t,X_{\mathrm{f}}(t),A(t))}{X_{\mathrm{f}}(t)}\cdot\sigma_{s,\mathrm{f}}(t,A(t))\cdot\mathrm{d}B(t).
\end{aligned}
\tag{7.2.20}
$$

So from (7.2.16), we have

$$
\begin{aligned}
&m_1 \mathrm{d}t + m_2.\mathrm{d}B(t) \\
&= \left(\frac{\Pi_{\mathrm{d}}^{\star}(t, Y_{\mathrm{d,f}}(t)X_{\mathrm{f}}(t), A(t))}{Y_{\mathrm{d,f}}(t)X_{\mathrm{f}}(t)} \cdot (a_{s,\mathrm{d}}(t,A(t)) - r_{\mathrm{d}}(t,A(t))) + r_{\mathrm{d}}(t,A(t)) \right) \mathrm{d}t \\
&\quad + \frac{\Pi_{\mathrm{d}}^{\star}(t, Y_{\mathrm{d,f}}(t)X_{\mathrm{f}}(t), A(t))}{Y_{\mathrm{d,f}}(t)X_{\mathrm{f}}(t)} \cdot \sigma_{s,\mathrm{d}}(t,A(t)) \cdot \mathrm{d}B(t) \\
&\quad - \left(\left(\frac{\Pi_{\mathrm{f}}^{\star}(t, X_{\mathrm{f}}(t), A(t))}{X_{\mathrm{f}}(t)} \cdot (a_{s,\mathrm{f}}(t,A(t)) - r_{\mathrm{f}}(t,A(t))) + r_{\mathrm{f}}(t,A(t)) \right) \mathrm{d}t \right. \\
&\qquad \left. + \frac{\Pi_{\mathrm{f}}^{\star}(t, X_{\mathrm{f}}(t), A(t))}{X_{\mathrm{f}}(t)} \cdot \sigma_{s,\mathrm{f}}(t,A(t)) \cdot \mathrm{d}B(t) \right) \\
&\quad - \frac{\Pi_{\mathrm{f}}^{\star}(t, X_{\mathrm{f}}(t), A(t))}{X_{\mathrm{f}}(t)} \cdot \sigma_{s,\mathrm{f}}(t,A(t)) \cdot m_2 \mathrm{d}t \\
&= \left(\frac{\Pi_{\mathrm{d}}^{\star}(t, Y_{\mathrm{d,f}}(t)X_{\mathrm{f}}(t), A(t))}{Y_{\mathrm{d,f}}(t)X_{\mathrm{f}}(t)} \cdot (a_{s,\mathrm{d}}(t,A(t)) - r_{\mathrm{d}}(t,A(t))) - \frac{\Pi_{\mathrm{f}}^{\star}(t, X_{\mathrm{f}}(t), A(t))}{X_{\mathrm{f}}(t)} \right. \\
&\qquad . (a_{s,\mathrm{f}}(t,A(t)) - r_{\mathrm{f}}(t,A(t))) + r_{\mathrm{d}}(t,A(t)) - r_{\mathrm{f}}(t,A(t)) \\
&\qquad \left. - \frac{\Pi_{\mathrm{f}}^{\star}(t, X_{\mathrm{f}}(t), A(t))}{X_{\mathrm{f}}(t)} \cdot \sigma_{s,\mathrm{f}}(t,A(t)) \cdot m_2 \right) \mathrm{d}t \\
&\quad + \left(\frac{\Pi_{\mathrm{d}}^{\star}(t, Y_{\mathrm{d,f}}(t)X_{\mathrm{f}}(t), A(t))}{Y_{\mathrm{d,f}}(t)X_{\mathrm{f}}(t)} \cdot \sigma_{s,\mathrm{d}}(t,A(t)) \right. \\
&\qquad \left. - \frac{\Pi_{\mathrm{f}}^{\star}(t, X_{\mathrm{f}}(t), A(t))}{X_{\mathrm{f}}(t)} \cdot \sigma_{s,\mathrm{f}}(t,A(t)) \right) \cdot \mathrm{d}B(t). \qquad (7.2.21)
\end{aligned}
$$

Therefore,

$$
m_2 = \frac{\Pi_{\mathrm{d}}^{\star}(t, Y_{\mathrm{d,f}}(t)X_{\mathrm{f}}(t), A(t))}{Y_{\mathrm{d,f}}(t)X_{\mathrm{f}}(t)} \cdot \sigma_{s,\mathrm{d}}(t,A(t)) - \frac{\Pi_{\mathrm{f}}^{\star}(t, X_{\mathrm{f}}(t), A(t))}{X_{\mathrm{f}}(t)} \cdot \sigma_{s,\mathrm{f}}(t,A(t)), \qquad (7.2.22)
$$

and then

$$
\begin{aligned}
m_1 &= r_{\mathrm{d}}(t,A(t)) - r_{\mathrm{f}}(t,A(t)) + \frac{\Pi_{\mathrm{d}}^{\star}(t, Y_{\mathrm{d,f}}(t)X_{\mathrm{f}}(t), A(t))}{Y_{\mathrm{d,f}}(t)X_{\mathrm{f}}(t)} \cdot (a_{s,\mathrm{d}}(t,A(t)) - r_{\mathrm{d}}(t,A(t))) \\
&\quad - \frac{\Pi_{\mathrm{f}}^{\star}(t, X_{\mathrm{f}}(t), A(t))}{X_{\mathrm{f}}(t)} \cdot (a_{s,\mathrm{f}}(t,A(t)) - r_{\mathrm{f}}(t,A(t)))
\end{aligned}
$$

$$-\frac{\Pi_{\mathrm{f}}^{\star}(t,X_{\mathrm{f}}(t),A(t))}{X_{\mathrm{f}}(t)}\cdot\sigma_{s,\mathrm{f}}(t,A(t))$$

$$\cdot\left(\frac{\Pi_{\mathrm{d}}^{\star}(t,Y_{\mathrm{d,f}}(t)X_{\mathrm{f}}(t),A(t))}{Y_{\mathrm{d,f}}(t)X_{\mathrm{f}}(t)}\cdot\sigma_{s,\mathrm{d}}(t,A(t))-\frac{\Pi_{\mathrm{f}}^{\star}(t,X_{\mathrm{f}}(t),A(t))}{X_{\mathrm{f}}(t)}\cdot\sigma_{s,\mathrm{f}}(t,A(t))\right). \tag{7.2.23}$$

Therefore, we have proved the following theorem.

Theorem 7.2.1. *(General FXR SDE.) In the above framework the FXR process $Y_{\mathrm{d,f}}(t)$ is a solution of the SDE*

$$\begin{aligned}\frac{\mathrm{d}Y_{\mathrm{d,f}}(t)}{Y_{\mathrm{d,f}}(t)}=&\left(r_{\mathrm{d}}-r_{\mathrm{f}}+\frac{\Pi_{\mathrm{d}}^{\star}(t,Y_{\mathrm{d,f}}(t)X_{\mathrm{f}}(t),A(t))}{Y_{\mathrm{d,f}}(t)X_{\mathrm{f}}(t)}\cdot(a_{s,\mathrm{d}}-r_{\mathrm{d}})-\frac{\Pi_{\mathrm{f}}^{\star}(t,X_{\mathrm{f}}(t),A(t))}{X_{\mathrm{f}}(t)}\right.\\&\cdot(a_{s,\mathrm{f}}-r_{\mathrm{f}})-\frac{\Pi_{\mathrm{f}}^{\star}(t,X_{\mathrm{f}}(t),A(t))}{X_{\mathrm{f}}(t)}\cdot\sigma_{s,\mathrm{f}}\cdot{\sigma_{s,\mathrm{d}}}^{\mathrm{T}}\cdot\frac{\Pi_{\mathrm{d}}^{\star}(t,Y_{\mathrm{d,f}}(t)X_{\mathrm{f}}(t),A(t))}{Y_{\mathrm{d,f}}(t)X_{\mathrm{f}}(t)}\\&\left.+\frac{\Pi_{\mathrm{f}}^{\star}(t,X_{\mathrm{f}}(t),A(t))}{X_{\mathrm{f}}(t)}\cdot\sigma_{s,\mathrm{f}}\cdot{\sigma_{s,\mathrm{f}}}^{\mathrm{T}}\cdot\frac{\Pi_{\mathrm{f}}^{\star}(t,X_{\mathrm{f}}(t),A(t))}{X_{\mathrm{f}}(t)}\right)\mathrm{d}t\\&+\left(\frac{\Pi_{\mathrm{d}}^{\star}(t,Y_{\mathrm{d,f}}(t)X_{\mathrm{f}}(t),A(t))}{Y_{\mathrm{d,f}}(t)X_{\mathrm{f}}(t)}\cdot\sigma_{s,\mathrm{d}}-\frac{\Pi_{\mathrm{f}}^{\star}(t,X_{\mathrm{f}}(t),A(t))}{X_{\mathrm{f}}(t)}\cdot\sigma_{s,\mathrm{f}}\right)\cdot\mathrm{d}B(t)\end{aligned} \tag{7.2.24}$$

coupled with SDEs (7.2.1) and (7.2.13), and where $\Pi_{\mathrm{d}}^{\star}$ and $\Pi_{\mathrm{f}}^{\star}$ are given in (7.2.8) and (7.2.9) respectively, and where φ_{d} and φ_{f} are solutions of PDEs (7.2.6) and (7.2.7), respectively.

Remark 7.2.3. We note that if no foreign investments are available or considered, i.e., if $S_{\mathrm{f}}=\emptyset$, by setting $\Pi_{\mathrm{f}}^{\star}(t,X_{\mathrm{f}}(t),A(t))=0$, we get

$$\begin{aligned}\frac{\mathrm{d}Y_{\mathrm{d,f}}(t)}{Y_{\mathrm{d,f}}(t)}=&\left(r_{\mathrm{d}}-r_{\mathrm{f}}+\frac{\Pi_{\mathrm{d}}^{\star}(t,Y_{\mathrm{d,f}}(t)X_{\mathrm{f}}(t),A(t))}{Y_{\mathrm{d,f}}(t)X_{\mathrm{f}}(t)}\cdot(a_{s,\mathrm{d}}-r_{\mathrm{d}})\right)\mathrm{d}t\\&+\frac{\Pi_{\mathrm{d}}^{\star}(t,Y_{\mathrm{d,f}}(t)X_{\mathrm{f}}(t),A(t))}{Y_{\mathrm{d,f}}(t)X_{\mathrm{f}}(t)}\cdot\sigma_{s,\mathrm{d}}\cdot\mathrm{d}B(t).\end{aligned} \tag{7.2.25}$$

Remark 7.2.4. In applications it is sometimes more natural to express the volatility term in the SDE (7.2.17) using the increments of the correlated Brownian motions $\mathrm{d}\mathbb{B}(t)$ instead of the independent ones $\mathrm{d}B(t)$. Since $\mathrm{d}\mathbb{B}(t)=\mathbb{S}_n.\mathrm{d}B(t)$, then $\mathrm{d}B(t)={\mathbb{S}_n}^{-1}\cdot\mathrm{d}\mathbb{B}(t)$, where the matrix $\mathbb{S}_n$ is as discussed in the appendix.

7.2.2 CRRA FXR SDE

We now specialize in the case of CRRA utility of wealth. In such a case, φ_{d} and φ_{f}, solutions of (7.2.6) and (7.2.7) respectively, can be found in the form (3.6.1). More precisely,

$$\varphi_{\mathrm{d}}(t,X,A) = \frac{X^{1-\gamma}e^{g_{\gamma_{\mathrm{d}},d,T}(t,A)} - 1}{1-\gamma}, \tag{7.2.26}$$

with $g_{\gamma_{\mathrm{d}},d,T}$ solving

$$\begin{aligned}
&\frac{\partial g_{\gamma_{\mathrm{d}},d,T}}{\partial t} + \frac{1}{2}\mathrm{Tr}\left(c.c^{\mathrm{T}} \cdot \nabla_A \nabla_A g_{\gamma_{\mathrm{d}},d,T}\right) \\
&\quad + \left(b - \frac{\gamma_{\mathrm{d}}-1}{\gamma_{\mathrm{d}}}(a_{s,\mathrm{d}} - r_{\mathrm{d}}) \cdot \left(\sigma_{s,\mathrm{d}} \cdot \sigma_{s,\mathrm{d}}{}^{\mathrm{T}}\right)^{-1} \cdot \sigma_{s,\mathrm{d}} \cdot c^{\mathrm{T}}\right) . \nabla_A g_{\gamma_{\mathrm{d}},d,T} \\
&\quad + \frac{1}{2}\nabla_A g_{\gamma_{\mathrm{d}},d,T} \cdot c \cdot \left(\mathbb{I}_n - \frac{\gamma_{\mathrm{d}}-1}{\gamma_{\mathrm{d}}}\sigma_{s,\mathrm{d}}{}^{\mathrm{T}} \cdot \left(\sigma_{s,\mathrm{d}} \cdot \sigma_{s,\mathrm{d}}{}^{\mathrm{T}}\right)^{-1} \cdot \sigma_{s,\mathrm{d}}\right) \cdot c^{\mathrm{T}} \cdot \nabla_A g_{\gamma_{\mathrm{d}},d,T} \\
&= \frac{\gamma_{\mathrm{d}}-1}{\gamma_{\mathrm{d}}}\left(\frac{1}{2}(a_{s,\mathrm{d}} - r_{\mathrm{d}}) \cdot \left(\sigma_{s,\mathrm{d}} \cdot \sigma_{s,\mathrm{d}}{}^{\mathrm{T}}\right)^{-1} \cdot (a_{s,\mathrm{d}} - r_{\mathrm{d}}) + r_{\mathrm{d}}\gamma\right)
\end{aligned} \tag{7.2.27}$$

for $t < T$, together with the terminal condition $g_{\gamma_{\mathrm{d}},d,T}(T,A) = 0$, and similarly for φ_{f} (expressed in terms of $g_{\gamma_{\mathrm{f}},f,T}$). The SDE (7.2.1) remains the same, while the SDE (7.2.13) is not needed, since under the assumption of a CRRA utility, the process $X_{\mathrm{f}}(t)$ is eliminated from the SDE (7.2.24). Indeed, using (3.6.21) and (3.6.22), we have

$$\begin{aligned}
&\frac{\Pi_{\mathrm{d}}^{\star}(t, Y_{\mathrm{d,f}}(t)X_{\mathrm{f}}(t), A(t))}{Y_{\mathrm{d,f}}(t)X_{\mathrm{f}}(t)} \\
&\quad = \frac{1}{\gamma_{\mathrm{d}}}\left(a_{s,\mathrm{d}} - r_{\mathrm{d}} + \left(\lim_{T\to\infty} \nabla_A g_{\gamma_{\mathrm{d}},d,T}\right) \cdot c \cdot \sigma_{s,\mathrm{d}}{}^{\mathrm{T}}\right) \cdot \left(\sigma_{s,\mathrm{d}} \cdot \sigma_{s,\mathrm{d}}{}^{\mathrm{T}}\right)^{-1}
\end{aligned} \tag{7.2.28}$$

and

$$\frac{\Pi_{\mathrm{f}}^{\star}(t, X_{\mathrm{f}}(t), A(t))}{X_{\mathrm{f}}(t)} = \frac{1}{\gamma_{\mathrm{f}}}\left(a_{s,\mathrm{f}} - r_{\mathrm{f}} + \left(\lim_{T\to\infty} \nabla_A g_{\gamma_{\mathrm{f}},f,T}\right) \cdot c \cdot \sigma_{s,\mathrm{f}}{}^{\mathrm{T}}\right) \cdot \left(\sigma_{s,\mathrm{f}} \cdot \sigma_{s,\mathrm{f}}{}^{\mathrm{T}}\right)^{-1}, \tag{7.2.29}$$

and consequently,

$$\begin{aligned}
m_1 &= r_{\mathrm{d}}(t,A(t)) - r_{\mathrm{f}}(t,A(t)) \\
&\quad + \frac{\Pi_{\mathrm{d}}^{\star}(t, Y_{\mathrm{d,f}}(t)X_{\mathrm{f}}(t), A(t))}{Y_{\mathrm{d,f}}(t)X_{\mathrm{f}}(t)} \cdot (a_{s,\mathrm{d}}(t,A(t)) - r_{\mathrm{d}}(t,A(t))) \\
&\quad - \frac{\Pi_{\mathrm{f}}^{\star}(t, X_{\mathrm{f}}(t), A(t))}{X_{\mathrm{f}}(t)} \cdot (a_{s,\mathrm{f}}(t,A(t)) - r_{\mathrm{f}}(t,A(t))) - \frac{\Pi_{\mathrm{f}}^{\star}(t, X_{\mathrm{f}}(t), A(t))}{X_{\mathrm{f}}(t)}
\end{aligned}$$

$$
\begin{aligned}
&\cdot\sigma_{s,\mathrm{f}}(t,A(t))\left(\frac{\Pi_{\mathrm{d}}^{\star}(t,Y_{\mathrm{d,f}}(t)X_{\mathrm{f}}(t),A(t))}{Y_{\mathrm{d,f}}(t)X_{\mathrm{f}}(t)}\cdot\sigma_{s,\mathrm{d}}(t,A(t))\right.\\
&\qquad\left.-\frac{\Pi_{\mathrm{f}}^{\star}(t,X_{\mathrm{f}}(t),A(t))}{X_{\mathrm{f}}(t)}\cdot\sigma_{s,\mathrm{f}}(t,A(t))\right)\\
&=r_{\mathrm{d}}-r_{\mathrm{f}}+\frac{1}{\gamma_{\mathrm{d}}}\left(a_{s,\mathrm{d}}-r_{\mathrm{d}}+\left(\lim_{T\to\infty}\nabla_A g_{\gamma_{\mathrm{d}},d,T}\right)\cdot c\cdot\sigma_{s,\mathrm{d}}{}^{\mathrm{T}}\right)\cdot\left(\sigma_{s,\mathrm{d}}\cdot\sigma_{s,\mathrm{d}}{}^{\mathrm{T}}\right)^{-1}\cdot(a_{s,\mathrm{d}}-r_{\mathrm{d}})\\
&\quad-\frac{1}{\gamma_{\mathrm{f}}}\left(a_{s,\mathrm{f}}-r_{\mathrm{f}}+\left(\lim_{T\to\infty}\nabla_A g_{\gamma_{\mathrm{f}},f,T}\right)\cdot c\cdot\sigma_{s,\mathrm{f}}{}^{\mathrm{T}}\right)\cdot\left(\sigma_{s,\mathrm{f}}\cdot\sigma_{s,\mathrm{f}}{}^{\mathrm{T}}\right)^{-1}\cdot(a_{s,\mathrm{f}}-r_{\mathrm{f}})\\
&\quad-\frac{1}{\gamma_{\mathrm{f}}}\left(a_{s,\mathrm{f}}-r_{\mathrm{f}}+\left(\lim_{T\to\infty}\nabla_A g_{\gamma_{\mathrm{f}},f,T}\right)\cdot c\cdot\sigma_{s,\mathrm{f}}{}^{\mathrm{T}}\right)\cdot\left(\sigma_{s,\mathrm{f}}\cdot\sigma_{s,\mathrm{f}}{}^{\mathrm{T}}\right)^{-1}\cdot\sigma_{s,\mathrm{f}}\\
&\quad\cdot\left(\frac{1}{\gamma_{\mathrm{d}}}\left(a_{s,\mathrm{d}}-r_{\mathrm{d}}+\left(\lim_{T\to\infty}\nabla_A g_{\gamma_{\mathrm{d}},d,T}\right)\cdot c\cdot\sigma_{s,\mathrm{d}}{}^{\mathrm{T}}\right)\cdot\left(\sigma_{s,\mathrm{d}}\cdot\sigma_{s,\mathrm{d}}{}^{\mathrm{T}}\right)^{-1}\right.\\
&\qquad\left.\cdot\sigma_{s,\mathrm{d}}-\frac{1}{\gamma_{\mathrm{f}}}\left(a_{s,\mathrm{f}}-r_{\mathrm{f}}\left(\lim_{T\to\infty}\nabla_A g_{\gamma_{\mathrm{f}},f,T}\right)\cdot c\cdot\sigma_{s,\mathrm{f}}{}^{\mathrm{T}}\right)\cdot\left(\sigma_{s,\mathrm{f}}\cdot\sigma_{s,\mathrm{f}}{}^{\mathrm{T}}\right)^{-1}\cdot\sigma_{s,\mathrm{f}}\right)\\
&=r_{\mathrm{d}}-r_{\mathrm{f}}+\frac{1}{\gamma_{\mathrm{d}}}\left(a_{s,\mathrm{d}}-r_{\mathrm{d}}+\left(\lim_{T\to\infty}\nabla_A g_{\gamma_{\mathrm{d}},d,T}\right)\cdot c\cdot\sigma_{s,\mathrm{d}}{}^{\mathrm{T}}\right)\cdot\left(\sigma_{s,\mathrm{d}}\cdot\sigma_{s,\mathrm{d}}{}^{\mathrm{T}}\right)^{-1}\\
&\quad\cdot(a_{s,\mathrm{d}}-r_{\mathrm{d}})-\frac{1}{\gamma_{\mathrm{f}}}\left(a_{s,\mathrm{f}}-r_{\mathrm{f}}+\left(\lim_{T\to\infty}\nabla_A g_{\gamma_{\mathrm{f}},f,T}\right)\cdot c\cdot\sigma_{s,\mathrm{f}}{}^{\mathrm{T}}\right)\cdot\left(\sigma_{s,\mathrm{f}}\cdot\sigma_{s,\mathrm{f}}{}^{\mathrm{T}}\right)^{-1}\\
&\quad\cdot(a_{s,\mathrm{f}}-r_{\mathrm{f}})-\frac{1}{\gamma_{\mathrm{d}}\gamma_{\mathrm{f}}}\left(a_{s,\mathrm{f}}-r_{\mathrm{f}}+\left(\lim_{T\to\infty}\nabla_A g_{\gamma_{\mathrm{f}},f,T}\right)\cdot c\cdot\sigma_{s,\mathrm{f}}{}^{\mathrm{T}}\right)\cdot\left(\sigma_{s,\mathrm{f}}\cdot\sigma_{s,\mathrm{f}}{}^{\mathrm{T}}\right)^{-1}\\
&\quad\cdot\sigma_{s,\mathrm{f}}\cdot\sigma_{s,\mathrm{d}}{}^{\mathrm{T}}\left(\sigma_{s,\mathrm{d}}\cdot\sigma_{s,\mathrm{d}}{}^{\mathrm{T}}\right)^{-1}\cdot\left(a_{s,\mathrm{d}}-r_{\mathrm{d}}+\left(\lim_{T\to\infty}\nabla_A g_{\gamma_{\mathrm{d}},d,T}\right)\cdot c\cdot\sigma_{s,\mathrm{d}}{}^{\mathrm{T}}\right)\\
&\quad+\frac{1}{\gamma_{\mathrm{f}}{}^2}\left(a_{s,\mathrm{f}}-r_{\mathrm{f}}+\left(\lim_{T\to\infty}\nabla_A g_{\gamma_{\mathrm{f}},f,T}\right)\cdot c\cdot\sigma_{s,\mathrm{f}}{}^{\mathrm{T}}\right)\cdot\left(\sigma_{s,\mathrm{f}}\cdot\sigma_{s,\mathrm{f}}{}^{\mathrm{T}}\right)^{-1}\\
&\quad\cdot\left(a_{s,\mathrm{f}}-r_{\mathrm{f}}+\left(\lim_{T\to\infty}\nabla_A g_{\gamma_{\mathrm{f}},f,T}\right)\cdot c\cdot\sigma_{s,\mathrm{f}}^{\mathrm{T}}\right)
\end{aligned}
\tag{7.2.30}
$$

and

$$
\begin{aligned}
m_2&=\frac{\Pi_{\mathrm{d}}^{\star}(t,Y_{\mathrm{d,f}}(t)X_{\mathrm{f}}(t),A(t))}{Y_{\mathrm{d,f}}(t)X_{\mathrm{f}}(t)}\cdot\sigma_{s,\mathrm{d}}(t,A(t))-\frac{\Pi_{\mathrm{f}}^{\star}(t,X_{\mathrm{f}}(t),A(t))}{X_{\mathrm{f}}(t)}\cdot\sigma_{s,\mathrm{f}}(t,A(t))\\
&=\frac{1}{\gamma_{\mathrm{d}}}\left(a_{s,\mathrm{d}}-r_{\mathrm{d}}+\left(\lim_{T\to\infty}\nabla_A g_{\gamma_{\mathrm{d}},d,T}\right)\cdot c\cdot\sigma_{s,\mathrm{d}}{}^{\mathrm{T}}\right)\cdot\left(\sigma_{s,\mathrm{d}}\cdot\sigma_{s,\mathrm{d}}{}^{\mathrm{T}}\right)^{-1}\cdot\sigma_{s,\mathrm{d}}\\
&\quad-\frac{1}{\gamma_{\mathrm{f}}}\left(a_{s,\mathrm{f}}-r_{\mathrm{f}}+\left(\lim_{T\to\infty}\nabla_A g_{\gamma_{\mathrm{f}},f,T}\right)\cdot c\cdot\sigma_{s,\mathrm{f}}{}^{\mathrm{T}}\right)\cdot\left(\sigma_{s,\mathrm{f}}\cdot\sigma_{s,\mathrm{f}}{}^{\mathrm{T}}\right)^{-1}\cdot\sigma_{s,\mathrm{f}},
\end{aligned}
\tag{7.2.31}
$$

which gives us the following theorem.

Theorem 7.2.2 (CRRA FXR SDE). *In the above framework, in the case of the CRRA utility of wealth, f or given relative risk-aversion parameters* $\gamma_d, \gamma_f \in (0,\infty]$, the FXR process $Y_{d,f}(t)$ *is a solution of the SDE*

$$\frac{dY_{d,f}(t)}{Y_{d,f}(t)} = \Bigg(r_d - r_f + \frac{1}{\gamma_d}\Big(a_{s,d} - r_d + \big(\lim_{T\to\infty} \nabla_A g_{\gamma_d,d,T}\big)\cdot c\cdot \sigma_{s,d}{}^T\Big)\cdot\big(\sigma_{s,d}\,\sigma_{s,d}{}^T\big)^{-1}$$
$$\cdot(a_{s,d} - r_d) - \frac{1}{\gamma_f}\Big(a_{s,f} - r_f + \big(\lim_{T\to\infty} \nabla_A g_{\gamma_f,f,T}\big)\cdot c\cdot \sigma_{s,f}{}^T\Big)$$
$$\big(\sigma_{s,f}\cdot\sigma_{s,f}{}^T\big)^{-1}\cdot(a_{s,f} - r_f) - \frac{1}{\gamma_d\gamma_f}\Big(a_{s,f} - r_f + \big(\lim_{T\to\infty} \nabla_A g_{\gamma_f,f,T}\big)$$
$$\cdot c\cdot\sigma_{s,f}{}^T\Big)\cdot\big(\sigma_{s,f}\cdot\sigma_{s,f}{}^T\big)^{-1}\cdot\sigma_{s,f}\cdot\sigma_{s,d}{}^T\cdot\big(\sigma_{s,d}\cdot\sigma_{s,d}{}^T\big)^{-1}$$
$$\cdot\Big(a_{s,d} - r_d + \big(\lim_{T\to\infty} \nabla_A g_{\gamma_d,d,T}\big)\cdot c\cdot\sigma_{s,d}{}^T\Big)$$
$$+\frac{1}{\gamma_f{}^2}\Big(a_{s,f} - r_f + \big(\lim_{T\to\infty} \nabla_A g_{\gamma_f,f,T}\big)\cdot c\cdot\sigma_{s,f}{}^T\Big)\cdot\big(\sigma_{s,f}\cdot\sigma_{s,f}{}^T\big)^{-1}$$
$$\cdot\Big(a_{s,f} - r_f + \big(\lim_{T\to\infty} \nabla_A g_{\gamma_f,f,T}\big)\cdot c\cdot\sigma_{s,f}{}^T\Big)\Bigg)dt + \Bigg(\frac{1}{\gamma_d}\Big(a_{s,d} - r_d$$
$$+\big(\lim_{T\to\infty} \nabla_A g_{\gamma_d,d,T}\big)\cdot c\cdot\sigma_{s,d}{}^T\Big)\cdot\big(\sigma_{s,d}\cdot\sigma_{s,d}{}^T\big)^{-1}\cdot\sigma_{s,d} - \frac{1}{\gamma_f}\Big(a_{s,f} - r_f$$
$$+\big(\lim_{T\to\infty} \nabla_A g_{\gamma_f,f,T}\big)\cdot c\cdot\sigma_{s,f}{}^T\Big)\cdot\big(\sigma_{s,f}\cdot\sigma_{s,f}{}^T\big)^{-1}\cdot\sigma_{s,f}\Bigg)\cdot dB(t) \quad (7.2.32)$$

coupled with SDE (7.2.1), and where $g_{\gamma_d,d}$ *and* $g_{\gamma_f,f}$ *are solutions of PDEs (7.2.27) and (7.2.28), respectively.*

Remark 7.2.5. If instead, CARA utility is used, the process $X_f(t)$ is not eliminated from the FXR SDE.

Remark 7.2.6. We remark that if no foreign investments are available or considered, i.e., if $S_f = \emptyset$, but $S_d \neq \emptyset$, then

$$\frac{dY_{d,f}(t)}{Y_{d,f}(t)} = \Bigg(r_d - r_f + \frac{1}{\gamma_d}\Big(a_{s,d} - r_d + \big(\lim_{T\to\infty} \nabla_A g_{\gamma_d,d,T}\big)\cdot c\cdot\sigma_{s,d}{}^T\Big)\cdot\big(\sigma_{s,d}\cdot\sigma_{s,d}{}^T\big)^{-1}$$
$$\cdot(a_{s,d} - r_d)\Bigg)dt + \Bigg(\frac{1}{\gamma_d}\Big(a_{s,d} - r_d + \big(\lim_{T\to\infty} \nabla_A g_{\gamma_d,d,T}\big)\cdot c\cdot\sigma_{s,d}{}^T\Big)$$
$$\cdot\big(\sigma_{s,d}\cdot\sigma_{s,d}{}^T\big)^{-1}\cdot\sigma_{s,d} - \frac{1}{\gamma_f}\Big(a_{s,f} - r_f + \big(\lim_{T\to\infty} \nabla_A g_{\gamma_f,f,T}\big)$$
$$\cdot c\cdot\sigma_{s,f}{}^T\Big)\cdot\big(\sigma_{s,f}\cdot\sigma_{s,f}{}^T\big)^{-1}\cdot\sigma_{s,f}\Bigg)\cdot dB(t). \quad (7.2.33)$$

If no domestic risky investments are considered, i.e., if $S_{\mathrm{d}}=\emptyset$ but $S_{\mathrm{f}}\neq\emptyset$, then

$$\frac{\mathrm{d}Y_{\mathrm{d,f}}(t)}{Y_{\mathrm{d,f}}(t)}=\Bigg(r_{\mathrm{d}}-r_{\mathrm{f}}-\frac{1}{\gamma_{\mathrm{f}}}\left(a_{s,\mathrm{f}}-r_{\mathrm{f}}+\left(\lim_{T\to\infty}\nabla_{A}g_{\gamma_{\mathrm{f}},f,T}\right)\cdot c\cdot\sigma_{s,\mathrm{f}}{}^{\mathrm{T}}\right)\cdot\left(\sigma_{s,\mathrm{f}}\cdot\sigma_{s,\mathrm{f}}{}^{\mathrm{T}}\right)^{-1}$$
$$\cdot(a_{s,\mathrm{f}}-r_{\mathrm{f}})+\frac{1}{\gamma_{\mathrm{f}}^{2}}\left(a_{s,\mathrm{f}}-r_{\mathrm{f}}+\left(\lim_{T\to\infty}\nabla_{A}g_{\gamma_{\mathrm{f}},f,T}\right)\cdot c\cdot\sigma_{s,\mathrm{f}}{}^{\mathrm{T}}\right)\cdot(\sigma_{s,\mathrm{f}}$$
$$.\sigma_{s,\mathrm{f}}{}^{\mathrm{T}})^{-1}\cdot\left(a_{s,\mathrm{f}}-r_{\mathrm{f}}+\left(\lim_{T\to\infty}\nabla_{A}g_{\gamma_{\mathrm{f}},f,T}\right)\cdot c\cdot\sigma_{s,\mathrm{f}}{}^{\mathrm{T}}\right)\Bigg)\mathrm{d}t \tag{7.2.34}$$
$$-\frac{1}{\gamma_{\mathrm{f}}}\left(a_{s,\mathrm{f}}-r_{\mathrm{f}}+\left(\lim_{T\to\infty}\nabla_{A}g_{\gamma_{\mathrm{f}},f,T}\right)\cdot c\cdot\sigma_{s,\mathrm{f}}{}^{\mathrm{T}}\right)\cdot\left(\sigma_{s,\mathrm{f}}\cdot\sigma_{s,\mathrm{f}}{}^{\mathrm{T}}\right)^{-1}\cdot\sigma_{s,\mathrm{f}}\cdot\mathrm{d}B(t),$$

and finally, if $S_{\mathrm{d}}=S_{\mathrm{f}}=\emptyset$, then

$$\frac{\mathrm{d}Y_{\mathrm{d,f}}(t)}{Y_{\mathrm{d,f}}(t)}=(r_{\mathrm{d}}-r_{\mathrm{f}})\,\mathrm{d}t. \tag{7.2.35}$$

Remark 7.2.7. Note in (7.2.32) that the relative risk-aversion parameter γ (domestic and/or foreign), i.e., the investor's risk-aversion, influences FX markets considerably, as is well known among practitioners. The presented theory captures this phenomenon quantitatively, a fact that we shall exploit in great detail later in this chapter.

7.2.3 *Multilateral FXR: The Cross-Currency Rule Fulfilled*

Any FXR model has to satisfy the so called cross-currency rule, which we shall discuss next.

The cross-currency rule being fulfilled shows that while solving bilateral FXR problems, we are simultaneously solving the multilateral FXR problem – the solutions obtained bilaterally are consistent with the multilateral cross-currency rule.

Proposition 7.2.1 (Cross-currency rule fulfilled). *Let $Y_{d,f_1}(t)$, $Y_{d,f_2}(t)$, and $Y_{f_1,f_2}(t)$ be the bilateral FXR processes, i.e., processes satisfying SDEs like (7.2.24). Then, in the above framework (assuming, for example, $X_{\mathrm{d}}=Y_{d,f_1}X_{f_1}=Y_{d,f_2}X_{f_2}$),*

$$\frac{\mathrm{d}Y_{f_1,f_2}(t)}{Y_{f_1,f_2}(t)}=\frac{d\left(Y_{d,f_2}(t)/Y_{d,f_1}(t)\right)}{Y_{d,f_2}(t)/Y_{d,f_1}(t)}. \tag{7.2.36}$$

Proof. Recall the Itô quotient rule

$$d\left(\frac{Y_{d,f_2}(t)}{Y_{d,f_1}(t)}\right)=\frac{\mathrm{d}Y_{d,f_2}(t)Y_{d,f_1}(t)-Y_{d,f_2}(t)\mathrm{d}Y_{d,f_1}(t)}{Y_{d,f_1}(t)^2}+\left((\mathrm{d}Y_{d,f_1}(t))^2Y_{d,f_2}(t)\right.$$
$$\left.-\mathrm{d}Y_{d,f_2}(t)\mathrm{d}Y_{d,f_1}(t)Y_{d,f_1}(t)\right)/Y_{d,f_1}(t)^3, \tag{7.2.37}$$

which implies

$$
\begin{aligned}
&\frac{d\left(Y_{d,f_2}(t)/Y_{d,f_1}(t)\right)}{Y_{d,f_2}(t)/Y_{d,f_1}(t)}\\
&=\frac{Y_{d,f_1}(t)}{Y_{d,f_2}(t)}\left(\mathrm{d}Y_{d,f_2}(t)Y_{d,f_1}(t)-Y_{d,f_2}(t)\mathrm{d}Y_{d,f_1}(t)\right)/Y_{d,f_1}(t)^2\\
&\quad+\frac{Y_{d,f_1}(t)}{Y_{d,f_2}(t)}\frac{1}{Y_{d,f_1}(t)^3}\left(\left(\mathrm{d}Y_{d,f_1}(t)\right)^2Y_{d,f_2}(t)-\mathrm{d}Y_{d,f_2}(t)\mathrm{d}Y_{d,f_1}(t)Y_{d,f_1}(t)\right)\\
&=\frac{\mathrm{d}Y_{d,f_2}(t)}{Y_{d,f_2}(t)}-\frac{\mathrm{d}Y_{d,f_1}(t)}{Y_{d,f_1}(t)}+\left(\frac{\mathrm{d}Y_{d,f_1}(t)}{Y_{d,f_1}(t)}\right)^2-\frac{\mathrm{d}Y_{d,f_1}(t)}{Y_{d,f_1}(t)}\frac{\mathrm{d}Y_{d,f_2}(t)}{Y_{d,f_2}(t)}\\
&=\Bigg(r_{\mathrm{d}}-r_{f_2}+\frac{\Pi_{\mathrm{d}}^{\star}\left(t,Y_{d,f_2}(t)X_{f_2}(t),A(t)\right)}{Y_{d,f_2}(t)X_{f_2}(t)}\cdot(a_{s,\mathrm{d}}-r_{\mathrm{d}})-\frac{\Pi_{f_2}^{\star}\left(t,X_{f_2}(t),A(t)\right)}{X_{f_2}(t)}\\
&\quad\cdot\left(a_{s,f_2}-r_{f_2}\right)-\frac{\Pi_{f_2}^{\star}\left(t,X_{f_2}(t),A(t)\right)}{X_{f_2}(t)}\\
&\qquad\cdot\sigma_{s,f_2}\cdot\sigma_{s,\mathrm{d}}{}^{\mathrm{T}}\cdot\Pi_{\mathrm{d}}^{\star}\left(t,Y_{d,f_2}(t)X_{f_2}(t),A(t)\right)/\left(Y_{d,f_2}(t)X_{f_2}(t)\right)\\
&\qquad+\frac{\Pi_{f_2}^{\star}\left(t,X_{f_2}(t),A(t)\right)}{X_{f_2}(t)}\cdot\sigma_{s,f_2}\cdot\sigma_{s,f_2}{}^{\mathrm{T}}\cdot\frac{\Pi_{f_2}^{\star}\left(t,X_{f_2}(t),A(t)\right)}{X_{f_2}(t)}\Bigg)\mathrm{d}t\\
&\quad+\left(\frac{\Pi_{\mathrm{d}}^{\star}\left(t,Y_{d,f_2}(t)X_{f_2}(t),A(t)\right)}{Y_{d,f_2}(t)X_{f_2}(t)}\cdot\sigma_{s,\mathrm{d}}-\frac{\Pi_{f_2}^{\star}\left(t,X_{f_2}(t),A(t)\right)}{X_{f_2}(t)}\cdot\sigma_{s,f_2}\right)\cdot\mathrm{d}B(t)\\
&\quad-\Bigg(r_{\mathrm{d}}-r_{f_1}+\frac{\Pi_{\mathrm{d}}^{\star}\left(t,Y_{d,f_1}(t)X_{f_1}(t),A(t)\right)}{Y_{d,f_1}(t)X_{f_1}(t)}\cdot(a_{s,\mathrm{d}}-r_{\mathrm{d}})-\frac{\Pi_{f_1}^{\star}\left(t,X_{f_1}(t),A(t)\right)}{X_{f_1}(t)}\\
&\qquad\cdot\left(a_{s,f_1}-r_{f_1}\right)-\frac{\Pi_{f_1}^{\star}\left(t,X_{f_1}(t),A(t)\right)}{X_{f_1}(t)}\\
&\qquad\cdot\sigma_{s,f_1}\cdot\sigma_{s,\mathrm{d}}{}^{\mathrm{T}}\cdot\Pi_{\mathrm{d}}^{\star}\left(t,Y_{d,f_1}(t)X_{f_1}(t),A(t)\right)/\left(Y_{d,f_1}(t)X_{f_1}(t)\right)\\
&\qquad+\frac{\Pi_{f_1}^{\star}\left(t,X_{f_1}(t),A(t)\right)}{X_{f_1}(t)}\cdot\sigma_{s,f_1}\cdot\sigma_{s,f_1}{}^{\mathrm{T}}\cdot\frac{\Pi_{f_1}^{\star}\left(t,X_{f_1}(t),A(t)\right)}{X_{f_1}(t)}\Bigg)\mathrm{d}t\\
&\quad+\left(\frac{\Pi_{\mathrm{d}}^{\star}\left(t,Y_{d,f_1}(t)X_{f_1}(t),A(t)\right)}{Y_{d,f_1}(t)X_{f_1}(t)}\cdot\sigma_{s,\mathrm{d}}-\frac{\Pi_{f_1}^{\star}\left(t,X_{f_1}(t),A(t)\right)}{X_{f_1}(t)}\cdot\sigma_{s,f_1}\right)\cdot\mathrm{d}B(t)\\
&\quad+\Big(\Big(\Pi_{\mathrm{d}}^{\star}\left(t,Y_{d,f_1}(t)X_{f_1}(t),A(t)\right)\Big/\left(Y_{d,f_1}(t)X_{f_1}(t)\right)\cdot\sigma_{s,\mathrm{d}}
\end{aligned}
$$

$$-\frac{\Pi_{f_1}^{\star}\left(t,X_{f_1}(t),A(t)\right)}{X_{f_1}(t)}\cdot\sigma_{s,f_1}\Bigg)\cdot\mathrm{d}B(t)\Bigg)^2-\Bigg(\Bigg(\Pi_{\mathrm{d}}^{\star}\left(t,Y_{d,f_1}(t)X_{f_1}(t),A(t)\right)/$$

$$-\left(Y_{d,f_1}(t)X_{f_1}(t)\right)\cdot\sigma_{s,\mathrm{d}}-\frac{\Pi_{f_1}^{\star}\left(t,X_{f_1}(t),A(t)\right)}{X_{f_1}(t)}\cdot\sigma_{s,f_1}\Bigg)\cdot\mathrm{d}B(t)\Bigg)$$

$$\Bigg(\Bigg(\Pi_{\mathrm{d}}^{\star}\left(t,Y_{d,f_2}(t)X_{f_2}(t),A(t)\right)/\left(Y_{d,f_2}(t)X_{f_2}(t)\right)\cdot\sigma_{s,\mathrm{d}}$$

$$-\frac{\Pi_{f_2}^{\star}\left(t,X_{f_2}(t),A(t)\right)}{X_{f_2}(t)}\cdot\sigma_{s,f_2}\Bigg)\cdot\mathrm{d}B(t)\Bigg)$$

$$=\Bigg(r_{f_1}-r_{f_2}+\frac{\Pi_{f_1}^{\star}\left(t,Y_{f_1,f_2}(t)X_{f_2}(t),A(t)\right)}{Y_{f_1,f_2}(t)X_{f_2}(t)}\cdot\left(a_{s,f_1}-r_{f_1}\right)$$

$$-\frac{\Pi_{f_2}^{\star}\left(t,X_{f_2}(t),A(t)\right)}{X_{f_2}(t)}\cdot\left(a_{s,f_2}-r_{f_2}\right)-\frac{\Pi_{f_2}^{\star}\left(t,X_{f_2}(t),A(t)\right)}{X_{f_2}(t)}$$

$$\cdot\sigma_{s,f_2}\cdot{\sigma_{s,f_1}}^{\mathrm{T}}\cdot\Pi_{f_1}^{\star}\left(t,Y_{f_1,f_2}(t)X_{f_2}(t),A(t)\right)/\left(Y_{f_1,f_2}(t)X_{f_2}(t)\right)$$

$$+\frac{\Pi_{f_2}^{\star}\left(t,X_{f_2}(t),A(t)\right)}{X_{f_2}(t)}\cdot\sigma_{s,f_2}\cdot{\sigma_{s,f_2}}^{\mathrm{T}}\cdot\frac{\Pi_{f_2}^{\star}\left(t,X_{f_2}(t),A(t)\right)}{X_{f_2}(t)}\Bigg)\mathrm{d}t$$

$$+\Bigg(\frac{\Pi_{f_1}^{\star}\left(t,Y_{f_1,f_2}(t)X_{f_2}(t),A(t)\right)}{Y_{f_1,f_2}(t)X_{f_2}(t)}.\sigma_{s,f_1}-\frac{\Pi_{f_2}^{\star}\left(t,X_{f_2}(t),A(t)\right)}{X_{f_2}(t)}\cdot\sigma_{s,f_2}\Bigg)\cdot\mathrm{d}B(t),\tag{7.2.38}$$

since

$$\Bigg(-\frac{\Pi_{f_2}^{\star}\left(t,X_{f_2}(t),A(t)\right)}{X_{f_2}(t)}\cdot\sigma_{s,f_2}\cdot{\sigma_{s,\mathrm{d}}}^{\mathrm{T}}\cdot\Pi_{\mathrm{d}}^{\star}\left(t,Y_{d,f_2}(t)X_{f_2}(t),A(t)\right)/\left(Y_{d,f_2}(t)X_{f_2}(t)\right)$$

$$+\frac{\Pi_{f_2}^{\star}\left(t,X_{f_2}(t),A(t)\right)}{X_{f_2}(t)}\cdot\sigma_{s,f_2}\cdot{\sigma_{s,f_2}}^{\mathrm{T}}\cdot\frac{\Pi_{f_2}^{\star}\left(t,X_{f_2}(t),A(t)\right)}{X_{f_2}(t)}\Bigg)\mathrm{d}t$$

$$-\Bigg(-\frac{\Pi_{f_1}^{\star}\left(t,X_{f_1}(t),A(t)\right)}{X_{f_1}(t)}\cdot\sigma_{s,f_1}\cdot{\sigma_{s,\mathrm{d}}}^{\mathrm{T}}\cdot\Pi_{\mathrm{d}}^{\star}\left(t,Y_{d,f_1}(t)X_{f_1}(t),A(t)\right)/$$

$$\left(Y_{d,f_1}(t)X_{f_1}(t)\right)+\frac{\Pi_{f_1}^{\star}\left(t,X_{f_1}(t),A(t)\right)}{X_{f_1}(t)}.\sigma_{s,f_1}\cdot{\sigma_{s,f_1}}^{\mathrm{T}}\cdot\frac{\Pi_{f_1}^{\star}\left(t,X_{f_1}(t),A(t)\right)}{X_{f_1}(t)}\Bigg)\mathrm{d}t$$

$$
\begin{aligned}
&+\left(\left(\Pi_{\mathrm{d}}^{\star}\left(t,Y_{d,f_1}(t)X_{f_1}(t),A(t)\right)/\left(Y_{d,f_1}(t)X_{f_1}(t)\right)\cdot\sigma_{s,\mathrm{d}}\right.\right.\\
&\left.\left.\qquad-\frac{\Pi_{f_1}^{\star}\left(t,X_{f_1}(t),A(t)\right)}{X_{f_1}(t)}\cdot\sigma_{s,f_1}\right)\cdot\mathrm{d}B(t)\right)^2\\
&-\left(\left(\frac{\Pi_{\mathrm{d}}^{\star}\left(t,Y_{d,f_1}(t)X_{f_1}(t),A(t)\right)}{Y_{d,f_1}(t)X_{f_1}(t)}.\sigma_{s,\mathrm{d}}-\frac{\Pi_{f_1}^{\star}\left(t,X_{f_1}(t),A(t)\right)}{X_{f_1}(t)}\cdot\sigma_{s,f_1}\right)\cdot\mathrm{d}B(t)\right)\\
&\left(\left(\frac{\Pi_{\mathrm{d}}^{\star}\left(t,Y_{d,f_2}(t)X_{f_2}(t),A(t)\right)}{Y_{d,f_2}(t)X_{f_2}(t)}\sigma_{s,\mathrm{d}}-\frac{\Pi_{f_2}^{\star}\left(t,X_{f_2}(t),A(t)\right)}{X_{f_2}(t)}\cdot\sigma_{s,f_2}\right)\cdot\mathrm{d}B(t)\right)\\
&=\frac{\Pi_{f_2}^{\star}\left(t,X_{f_2}(t),A(t)\right)}{X_{f_2}(t)}\cdot\sigma_{s,f_2}\cdot{\sigma_{s,f_2}}^{\mathrm{T}}\cdot\frac{\Pi_{f_2}^{\star}\left(t,X_{f_2}(t),A(t)\right)}{X_{f_2}(t)}\mathrm{d}t\\
&\quad-\frac{\Pi_{f_1}^{\star}\left(t,X_{f_1}(t),A(t)\right)}{X_{f_1}(t)}\sigma_{s,f_1}\cdot{\sigma_{s,f_2}}^{\mathrm{T}}\cdot\frac{\Pi_{f_2}^{\star}\left(t,X_{f_2}(t),A(t)\right)}{X_{f_2}(t)}\mathrm{d}t,
\end{aligned}
\tag{7.2.39}
$$

which completes the proof of the proposition. □

As a special case of the cross-currency rule, we have that the reciprocal of an FXR is an FXR, as stated in the following corollary.

Corollary 7.2.1. *("$Y_{\mathrm{f,d}}(t)=1/Y_{\mathrm{d,f}}(t)$.") In the above framework,*

$$
\frac{\mathrm{d}Y_{\mathrm{f,d}}(t)}{Y_{\mathrm{f,d}}(t)}=\frac{d\left(1/Y_{\mathrm{d,f}}(t)\right)}{1/Y_{\mathrm{d,f}}(t)}.
\tag{7.2.40}
$$

Proof. Setting $f_1=f$, $f_2=d$ in (1.1.24), we get

$$
\frac{\mathrm{d}Y_{\mathrm{f,d}}(t)}{Y_{\mathrm{f,d}}(t)}=\frac{d\left(Y_{d,d}(t)/Y_{\mathrm{d,f}}(t)\right)}{Y_{d,d}(t)/Y_{\mathrm{d,f}}(t)}=\frac{d\left(1/Y_{\mathrm{d,f}}(t)\right)}{1/Y_{\mathrm{d,f}}(t)}.
\tag{7.2.41}
$$

□

Remark 7.2.9. ("Siegel's paradox.") Observe that

$$
\mathrm{d}Y_{\mathrm{f,d}}(t)=d\frac{1}{Y_{\mathrm{d,f}}(t)}=-\frac{1}{Y_{\mathrm{d,f}}(t)^2}\mathrm{d}Y_{\mathrm{d,f}}(t)+\frac{1}{2}\frac{2}{Y_{\mathrm{d,f}}(t)^3}\left(\mathrm{d}Y_{\mathrm{d,f}}(t)\right)^2,
\tag{7.2.42}
$$

and therefore, since $1/Y_{\mathrm{f,d}}(t)=Y_{\mathrm{d,f}}(t)$,

$$
\frac{\mathrm{d}Y_{\mathrm{f,d}}(t)}{Y_{\mathrm{f,d}}(t)}=-\frac{\mathrm{d}Y_{\mathrm{d,f}}(t)}{Y_{\mathrm{d,f}}(t)}+\left(\frac{\mathrm{d}Y_{\mathrm{d,f}}(t)}{Y_{\mathrm{d,f}}(t)}\right)^2
\tag{7.2.43}
$$

and

$$\frac{dY_{d,f}(t)}{Y_{d,f}(t)} + \frac{dY_{f,d}(t)}{Y_{f,d}(t)} = \left(\frac{dY_{f,d}(t)}{Y_{f,d}(t)}\right)^2 = \left(\frac{dY_{d,f}(t)}{Y_{d,f}(t)}\right)^2$$
$$= \left(\left(\frac{\Pi_d^\star(t, Y_{d,f}(t)X_f(t), A(t))}{Y_{d,f}(t)X_f(t)} \cdot \sigma_{s,d} - \frac{\Pi_f^\star(t, X_f(t), A(t))}{X_f(t)} \cdot \sigma_{s,f}\right) \cdot dB(t)\right)^2$$
$$= \left|\frac{\Pi_d^\star(t, Y_{d,f}(t)X_f(t), A(t))}{Y_{d,f}(t)X_f(t)} \cdot \sigma_{s,d} - \frac{\Pi_f^\star(t, X_f(t), A(t))}{X_f(t)} \cdot \sigma_{s,f}\right|^2 dt \neq 0. \tag{7.2.44}$$

So "if two investors from different countries have the same expectation of the probable distribution of future exchange rates, the expected returns of the two currencies are not offsetting." This is so-called "Siegel's paradox".

Remark 7.2.10. It is also useful to state the FXR SDE for $Y_{f,d}(t) = 1/Y_{d,f}(t)$. From (7.2.44) and (7.2.24), we get

$$\frac{dY_{f,d}(t)}{Y_{f,d}(t)} = -\frac{dY_{d,f}(t)}{Y_{d,f}(t)} + \left(\frac{dY_{d,f}(t)}{Y_{d,f}(t)}\right)^2 = -\Big(\Big(r_d - r_f + \Pi_d^\star(t, Y_{d,f}(t)X_f(t), A(t))/$$
$$(Y_{d,f}(t)X_f(t)) \cdot (a_{s,d} - r_d) - \frac{\Pi_f^\star(t, X_f(t), A(t))}{X_f(t)} \cdot (a_{s,f} - r_f) - \frac{\Pi_f^\star(t, X_f(t), A(t))}{X_f(t)}$$
$$\cdot \sigma_{s,f} \cdot \sigma_{s,d}{}^T \cdot \Pi_d^\star(t, Y_{d,f}(t)X_f(t), A(t)) \big/ (Y_{d,f}(t)X_f(t)) + \frac{\Pi_f^\star(t, X_f(t), A(t))}{X_f(t)}$$
$$\cdot \sigma_{s,f} \cdot \sigma_{s,f}{}^T \cdot \frac{\Pi_f^\star(t, X_f(t), A(t))}{X_f(t)}\Big) dt + \Big(\Pi_d^\star(t, Y_{d,f}(t)X_f(t), A(t))/$$
$$(Y_{d,f}(t)X_f(t)) \cdot \sigma_{s,d} - \frac{\Pi_f^\star(t, X_f(t), A(t))}{X_f(t)} \cdot \sigma_{s,f}\Big) \cdot dB(t)\Big)$$
$$+ \left|\frac{\Pi_d^\star(t, Y_{d,f}(t)X_f(t), A(t))}{Y_{d,f}(t)X_f(t)} \cdot \sigma_{s,d} - \frac{\Pi_f^\star(t, X_f(t), A(t))}{X_f(t)} \cdot \sigma_{s,f}\right|^2 dt, \tag{7.2.45}$$

so that the reciprocal FXR SDE reads

$$\frac{dY_{f,d}(t)}{Y_{f,d}(t)} = \Big(r_f - r_d + \frac{\Pi_f^\star(t, X_f(t), A(t))}{X_f(t)} \cdot (a_{s,f} - r_f) - \frac{\Pi_d^\star(t, Y_{d,f}(t)X_f(t), A(t))}{Y_{d,f}(t)X_f(t)}$$
$$\cdot (a_{s,d} - r_d) - \frac{\Pi_f^\star(t, X_f(t), A(t))}{X_f(t)} \cdot \sigma_{s,f} \cdot \sigma_{s,d}{}^T \cdot \frac{\Pi_d^\star(t, Y_{d,f}(t)X_f(t), A(t))}{Y_{d,f}(t)X_f(t)}$$

$$+\frac{\Pi_{\mathrm{d}}^{\star}(t,X_{\mathrm{d}}(t),A(t))}{X_{\mathrm{d}}(t)}\cdot\sigma_{s,\mathrm{d}}\cdot\sigma_{s,\mathrm{d}}^{\mathrm{T}}\cdot\frac{\Pi_{\mathrm{d}}^{\star}(t,X_{\mathrm{d}}(t),A(t))}{X_{\mathrm{d}}(t)}\Bigg)\mathrm{d}t$$
$$+\left(\frac{\Pi_{\mathrm{f}}^{\star}(t,X_{\mathrm{f}}(t),A(t))}{X_{\mathrm{f}}(t)}.\sigma_{s,\mathrm{f}}-\frac{\Pi_{\mathrm{d}}^{\star}(t,Y_{\mathrm{d,f}}(t)X_{\mathrm{f}}(t),A(t))}{Y_{\mathrm{d,f}}(t)X_{\mathrm{f}}(t)}\cdot\sigma_{s,\mathrm{d}}\right)\cdot\mathrm{d}B(t). \tag{7.2.46}$$

Remark 7.2.11 (FXR for an economically unbalanced world). What if the postulated equivalence between investment opportunities in two economies, i.e., condition (7.2.14) is disturbed? For example, we can try to quantify/model such a situation by extending the condition (7.2.14) to

$$\mathrm{d}X_{\mathrm{d}}(t)-X_{\mathrm{d}}(t)s_{d,\mathrm{f}}\mathrm{d}t=d\left(Y_{\mathrm{d,f}}(t)X_{\mathrm{f}}(t)\right), \tag{7.2.47}$$

where $s_{d,f}$ (if positive, and analogously if negative) quantifies how much ("percentagewise") superior, overall, it is to invest in the domestic economy than in the foreign economy. It is easy to see that SDE (7.2.24) is generalized to

$$\frac{\mathrm{d}Y_{\mathrm{d,f}}(t)}{Y_{\mathrm{d,f}}(t)}=\Bigg(r_{\mathrm{d}}-r_{\mathrm{f}}-s_{d,\mathrm{f}}+\frac{\Pi_{\mathrm{d}}^{\star}(t,Y_{\mathrm{d,f}}(t)X_{\mathrm{f}}(t),A(t))}{Y_{\mathrm{d,f}}(t)X_{\mathrm{f}}(t)}.(a_{s,\mathrm{d}}-r_{\mathrm{d}})$$
$$\cdot-\frac{\Pi_{\mathrm{f}}^{\star}(t,X_{\mathrm{f}}(t),A(t))}{X_{\mathrm{f}}(t)}\cdot(a_{s,\mathrm{f}}-r_{\mathrm{f}})-\frac{\Pi_{\mathrm{f}}^{\star}(t,X_{\mathrm{f}}(t),A(t))}{X_{\mathrm{f}}(t)}\cdot\sigma_{s,\mathrm{f}}\cdot\sigma_{s,\mathrm{d}}^{\mathrm{T}}$$
$$\cdot\frac{\Pi_{\mathrm{d}}^{\star}(t,Y_{\mathrm{d,f}}(t)X_{\mathrm{f}}(t),A(t))}{Y_{\mathrm{d,f}}(t)X_{\mathrm{f}}(t)}+\frac{\Pi_{\mathrm{f}}^{\star}(t,X_{\mathrm{f}}(t),A(t))}{X_{\mathrm{f}}(t)}\cdot\sigma_{s,\mathrm{f}}\cdot\sigma_{s,\mathrm{f}}^{\mathrm{T}}$$
$$\cdot\frac{\Pi_{\mathrm{f}}^{\star}(t,X_{\mathrm{f}}(t),A(t))}{X_{\mathrm{f}}(t)}\Bigg)\mathrm{d}t$$
$$+\left(\frac{\Pi_{\mathrm{d}}^{\star}(t,Y_{\mathrm{d,f}}(t)X_{\mathrm{f}}(t),A(t))}{Y_{\mathrm{d,f}}(t)X_{\mathrm{f}}(t)}\cdot\sigma_{s,\mathrm{d}}-\frac{\Pi_{\mathrm{f}}^{\star}(t,X_{\mathrm{f}}(t),A(t))}{X_{\mathrm{f}}(t)}\cdot\sigma_{s,\mathrm{f}}\right)\cdot\mathrm{d}B(t). \tag{7.2.48}$$

Obviously, $s_{d,f}=-s_{f,d}$, and an analogous condition is needed if more then two economies are considered and the cross-currency rule is required.

7.3 Applications

7.3.1 FXR for Log-Normal Markets

The simplest stochastic example of two economies and the corresponding FXR process is the case in which the risky asset prices in both the domestic and foreign economies are modeled as log-normal processes. So let $S_1(t)$ be, for example, a market index in the domestic economy, and let $S_2(t)$ be a market index in the foreign economy. Let the (predividend) appreciation rates for domestic and foreign markets be $\mathbf{a}_\mathrm{d}$ and $\mathbf{a}_\mathrm{f}$, and let the dividend rates be $\mathbb{D}_\mathrm{d}$ and $\mathbb{D}_\mathrm{f}$. So assume that the two economies considered are represented via two factors $A = \{S_1, S_2\}$ and tradables $S_\mathrm{d} = \{S_1\}$ and $S_\mathrm{f} = \{S_2\}$ obeying

$$\begin{aligned}
\mathrm{d}S_1(t) &= S_1(t)\left(\mathbf{a}_\mathrm{d} - \mathbb{D}_\mathrm{d}\right)\mathrm{d}t + S_1(t)\sigma_\mathrm{d}\mathrm{d}B_1(t), \\
\mathrm{d}S_2(t) &= S_2(t)\left(\mathbf{a}_\mathrm{f} - \mathbb{D}_\mathrm{f}\right)\mathrm{d}t + S_2(t)\sigma_\mathrm{f}\mathrm{d}\mathbb{B}_2(t),
\end{aligned} \tag{7.3.1}$$

where $\mathrm{d}B_1(t)\mathrm{d}\mathbb{B}_2(t) = \rho_{2,1}\mathrm{d}t$.

The market coefficients are then equal to

$$\begin{aligned}
a_{s,\mathrm{d}} &= \{\mathbf{a}_\mathrm{d}\}, a_{s,\mathrm{f}} = \{\mathbf{a}_\mathrm{f}\}, \\
\sigma_{s,\mathrm{d}} &= \begin{pmatrix} \sigma_\mathrm{d} & 0 \end{pmatrix}, \sigma_{s,\mathrm{f}} = \begin{pmatrix} \sigma_\mathrm{f}\rho_{2,1} & \sigma_\mathrm{f}\sqrt{1-\rho_{2,1}^2} \end{pmatrix}, \\
b &= \{S_1\left(\mathbf{a}_\mathrm{d} - \mathbb{D}_\mathrm{d}\right), S_2\left(\mathbf{a}_\mathrm{f} - \mathbb{D}_\mathrm{f}\right)\}, c = \begin{pmatrix} S_1\sigma_\mathrm{d} & 0 \\ S_2\sigma_\mathrm{f}\rho_{2,1} & S_2\sigma_\mathrm{f}\sqrt{1-\rho_{2,1}^2} \end{pmatrix}.
\end{aligned} \tag{7.3.2}$$

In such a case (and for CRRA utility), as discussed before, we have

$$\varphi_\mathrm{d}(t,X,A) = \varphi_\mathrm{d}(t,X) = \frac{X^{1-\gamma}e^{g_{\gamma_\mathrm{d},d,T}(t)} - 1}{1-\gamma} \tag{7.3.3}$$

for $t < T$, and therefore $\nabla_A g_{\gamma_\mathrm{d},d,T} = 0$,

$$\frac{\Pi_\mathrm{d}^\star\left(t, Y_{\mathrm{d,f}}X_\mathrm{f}, A\right)}{Y_{\mathrm{d,f}}X_\mathrm{f}} = \frac{1}{\gamma_\mathrm{d}}\left(a_{s,\mathrm{d}} - r_\mathrm{d}\right)\cdot\left(\sigma_{s,\mathrm{d}}\cdot\sigma_{s,\mathrm{d}}^\mathrm{T}\right)^{-1}, \tag{7.3.4}$$

and

$$\frac{\Pi_\mathrm{f}^\star\left(t, X_\mathrm{f}, A\right)}{X_\mathrm{f}} = \frac{1}{\gamma_\mathrm{f}}\left(a_{s,\mathrm{f}} - r_\mathrm{f}\right)\cdot\left(\sigma_{s,\mathrm{f}}\cdot\sigma_{s,\mathrm{f}}^\mathrm{T}\right)^{-1}, \tag{7.3.5}$$

and consequently,

$$\begin{aligned}
\frac{\mathrm{d}Y_{\mathrm{d,f}}(t)}{Y_{\mathrm{d,f}}(t)} = \Bigg(& r_\mathrm{d} - r_\mathrm{f} + \frac{1}{\gamma_\mathrm{d}}\left(a_{s,\mathrm{d}} - r_\mathrm{d}\right)\cdot\left(\sigma_{s,\mathrm{d}}\cdot\sigma_{s,\mathrm{d}}^\mathrm{T}\right)^{-1}\cdot\left(a_{s,\mathrm{d}} - r_\mathrm{d}\right) - \frac{1}{\gamma_\mathrm{f}}\left(a_{s,\mathrm{f}} - r_\mathrm{f}\right) \\
&\cdot\left(\sigma_{s,\mathrm{f}}\cdot\sigma_{s,\mathrm{f}}^\mathrm{T}\right)^{-1}\cdot\left(a_{s,\mathrm{f}} - r_\mathrm{f}\right) - \frac{1}{\gamma_\mathrm{d}\gamma_\mathrm{f}}\left(a_{s,\mathrm{f}} - r_\mathrm{f}\right)\cdot\left(\sigma_{s,\mathrm{f}}\cdot\sigma_{s,\mathrm{f}}^\mathrm{T}\right)^{-1} \\
&\cdot\sigma_{s,\mathrm{f}}\cdot\sigma_{s,\mathrm{d}}^\mathrm{T}\cdot\left(\sigma_{s,\mathrm{d}}\cdot\sigma_{s,\mathrm{d}}^\mathrm{T}\right)^{-1}\cdot\left(a_{s,\mathrm{d}} - r_\mathrm{d}\right) + \frac{1}{\gamma_\mathrm{f}^2}\left(a_{s,\mathrm{f}} - r_\mathrm{f}\right)\cdot
\end{aligned}$$

$$\left(\sigma_{s,\mathrm{f}}\cdot\sigma_{s,\mathrm{f}}{}^{\mathrm{T}}\right)^{-1}\cdot(a_{s,\mathrm{f}}-r_{\mathrm{f}})\Big)\mathrm{d}t+\left(\frac{1}{\gamma_{\mathrm{d}}}(a_{s,\mathrm{d}}-r_{\mathrm{d}})\cdot\left(\sigma_{s,\mathrm{d}}\cdot\sigma_{s,\mathrm{d}}{}^{\mathrm{T}}\right)^{-1}\right.$$
$$\left.\cdot\sigma_{s,\mathrm{d}}-\frac{1}{\gamma_{\mathrm{f}}}(a_{s,\mathrm{f}}-r_{\mathrm{f}})\cdot\left(\sigma_{s,\mathrm{f}}\cdot\sigma_{s,\mathrm{f}}{}^{\mathrm{T}}\right)^{-1}\cdot\sigma_{s,\mathrm{f}}\right)\cdot\mathrm{d}B(t), \tag{7.3.6}$$

or if $\gamma_{\mathrm{d}}=\gamma_{\mathrm{f}}=\gamma$,

$$\begin{aligned}\frac{\mathrm{d}Y_{\mathrm{d,f}}(t)}{Y_{\mathrm{d,f}}(t)}&=\left(r_{\mathrm{d}}-r_{\mathrm{f}}+\frac{1}{\gamma}\left(\left(\frac{\mathbf{a}_{\mathrm{d}}-r_{\mathrm{d}}}{\sigma_{\mathrm{d}}}\right)^2-\left(\frac{\mathbf{a}_{\mathrm{f}}-r_{\mathrm{f}}}{\sigma_{\mathrm{f}}}\right)^2\right)\right.\\&\quad\left.-\frac{1}{\gamma^2}\frac{\mathbf{a}_{\mathrm{d}}-r_{\mathrm{d}}}{\sigma_{\mathrm{d}}}\frac{\mathbf{a}_{\mathrm{f}}-r_{\mathrm{f}}}{\sigma_{\mathrm{f}}}\rho_{2,1}+\frac{1}{\gamma^2}\left(\frac{\mathbf{a}_{\mathrm{f}}-r_{\mathrm{f}}}{\sigma_{\mathrm{f}}}\right)^2\right)\mathrm{d}t\\&\quad+\frac{1}{\gamma}\frac{\mathbf{a}_{\mathrm{d}}-r_{\mathrm{d}}}{\sigma_{\mathrm{d}}}\mathrm{d}B_1(t)-\frac{1}{\gamma}\frac{\mathbf{a}_{\mathrm{f}}-r_{\mathrm{f}}}{\sigma_{\mathrm{f}}}\left(\rho_{2,1}\mathrm{d}B_1(t)+\sqrt{1-\rho_{2,1}^2}\mathrm{d}B_2(t)\right)\\&=\left(r_{\mathrm{d}}-r_{\mathrm{f}}+\frac{1}{\gamma}\left(\left(\frac{\mathbf{a}_{\mathrm{d}}-r_{\mathrm{d}}}{\sigma_{\mathrm{d}}}\right)^2-\left(\frac{\mathbf{a}_{\mathrm{f}}-r_{\mathrm{f}}}{\sigma_{\mathrm{f}}}\right)^2\right)-\frac{1}{\gamma^2}\frac{\mathbf{a}_{\mathrm{d}}-r_{\mathrm{d}}}{\sigma_{\mathrm{d}}}\frac{\mathbf{a}_{\mathrm{f}}-r_{\mathrm{f}}}{\sigma_{\mathrm{f}}}\rho_{2,1}\right.\\&\quad\left.+\frac{1}{\gamma^2}\left(\frac{\mathbf{a}_{\mathrm{f}}-r_{\mathrm{f}}}{\sigma_{\mathrm{f}}}\right)^2\right)\mathrm{d}t+\frac{1}{\gamma}\left(\frac{\mathbf{a}_{\mathrm{d}}-r_{\mathrm{d}}}{\sigma_{\mathrm{d}}}\mathrm{d}\mathbb{B}_1(t)-\frac{\mathbf{a}_{\mathrm{f}}-r_{\mathrm{f}}}{\sigma_{\mathrm{f}}}\mathrm{d}\mathbb{B}_2(t)\right).\end{aligned} \tag{7.3.7}$$

Exercise 7.3.1. Show that

$$\frac{\mathrm{d}Y_{\mathrm{d,f}}(t)}{Y_{\mathrm{d,f}}(t)}+\frac{\mathrm{d}Y_{\mathrm{f,d}}(t)}{Y_{\mathrm{f,d}}(t)}=\frac{1}{\gamma^2}\left(\left(\frac{\mathbf{a}_{\mathrm{d}}-r_{\mathrm{d}}}{\sigma_{\mathrm{d}}}\right)^2-2\frac{\mathbf{a}_{\mathrm{f}}-r_{\mathrm{f}}}{\sigma_{\mathrm{f}}}\frac{\mathbf{a}_{\mathrm{d}}-r_{\mathrm{d}}}{\sigma_{\mathrm{d}}}\rho_{2,1}+\left(\frac{\mathbf{a}_{\mathrm{f}}-r_{\mathrm{f}}}{\sigma_{\mathrm{f}}}\right)^2\right)\mathrm{d}t.$$

Under what scenarios do we have $\frac{\mathrm{d}Y_{\mathrm{d,f}}(t)}{Y_{\mathrm{d,f}}(t)}+\frac{\mathrm{d}Y_{\mathrm{f,d}}(t)}{Y_{\mathrm{f,d}}(t)}=0$?

The FXR process $Y_{\mathrm{d,f}}(t)$, has the (conditional) expectation

$$E_{t_0,Y}Y_{\mathrm{d,f}}(t)=Y\mathrm{e}^{(t-t_0)\left(r_{\mathrm{d}}-r_{\mathrm{f}}+\frac{1}{\gamma}\left(\left(\frac{\mathbf{a}_{\mathrm{d}}-r_{\mathrm{d}}}{\sigma_{\mathrm{d}}}\right)^2-\left(\frac{\mathbf{a}_{\mathrm{f}}-r_{\mathrm{f}}}{\sigma_{\mathrm{f}}}\right)^2\right)-\frac{1}{\gamma^2}\frac{\mathbf{a}_{\mathrm{d}}-r_{\mathrm{d}}}{\sigma_{\mathrm{d}}}\frac{\mathbf{a}_{\mathrm{f}}-r_{\mathrm{f}}}{\sigma_{\mathrm{f}}}\rho_{2,1}+\frac{1}{\gamma^2}\left(\frac{\mathbf{a}_{\mathrm{f}}-r_{\mathrm{f}}}{\sigma_{\mathrm{f}}}\right)^2\right)}, \tag{7.3.8}$$

which might possibly be used for FXR projections into the future. More importantly, denote by σ_{FXR} the FXR volatility, i.e., let

$$\sigma_{\mathrm{FXR}}=\sqrt{\frac{1}{\mathrm{d}t}\left(\frac{\mathrm{d}Y_{\mathrm{d,f}}(t)}{Y_{\mathrm{d,f}}(t)}\right)^2}=\sqrt{\frac{1}{\mathrm{d}t}\left(\frac{\mathrm{d}Y_{\mathrm{f,d}}(t)}{Y_{\mathrm{f,d}}(t)}\right)^2}. \tag{7.3.9}$$

Then

$$\sigma_{\mathrm{FXR}}=\frac{1}{\gamma}\surd\left(\left(\frac{\mathbf{a}_{\mathrm{d}}-r_{\mathrm{d}}}{\sigma_{\mathrm{d}}}\right)^2-2\frac{\mathbf{a}_{\mathrm{d}}-r_{\mathrm{d}}}{\sigma_{\mathrm{d}}}\frac{\mathbf{a}_{\mathrm{f}}-r_{\mathrm{f}}}{\sigma_{\mathrm{f}}}\rho_{2,1}+\left(\frac{\mathbf{a}_{\mathrm{f}}-r_{\mathrm{f}}}{\sigma_{\mathrm{f}}}\right)^2\right), \tag{7.3.10}$$

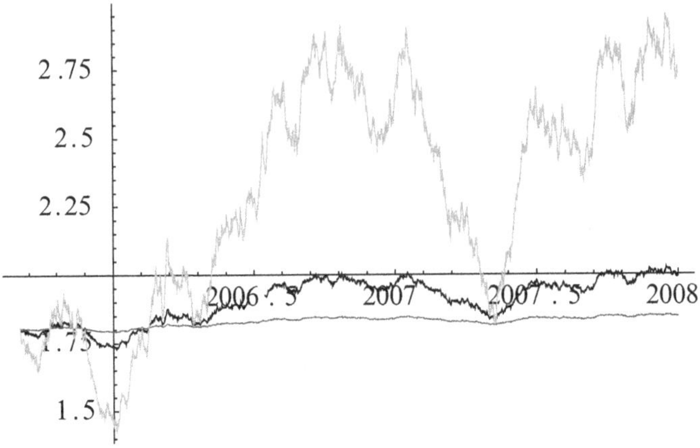

Fig. 7.1 Simulated USD/GBP price trajectories for realistic and non-realistic values of risk aversion

which, solving for γ, yields

$$\gamma = \frac{1}{\sigma_{\mathrm{FXR}}}\sqrt{\left(\left(\frac{\mathbf{a}_{\mathrm{d}} - r_{\mathrm{d}}}{\sigma_{\mathrm{d}}}\right)^2 - 2\frac{\mathbf{a}_{\mathrm{d}} - r_{\mathrm{d}}}{\sigma_{\mathrm{d}}}\frac{\mathbf{a}_{\mathrm{f}} - r_{\mathrm{f}}}{\sigma_{\mathrm{f}}}\rho_{2,1} + \left(\frac{\mathbf{a}_{\mathrm{f}} - r_{\mathrm{f}}}{\sigma_{\mathrm{f}}}\right)^2\right)}. \tag{7.3.11}$$

Since $\mathbf{a}_{\mathrm{d}}, \mathbf{a}_{\mathrm{f}}, r_{\mathrm{d}}, r_{\mathrm{f}}, \sigma_{\mathrm{d}}, \sigma_{\mathrm{f}}$, and $\rho_{2,1}$ can be more or less efficiently estimated from the market data, formula (7.3.11) provides a way of estimating the *(aggregate market) relative risk-aversion (for the two economies)*. A difficulty is that estimating $\mathbf{a}_{\mathrm{d}}, \mathbf{a}_{\mathrm{f}}$ is inefficient.

Remark 7.3.1. Throughout this book, a single parameter that has profoundly affected most of the results presented was the *(relative) risk-aversion* parameter γ. It is therefore very satisfying now to have an explicit and efficient estimate for γ, i.e., formula (7.3.11).

Example 7.3.1. Monthly data were used for USD and GBP, between December 31, 1985, and August 31, 2005 (DJI and FTSE were taken as market representatives; interest rates were averaged). The corresponding risk-aversion is estimated using (7.3.11) to be

$$\gamma_{\mathrm{US/GB}} = 4.16761. \tag{7.3.12}$$

This suggests that assuming $\gamma = 1$, i.e., using logarithmic utility in financial mathematics, which seems still to be a popular choice due to its simplicity, is not realistic. See also Fig. 7.1, which shows simulated FXR trajectories modeled after the USD/GBP exchange rate, corresponding to $\gamma = 1$ (too aggressive), $\gamma = 4.16$, and $\gamma = 4.16^2 = 17.3056$ (too conservative).

Remark 7.3.2. Alternatively, if one has a sense of what the market risk-aversion is, then taking the lead from the (overseas) foreign market performance, (7.3.10) can be used to gauge and quantify the direction of the domestic market by solving (7.3.10) for $\mathbf{a}_\mathrm{d}$:

$$\mathbf{a}_\mathrm{d} = r_\mathrm{d} + \frac{1}{\sigma_\mathrm{f}^2}\left(\left(\mathbf{a}_\mathrm{f} - r_\mathrm{f}\right)\sigma_\mathrm{d}\sigma_\mathrm{f}\rho_{2,1} \pm \sqrt{\left(\sigma_\mathrm{d}^2\sigma_\mathrm{f}^2\left(\gamma^2\sigma_\mathrm{f}^2\sigma_\mathrm{FXR}^2 - \left(\mathbf{a}_\mathrm{f} - r_\mathrm{f}\right)^2\left(1 - \rho_{2,1}^2\right)\right)\right)}\right). \tag{7.3.13}$$

Remark 7.3.3. (Why did the JPY fall whenever investors ignored risks (in 2007)?)

Financial Times, By Neil Dennis:

The yen fell sharply as investors ignored the risks (Published: Jun 06, 2007) ... earlier this year, the yen rose amid a general increase in risk-aversion (Published: Jun 05, 2007).

Is the FXR SDE (7.3.14) consistent with the empirical observation (in 2007) that the Japanese yen usually falls/rises when market risk-aversion falls/rises?

The answer is affirmative. Indeed, taking the conditional expectation of (7.3.14), we get

$$E_{t,Y_{\mathrm{d,f}}}\mathrm{d}Y_{\mathrm{d,f}}(t) = Y_{\mathrm{d,f}}\left(r_\mathrm{d} - r_\mathrm{f} + \frac{1}{\gamma}\left(\left(\frac{\mathbf{a}_\mathrm{d} - r_\mathrm{d}}{\sigma_\mathrm{d}}\right)^2 - \left(\frac{\mathbf{a}_\mathrm{f} - r_\mathrm{f}}{\sigma_\mathrm{f}}\right)^2\right) - \frac{1}{\gamma^2}\frac{\mathbf{a}_\mathrm{d} - r_\mathrm{d}}{\sigma_\mathrm{d}}\right.$$
$$\left.\frac{\mathbf{a}_\mathrm{f} - r_\mathrm{f}}{\sigma_\mathrm{f}}\rho_{2,1} + \frac{1}{\gamma^2}\left(\frac{\mathbf{a}_\mathrm{f} - r_\mathrm{f}}{\sigma_\mathrm{f}}\right)^2\right)\mathrm{d}t. \tag{7.3.14}$$

Also, since $\gamma > 1$, we shall ignore $1/\gamma^2$ terms, arriving at

$$E_{t,Y_{\mathrm{d,f}}}\mathrm{d}Y_{\mathrm{d,f}}(t) \approx Y_{\mathrm{d,f}}\left(r_\mathrm{d} - r_\mathrm{f} + \frac{1}{\gamma}\left(\left(\frac{\mathbf{a}_\mathrm{d} - r_\mathrm{d}}{\sigma_\mathrm{d}}\right)^2 - \left(\frac{\mathbf{a}_\mathrm{f} - r_\mathrm{f}}{\sigma_\mathrm{f}}\right)^2\right)\right)\mathrm{d}t. \tag{7.3.15}$$

Back in 2007, the interest rates in Japan were lower than in the United States, so that $r_\mathrm{d} - r_\mathrm{f} > 0$, and therefore, the FXR balance was necessarily maintained by the difference in Sharpe ratios squared:

$$\left(\frac{\mathbf{a}_\mathrm{d} - r_\mathrm{d}}{\sigma_\mathrm{d}}\right)^2 - \left(\frac{\mathbf{a}_\mathrm{f} - r_\mathrm{f}}{\sigma_\mathrm{f}}\right)^2 < 0, \tag{7.3.16}$$

yielding ("FXR balance")

$$r_\mathrm{d} - r_\mathrm{f} + \frac{1}{\gamma}\left(\left(\frac{\mathbf{a}_\mathrm{d} - r_\mathrm{d}}{\sigma_\mathrm{d}}\right)^2 - \left(\frac{\mathbf{a}_\mathrm{f} - r_\mathrm{f}}{\sigma_\mathrm{f}}\right)^2\right) \approx 0. \tag{7.3.17}$$

If there was an increase in risk-aversion, say from γ to γ_1, with $\gamma_1 > \gamma$, then

$$E_{t,Y_{\mathrm{d,f}}}\mathrm{d}Y_{\mathrm{d,f}}(t) \approx Y_{\mathrm{d,f}}\left(r_\mathrm{d} - r_\mathrm{f} + \frac{1}{\gamma_1}\left(\left(\frac{\mathbf{a}_\mathrm{d} - r_\mathrm{d}}{\sigma_\mathrm{d}}\right)^2 - \left(\frac{\mathbf{a}_\mathrm{f} - r_\mathrm{f}}{\sigma_\mathrm{f}}\right)^2\right)\right)\mathrm{d}t$$
$$> Y_{\mathrm{d,f}}\left(r_\mathrm{d} - r_\mathrm{f} + \frac{1}{\gamma}\left(\left(\frac{\mathbf{a}_\mathrm{d} - r_\mathrm{d}}{\sigma_\mathrm{d}}\right)^2 - \left(\frac{\mathbf{a}_\mathrm{f} - r_\mathrm{f}}{\sigma_\mathrm{f}}\right)^2\right)\right)\mathrm{d}t \approx 0, \tag{7.3.18}$$

so that an increase in investors' risk-aversion causes JPY to rise (and conversely).

7.3.2 FXR for Economies with Stochastic Interest Rates

Now we modify the previous example to introduce a possibility of stochastic interest rates.

As in Sects. 2.4.5 and 3.11, we assume the interest rate model (2.4.23), with $S = \{S_1\}$, $A = \{S_1, r\}$, and market coefficients (2.4.27), but now extended to two economies. Therefore, we have $A = \{S_1, S_2, r_{\mathrm{d}}, r_{\mathrm{f}}\}$, $S_{\mathrm{d}} = \{S_1\}$, $S_{\mathrm{f}} = \{S_2\}$, and the market coefficients are

$$\begin{aligned} a_{s,\mathrm{d}} &= \{\mathbf{a}_{\mathrm{d}} + r_{\mathrm{d}}\beta_{\mathrm{d}}\}, \\ \sigma_{s,\mathrm{d}} &= \begin{pmatrix} \sigma_{\mathrm{d}} & 0 & 0 & 0 \end{pmatrix}, \end{aligned} \tag{7.3.19}$$

$$\begin{aligned} a_{s,\mathrm{f}} &= \{\mathbf{a}_{\mathrm{f}} + r_{\mathrm{f}}\beta_{\mathrm{f}}\}, \\ \sigma_{s,\mathrm{f}} &= \left(\sigma_{\mathrm{f}}\rho_{2,1} \quad \sigma_{\mathrm{f}}\sqrt{1-\rho_{2,1}^2} \quad 0 \quad 0\right), \end{aligned} \tag{7.3.20}$$

$$b = \left\{S_1\left(\mathbf{a}_{\mathrm{d}} + r_{\mathrm{d}}\beta_{\mathrm{d}} - \mathbb{D}_{\mathrm{d}}\right), S_2\left(\mathbf{a}_{\mathrm{f}} + r_{\mathrm{f}}\beta_{\mathrm{f}} - \mathbb{D}_{\mathrm{f}}\right), q_{d,0} + r_{\mathrm{d}}q_{d,1}, q_{f,0} + r_{\mathrm{f}}q_{f,1}\right\}, \tag{7.3.21}$$

$$c = \begin{pmatrix} S_1\sigma_{\mathrm{d}} & 0 & 0 & 0 \\ S_2\sigma_{\mathrm{f}}\rho_{2,1} & S_2\sigma_{\mathrm{f}}\sqrt{1-\rho_{2,1}^2} & 0 & 0 \\ w_{\mathrm{d}}\rho_{3,1} & \frac{w_{\mathrm{d}}\left(\rho_{3,2}-\rho_{2,1}\rho_{3,1}\right)}{\sqrt{1-\rho_{2,1}^2}} & w_{\mathrm{d}}\sqrt{1-\rho_{3,1}^2 + \frac{\left(-\rho_{2,1}\rho_{3,1}+\rho_{3,2}\right)^2}{-1+\rho_{2,1}^2}} & 0 \\ w_{\mathrm{f}}\rho_{4,1} & \frac{w_{\mathrm{f}}\left(\rho_{4,2}-\rho_{2,1}\rho_{4,1}\right)}{\sqrt{1-\rho_{2,1}^2}} & w_{\mathrm{f}}\mathbf{s}_{4,3} & w_{\mathrm{f}}\mathbf{s}_{4,4} \end{pmatrix}, \tag{7.3.22}$$

where $\mathbf{s}_{4,3}$ and $\mathbf{s}_{4,4}$ are given in (A.2.30) and (A.2.31), respectively. In other words, we assume that the domestic and foreign markets obey SDEs

$$\begin{aligned} \mathrm{d}S_1(t) &= S_1(t)\left(\mathbf{a}_{\mathrm{d}} + r_{\mathrm{d}}(t)\beta_{\mathrm{d}} - \mathbb{D}_{\mathrm{d}}\right)\mathrm{d}t + S_1(t)\sigma_{\mathrm{d}}\mathrm{d}B_1(t), \\ \mathrm{d}S_2(t) &= S_2(t)\left(\mathbf{a}_{\mathrm{f}} + r_{\mathrm{f}}(t)\beta_{\mathrm{f}} - \mathbb{D}_{\mathrm{f}}\right)\mathrm{d}t + S_2(t)\sigma_{\mathrm{f}}\mathrm{d}\mathbb{B}_2(t), \end{aligned} \tag{7.3.23}$$

while domestic and foreign (short) interest rates obey

$$\begin{aligned} \mathrm{d}r_{\mathrm{d}}(t) &= \left(q_{d,0} + r_{\mathrm{d}}(t)q_{d,1}\right)\mathrm{d}t + w_{\mathrm{d}}\mathrm{d}\mathbb{B}_3(t), \\ \mathrm{d}r_{\mathrm{f}}(t) &= \left(q_{f,0} + r_{\mathrm{f}}(t)q_{f,1}\right)\mathrm{d}t + w_{\mathrm{f}}\mathrm{d}\mathbb{B}_4(t), \end{aligned} \tag{7.3.24}$$

where $B_1(t), \mathbb{B}_2(t), \mathbb{B}_3(t)$, and $\mathbb{B}_4(t)$ are correlated Brownian motions, and where the (predividend) appreciation rates for domestic and foreign markets are equal to $\mathbf{a}_{\mathrm{d}} + r_{\mathrm{d}}\beta_{\mathrm{d}}$ and $\mathbf{a}_{\mathrm{f}} + r_{\mathrm{f}}\beta_{\mathrm{f}}$, respectively, with corresponding dividend rates denoted by $\mathbb{D}_{\mathrm{d}}$ and $\mathbb{D}_{\mathrm{f}}$, with the corresponding volatilities denoted by σ_{d} and σ_{f}. We assume mean reversion for domestic and foreign interest rates: $q_{d,1} < 0$ and $q_{f,1} < 0$.

As can be confirmed by direct calculation here, from (3.11.18), we conclude that in the case of CRRA utility of wealth with relative risk-aversion $\gamma_{\mathrm{d}} = \gamma_{\mathrm{f}} = \gamma$,

$$\frac{\Pi_{\mathrm{d}}^{\star}(t, Y_{\mathrm{d,f}} X_{\mathrm{f}}, A)}{Y_{\mathrm{d,f}} X_{\mathrm{f}}} = \left\{ \frac{1}{\gamma} \left(\frac{\mathbf{a}_{\mathrm{d}} + (\beta_{\mathrm{d}} - 1) r_{\mathrm{d}}}{\sigma_{\mathrm{d}}^2} + \frac{w_{\mathrm{d}} \left(\mathbb{H}_{d,0} + r_{\mathrm{d}} \mathbb{H}_{d,1} \right) \rho_{3,1}}{\sigma_{\mathrm{d}}} \right) \right\}, \tag{7.3.25}$$

with

$$\frac{w_{\mathrm{d}}^2 \left(\gamma - (\gamma - 1) \rho_{3,1}^2 \right)}{\gamma} \mathbb{H}_{d,1}^2 + \frac{1}{\gamma \sigma_{\mathrm{d}}} 2 \left(\gamma q_{d,0} \sigma_{\mathrm{d}} - w_{\mathrm{d}} (\beta_{\mathrm{d}} - 1)(\gamma - 1) \rho_{3,1} \right) \mathbb{H}_{d,1} - \frac{(\beta_{\mathrm{d}} - 1)^2 (\gamma - 1)}{\gamma \sigma_{\mathrm{d}}^2} = 0 r \tag{7.3.26}$$

$$\left(q_{d,1} + \frac{1}{\gamma \sigma_{\mathrm{d}}} w_{\mathrm{d}} \left(w_{\mathrm{d}} \mathbb{H}_{d,1} \sigma_{\mathrm{d}} \left(\gamma - (\gamma - 1) \rho_{3,1}^2 \right) - (\beta_{\mathrm{d}} - 1)(\gamma - 1) \rho_{3,1} \right) \right) \mathbb{H}_{d,0} - \gamma + q_{d,0} \mathbb{H}_{d,1} - \frac{1}{\gamma \sigma_{\mathrm{d}}^2} (\gamma - 1) \mathbf{a}_{\mathrm{d}} \left(\beta_{\mathrm{d}} - 1 + w_{\mathrm{d}} \mathbb{H}_{d,1} \sigma_{\mathrm{d}} \rho_{3,1} \right) + 1 = 0$$

and

$$\frac{\Pi_{\mathrm{f}}^{\star}(t, X_{\mathrm{f}}, A)}{X_{\mathrm{f}}} = \left\{ \frac{1}{\gamma} \left(\frac{\mathbf{a}_{\mathrm{f}} + (\beta_{\mathrm{f}} - 1) r_{\mathrm{f}}}{\sigma_{\mathrm{f}}^2} + \frac{w_{\mathrm{f}} \left(\mathbb{H}_{f,0} + r_{\mathrm{f}} \mathbb{H}_{f,1} \right) \rho_{4,2}}{\sigma_{\mathrm{f}}} \right) \right\} \tag{7.3.27}$$

with

$$\frac{w_{\mathrm{f}}^2 \left(\gamma - (\gamma - 1) \rho_{4,2}^2 \right)}{\gamma} \mathbb{H}_{f,1}^2 + \frac{1}{\gamma \sigma_{\mathrm{f}}} 2 \left(\gamma q_{f,0} \sigma_{\mathrm{f}} - w_{\mathrm{f}} (\beta_{\mathrm{f}} - 1)(\gamma - 1) \rho_{4,2} \right) \mathbb{H}_{f,1} - \frac{(\beta_{\mathrm{f}} - 1)^2 (\gamma - 1)}{\gamma \sigma_{\mathrm{f}}^2} = 0 \tag{7.3.28}$$

$$\left(q_{f,1} + \frac{1}{\gamma \sigma_{\mathrm{f}}} w_{\mathrm{f}} \left(w_{\mathrm{f}} \mathbb{H}_{f,1} \sigma_{\mathrm{f}} \left(\gamma - (\gamma - 1) \rho_{4,2}^2 \right) - (\beta_{\mathrm{f}} - 1)(\gamma - 1) \rho_{4,2} \right) \right) \mathbb{H}_{f,0} - \gamma + q_{f,0} \mathbb{H}_{f,1} - \frac{1}{\gamma \sigma_{\mathrm{f}}^2} (\gamma - 1) \mathbf{a}_{\mathrm{f}} \left(\beta_{\mathrm{f}} - 1 + w_{\mathrm{f}} \mathbb{H}_{f,1} \sigma_{\mathrm{f}} \rho_{4,2} \right) + 1 = 0,$$

and therefore, the FXR SDE (7.2.24) becomes

$$\frac{dY_{\mathrm{d,f}}(t)}{Y_{\mathrm{d,f}}(t)} = \Bigg(r_{\mathrm{d}} - r_{\mathrm{f}} + \left\{ \frac{1}{\gamma} \left(\frac{\mathbf{a}_{\mathrm{d}} + (\beta_{\mathrm{d}} - 1) r_{\mathrm{d}}}{\sigma_{\mathrm{d}}^2} + \frac{1}{\sigma_{\mathrm{d}}} w_{\mathrm{d}} \left(\mathbb{H}_{d,0} + r_{\mathrm{d}} \mathbb{H}_{d,1} \right) \rho_{3,1} \right) \right\} \cdot (a_{s,\mathrm{d}} - r_{\mathrm{d}}) - \left\{ \frac{1}{\gamma} \left(\frac{\mathbf{a}_{\mathrm{f}} + (\beta_{\mathrm{f}} - 1) r_{\mathrm{f}}}{\sigma_{\mathrm{f}}^2} + \frac{1}{\sigma_{\mathrm{f}}} w_{\mathrm{f}} \left(\mathbb{H}_{f,0} + r_{\mathrm{f}} \mathbb{H}_{f,1} \right) \rho_{4,2} \right) \right\}$$

$$
\begin{aligned}
&\cdot (a_{s,\mathrm{f}} - r_\mathrm{f}) - \left\{ \frac{1}{\gamma} \left(\frac{\mathbf{a}_\mathrm{f} + (\beta_\mathrm{f} - 1)\, r_\mathrm{f}}{\sigma_\mathrm{f}^2} + \frac{1}{\sigma_\mathrm{f}} w_\mathrm{f} \left(\mathbb{H}_{f,0} + r_\mathrm{f} \mathbb{H}_{f,1} \right) \rho_{4,2} \right) \right\} \\
&\cdot \sigma_{s,\mathrm{f}} \cdot \sigma_{s,\mathrm{d}}{}^\mathrm{T} \left\{ \frac{1}{\gamma} \left(\frac{\mathbf{a}_\mathrm{d} + (\beta_\mathrm{d} - 1)\, r_\mathrm{d}}{\sigma_\mathrm{d}^2} + \frac{1}{\sigma_\mathrm{d}} w_\mathrm{d} \left(\mathbb{H}_{d,0} + r_\mathrm{d} \mathbb{H}_{d,1} \right) \rho_{3,1} \right) \right\} \\
&+ \left\{ \frac{1}{\gamma} \left(\frac{\mathbf{a}_\mathrm{f} + (\beta_\mathrm{f} - 1)\, r_\mathrm{f}}{\sigma_\mathrm{f}^2} + \frac{1}{\sigma_\mathrm{f}} w_\mathrm{f} \left(\mathbb{H}_{f,0} + r_\mathrm{f} \mathbb{H}_{f,1} \right) \rho_{4,2} \right) \right\} \cdot \sigma_{s,\mathrm{f}} \cdot \sigma_{s,\mathrm{f}}{}^\mathrm{T} \\
&\cdot \left\{ \frac{1}{\gamma} \left(\frac{\mathbf{a}_\mathrm{f} + (\beta_\mathrm{f} - 1)\, r_\mathrm{f}}{\sigma_\mathrm{f}^2} + \frac{1}{\sigma_\mathrm{f}} w_\mathrm{f} \left(\mathbb{H}_{f,0} + r_\mathrm{f} \mathbb{H}_{f,1} \right) \rho_{4,2} \right) \right\} \Bigg) \mathrm{d}t \\
&+ \Bigg(\left\{ \frac{1}{\gamma} \left(\frac{\mathbf{a}_\mathrm{d} + (\beta_\mathrm{d} - 1)\, r_\mathrm{d}}{\sigma_\mathrm{d}^2} + \frac{1}{\sigma_\mathrm{d}} w_\mathrm{d} \left(\mathbb{H}_{d,0} + r_\mathrm{d} \mathbb{H}_{d,1} \right) \rho_{3,1} \right) \right\} \\
&\quad \cdot \sigma_{s,\mathrm{d}} - \left\{ \frac{1}{\gamma} \left(\frac{\mathbf{a}_\mathrm{f} + (\beta_\mathrm{f} - 1)\, r_\mathrm{f}}{\sigma_\mathrm{f}^2} + \frac{1}{\sigma_\mathrm{f}} w_\mathrm{f} \left(\mathbb{H}_{f,0} + r_\mathrm{f} \mathbb{H}_{f,1} \right) \rho_{4,2} \right) \right\} \\
&\quad \cdot \sigma_{s,\mathrm{f}} \Bigg) \cdot \mathrm{d}B(t),
\end{aligned}
\tag{7.3.29}
$$

and since

$$
\begin{aligned}
a_{s,\mathrm{d}} - r_\mathrm{d} &= \{\mathbf{a}_\mathrm{d} + r_\mathrm{d}\beta_\mathrm{d}\} - r_\mathrm{d} = \{\mathbf{a}_\mathrm{d} + r_\mathrm{d}\,(\beta_\mathrm{d} - 1)\}, \\
a_{s,\mathrm{f}} - r_\mathrm{f} &= \{\mathbf{a}_\mathrm{f} + r_\mathrm{f}\beta_\mathrm{f}\} - r_\mathrm{f} = \{\mathbf{a}_\mathrm{f} + r_\mathrm{f}\,(\beta_\mathrm{f} - 1)\}, \\
\sigma_{s,\mathrm{f}} \cdot \sigma_{s,\mathrm{d}}{}^\mathrm{T} &= \left(\sigma_\mathrm{f}\rho_{2,1} \;\; \sigma_\mathrm{f}\sqrt{1-\rho_{2,1}^2} \;\; 0 \;\; 0 \right) \cdot \left(\sigma_\mathrm{d} \;\; 0 \;\; 0 \;\; 0 \right)^\mathrm{T} = \left(\sigma_\mathrm{d}\sigma_\mathrm{f}\rho_{2,1} \right), \\
\sigma_{s,\mathrm{f}} \cdot \sigma_{s,\mathrm{f}}{}^\mathrm{T} &= \left(\sigma_\mathrm{f}\rho_{2,1} \;\; \sigma_\mathrm{f}\sqrt{1-\rho_{2,1}^2} \;\; 0 \;\; 0 \right) \cdot \left(\sigma_\mathrm{f}\rho_{2,1} \;\; \sigma_\mathrm{f}\sqrt{1-\rho_{2,1}^2} \;\; 0 \;\; 0 \right)^\mathrm{T} = \left(\sigma_\mathrm{f}^2 \right),
\end{aligned}
\tag{7.3.30}
$$

we conclude that

$$
\begin{aligned}
\frac{\mathrm{d}Y_{\mathrm{d,f}}(t)}{Y_{\mathrm{d,f}}(t)} = &\Bigg(r_\mathrm{d} - r_\mathrm{f} + \frac{1}{\gamma} \left(\frac{\mathbf{a}_\mathrm{d} + (\beta_\mathrm{d} - 1)\, r_\mathrm{d}}{\sigma_\mathrm{d}^2} + \frac{w_\mathrm{d} \left(\mathbb{H}_{d,0} + r_\mathrm{d} \mathbb{H}_{d,1} \right) \rho_{3,1}}{\sigma_\mathrm{d}} \right) \\
&(\mathbf{a}_\mathrm{d} + r_\mathrm{d}\,(\beta_\mathrm{d} - 1)) - \frac{1}{\gamma} \left(\frac{\mathbf{a}_\mathrm{f} + (\beta_\mathrm{f} - 1)\, r_\mathrm{f}}{\sigma_\mathrm{f}^2} + \frac{w_\mathrm{f} \left(\mathbb{H}_{f,0} + r_\mathrm{f} \mathbb{H}_{f,1} \right) \rho_{4,2}}{\sigma_\mathrm{f}} \right) \\
&(\mathbf{a}_\mathrm{f} + r_\mathrm{f}\,(\beta_\mathrm{f} - 1)) - \frac{1}{\gamma^2} \left(\frac{\mathbf{a}_\mathrm{f} + (\beta_\mathrm{f} - 1)\, r_\mathrm{f}}{\sigma_\mathrm{f}} + w_\mathrm{f} \left(\mathbb{H}_{f,0} + r_\mathrm{f} \mathbb{H}_{f,1} \right) \rho_{4,2} \right) \\
&\rho_{2,1} \left(\frac{\mathbf{a}_\mathrm{d} + (\beta_\mathrm{d} - 1)\, r_\mathrm{d}}{\sigma_\mathrm{d}} + w_\mathrm{d} \left(\mathbb{H}_{d,0} + r_\mathrm{d} \mathbb{H}_{d,1} \right) \rho_{3,1} \right) \\
&+ \frac{1}{\gamma^2} \left(\frac{\mathbf{a}_\mathrm{f} + (\beta_\mathrm{f} - 1)\, r_\mathrm{f}}{\sigma_\mathrm{f}} + w_\mathrm{f} \left(\mathbb{H}_{f,0} + r_\mathrm{f} \mathbb{H}_{f,1} \right) \rho_{4,2} \right)^2 \Bigg) \mathrm{d}t
\end{aligned}
$$

$$+\left(\left\{\frac{1}{\gamma}\left(\frac{\mathbf{a}_\mathrm{d}+(\beta_\mathrm{d}-1)\,r_\mathrm{d}}{\sigma_\mathrm{d}}+w_\mathrm{d}\left(\mathbb{H}_{d,0}+r_\mathrm{d}\mathbb{H}_{d,1}\right)\rho_{3,1}\right),0,0,0\right\}\right.$$
$$-\left\{\frac{1}{\gamma}\left(\frac{\mathbf{a}_\mathrm{f}+(\beta_\mathrm{f}-1)\,r_\mathrm{f}}{\sigma_\mathrm{f}}+w_\mathrm{f}\left(\mathbb{H}_{f,0}+r_\mathrm{f}\mathbb{H}_{f,1}\right)\rho_{4,2}\right)\rho_{2,1},\right.$$
$$\frac{1}{\gamma}\left(\frac{\mathbf{a}_\mathrm{f}+(\beta_\mathrm{f}-1)\,r_\mathrm{f}}{\sigma_\mathrm{f}}+w_\mathrm{f}\left(\mathbb{H}_{f,0}\right.\right.$$
$$\left.\left.\left.\left.+r_\mathrm{f}\mathbb{H}_{f,1}\right)\rho_{4,2}\right)\sqrt{1-\rho_{2,1}^2},0,0\right\}\right)\cdot \mathrm{d}B(t), \tag{7.3.31}$$

i.e.,

$$\frac{\mathrm{d}Y_{\mathrm{d,f}}(t)}{Y_{\mathrm{d,f}}(t)}=\left(r_\mathrm{d}-r_\mathrm{f}+\frac{1}{\gamma}\left(\frac{\mathbf{a}_\mathrm{d}+(\beta_\mathrm{d}-1)\,r_\mathrm{d}}{\sigma_\mathrm{d}^2}+\frac{w_\mathrm{d}\left(\mathbb{H}_{d,0}+r_\mathrm{d}\mathbb{H}_{d,1}\right)\rho_{3,1}}{\sigma_\mathrm{d}}\right)\right.$$
$$(\mathbf{a}_\mathrm{d}+r_\mathrm{d}(\beta_\mathrm{d}-1))-\frac{1}{\gamma}\left(\frac{\mathbf{a}_\mathrm{f}+(\beta_\mathrm{f}-1)\,r_\mathrm{f}}{\sigma_\mathrm{f}^2}+\frac{w_\mathrm{f}\left(\mathbb{H}_{f,0}+r_\mathrm{f}\mathbb{H}_{f,1}\right)\rho_{4,2}}{\sigma_\mathrm{f}}\right)$$
$$(\mathbf{a}_\mathrm{f}+r_\mathrm{f}(\beta_\mathrm{f}-1))-\frac{1}{\gamma^2}\left(\frac{\mathbf{a}_\mathrm{f}+(\beta_\mathrm{f}-1)\,r_\mathrm{f}}{\sigma_\mathrm{f}}+w_\mathrm{f}\left(\mathbb{H}_{f,0}+r_\mathrm{f}\mathbb{H}_{f,1}\right)\rho_{4,2}\right)$$
$$\rho_{2,1}\left(\frac{\mathbf{a}_\mathrm{d}+(\beta_\mathrm{d}-1)\,r_\mathrm{d}}{\sigma_\mathrm{d}}+w_\mathrm{d}\left(\mathbb{H}_{d,0}+r_\mathrm{d}\mathbb{H}_{d,1}\right)\rho_{3,1}\right)$$
$$\left.+\frac{1}{\gamma^2}\left(\frac{\mathbf{a}_\mathrm{f}+(\beta_\mathrm{f}-1)\,r_\mathrm{f}}{\sigma_\mathrm{f}}+w_\mathrm{f}\left(\mathbb{H}_{f,0}+r_\mathrm{f}\mathbb{H}_{f,1}\right)\rho_{4,2}\right)^2\right)\mathrm{d}t$$
$$+\frac{1}{\gamma}\left(\frac{\mathbf{a}_\mathrm{d}+(\beta_\mathrm{d}-1)\,r_\mathrm{d}}{\sigma_\mathrm{d}}+w_\mathrm{d}\left(\mathbb{H}_{d,0}+r_\mathrm{d}\mathbb{H}_{d,1}\right)\rho_{3,1}\right)\mathrm{d}B_1(t)$$
$$-\frac{1}{\gamma}\left(\frac{\mathbf{a}_\mathrm{f}+(\beta_\mathrm{f}-1)\,r_\mathrm{f}}{\sigma_\mathrm{f}}+w_\mathrm{f}\left(\mathbb{H}_{f,0}+r_\mathrm{f}\mathbb{H}_{f,1}\right)\rho_{4,2}\right)\mathrm{d}\mathbb{B}_2(t), \tag{7.3.32}$$

coupled with (7.3.24). In the more convenient special case $\beta_\mathrm{d}=\beta_\mathrm{f}=1$, we have $\mathbb{H}_{d,1}=\mathbb{H}_{f,1}=0$, $\mathbb{H}_{d,0}=(\gamma-1)/q_{d,1}$, and $\mathbb{H}_{f,0}=(\gamma-1)/q_{f,1}$, and consequently,

$$\frac{\mathrm{d}Y_{\mathrm{d,f}}(t)}{Y_{\mathrm{d,f}}(t)}=\left(r_\mathrm{d}-r_\mathrm{f}+\frac{1}{\gamma}\left(\frac{\mathbf{a}_\mathrm{d}}{\sigma_\mathrm{d}^2}+\frac{w_\mathrm{d}(\gamma-1)\rho_{3,1}}{\sigma_\mathrm{d}q_{d,1}}\right)\mathbf{a}_\mathrm{d}-\frac{1}{\gamma}\left(\frac{\mathbf{a}_\mathrm{f}}{\sigma_\mathrm{f}^2}+\frac{w_\mathrm{f}(\gamma-1)\rho_{4,2}}{\sigma_\mathrm{f}q_{f,1}}\right)\right.$$
$$\mathbf{a}_\mathrm{f}-\frac{1}{\gamma^2}\left(\frac{\mathbf{a}_\mathrm{f}}{\sigma_\mathrm{f}}+w_\mathrm{f}\frac{\gamma-1}{q_{f,1}}\rho_{4,2}\right)\rho_{2,1}\left(\frac{\mathbf{a}_\mathrm{d}}{\sigma_\mathrm{d}}+w_\mathrm{d}\frac{\gamma-1}{q_{d,1}}\rho_{3,1}\right)$$
$$\left.+\frac{1}{\gamma^2}\left(\frac{\mathbf{a}_\mathrm{f}}{\sigma_\mathrm{f}}+w_\mathrm{f}\frac{\gamma-1}{q_{f,1}}\rho_{4,2}\right)^2\right)\mathrm{d}t+\frac{1}{\gamma}\left(\frac{\mathbf{a}_\mathrm{d}}{\sigma_\mathrm{d}}+w_\mathrm{d}\frac{\gamma-1}{q_{d,1}}\rho_{3,1}\right)\mathrm{d}B_1(t)$$
$$-\frac{1}{\gamma}\left(\frac{\mathbf{a}_\mathrm{f}}{\sigma_\mathrm{f}}+w_\mathrm{f}\frac{\gamma-1}{q_{f,1}}\rho_{4,2}\right)\mathrm{d}\mathbb{B}_2(t), \tag{7.3.33}$$

coupled with (7.3.24).

Table 7.1 Stochastic Interest Rate FXR Model parameter estimates for various currencies, based on 10 and 20 years of data

	USD/JPY		USD/CAD		USD/GBP	
	86/06	96/06	86/06	96/06	86/06	96/06
a_d	0.0553827	0.0432591	0.0553827	0.0432591	0.0553827	0.0432591
σ_d	0.126669	0.130523	0.126669	0.130523	0.126669	0.130523
w_d	0.00934253	0.00638909	0.00934253	0.00638909	0.00934253	0.00638909
$q_{d,0}$	0.00444408	0.00224622	0.00444408	0.00224622	0.00444408	0.00224622
$q_{d,1}$	−0.0876372	−0.056621	−0.0876372	−0.056621	−0.0876372	−0.056621
$\mathbf{a}_f$	0.0116198	−0.00908073	0.0208129	0.059647	0.00994607	0.00124463
σ_f	0.218095	0.191967	0.156741	0.168861	0.166314	0.139327
w_f	0.00675006	0.00109607	0.0178495	0.00815306	0.0251519	0.0210506
$q_{f,0}$	0.00079536	0.000235767	0.0098136	0.00862889	0.0213267	0.0784621
$q_{f,1}$	−0.0351943	−0.150686	−0.162631	−0.232491	−0.287016	−1.50054
$\rho_{2,1}$	0.335031	0.359153	0.573433	0.534128	0.577589	0.538008
$\rho_{3,1}$	−0.110615	−0.116515	−0.110615	−0.116515	−0.110615	−0.116515
$\rho_{4,1}$	−0.132049	0.00385812	−0.0792655	−0.126492	−0.0202239	−0.099886
$\rho_{3,2}$	−0.0709149	0.0278757	−0.0478924	0.0745951	−0.0053297	0.086821
$\rho_{4,2}$	−0.11823	−0.0468404	0.010382	−0.0225247	−0.0590464	−0.202874
$\rho_{4,3}$	0.26569	0.187291	0.238046	0.469543	0.0317397	−0.109456
σ_{FXR}	0.0965982	0.0887357	0.0438097	0.0507955	0.0832688	0.0635197
γ	4.69634	4.46095	11.431	7.42247	5.35781	6.07881

Notice that the Brownian motions $\mathbb{B}_3(t)$ and $\mathbb{B}_4(t)$ affect the FXR processes only through the drift terms by affecting r_d and r_f.

We focus now on the case $\beta_d = \beta_f = 1$. Denoting by σ_{FXR} the volatility of the process $Y_{d,f}(t)$, i.e., $\sigma_{FXR} = \sqrt{(dY_{d,f}(t)/Y_{d,f}(t))^2/dt}$, from (7.3.33), we have

$$\gamma^2 = \frac{1}{\sigma_{FXR}{}^2}\left(\left(\frac{\mathbf{a}_d}{\sigma_d} + w_d\frac{\gamma-1}{q_{d,1}}\rho_{3,1}\right)^2 - 2\left(\frac{\mathbf{a}_d}{\sigma_d} + w_d\frac{\gamma-1}{q_{d,1}}\rho_{3,1}\right)\right.$$
$$\left.\left(\frac{\mathbf{a}_f}{\sigma_f} + w_f\frac{\gamma-1}{q_{f,1}}\rho_{4,2}\right)\rho_{2,1} + \left(\frac{\mathbf{a}_f}{\sigma_f} + w_f\frac{\gamma-1}{q_{f,1}}\rho_{4,2}\right)^2\right), \tag{7.3.34}$$

and the market risk-aversion is characterized as *the positive solution* of the quadratic equation (7.3.34).

Example 7.3.2. Parameters of the above stochastic interest rate FXR model are estimated on the data for the US dollar (USD) paired with three foreign currencies: Japanese yen (JPY), Canadian dollar (CAD), and British pound (GBP). Two sets of data for each of the currencies was considered: data from 1986 to 2006 and data from 1996 to 2006. The model parameters, as well as the FXR volatilities, are estimated using the methodology described in Sect. 2.5. The risk-aversion parameter γ is then calculated as a positive solution of the quadratic equation (7.3.34). The results are listed in the Table 7.1.

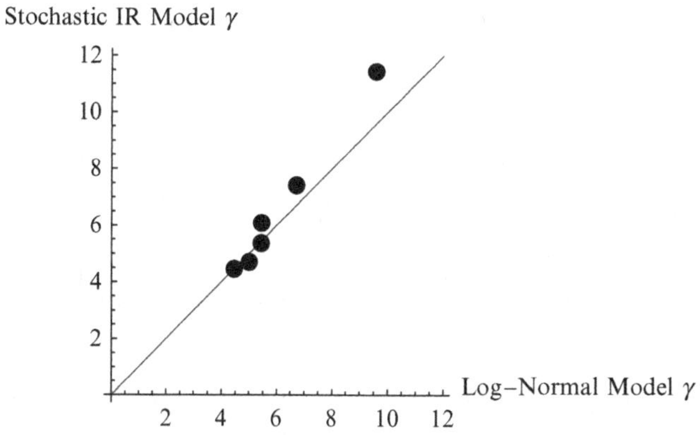

Fig. 7.2 Comparing the risk aversion parameter estimates

Remark 7.3.4. (The approximate model-independence of risk-aversion estimation.) Using the above six data sets, the relative risk-aversion parameter γ was estimated using two methods: by applying formula (7.3.11) and by finding the positive solution of the quadratic equation (7.3.34). The results are plotted in Fig. 7.2.

7.3.3 FXR Under Stochastic Volatility

At least in short-term FX risk management, the interest rate risk, domestic or foreign, may be dwarfed by the market volatility risk. Such a scenario motivates the present study. We consider the case of Heston's stochastic volatility for risky assets in the underlying economies. More precisely, let the four factors be (with the first two being the tradables in the domestic and foreign economies, respectively)

$$\mathrm{d}S_1(t) = S_1(t)\left(\mathbf{a}_\mathrm{d} + \lambda_\mathrm{d}\mathbf{s}_\mathrm{d}(t) - \mathbb{D}_\mathrm{d}\right)\mathrm{d}t + S_1(t)\sqrt{\mathbf{s}_\mathrm{d}(t)}\mathrm{d}\mathbb{B}_1(t), \tag{7.3.35}$$

$$\mathrm{d}S_2(t) = S_2(t)\left(\mathbf{a}_\mathrm{f} + \lambda_f\mathbf{s}_\mathrm{f}(t) - \mathbb{D}_\mathrm{f}\right)\mathrm{d}t + S_2(t)\sqrt{\mathbf{s}_\mathrm{f}(t)}\mathrm{d}\mathbb{B}_2(t),$$

$$\mathrm{d}\mathbf{s}_\mathrm{d}(t) = \left(q_{d,0} + q_{d,1}\mathbf{s}_\mathrm{d}(t)\right)\mathrm{d}t + w_\mathrm{d}\sqrt{\mathbf{s}_\mathrm{d}(t)}\mathrm{d}\mathbb{B}_3(t), \tag{7.3.36}$$

$$\mathrm{d}\mathbf{s}_\mathrm{f}(t) = \left(q_{f,0} + q_{f,1}\mathbf{s}_\mathrm{f}(t)\right)\mathrm{d}t + w_\mathrm{f}\sqrt{\mathbf{s}_\mathrm{f}(t)}\mathrm{d}\mathbb{B}_4(t) \tag{7.3.37}$$

As before, $\mathbb{B}_j(t), j = 1,\ldots,4$, are dependent Brownian motions with correlations $\rho_{j,i}$, $1 \le i \le j \le 4$. Therefore, the market coefficients are

$$a_{s,\mathrm{d}} = \{\mathbf{a}_\mathrm{d} + \lambda_\mathrm{d}\mathbf{s}_\mathrm{d}\}, a_{s,\mathrm{f}} = \left\{\mathbf{a}_\mathrm{f} + \lambda_f\mathbf{s}_\mathrm{f}\right\},$$

$$\sigma_{s,\mathrm{d}} = \left(\sqrt{\mathbf{s}_\mathrm{d}}\ 0\ 0\ 0\right), \sigma_{s,\mathrm{f}} = \left(\sqrt{\mathbf{s}_\mathrm{f}}\rho_{2,1}\ \sqrt{\mathbf{s}_\mathrm{f}}\sqrt{1-\rho_{2,1}^2}\ 0\ 0\right), \tag{7.3.38}$$

$$b=\left\{S_1\left(\lambda_{\mathrm{d}}\mathbf{s}_{\mathrm{d}}+\mathbf{a}_{\mathrm{d}}-\mathbb{D}_{\mathrm{d}}\right),S_2\left(\lambda_f\mathbf{s}_{\mathrm{f}}+\mathbf{a}_{\mathrm{f}}-\mathbb{D}_{\mathrm{f}}\right),q_{d,0}+q_{d,1}\mathbf{s}_{\mathrm{d}},q_{f,0}+q_{f,1}\mathbf{s}_{\mathrm{f}}\right\}, \quad (7.3.39)$$

$$c=\begin{pmatrix} S_1\sqrt{\mathbf{s}_{\mathrm{d}}} & 0 & 0 & 0 \\ S_2\sqrt{\mathbf{s}_{\mathrm{f}}}\rho_{2,1} & S_2\sqrt{\mathbf{s}_{\mathrm{f}}}\sqrt{1-\rho_{2,1}^2} & 0 & 0 \\ \sqrt{\mathbf{s}_{\mathrm{d}}}w_{\mathrm{d}}\rho_{3,1} & \sqrt{\mathbf{s}_{\mathrm{d}}}w_{\mathrm{d}}\frac{-\rho_{2,1}\rho_{3,1}+\rho_{3,2}}{\sqrt{1-\rho_{2,1}^2}} & \sqrt{\mathbf{s}_{\mathrm{d}}}w_{\mathrm{d}}\mathbf{s}_{3,3} & 0 \\ \sqrt{\mathbf{s}_{\mathrm{f}}}w_{\mathrm{f}}\rho_{4,1} & \sqrt{\mathbf{s}_{\mathrm{f}}}w_{\mathrm{f}}\frac{-\rho_{2,1}\rho_{4,1}+\rho_{4,2}}{\sqrt{1-\rho_{2,1}^2}} & \sqrt{\mathbf{s}_{\mathrm{f}}}w_{\mathrm{f}}\mathbf{s}_{4,3} & \sqrt{\mathbf{s}_{\mathrm{f}}}w_{\mathrm{f}}\mathbf{s}_{4,4} \end{pmatrix}, \quad (7.3.40)$$

with $\mathbf{s}_{3,3}$, $\mathbf{s}_{4,3}$, and $\mathbf{s}_{4,4}$ given in (A.2.27), (A.2.30), and (A.2.31), respectively.

As can be confirmed by direct calculation here, from (3.10.20), we conclude that in the case of CRRA utility of wealth with relative risk-aversion $\gamma_{\mathrm{d}}=\gamma_{\mathrm{f}}=\gamma$,

$$\frac{\Pi_{\mathrm{d}}^{\star}\left(t,Y_{\mathrm{d,f}}X_{\mathrm{f}},A\right)}{Y_{\mathrm{d,f}}X_{\mathrm{f}}}=\left\{\frac{1}{\gamma}\left(\frac{\mathbf{a}_{\mathrm{d}}-r_{\mathrm{d}}}{\mathbf{s}_{\mathrm{d}}}+\lambda_d+w_{\mathrm{d}}\rho_{3,1}\left(\mathbb{H}_{d,0}+\frac{\mathbb{H}_{d,-1}}{\mathbf{s}_{\mathrm{d}}}\right)\right)\right\}, \quad (7.3.41)$$

with

$$\begin{aligned}&\mathbb{H}_{d,-1}^2\frac{w_{\mathrm{d}}^2}{2}\left(1-\frac{\gamma-1}{\gamma}\rho_{3,1}^2\right)-\mathbb{H}_{d,-1}\left(\frac{w_{\mathrm{d}}^2}{2}+\frac{\gamma-1}{\gamma}w_{\mathrm{d}}\left(\mathbf{a}_{\mathrm{d}}-r_{\mathrm{d}}\right)\rho_{3,1}-q_{d,0}\right)\\&-\frac{1}{2}\frac{\gamma-1}{\gamma}\left(\mathbf{a}_{\mathrm{d}}-r_{\mathrm{d}}\right)^2=0,\\&\mathbb{H}_{d,0}^2\frac{w_{\mathrm{d}}^2}{2}\left(1-\frac{\gamma-1}{\gamma}\rho_{3,1}^2\right)-\mathbb{H}_{d,0}\left(-q_{d,1}+\frac{\gamma-1}{\gamma}w_{\mathrm{d}}\rho_{3,1}\lambda_d\right)-\frac{1}{2}\frac{\gamma-1}{\gamma}\lambda_d^2=0,\end{aligned} \quad (7.3.42)$$

and

$$\frac{\Pi_{\mathrm{f}}^{\star}\left(t,X_{\mathrm{f}},A\right)}{X_{\mathrm{f}}}=\left\{\frac{1}{\gamma}\left(\frac{\mathbf{a}_{\mathrm{f}}-r_{\mathrm{f}}}{\mathbf{s}_{\mathrm{f}}}+\lambda_f+w_{\mathrm{f}}\rho_{4,2}\left(\mathbb{H}_{f,0}+\frac{\mathbb{H}_{f,-1}}{\mathbf{s}_{\mathrm{f}}}\right)\right)\right\}, \quad (7.3.43)$$

with

$$\begin{aligned}&\mathbb{H}_{f,-1}^2\frac{w_{\mathrm{f}}^2}{2}\left(1-\frac{\gamma-1}{\gamma}\rho_{4,2}^2\right)-\mathbb{H}_{f,-1}\left(\frac{w_{\mathrm{f}}^2}{2}+\frac{\gamma-1}{\gamma}w_{\mathrm{f}}\left(\mathbf{a}_{\mathrm{f}}-r_{\mathrm{f}}\right)\rho_{4,2}-q_{f,0}\right)\\&-\frac{1}{2}\frac{\gamma-1}{\gamma}\left(\mathbf{a}_{\mathrm{f}}-r_{\mathrm{f}}\right)^2=0,\\&\mathbb{H}_{f,0}^2\frac{w_{\mathrm{f}}^2}{2}\left(1-\frac{\gamma-1}{\gamma}\rho_{4,2}^2\right)-\mathbb{H}_{f,0}\left(-q_{f,1}+\frac{\gamma-1}{\gamma}w_{\mathrm{f}}\rho_{4,2}\lambda_f\right)-\frac{1}{2}\frac{\gamma-1}{\gamma}\lambda_f^2=0.\end{aligned} \quad (7.3.44)$$

Therefore, the FXR SDE (7.2.24) becomes

$$\begin{aligned}\frac{\mathrm{d}Y_{\mathrm{d,f}}(t)}{Y_{\mathrm{d,f}}(t)} = &\left(r_{\mathrm{d}} - r_{\mathrm{f}} + \left\{\frac{1}{\gamma}\left(\frac{\mathbf{a}_{\mathrm{d}} - r_{\mathrm{d}}}{\mathbf{s}_{\mathrm{d}}} + \lambda_d + w_{\mathrm{d}}\rho_{3,1}\left(\mathbb{H}_{d,0} + \frac{\mathbb{H}_{d,-1}}{\mathbf{s}_{\mathrm{d}}}\right)\right)\right\}\right.\\
&\cdot (a_{s,\mathrm{d}} - r_{\mathrm{d}}) - \left\{\frac{1}{\gamma}\left(\frac{\mathbf{a}_{\mathrm{f}} - r_{\mathrm{f}}}{\mathbf{s}_{\mathrm{f}}} + \lambda_f + w_{\mathrm{f}}\rho_{4,2}\left(\mathbb{H}_{f,0} + \frac{\mathbb{H}_{f,-1}}{\mathbf{s}_{\mathrm{f}}}\right)\right)\right\}\\
&\cdot (a_{s,\mathrm{f}} - r_{\mathrm{f}}) - \left\{\frac{1}{\gamma}\left(\frac{\mathbf{a}_{\mathrm{f}} - r_{\mathrm{f}}}{\mathbf{s}_{\mathrm{f}}} + \lambda_f + w_{\mathrm{f}}\rho_{4,2}\left(\mathbb{H}_{f,0} + \frac{\mathbb{H}_{f,-1}}{\mathbf{s}_{\mathrm{f}}}\right)\right)\right\}\\
&\cdot \sigma_{s,\mathrm{f}} \cdot {\sigma_{s,\mathrm{d}}}^{\mathrm{T}} \cdot \left\{\frac{1}{\gamma}\left(\frac{\mathbf{a}_{\mathrm{d}} - r_{\mathrm{d}}}{\mathbf{s}_{\mathrm{d}}} + \lambda_d + w_{\mathrm{d}}\rho_{3,1}\left(\mathbb{H}_{d,0} + \frac{\mathbb{H}_{d,-1}}{\mathbf{s}_{\mathrm{d}}}\right)\right)\right\}\\
&+ \left\{\frac{1}{\gamma}\left(\frac{\mathbf{a}_{\mathrm{f}} - r_{\mathrm{f}}}{\mathbf{s}_{\mathrm{f}}} + \lambda_f + w_{\mathrm{f}}\rho_{4,2}\left(\mathbb{H}_{f,0} + \frac{\mathbb{H}_{f,-1}}{\mathbf{s}_{\mathrm{f}}}\right)\right)\right\} \cdot \sigma_{s,\mathrm{f}} \cdot {\sigma_{s,\mathrm{f}}}^{\mathrm{T}}\\
&\left.\cdot \left\{\frac{1}{\gamma}\left(\frac{\mathbf{a}_{\mathrm{f}} - r_{\mathrm{f}}}{\mathbf{s}_{\mathrm{f}}} + \lambda_f + w_{\mathrm{f}}\rho_{4,2}\left(\mathbb{H}_{f,0} + \frac{\mathbb{H}_{f,-1}}{\mathbf{s}_{\mathrm{f}}}\right)\right)\right\}\right) \mathrm{d}t\\
&+ \left(\left\{\frac{1}{\gamma}\left(\frac{\mathbf{a}_{\mathrm{d}} - r_{\mathrm{d}}}{\mathbf{s}_{\mathrm{d}}} + \lambda_d + w_{\mathrm{d}}\rho_{3,1}\left(\mathbb{H}_{d,0} + \frac{\mathbb{H}_{d,-1}}{\mathbf{s}_{\mathrm{d}}}\right)\right)\right\} \cdot \sigma_{s,\mathrm{d}}\right.\\
&\left. - \left\{\frac{1}{\gamma}\left(\frac{\mathbf{a}_{\mathrm{f}} - r_{\mathrm{f}}}{\mathbf{s}_{\mathrm{f}}} + \lambda_f + w_{\mathrm{f}}\rho_{4,2}\left(\mathbb{H}_{f,0} + \frac{\mathbb{H}_{f,-1}}{\mathbf{s}_{\mathrm{f}}}\right)\right)\right\} \cdot \sigma_{s,\mathrm{f}}\right) \cdot \mathrm{d}B(t),\end{aligned} \tag{7.3.45}$$

and since

$$\begin{aligned}a_{s,\mathrm{d}} - r_{\mathrm{d}} &= \{\mathbf{a}_{\mathrm{d}} + \lambda_{\mathrm{d}}\mathbf{s}_{\mathrm{d}}\} - r_{\mathrm{d}} = \{\mathbf{a}_{\mathrm{d}} + \lambda_{\mathrm{d}}\mathbf{s}_{\mathrm{d}} - r_{\mathrm{d}}\},\\
a_{s,\mathrm{f}} - r_{\mathrm{f}} &= \{\mathbf{a}_{\mathrm{f}} + \lambda_f\mathbf{s}_{\mathrm{f}}\} - r_{\mathrm{f}} = \{\mathbf{a}_{\mathrm{f}} + \lambda_f\mathbf{s}_{\mathrm{f}} - r_{\mathrm{f}}\},\\
\sigma_{s,\mathrm{f}} \cdot {\sigma_{s,\mathrm{d}}}^{\mathrm{T}} &= \left(\sqrt{\mathbf{s}_{\mathrm{f}}}\rho_{2,1}\ \sqrt{\mathbf{s}_{\mathrm{f}}}\sqrt{1-\rho_{2,1}^2}\ 0\ 0\right) \cdot \left(\sqrt{\mathbf{s}_{\mathrm{d}}}\ 0\ 0\ 0\right)^{\mathrm{T}} = \left(\sqrt{\mathbf{s}_{\mathrm{d}}}\sqrt{\mathbf{s}_{\mathrm{f}}}\rho_{2,1}\right),\\
\sigma_{s,\mathrm{f}} \cdot {\sigma_{s,\mathrm{f}}}^{\mathrm{T}} &= \left(\sqrt{\mathbf{s}_{\mathrm{f}}}\rho_{2,1}\ \sqrt{\mathbf{s}_{\mathrm{f}}}\sqrt{1-\rho_{2,1}^2}\ 0\ 0\right)\\
&\quad \cdot \left(\sqrt{\mathbf{s}_{\mathrm{f}}}\rho_{2,1}\ \sqrt{\mathbf{s}_{\mathrm{f}}}\sqrt{1-\rho_{2,1}^2}\ 0\ 0\right)^{\mathrm{T}} = \left(\mathbf{s}_{\mathrm{f}}\right),\end{aligned} \tag{7.3.46}$$

we conclude that

$$\begin{aligned}\frac{\mathrm{d}Y_{\mathrm{d,f}}(t)}{Y_{\mathrm{d,f}}(t)} = &\left(r_{\mathrm{d}} - r_{\mathrm{f}} + \frac{1}{\gamma}\left(\frac{\mathbf{a}_{\mathrm{d}} - r_{\mathrm{d}}}{\mathbf{s}_{\mathrm{d}}} + \lambda_d + w_{\mathrm{d}}\rho_{3,1}\left(\mathbb{H}_{d,0} + \frac{\mathbb{H}_{d,-1}}{\mathbf{s}_{\mathrm{d}}}\right)\right)\right.\\
&(\mathbf{a}_{\mathrm{d}} + \lambda_{\mathrm{d}}\mathbf{s}_{\mathrm{d}} - r_{\mathrm{d}}) - \frac{1}{\gamma}\left(\frac{\mathbf{a}_{\mathrm{f}} - r_{\mathrm{f}}}{\mathbf{s}_{\mathrm{f}}} + \lambda_f + w_{\mathrm{f}}\rho_{4,2}\left(\mathbb{H}_{f,0} + \frac{\mathbb{H}_{f,-1}}{\mathbf{s}_{\mathrm{f}}}\right)\right)\end{aligned}$$

$$
\begin{aligned}
&(\mathbf{a}_{\mathrm{f}}+\lambda_f \mathbf{s}_{\mathrm{f}}-r_{\mathrm{f}})-\frac{\sqrt{\mathbf{s}_{\mathrm{d}}}\sqrt{\mathbf{s}_{\mathrm{f}}}\rho_{2,1}}{\gamma^2}\left(\frac{\mathbf{a}_{\mathrm{f}}-r_{\mathrm{f}}}{\mathbf{s}_{\mathrm{f}}}+\lambda_f+w_{\mathrm{f}}\rho_{4,2}\left(\mathbb{H}_{f,0}+\frac{\mathbb{H}_{f,-1}}{\mathbf{s}_{\mathrm{f}}}\right)\right)\\
&\left(\frac{\mathbf{a}_{\mathrm{d}}-r_{\mathrm{d}}}{\mathbf{s}_{\mathrm{d}}}+\lambda_d+w_{\mathrm{d}}\rho_{3,1}\left(\mathbb{H}_{d,0}+\frac{\mathbb{H}_{d,-1}}{\mathbf{s}_{\mathrm{d}}}\right)\right)+\frac{\mathbf{s}_{\mathrm{f}}}{\gamma^2}\left(\frac{\mathbf{a}_{\mathrm{f}}-r_{\mathrm{f}}}{\mathbf{s}_{\mathrm{f}}}+\lambda_f+w_{\mathrm{f}}\rho_{4,2}\right.\\
&\left.\left(\mathbb{H}_{f,0}+\frac{\mathbb{H}_{f,-1}}{\mathbf{s}_{\mathrm{f}}}\right)\right)^2\Bigg)\,\mathrm{d}t+\Bigg(\left\{\frac{\sqrt{\mathbf{s}_{\mathrm{d}}}}{\gamma}\left(\frac{\mathbf{a}_{\mathrm{d}}-r_{\mathrm{d}}}{\mathbf{s}_{\mathrm{d}}}+\lambda_d+w_{\mathrm{d}}\rho_{3,1}\right.\right.\\
&\left.\left.\left(\mathbb{H}_{d,0}+\frac{\mathbb{H}_{d,-1}}{\mathbf{s}_{\mathrm{d}}}\right)\right),0,0,0\right\}-\left\{\frac{\sqrt{\mathbf{s}_{\mathrm{f}}}\rho_{2,1}}{\gamma}\left(\frac{\mathbf{a}_{\mathrm{f}}-r_{\mathrm{f}}}{\mathbf{s}_{\mathrm{f}}}+\lambda_f\right.\right.\\
&\left.+w_{\mathrm{f}}\rho_{4,2}\left(\mathbb{H}_{f,0}+\frac{\mathbb{H}_{f,-1}}{\mathbf{s}_{\mathrm{f}}}\right)\right),\frac{\sqrt{\mathbf{s}_{\mathrm{f}}}\sqrt{1-\rho_{2,1}^2}}{\gamma}\\
&\left.\left(\frac{\mathbf{a}_{\mathrm{f}}-r_{\mathrm{f}}}{\mathbf{s}_{\mathrm{f}}}+\lambda_f+w_{\mathrm{f}}\rho_{4,2}\left(\mathbb{H}_{f,0}+\frac{\mathbb{H}_{f,-1}}{\mathbf{s}_{\mathrm{f}}}\right)\right),0,0\right\}\Bigg)\cdot\mathrm{d}B(t),
\end{aligned}
\tag{7.3.47}
$$

i.e.,

$$
\begin{aligned}
\frac{\mathrm{d}Y_{\mathrm{d,f}}(t)}{Y_{\mathrm{d,f}}(t)}=&\Bigg(r_{\mathrm{d}}-r_{\mathrm{f}}+\frac{1}{\gamma}\left(\frac{\mathbf{a}_{\mathrm{d}}-r_{\mathrm{d}}}{\mathbf{s}_{\mathrm{d}}}+\lambda_d+w_{\mathrm{d}}\rho_{3,1}\left(\mathbb{H}_{d,0}+\frac{\mathbb{H}_{d,-1}}{\mathbf{s}_{\mathrm{d}}}\right)\right)\\
&(\mathbf{a}_{\mathrm{d}}+\lambda_{\mathrm{d}}\mathbf{s}_{\mathrm{d}}-r_{\mathrm{d}})-\frac{1}{\gamma}\left(\frac{\mathbf{a}_{\mathrm{f}}-r_{\mathrm{f}}}{\mathbf{s}_{\mathrm{f}}}+\lambda_f+w_{\mathrm{f}}\rho_{4,2}\left(\mathbb{H}_{f,0}+\frac{\mathbb{H}_{f,-1}}{\mathbf{s}_{\mathrm{f}}}\right)\right)\\
&(\mathbf{a}_{\mathrm{f}}+\lambda_f\mathbf{s}_{\mathrm{f}}-r_{\mathrm{f}})-\frac{\sqrt{\mathbf{s}_{\mathrm{d}}}\sqrt{\mathbf{s}_{\mathrm{f}}}\rho_{2,1}}{\gamma^2}\left(\frac{\mathbf{a}_{\mathrm{f}}-r_{\mathrm{f}}}{\mathbf{s}_{\mathrm{f}}}+\lambda_f+w_{\mathrm{f}}\rho_{4,2}\left(\mathbb{H}_{f,0}+\frac{\mathbb{H}_{f,-1}}{\mathbf{s}_{\mathrm{f}}}\right)\right)\\
&\left(\frac{\mathbf{a}_{\mathrm{d}}-r_{\mathrm{d}}}{\mathbf{s}_{\mathrm{d}}}+\lambda_d+w_{\mathrm{d}}\rho_{3,1}\left(\mathbb{H}_{d,0}+\frac{\mathbb{H}_{d,-1}}{\mathbf{s}_{\mathrm{d}}}\right)\right)+\frac{\mathbf{s}_{\mathrm{f}}}{\gamma^2}\left(\frac{\mathbf{a}_{\mathrm{f}}-r_{\mathrm{f}}}{\mathbf{s}_{\mathrm{f}}}+\lambda_f+w_{\mathrm{f}}\rho_{4,2}\right.\\
&\left.\left(\mathbb{H}_{f,0}+\frac{\mathbb{H}_{f,-1}}{\mathbf{s}_{\mathrm{f}}}\right)\right)^2\Bigg)\,\mathrm{d}t+\frac{\sqrt{\mathbf{s}_{\mathrm{d}}}}{\gamma}\left(\frac{\mathbf{a}_{\mathrm{d}}-r_{\mathrm{d}}}{\mathbf{s}_{\mathrm{d}}}+\lambda_d+w_{\mathrm{d}}\rho_{3,1}\left(\mathbb{H}_{d,0}\right.\right.\\
&\left.\left.+\frac{\mathbb{H}_{d,-1}}{\mathbf{s}_{\mathrm{d}}}\right)\right)\mathrm{d}B_1(t)-\frac{\sqrt{\mathbf{s}_{\mathrm{f}}}}{\gamma}\left(\frac{\mathbf{a}_{\mathrm{f}}-r_{\mathrm{f}}}{\mathbf{s}_{\mathrm{f}}}+\lambda_f\right.\\
&\left.+w_{\mathrm{f}}\rho_{4,2}\left(\mathbb{H}_{f,0}+\frac{\mathbb{H}_{f,-1}}{\mathbf{s}_{\mathrm{f}}}\right)\right)\mathrm{d}\mathbb{B}_2(t),
\end{aligned}
\tag{7.3.48}
$$

coupled with (7.3.36)–(7.3.37). If $\mathbf{a}_{\mathrm{d}} = r_{\mathrm{d}}$ and $\mathbf{a}_{\mathrm{f}} = r_{\mathrm{f}}$, and consequently $\mathbb{H}_{d,-1} = \mathbb{H}_{f,-1} = 0$, then

$$
\begin{aligned}
\frac{\mathrm{d}Y_{\mathrm{d,f}}(t)}{Y_{\mathrm{d,f}}(t)}=&\Bigg(r_{\mathrm{d}}-r_{\mathrm{f}}+\frac{1}{\gamma}\left(\lambda_d+w_{\mathrm{d}}\rho_{3,1}\mathbb{H}_{d,0}\right)\lambda_{\mathrm{d}}\mathbf{s}_{\mathrm{d}}-\frac{1}{\gamma}\left(\lambda_f+w_{\mathrm{f}}\rho_{4,2}\mathbb{H}_{f,0}\right)\lambda_f\mathbf{s}_{\mathrm{f}}\\
&-\frac{\sqrt{\mathbf{s}_{\mathrm{d}}}\sqrt{\mathbf{s}_{\mathrm{f}}}\rho_{2,1}}{\gamma^2}\left(\lambda_f+w_{\mathrm{f}}\rho_{4,2}\mathbb{H}_{f,0}\right)\left(\lambda_d+w_{\mathrm{d}}\rho_{3,1}\mathbb{H}_{d,0}\right)
\end{aligned}
$$

$$+\frac{\mathrm{s_f}}{\gamma^2}\left(\lambda_f + w_\mathrm{f}\rho_{4,2}\mathbb{H}_{f,0}\right)^2\Big)\,\mathrm{d}t + \frac{\sqrt{\mathrm{s_d}}}{\gamma}\left(\lambda_d + w_\mathrm{d}\rho_{3,1}\mathbb{H}_{d,0}\right)\mathrm{d}B_1(t)$$

$$-\frac{\sqrt{\mathrm{s_f}}}{\gamma}\left(\lambda_f + w_\mathrm{f}\rho_{4,2}\mathbb{H}_{f,0}\right)\mathrm{d}\mathbb{B}_2(t), \tag{7.3.49}$$

coupled with (7.3.36)–(7.3.37). If, additionally, no foreign risky investments are considered, that is, by setting

$$0 = \frac{\Pi^\star_\mathrm{f}(t, X_\mathrm{f}, A)}{X_\mathrm{f}} = \left\{\frac{1}{\gamma}\left(\lambda_f + w_\mathrm{f}\rho_{4,2}\mathbb{H}_{f,0}\right)\right\}, \tag{7.3.50}$$

we get

$$\frac{\mathrm{d}Y_\mathrm{d,f}(t)}{Y_\mathrm{d,f}(t)} = \left(r_\mathrm{d} - r_\mathrm{f} + \frac{1}{\gamma}\left(\lambda_d + w_\mathrm{d}\rho_{3,1}\mathbb{H}_{d,0}\right)\lambda_\mathrm{d}\mathrm{s_d}\right)\mathrm{d}t$$

$$+\frac{\sqrt{\mathrm{s_d}}}{\gamma}\left(\lambda_d + w_\mathrm{d}\rho_{3,1}\mathbb{H}_{d,0}\right)\mathrm{d}B_1(t), \tag{7.3.51}$$

coupled with (7.3.36), which, finally, is a Heston-like FXR model.

7.4 FX Derivatives

Having developed the underlying FXR models, and having developed the general theory of neutral and indifference pricing and hedging, one would expect that the pricing and hedging of FX derivatives is just an application of the two. To some extent that is true, except for one detail.

Indeed, in the FX derivatives literature, following Garman and Kohlhagen [11], the foreign currency "dividend" is set equal to the foreign interest rate r_d. Equivalently, it is assumed that the foreign currency is used solely for foreign money-market investments. This is certainly a legitimate point of view: by holding, say, a *call option* contract, instead of owning the currency, an investor is forfeiting (at least) the interest to be accumulated in the foreign bank.

Our results will include such a point of view. On the other hand, such a point of view will be only a special case of the presented results, since our main result is broader and includes also a possibility of considering risky foreign assets a generator of a foreign currency "dividend," and henceforth also one of the drivers for the FXR SDE dynamics (as discussed above).

Recall that throughout (see Remark 2.2.1; see also (2.2.3), (4.5.4), etc.) our discussions it has been important to distinguish between predividend and postdividend underlying dynamics. In that context, observe that all the FXR SDEs derived above should be considered postdividend dynamics, and therefore, to apply the general pricing theory, we also need to identify the predividend FXR dynamics.

Foreign currency when invested in foreign risky assets and foreign money markets yields gains or losses according to the wealth evolution SDE ($X_f(t)$ is expressed in foreign currency)

$$\begin{aligned}\frac{dX_f(t)}{X_f(t)} &= \left(\frac{\Pi_f^\star(t,X_f(t),A(t))}{X_f(t)}\cdot(a_{s,f}(t,A(t))-r_f(t,A(t)))+r_f(t,A(t))\right)dt\\&\quad+\frac{\Pi_f^\star(t,X_f(t),A(t))}{X_f(t)}\cdot\sigma_{s,f}(t,A(t))\cdot dB(t)\\&=\left(r_f+\frac{\Pi_f^\star(t,X_f(t),A(t))}{X_f(t)}\cdot(a_{s,f}-r_f+\sigma_{s,f}\cdot W(t))\right)dt,\end{aligned}\tag{7.4.1}$$

where $W(t)$ is white noise (recall Fig. 1.1), i.e., a measure-valued stochastic process such that $dB(t)=W(t)dt$. Therefore,

$$dX_f(t)=\mathbb{D}_{fx}X_f(t)dt,\tag{7.4.2}$$

where $\mathbb{D}_{fx}$ is the *"dividend" rate on the foreign currency*, given by

$$\mathbb{D}_{fx}=r_f+\frac{\Pi_f^\star(t,X_f(t),A(t))}{X_f(t)}\cdot(a_{s,f}-r_f+\sigma_{s,f}\cdot W(t)).\tag{7.4.3}$$

Of course, if no foreign risky assets are considered for investing, then simply by setting $\Pi_f^\star=0$, we get

$$\mathbb{D}_{fx}=r_f,\tag{7.4.4}$$

as in Garman and Kohlhagen [11] and the usual FX derivative literature.

We now proceed to establish the predividend FXR dynamics. Starting from the postdividend FXR dynamics (7.2.24), using (7.4.3), and abusing notation somewhat by using the same notation $dY_{d,f}(t)/Y_{d,f}(t)$ for pre- and postdividend dynamics, we have (the predividend FXR dynamics)

$$\begin{aligned}\frac{dY_{d,f}(t)}{Y_{d,f}(t)} &= \Bigg(\mathbb{D}_{fx}+r_d-r_f+\frac{\Pi_d^\star(t,Y_{d,f}(t)X_f(t),A(t))}{Y_{d,f}(t)X_f(t)}\cdot(a_{s,d}-r_d)-\frac{\Pi_f^\star(t,X_f(t),A(t))}{X_f(t)}\cdot\\&\quad(a_{s,f}-r_f)-\frac{\Pi_f^\star(t,X_f(t),A(t))}{X_f(t)}\cdot\sigma_{s,f}\cdot{\sigma_{s,d}}^T\cdot\frac{\Pi_d^\star(t,Y_{d,f}(t)X_f(t),A(t))}{Y_{d,f}(t)X_f(t)}\\&\quad+\frac{\Pi_f^\star(t,X_f(t),A(t))}{X_f(t)}\cdot\sigma_{s,f}\cdot{\sigma_{s,f}}^T\cdot\frac{\Pi_f^\star(t,X_f(t),A(t))}{X_f(t)}\Bigg)dt\\&\quad+\left(\frac{\Pi_d^\star(t,Y_{d,f}(t)X_f(t),A(t))}{Y_{d,f}(t)X_f(t)}\cdot\sigma_{s,d}-\frac{\Pi_f^\star(t,X_f(t),A(t))}{X_f(t)}\cdot\sigma_{s,f}\right)\cdot dB(t)\end{aligned}$$

$$= \Bigg(r_{\mathrm{f}} + \frac{\Pi_{\mathrm{f}}^{\star}(t, X_{\mathrm{f}}(t), A(t))}{X_{\mathrm{f}}(t)} \cdot (a_{s,\mathrm{f}} - r_{\mathrm{f}} + \sigma_{s,\mathrm{f}} \cdot W(t)) + r_{\mathrm{d}} - r_{\mathrm{f}}$$

$$+ \frac{\Pi_{\mathrm{d}}^{\star}(t, Y_{\mathrm{d,f}}(t)X_{\mathrm{f}}(t), A(t))}{Y_{\mathrm{d,f}}(t)X_{\mathrm{f}}(t)} \cdot (a_{s,\mathrm{d}} - r_{\mathrm{d}}) - \frac{\Pi_{\mathrm{f}}^{\star}(t, X_{\mathrm{f}}(t), A(t))}{X_{\mathrm{f}}(t)} \cdot (a_{s,\mathrm{f}} - r_{\mathrm{f}})$$

$$- \frac{\Pi_{\mathrm{f}}^{\star}(t, X_{\mathrm{f}}(t), A(t))}{X_{\mathrm{f}}(t)} \cdot \sigma_{s,\mathrm{f}} \cdot \sigma_{s,\mathrm{d}}^{\mathrm{T}} \cdot \frac{\Pi_{\mathrm{d}}^{\star}(t, Y_{\mathrm{d,f}}(t)X_{\mathrm{f}}(t), A(t))}{Y_{\mathrm{d,f}}(t)X_{\mathrm{f}}(t)}$$

$$+ \frac{\Pi_{\mathrm{f}}^{\star}(t, X_{\mathrm{f}}(t), A(t))}{X_{\mathrm{f}}(t)} \cdot \sigma_{s,\mathrm{f}} \cdot \sigma_{s,\mathrm{f}}^{\mathrm{T}} \cdot \frac{\Pi_{\mathrm{f}}^{\star}(t, X_{\mathrm{f}}(t), A(t))}{X_{\mathrm{f}}(t)} \Bigg) \mathrm{d}t$$

$$+ \left(\frac{\Pi_{\mathrm{d}}^{\star}(t, Y_{\mathrm{d,f}}(t)X_{\mathrm{f}}(t), A(t))}{Y_{\mathrm{d,f}}(t)X_{\mathrm{f}}(t)} \cdot \sigma_{s,\mathrm{d}} - \frac{\Pi_{\mathrm{f}}^{\star}(t, X_{\mathrm{f}}(t), A(t))}{X_{\mathrm{f}}(t)} \cdot \sigma_{s,\mathrm{f}} \right) \cdot \mathrm{d}B(t), \tag{7.4.5}$$

yielding the general predividend FXR dynamics

$$\frac{\mathrm{d}Y_{\mathrm{d,f}}(t)}{Y_{\mathrm{d,f}}(t)} = \Bigg(r_{\mathrm{d}} + \frac{\Pi_{\mathrm{d}}^{\star}(t, Y_{\mathrm{d,f}}(t)X_{\mathrm{f}}(t), A(t))}{Y_{\mathrm{d,f}}(t)X_{\mathrm{f}}(t)} \cdot (a_{s,\mathrm{d}} - r_{\mathrm{d}}) - \frac{\Pi_{\mathrm{f}}^{\star}(t, X_{\mathrm{f}}(t), A(t))}{X_{\mathrm{f}}(t)}$$

$$\cdot \sigma_{s,\mathrm{f}} \cdot \sigma_{s,\mathrm{d}}^{\mathrm{T}} \cdot \frac{\Pi_{\mathrm{d}}^{\star}(t, Y_{\mathrm{d,f}}(t)X_{\mathrm{f}}(t), A(t))}{Y_{\mathrm{d,f}}(t)X_{\mathrm{f}}(t)} + \frac{\Pi_{\mathrm{f}}^{\star}(t, X_{\mathrm{f}}(t), A(t))}{X_{\mathrm{f}}(t)} \cdot \sigma_{s,\mathrm{f}}$$

$$\cdot \sigma_{s,\mathrm{f}}^{\mathrm{T}} \cdot \frac{\Pi_{\mathrm{f}}^{\star}(t, X_{\mathrm{f}}(t), A(t))}{X_{\mathrm{f}}(t)} \Bigg) \mathrm{d}t + \frac{\Pi_{\mathrm{d}}^{\star}(t, Y_{\mathrm{d,f}}(t)X_{\mathrm{f}}(t), A(t))}{Y_{\mathrm{d,f}}(t)X_{\mathrm{f}}(t)} \cdot \sigma_{s,\mathrm{d}} \cdot \mathrm{d}B(t). \tag{7.4.6}$$

On the other hand, by setting $\Pi_{\mathrm{f}}^{\star}(t, X_{\mathrm{f}}(t), A(t)) = 0$ (corresponding to the economic scenario in which no foreign risky investments are deemed relevant for consideration), we obtain the postdividend FXR dynamics (7.2.25) and the predividend FXR dynamics

$$\frac{\mathrm{d}Y_{\mathrm{d,f}}(t)}{Y_{\mathrm{d,f}}(t)} = \left(r_{\mathrm{d}} + \frac{\Pi_{\mathrm{d}}^{\star}(t, Y_{\mathrm{d,f}}(t)X_{\mathrm{f}}(t), A(t))}{Y_{\mathrm{d,f}}(t)X_{\mathrm{f}}(t)} \cdot (a_{s,\mathrm{d}} - r_{\mathrm{d}}) \right) \mathrm{d}t$$

$$+ \frac{\Pi_{\mathrm{d}}^{\star}(t, Y_{\mathrm{d,f}}(t)X_{\mathrm{f}}(t), A(t))}{Y_{\mathrm{d,f}}(t)X_{\mathrm{f}}(t)} \cdot \sigma_{s,\mathrm{d}} \cdot \mathrm{d}B(t), \tag{7.4.7}$$

which together will amount to an analogue of the Garman–Kohlhagen framework.

Definition 7.4.1. Equations (7.2.24) and (7.4.6), possibly together with some other equations to close the system, will be referred to as the extended Garman–Kohlhagen framework for pricing FX derivatives, while equations (7.2.25) and (7.4.7), possibly together with some other equations to obtain a closed system, will be referred to as the Garman–Kohlhagen framework for pricing FX derivatives.

Definition 7.4.2 (FXR market coefficients for the extended Garman–Kohlhagen framework).
The predividend FXR appreciation rate:

$$
\begin{aligned}
a_{\mathrm{FX}} = r_{\mathrm{d}} &+ \frac{\Pi_{\mathrm{d}}^{\star}(t, Y_{\mathrm{d,f}}(t)X_{\mathrm{f}}(t), A(t))}{Y_{\mathrm{d,f}}(t)X_{\mathrm{f}}(t)} \cdot (a_{s,\mathrm{d}} - r_{\mathrm{d}}) \\
&- \frac{\Pi_{\mathrm{f}}^{\star}(t, X_{\mathrm{f}}(t), A(t))}{X_{\mathrm{f}}(t)} \cdot \sigma_{s,\mathrm{f}} \cdot \sigma_{s,\mathrm{d}}^{\mathrm{T}} \cdot \frac{\Pi_{\mathrm{d}}^{\star}(t, Y_{\mathrm{d,f}}(t)X_{\mathrm{f}}(t), A(t))}{Y_{\mathrm{d,f}}(t)X_{\mathrm{f}}(t)} \\
&+ \frac{\Pi_{\mathrm{f}}^{\star}(t, X_{\mathrm{f}}(t), A(t))}{X_{\mathrm{f}}(t)} \cdot \sigma_{s,\mathrm{f}} \cdot \sigma_{s,\mathrm{f}}^{\mathrm{T}} \cdot \frac{\Pi_{\mathrm{f}}^{\star}(t, X_{\mathrm{f}}(t), A(t))}{X_{\mathrm{f}}(t)}.
\end{aligned} \tag{7.4.8}
$$

The predividend FXR volatility:

$$
\sigma_{\mathrm{FX}} = \frac{\Pi_{\mathrm{d}}^{\star}(t, Y_{\mathrm{d,f}}(t)X_{\mathrm{f}}(t), A(t))}{Y_{\mathrm{d,f}}(t)X_{\mathrm{f}}(t)} \cdot \sigma_{s,\mathrm{d}}. \tag{7.4.9}
$$

The postdividend FXR drift:

$$
\begin{aligned}
b_{\mathrm{FX}} = Y_{\mathrm{d,f}} \Bigg(r_{\mathrm{d}} - r_{\mathrm{f}} &+ \frac{\Pi_{\mathrm{d}}^{\star}(t, Y_{\mathrm{d,f}}(t)X_{\mathrm{f}}(t), A(t))}{Y_{\mathrm{d,f}}(t)X_{\mathrm{f}}(t)} \cdot (a_{s,\mathrm{d}} - r_{\mathrm{d}}) - \frac{\Pi_{\mathrm{f}}^{\star}(t, X_{\mathrm{f}}(t), A(t))}{X_{\mathrm{f}}(t)} \cdot \\
&(a_{s,\mathrm{f}} - r_{\mathrm{f}}) - \frac{\Pi_{\mathrm{f}}^{\star}(t, X_{\mathrm{f}}(t), A(t))}{X_{\mathrm{f}}(t)} \cdot \sigma_{s,\mathrm{f}} \cdot \sigma_{s,\mathrm{d}}^{\mathrm{T}} \cdot \frac{\Pi_{\mathrm{d}}^{\star}(t, Y_{\mathrm{d,f}}(t)X_{\mathrm{f}}(t), A(t))}{Y_{\mathrm{d,f}}(t)X_{\mathrm{f}}(t)} \\
&+ \frac{\Pi_{\mathrm{f}}^{\star}(t, X_{\mathrm{f}}(t), A(t))}{X_{\mathrm{f}}(t)} \cdot \sigma_{s,\mathrm{f}} \cdot \sigma_{s,\mathrm{f}}^{\mathrm{T}} \cdot \frac{\Pi_{\mathrm{f}}^{\star}(t, X_{\mathrm{f}}(t), A(t))}{X_{\mathrm{f}}(t)} \Bigg).
\end{aligned} \tag{7.4.10}
$$

The postdividend FXR diffusion:

$$
c_{\mathrm{FX}} = Y_{\mathrm{d,f}} \left(\frac{\Pi_{\mathrm{d}}^{\star}(t, Y_{\mathrm{d,f}}(t)X_{\mathrm{f}}(t), A(t))}{Y_{\mathrm{d,f}}(t)X_{\mathrm{f}}(t)} \cdot \sigma_{s,\mathrm{d}} - \frac{\Pi_{\mathrm{f}}^{\star}(t, X_{\mathrm{f}}(t), A(t))}{X_{\mathrm{f}}(t)} \cdot \sigma_{s,\mathrm{f}} \right). \tag{7.4.11}
$$

Also, by setting $\Pi_{\mathrm{f}}^{\star}(t, X_{\mathrm{f}}(t), A(t)) = 0$ in the above, we have the following definition.

Definition 7.4.3 (FXR market coefficients for the Garman–Kohlhagen framework).
The predividend FXR appreciation rate:

$$
a_{\mathrm{FX}} = r_{\mathrm{d}} + \frac{\Pi_{\mathrm{d}}^{\star}(t, Y_{\mathrm{d,f}}(t)X_{\mathrm{f}}(t), A(t))}{Y_{\mathrm{d,f}}(t)X_{\mathrm{f}}(t)} \cdot (a_{s,\mathrm{d}} - r_{\mathrm{d}}). \tag{7.4.12}
$$

The predividend FXR volatility:

$$
\sigma_{\mathrm{FX}} = \frac{\Pi_{\mathrm{d}}^{\star}(t, Y_{\mathrm{d,f}}(t)X_{\mathrm{f}}(t), A(t))}{Y_{\mathrm{d,f}}(t)X_{\mathrm{f}}(t)} \cdot \sigma_{s,\mathrm{d}}. \tag{7.4.13}
$$

The postdividend FXR drift:

$$b_{\mathrm{FX}} = Y_{\mathrm{d,f}}\left(r_{\mathrm{d}} - r_{\mathrm{f}} + \frac{\Pi_{\mathrm{d}}^{\star}\left(t, Y_{\mathrm{d,f}}(t)X_{\mathrm{f}}(t), A(t)\right)}{Y_{\mathrm{d,f}}(t)X_{\mathrm{f}}(t)} \cdot \left(a_{s,\mathrm{d}} - r_{\mathrm{d}}\right)\right). \tag{7.4.14}$$

The postdividend FXR diffusion:

$$c_{\mathrm{FX}} = Y_{\mathrm{d,f}}\frac{\Pi_{\mathrm{d}}^{\star}\left(t, Y_{\mathrm{d,f}}(t)X_{\mathrm{f}}(t), A(t)\right)}{Y_{\mathrm{d,f}}(t)X_{\mathrm{f}}(t)} \cdot \sigma_{s,\mathrm{d}}. \tag{7.4.15}$$

Remark 7.4.1. The FX market coefficients above may, and often will, be augmented by some other factor coefficients.

7.5 FX Derivatives Under Stochastic Interest Rates

7.5.1 $S = \left\{Y_{\mathrm{d,f}}\right\}$

We continue Sect. 7.3.2. Recall the market coefficients (7.3.19)–(7.3.22), and assume that

$$\beta_{\mathrm{d}} = \beta_{\mathrm{f}} = 1, \tag{7.5.1}$$

which implies $\mathbb{H}_{d,1} = 0$, $\mathbb{H}_{f,1} = 0$, $\mathbb{H}_{d,0} = (\gamma - 1)/q_{d,1}$, $\mathbb{H}_{f,0} = (\gamma - 1)/q_{f,1}$. To fix ideas, we shall consider CRRA pricing of FX futures with zero position held ($\kappa = 0$). Consequently, when we refer to the pricing system (4.9.6)–(4.9.7), it will be understood that the zero-order term in (4.9.7) is omitted (and that the terminal condition is equal to $\upsilon\left(Y_{\mathrm{d,f}}\right) = Y_{\mathrm{d,f}}$).

In the case $S = \left\{Y_{\mathrm{d,f}}\right\}$, i.e., in the case that a contract position is hedged only by means of trading in the FX currency (with price $Y_{\mathrm{d,f}}$), using (7.4.8)–(7.4.11) the FX market coefficients are as follows:

- *The predividend FXR appreciation rate*

$$a_{\mathrm{FX}} = \left\{ \frac{\left(\mathbf{a}_{\mathrm{f}} + \frac{(\gamma-1)w_{\mathrm{f}}\sigma_{\mathrm{f}}\rho_{4,2}}{q_{f,1}}\right)^2}{\gamma^2\sigma_{\mathrm{f}}^2} - \left(\rho_{2,1}\left(\mathbf{a}_{\mathrm{d}} + \frac{(\gamma-1)w_{\mathrm{d}}\sigma_{\mathrm{d}}\rho_{3,1}}{q_{d,1}}\right)\right.\right.$$
$$\left.\left.\left(\mathbf{a}_{\mathrm{f}} + \frac{(\gamma-1)w_{\mathrm{f}}\sigma_{\mathrm{f}}\rho_{4,2}}{q_{f,1}}\right)\right)\Big/\left(\gamma^2\sigma_{\mathrm{d}}\sigma_{\mathrm{f}}\right) + r_{\mathrm{d}} + \frac{\mathbf{a}_{\mathrm{d}}\left(\mathbf{a}_{\mathrm{d}} + \frac{(\gamma-1)w_{\mathrm{d}}\sigma_{\mathrm{d}}\rho_{3,1}}{q_{d,1}}\right)}{\gamma\sigma_{\mathrm{d}}^2}\right\}; \tag{7.5.2}$$

- *The predividend FXR volatility*

$$\sigma_{\mathrm{FX}} = \left(\frac{\mathbf{a}_{\mathrm{d}} + \frac{(\gamma-1)w_{\mathrm{d}}\sigma_{\mathrm{d}}\rho_{3,1}}{q_{d,1}}}{\gamma\sigma_{\mathrm{d}}}\ 0\ 0\ 0\right); \tag{7.5.3}$$

- *The postdividend FXR drift* ((7.4.10) is augmented with four factor drifts)

$$b_{\text{FX}} = \Bigg\{ S_1\left(r_{\text{d}} + \mathbf{a}_{\text{d}} - \mathbb{D}_{\text{d}}\right), S_2\left(r_{\text{f}} + \mathbf{a}_{\text{f}} - \mathbb{D}_{\text{f}}\right), q_{d,0} + r_{\text{d}} q_{d,1}, q_{f,0} + r_{\text{f}} q_{f,1}, Y_{\text{d,f}}$$

$$\left(\frac{\left(\mathbf{a}_{\text{f}} + \frac{(\gamma-1)w_{\text{f}}\sigma_{\text{f}}\rho_{4,2}}{q_{f,1}}\right)^2}{\gamma^2\sigma_{\text{f}}^2} - \left(\rho_{2,1}\left(\mathbf{a}_{\text{d}} + \frac{(\gamma-1)w_{\text{d}}\sigma_{\text{d}}\rho_{3,1}}{q_{d,1}}\right)\right.\right.$$

$$\left.\left.\left(\mathbf{a}_{\text{f}} + \frac{(\gamma-1)w_{\text{f}}\sigma_{\text{f}}\rho_{4,2}}{q_{f,1}}\right)\right)\Big/\left(\gamma^2\sigma_{\text{d}}\sigma_{\text{f}}\right) + r_{\text{d}} - r_{\text{f}} + \frac{1}{\gamma}\right.$$

$$\left.\left.\left(\frac{\mathbf{a}_{\text{d}}\left(\mathbf{a}_{\text{d}} + \frac{(\gamma-1)w_{\text{d}}\sigma_{\text{d}}\rho_{3,1}}{q_{d,1}}\right)}{\sigma_{\text{d}}^2} - \frac{1}{\sigma_{\text{f}}^2}\mathbf{a}_{\text{f}}\left(\mathbf{a}_{\text{f}} + \frac{(\gamma-1)w_{\text{f}}\sigma_{\text{f}}\rho_{4,2}}{q_{f,1}}\right)\right)\right)\right\}; \quad (7.5.4)$$

- *The postdividend FXR diffusion* ((7.4.11) is augmented with four factor diffusions)

$$c_{\text{FX}} =$$

$$\begin{pmatrix} S_1\sigma_{\text{d}} & 0 & 0 & 0 \\ S_2\sigma_{\text{f}}\rho_{2,1} & S_2\sigma_{\text{f}}\sqrt{1-\rho_{2,1}^2} & 0 & 0 \\ w_{\text{d}}\rho_{3,1} & w_{\text{d}}\mathbf{s}_{3,2} & w_{\text{d}}\mathbf{s}_{3,3} & 0 \\ w_{\text{f}}\rho_{4,1} & w_{\text{f}}\mathbf{s}_{4,2} & w_{\text{f}}\mathbf{s}_{4,3} & w_{\text{f}}\mathbf{s}_{4,4} \\ \frac{Y_{\text{d,f}}}{\gamma}\left(\frac{\mathbf{a}_{\text{d}} + \frac{(\gamma-1)w_{\text{d}}\sigma_{\text{d}}\rho_{3,1}}{q_{d,1}}}{\sigma_{\text{d}}} - \frac{\rho_{2,1}\left(\mathbf{a}_{\text{f}} + \frac{(\gamma-1)w_{\text{f}}\sigma_{\text{f}}\rho_{4,2}}{q_{f,1}}\right)}{\sigma_{\text{f}}}\right) & -\frac{Y_{\text{d,f}}}{\gamma\sigma_{\text{f}}}\sqrt{1-\rho_{2,1}^2}\left(\mathbf{a}_{\text{f}} + \frac{(\gamma-1)w_{\text{f}}\sigma_{\text{f}}\rho_{4,2}}{q_{f,1}}\right) & 0 & 0 \end{pmatrix},$$

$$(7.5.5)$$

where $\mathbf{s}_{3,2}$, $\mathbf{s}_{3,3}$, $\mathbf{s}_{4,2}$, $\mathbf{s}_{4,3}$, and $\mathbf{s}_{4,4}$ are given in (A.2.26), (A.2.27), (A.2.29), (A.2.30) and (A.2.31), respectively.

We look for the solution of (4.9.6) in the form

$$g_{\gamma,\text{fx}}\left(t, S_1, S_2, r_{\text{d}}, r_{\text{f}}\right) = \mathbb{G}_0(t) + r_{\text{d}}\mathbb{G}_1(t), \quad (7.5.6)$$

obtaining two ODEs characterizing $\mathbb{G}_0(t)$ and $\mathbb{G}_1(t)$. Since $\mathbb{G}_0(t)$ is not needed, we state only the ODE characterizing $\mathbb{G}_1(t)$, which reads

$$-\gamma + q_{d,1}\mathbb{G}_1(t) + \mathbb{G}_1'(t) + 1 = 0 \quad (7.5.7)$$

for $t < T$, with the terminal condition $\mathbb{G}_1(T) = 0$, and it is solved by

$$\mathbb{G}_1(t) = \frac{\left(1 - \mathrm{e}^{(T-t)q_{d,1}}\right)(\gamma-1)}{q_{d,1}}. \quad (7.5.8)$$

The pricing PDE (7.4.11), after substituting (7.5.8), and with the terminal condition $V_\gamma(T, r_{\text{d}}, r_{\text{f}}, Y_{\text{d,f}}) = \upsilon\left(Y_{\text{d,f}}\right) = Y_{\text{d,f}}$, can be solved in the form

$$V_\gamma(t, r_d, r_f, Y_{d,f}) = e^{\mathbb{V}_0(t) + r_d \mathbb{V}_d(t) + r_f \mathbb{V}_f(t)} Y_{d,f}. \quad (7.5.9)$$

Indeed, plugging (7.5.9) into it, we deduce three ODEs:

$$\begin{aligned}
&\Big(\Big(\Big((\gamma-1)\left(2(\gamma-1)q_{d,1}^2\rho_{4,2}^2\left((\gamma-1)\rho_{2,1}\rho_{4,2}-\gamma q_{f,1}\rho_{4,1}\mathbb{V}_f(t)\right)w_f^3\right.\\
&+2\gamma w_d^2 q_{f,1}^2\rho_{3,1}\left(-(\gamma-1)\rho_{2,1}\rho_{3,1}\rho_{4,2}\left(\left(-1+e^{(T-t)q_{d,1}}\right)(\gamma-1)\right.\right.\\
&\left.-q_{d,1}\mathbb{V}_d(t)\right)+(\gamma-1)\rho_{3,2}\rho_{4,2}\left(\left(-1+e^{(T-t)q_{d,1}}\right)(\gamma-1)-q_{d,1}\mathbb{V}_d(t)\right)\\
&+q_{f,1}\left((\gamma-1)\left(\left(-1+e^{(T-t)q_{d,1}}\right)\gamma+1\right)\rho_{3,1}\rho_{4,1}-\gamma\rho_{4,3}\right.\\
&\left.\left.\left(\left(-1+e^{(T-t)q_{d,1}}\right)(\gamma-1)-q_{d,1}\mathbb{V}_d(t)\right)\right)\mathbb{V}_f(t)\right)w_f\\
&+\gamma w_d^3 q_{f,1}^3\rho_{3,1}\mathbb{V}_d(t)\left(2(\gamma-1)\left(\left(-1+e^{(T-t)q_{d,1}}\right)\gamma+1\right)\rho_{3,1}^2\right.\\
&\left.-\gamma\left(2\left(-1+e^{(T-t)q_{d,1}}\right)(\gamma-1)-q_{d,1}\mathbb{V}_d(t)\right)\right)\\
&+w_d q_{d,1}q_{f,1}\rho_{3,1}\left(\left(-2(\gamma-1)^2\rho_{2,1}^2\rho_{4,2}^2+2(\gamma-1)\gamma q_{f,1}\rho_{2,1}\rho_{4,1}\mathbb{V}_f(t)\rho_{4,2}\right.\right.\\
&+\gamma\left(q_{f,1}\mathbb{V}_f(t)\left(\gamma q_{f,1}\mathbb{V}_f(t)-2(\gamma-1)\rho_{4,2}^2\right)\right.\\
&\left.\left.-2(\gamma-1)q_{d,1}\rho_{4,2}^2\mathbb{V}_d(t)\right)\right)w_f^2+2\gamma^2 q_{f,1}^2\left(q_{d,0}\mathbb{V}_d(t)\right.\\
&\left.\left.+q_{f,0}\mathbb{V}_f(t)+\mathbb{V}_0'(t)\right)\right)\Big)\sigma_f^3-2(\gamma-1)\mathbf{a}_f q_{f,1}\left(w_f^2\rho_{4,2}\left(2\gamma q_{f,1}\rho_{4,1}\mathbb{V}_f(t)\right.\right.\\
&\left.-3(\gamma-1)\rho_{2,1}\rho_{4,2}\right)q_{d,1}^2+w_d w_f q_{f,1}\rho_{3,1}\left(\rho_{4,2}\left(2(\gamma-1)\rho_{2,1}^2\right.\right.\\
&\left.\left.+\gamma\left(\gamma+2q_{d,1}\mathbb{V}_d(t)+q_{f,1}\mathbb{V}_f(t)-1\right)\right)-\gamma q_{f,1}\rho_{2,1}\rho_{4,1}\mathbb{V}_f(t)\right)q_{d,1}\\
&+\gamma w_d^2 q_{f,1}^2\rho_{3,1}\left(\rho_{2,1}\rho_{3,1}-\rho_{3,2}\right)\left(\left(-1+e^{(T-t)q_{d,1}}\right)(\gamma-1)\right.\\
&\left.\left.-q_{d,1}\mathbb{V}_d(t)\right)\right)\sigma_f^2-2\mathbf{a}_f^2 q_{d,1}q_{f,1}^2\left(w_d q_{f,1}\rho_{3,1}\left((\gamma-1)\right.\right.\\
&\left.\rho_{2,1}^2+\gamma\left(\gamma+q_{d,1}\mathbb{V}_d(t)-1\right)\right)+w_f q_{d,1}\left(\gamma q_{f,1}\rho_{4,1}\right.\\
&\left.\left.\mathbb{V}_f(t)-3(\gamma-1)\rho_{2,1}\rho_{4,2}\right)\right)\sigma_f+2\mathbf{a}_f^3 q_{d,1}^2 q_{f,1}^3\rho_{2,1}\Big)\\
&\sigma_d^2+\mathbf{a}_d\sigma_f q_{d,1}q_{f,1}\left(\left(\gamma w_d^2\mathbb{V}_d(t)\left(2(\gamma-1)\left(\left(-2+e^{(T-t)q_{d,1}}\right)\gamma+2\right)\rho_{3,1}^2\right.\right.\right.\\
&\left.\left.-\gamma\left(2\left(-1+e^{(T-t)q_{d,1}}\right)(\gamma-1)-q_{d,1}\mathbb{V}_d(t)\right)\right)q_{f,1}^2\right.\\
&+2\gamma w_d w_f\left(-(\gamma-1)\rho_{2,1}\rho_{3,1}\rho_{4,2}\left(\left(-2+e^{(T-t)q_{d,1}}\right)(\gamma-1)-q_{d,1}\mathbb{V}_d(t)\right)\right.\\
&+(\gamma-1)\rho_{3,2}\rho_{4,2}\left(\left(-1+e^{(T-t)q_{d,1}}\right)(\gamma-1)-q_{d,1}\mathbb{V}_d(t)\right)
\end{aligned}$$

$$
\begin{aligned}
&+q_{f,1}\left((\gamma-1)\left(\left(-2+\mathrm{e}^{(T-t)q_{d,1}}\right)\gamma+2\right)\rho_{3,1}\rho_{4,1}\right.\\
&-\gamma\,\rho_{4,3}\left(\left(-1+\mathrm{e}^{(T-t)q_{d,1}}\right)(\gamma-1)-q_{d,1}\mathbb{V}_{\mathrm{d}}(t)\right)\Big)\,\mathbb{V}_f(t)\Big)\,q_{f,1}\\
&+q_{d,1}\left(\left(-2(\gamma-1)^2\rho_{2,1}^2\rho_{4,2}^2+2(\gamma-1)\gamma q_{f,1}\rho_{2,1}\rho_{4,1}\mathbb{V}_f(t)\rho_{4,2}\right.\right.\\
&+\gamma q_{f,1}\mathbb{V}_f(t)\left(\gamma q_{f,1}\mathbb{V}_f(t)-2(\gamma-1)\rho_{4,2}^2\right)\big)\,w_{\mathrm{f}}^2\\
&+2\gamma^2q_{f,1}^2\left(q_{d,0}\mathbb{V}_{\mathrm{d}}(t)+q_{f,0}\mathbb{V}_f(t)+\mathbb{V}_0'(t)\right)\big)\big)\,\sigma_{\mathrm{f}}^2\\
&-2\mathbf{a}_{\mathrm{f}}q_{f,1}\left(\gamma w_{\mathrm{d}}q_{f,1}\left(\rho_{2,1}\rho_{3,1}\left(\left(-2+\mathrm{e}^{(T-t)q_{d,1}}\right)(\gamma-1)-q_{d,1}\mathbb{V}_{\mathrm{d}}(t)\right)\right.\right.\\
&-\rho_{3,2}\left(\left(-1+\mathrm{e}^{(T-t)q_{d,1}}\right)(\gamma-1)-q_{d,1}\mathbb{V}_{\mathrm{d}}(t)\right)\Big)\\
&+w_{\mathrm{f}}q_{d,1}\left(\rho_{4,2}\left(2(\gamma-1)\rho_{2,1}^2+\gamma\left(\gamma+q_{f,1}\mathbb{V}_f(t)-1\right)\right)\right.\\
&-\gamma q_{f,1}\rho_{2,1}\ \rho_{4,1}\mathbb{V}_f(t)\big)\big)\,\sigma_{\mathrm{f}}-2\mathbf{a}_{\mathrm{f}}^2q_{d,1}q_{f,1}^2\left(\rho_{2,1}^2+\gamma\right)\big)\,\sigma_{\mathrm{d}}\\
&-2\gamma\mathbf{a}_{\mathrm{d}}^2\sigma_{\mathrm{f}}^2q_{d,1}^2q_{f,1}^2\left((\gamma-1)\sigma_{\mathrm{f}}\left(w_{\mathrm{d}}q_{f,1}\rho_{3,1}\mathbb{V}_{\mathrm{d}}(t)+w_{\mathrm{f}}\left(q_{f,1}\rho_{4,1}\mathbb{V}_f(t)\right.\right.\right.\\
&-\rho_{2,1}\ \rho_{4,2})\big)-\mathbf{a}_{\mathrm{f}}q_{f,1}\rho_{2,1})\big)\Big)\,/\,\left(\gamma\sigma_{\mathrm{d}}\sigma_{\mathrm{f}}q_{d,1}q_{f,1}\left(\mathbf{a}_{\mathrm{d}}q_{d,1}\right.\right.\\
&+(\gamma-1)\ w_{\mathrm{d}}\sigma_{\mathrm{d}}\rho_{3,1})\big)=0, \qquad (7.5.10)
\end{aligned}
$$

$$q_{f,1}\mathbb{V}_f(t)+\mathbb{V}_f'(t)=1, \qquad (7.5.11)$$

$$q_{d,1}\mathbb{V}_{\mathrm{d}}(t)+\mathbb{V}_{\mathrm{d}}'(t)+1=0, \qquad (7.5.12)$$

for $t<T$, together with the terminal condition $\mathbb{V}_0(T)=\mathbb{V}_{\mathrm{d}}(T)=\mathbb{V}_f(T)=0$. Solving the above ODEs yields the following result.

Proposition 7.5.1. *Under the stochastic interest rate model, assuming $\beta_{\mathrm{d}}=\beta_{\mathrm{f}}=0$ and assuming that the hedging is done only via trading in the foreign currency ($S=\{Y_{\mathrm{d,f}}\}$), an FX futures contract expiring at T has CRRA ($\kappa=0$)-neutral (or equivalently, ($\kappa=0$)-indifference) price given by (7.5.9), for $t<T$, where*

$$\mathbb{V}_{\mathrm{d}}(t)=-\frac{1-\mathrm{e}^{(T-t)q_{d,1}}}{q_{d,1}}, \qquad (7.5.13)$$

$$\mathbb{V}_f(t)=\frac{1-\mathrm{e}^{(T-t)q_{f,1}}}{q_{f,1}}, \qquad (7.5.14)$$

and where $\mathbb{V}_0(t)$ is the solution of the ODE (7.5.10). (To save space, we do not provide an explicit formula for $\mathbb{V}_0(t)$, although it is easily obtained using symbolic calculation.)

Furthermore, the (most conservative) hedging formula for the position $\kappa = \{\kappa_1\}$ (see (5.2.3)) reads

$$\begin{aligned}
\Pi^{\mathbf{h}}(t, r_{\mathrm{d}}, r_{\mathrm{f}}, Y_{\mathrm{d,f}}) = \Bigg\{ -\kappa_1 \Bigg(\Bigg(\sigma_{\mathrm{d}} \Bigg(\frac{1}{\sigma_{\mathrm{d}}} \big(\beta_{\mathrm{d}} r_{\mathrm{d}} - r_{\mathrm{d}} + \mathbf{a}_{\mathrm{d}} + w_{\mathrm{d}}\sigma_{\mathrm{d}} \big(\mathbb{H}_{d,0} \\
+ r_{\mathrm{d}}\mathbb{H}_{d,1}\big)\rho_{3,1}\big) - \frac{1}{\sigma_{\mathrm{f}}}\rho_{2,1}\big(\big(\beta_{\mathrm{f}} + w_{\mathrm{f}}\sigma_{\mathrm{f}}\mathbb{H}_{f,1}\rho_{4,2}\big) r_{\mathrm{f}} - r_{\mathrm{f}} + \mathbf{a}_{\mathrm{f}} \\
+ w_{\mathrm{f}}\sigma_{\mathrm{f}}\mathbb{H}_{f,0}\rho_{4,2}\big) \Bigg) Y_{\mathrm{d,f}} V_{\gamma}{}^{(0,0,0,1)}(t, r_{\mathrm{d}}, r_{\mathrm{f}}, Y_{\mathrm{d,f}}) \Bigg) \Big/ \big(\beta_{\mathrm{d}} r_{\mathrm{d}} - r_{\mathrm{d}} + \mathbf{a}_{\mathrm{d}} \\
+ w_{\mathrm{d}}\sigma_{\mathrm{d}}\big(\mathbb{H}_{d,0} + r_{\mathrm{d}}\mathbb{H}_{d,1}\big)\rho_{3,1}\big) + \Big(\gamma w_{\mathrm{f}}\sigma_{\mathrm{d}}\rho_{4,1} V_{\gamma}{}^{(0,0,1,0)} \\
(t, r_{\mathrm{d}}, r_{\mathrm{f}}, Y_{\mathrm{d,f}})\Big) \Big/ \big(\beta_{\mathrm{d}} r_{\mathrm{d}} - r_{\mathrm{d}} + \mathbf{a}_{\mathrm{d}} + w_{\mathrm{d}}\sigma_{\mathrm{d}}\big(\mathbb{H}_{d,0} + r_{\mathrm{d}}\mathbb{H}_{d,1}\big)\rho_{3,1}\big) \\
+ \Big(\gamma w_{\mathrm{d}}\sigma_{\mathrm{d}}\rho_{3,1} V_{\gamma}{}^{(0,1,0,0)}(t, r_{\mathrm{d}}, r_{\mathrm{f}}, Y_{\mathrm{d,f}})\Big) \Big/ \big(\beta_{\mathrm{d}} r_{\mathrm{d}} - r_{\mathrm{d}} \\
+ \mathbf{a}_{\mathrm{d}} + w_{\mathrm{d}}\sigma_{\mathrm{d}}\big(\mathbb{H}_{d,0} + r_{\mathrm{d}}\mathbb{H}_{d,1}\big)\rho_{3,1}\big) \Bigg) \Bigg\} \\
= \Bigg\{ -\kappa_1 \Bigg(\Big(\mathrm{e}^{\mathbb{V}_0(t) + S_1\mathbb{V}_{\mathrm{d}}(t) + S_2\mathbb{V}_f(t)} \gamma r_{\mathrm{d}} S_1 \mathbb{V}_{\mathrm{d}}(t)\sigma_{\mathrm{d}}^2 \Big) \Big/ \\
\left(\mathbf{a}_{\mathrm{d}} + \frac{(\gamma - 1) w_{\mathrm{d}}\sigma_{\mathrm{d}}\rho_{3,1}}{q_{d,1}} \right) + \Big(\mathrm{e}^{\mathbb{V}_0(t) + S_1\mathbb{V}_{\mathrm{d}}(t) + S_2\mathbb{V}_f(t)} \gamma w_{\mathrm{d}}\rho_{3,1}\sigma_{\mathrm{d}} \Big) \Big/ \\
\left(\mathbf{a}_{\mathrm{d}} + \frac{(\gamma - 1) w_{\mathrm{d}}\sigma_{\mathrm{d}}\rho_{3,1}}{q_{d,1}} \right) + \Big(\mathrm{e}^{\mathbb{V}_0(t) + S_1\mathbb{V}_{\mathrm{d}}(t) + S_2\mathbb{V}_f(t)} \\
\gamma r_{\mathrm{d}} S_2 \sigma_{\mathrm{f}} \rho_{2,1} \mathbb{V}_f(t)\sigma_{\mathrm{d}} \Big) \Big/ \left(\mathbf{a}_{\mathrm{d}} + \frac{(\gamma - 1) w_{\mathrm{d}}\sigma_{\mathrm{d}}\rho_{3,1}}{q_{d,1}} \right) \Bigg) \Bigg\},
\end{aligned} \tag{7.5.15}$$

which represents the hedging position in the foreign currency (expressed in the domestic currency).

7.5.2 $S = \{Y_{\mathbf{d,f}}, S_2\}$

Again, we shall assume that $\beta_{\mathrm{d}} = \beta_{\mathrm{f}} = 0$ and $\kappa = 0$.

What changes in the case $\{Y_{\mathrm{d,f}}, S_2\}$, i.e., in the case that a futures contract is hedged not only by means of trading in the FX currency (with price $Y_{\mathrm{d,f}}$) but also by trading in the foreign risky assets or their proxy, the market index? Again using (7.4.8)–(7.4.11), the predividend FX market coefficients (7.5.2) and (7.5.3) are modified/appended as follows:

- *The predividend FXR appreciation rate*

$$a_{\mathrm{FX}} = \left\{ \frac{\left(\mathbf{a}_{\mathrm{f}} + \frac{(\gamma-1)w_{\mathrm{f}}\sigma_{\mathrm{f}}\rho_{4,2}}{q_{f,1}}\right)^2}{\gamma^2\sigma_{\mathrm{f}}^2} - \left(\rho_{2,1}\left(\mathbf{a}_{\mathrm{d}} + \frac{(\gamma-1)w_{\mathrm{d}}\sigma_{\mathrm{d}}\rho_{3,1}}{q_{d,1}}\right)\right.\right.$$
$$\left.\left.\left(\mathbf{a}_{\mathrm{f}} + \frac{(\gamma-1)w_{\mathrm{f}}\sigma_{\mathrm{f}}\rho_{4,2}}{q_{f,1}}\right)\right)\Big/\left(\gamma^2\sigma_{\mathrm{d}}\sigma_{\mathrm{f}}\right) + r_{\mathrm{d}} + \frac{\mathbf{a}_{\mathrm{d}}\left(\mathbf{a}_{\mathrm{d}} + \frac{(\gamma-1)w_{\mathrm{d}}\sigma_{\mathrm{d}}\rho_{3,1}}{q_{d,1}}\right)}{\gamma\sigma_{\mathrm{d}}^2}, r_{\mathrm{f}} + \mathbf{a}_{\mathrm{f}} \right\}; \tag{7.5.16}$$

- *The predividend FXR volatility*

$$\sigma_{\mathrm{FX}} = \begin{pmatrix} \frac{\mathbf{a}_{\mathrm{d}} + \frac{(\gamma-1)w_{\mathrm{d}}\sigma_{\mathrm{d}}\rho_{3,1}}{q_{d,1}}}{\gamma\sigma_{\mathrm{d}}} & 0 & 0\;0 \\ \sigma_{\mathrm{f}}\rho_{2,1} & \sigma_{\mathrm{f}}\sqrt{1-\rho_{2,1}^2} & 0\;0 \end{pmatrix}, \tag{7.5.17}$$

while the postdividend FX market coefficients (7.5.2) and (7.5.3) remain the same – indeed, only tradables were modified/appended, not the factors.

We look for the solution of (4.9.6), instead of (7.5.6), in the form

$$\begin{aligned} &g_{\gamma,\mathrm{fx}}(t, S_1, S_2, r_{\mathrm{d}}, r_{\mathrm{f}}, Y_{\mathrm{d,f}}) \\ &= \mathbb{G}_{d,2}(t) r_{\mathrm{d}}^2 + \mathbb{G}_{d,1}(t) r_{\mathrm{d}} + r_{\mathrm{f}}\mathbb{G}_{d,f,1}(t) r_{\mathrm{d}} + \mathbb{G}_0(t) + r_{\mathrm{f}}\mathbb{G}_{f,1}(t) + r_{\mathrm{f}}^2\mathbb{G}_{f,2}(t), \end{aligned} \tag{7.5.18}$$

arriving at six ODEs. Since $\mathbb{G}_0(t)$ is not needed, we state only the five ODEs that are needed:

$$\begin{aligned} &(-(\gamma-1)((\gamma\sigma_{\mathrm{f}}q_{f,1}^2\rho_{3,1}(-\gamma\rho_{2,1}^2 + 2(\gamma-1)\rho_{3,1}\rho_{3,2}\rho_{2,1} - (\gamma-1)\rho_{3,1}^2 - \gamma\rho_{3,2}^2 \\ &+\rho_{3,2}^2 + \gamma)\mathbb{G}_{d,1}(t)\mathbb{G}_{d,f}(t)w_{\mathrm{d}}^3 + q_{f,1}\rho_{3,1}(w_{\mathrm{f}}\sigma_{\mathrm{f}}\rho_{2,1}(\rho_{3,1} - \rho_{2,1}\rho_{3,2})\rho_{4,2}\mathbb{G}_{d,f}(t) \\ &(\gamma-1)^2 + \gamma q_{f,1}(-\gamma w_{\mathrm{f}}\sigma_{\mathrm{f}}\rho_{4,3}(\mathbb{G}_{d,f}(t)\mathbb{G}_{f,1}(t) + 2\mathbb{G}_{d,1}(t)\mathbb{G}_{f,2}(t)) \\ &\rho_{2,1}^2 + (\gamma-1)(w_{\mathrm{f}}\sigma_{\mathrm{f}}\rho_{3,2}\rho_{4,1}(\mathbb{G}_{d,f}(t)\mathbb{G}_{f,1}(t) + 2\mathbb{G}_{d,1}(t)\mathbb{G}_{f,2}(t)) \\ &+\rho_{3,1}(w_{\mathrm{f}}\sigma_{\mathrm{f}}\rho_{4,2}\mathbb{G}_{d,f}(t)\mathbb{G}_{f,1}(t) + \mathbb{G}_{d,1}(t)(2w_{\mathrm{f}}\sigma_{\mathrm{f}}\rho_{4,2}\mathbb{G}_{f,2}(t) + 1)))\rho_{2,1} \\ &-w_{\mathrm{f}}\sigma_{\mathrm{f}}((\gamma-1)\rho_{3,1}\rho_{4,1} - \gamma\rho_{4,3})(\mathbb{G}_{d,f}(t)\mathbb{G}_{f,1}(t) + 2\mathbb{G}_{d,1}(t)\mathbb{G}_{f,2}(t)) \\ &-(\gamma-1)\rho_{3,2}(w_{\mathrm{f}}\sigma_{\mathrm{f}}\rho_{4,2}\mathbb{G}_{d,f}(t)\mathbb{G}_{f,1}(t) + \mathbb{G}_{d,1}(t)(2w_{\mathrm{f}}\sigma_{\mathrm{f}}\rho_{4,2}\mathbb{G}_{f,2}(t) + 1))))w_{\mathrm{d}}^2 \\ &-(\sigma_{\mathrm{f}}((\gamma-1)^2 q_{d,1}(\rho_{3,1} - \rho_{2,1}\rho_{3,2})\mathbb{G}_{d,f}(t)\rho_{4,2}^2 + 2q_{f,1}\rho_{3,1}((q_{f,1}\mathbb{G}_{f,1}(t)\gamma^2 \\ &+(\gamma-1)^2\rho_{4,2}^2)\rho_{2,1}^2 - (\gamma-1)\rho_{4,1}\rho_{4,2}(2q_{f,1}\mathbb{G}_{f,1}(t)\gamma + \gamma - 1)\rho_{2,1} \end{aligned}$$

$$+\gamma q_{f,1}((\gamma-1)\rho_{4,1}^2+(\gamma-1)\rho_{4,2}^2-\gamma)\mathbb{G}_{f,1}(t))\mathbb{G}_{f,2}(t))w_{\mathrm{f}}^2+(\gamma-1)q_{f,1}$$
$$\rho_{3,1}((\gamma-1)\rho_{4,2}\rho_{2,1}^2-\gamma q_{f,1}\rho_{4,1}\mathbb{G}_{f,1}(t)\rho_{2,1}$$
$$+\gamma q_{f,1}\rho_{4,2}\mathbb{G}_{f,1}(t))w_{\mathrm{f}}+\gamma^2\sigma_{\mathrm{f}}q_{f,1}^2(\rho_{2,1}^2-1)\rho_{3,1}(q_{d,0}\mathbb{G}_{d,f}(t)+q_{f,1}\mathbb{G}_{f,1}(t)$$
$$+2q_{f,0}\mathbb{G}_{f,2}(t)+\mathbb{G}_{f,1}'(t)))w_{\mathrm{d}}+(\gamma-1)^2w_{\mathrm{f}}^2q_{d,1}\rho_{4,2}^2(\rho_{2,1}(2w_{\mathrm{f}}\sigma_{\mathrm{f}}$$
$$\rho_{4,2}\mathbb{G}_{f,2}(t)+1)-2w_{\mathrm{f}}\sigma_{\mathrm{f}}\rho_{4,1}\mathbb{G}_{f,2}(t)))\sigma_{\mathrm{f}}^2$$
$$+(\gamma-1)\mathbf{a}_{\mathrm{f}}q_{f,1}(\sigma_{\mathrm{f}}q_{f,1}\rho_{3,1}(-\rho_{3,2}\rho_{2,1}^2+(\gamma+1)\rho_{3,1}\rho_{2,1}-\gamma\rho_{3,2})$$
$$\mathbb{G}_{d,f}(t)w_{\mathrm{d}}^2+(q_{f,1}\rho_{3,1}(-(2w_{\mathrm{f}}\sigma_{\mathrm{f}}\rho_{4,2}\mathbb{G}_{f,2}(t)+1)\rho_{2,1}^2$$
$$+2(\gamma+1)w_{\mathrm{f}}\sigma_{\mathrm{f}}\rho_{4,1}\mathbb{G}_{f,2}(t)\rho_{2,1}-\gamma(2w_{\mathrm{f}}\sigma_{\mathrm{f}}\rho_{4,2}\mathbb{G}_{f,2}(t)+1))$$
$$-2w_{\mathrm{f}}\sigma_{\mathrm{f}}q_{d,1}(\rho_{3,1}-\rho_{2,1}\rho_{3,2})\rho_{4,2}\mathbb{G}_{d,f}(t))w_{\mathrm{d}}$$
$$+2w_{\mathrm{f}}q_{d,1}\rho_{4,2}(\rho_{2,1}(2w_{\mathrm{f}}\sigma_{\mathrm{f}}\rho_{4,2}\mathbb{G}_{f,2}(t)+1)-2w_{\mathrm{f}}\sigma_{\mathrm{f}}\rho_{4,1}\mathbb{G}_{f,2}(t)))\sigma_{\mathrm{f}}$$
$$+\mathbf{a}_{\mathrm{f}}^2q_{d,1}q_{f,1}^2(\rho_{2,1}(w_{\mathrm{d}}\sigma_{\mathrm{f}}\rho_{3,2}\mathbb{G}_{d,f}(t)+2w_{\mathrm{f}}\sigma_{\mathrm{f}}\rho_{4,2}\mathbb{G}_{f,2}(t)+1)$$
$$-\sigma_{\mathrm{f}}(w_{\mathrm{d}}\rho_{3,1}\mathbb{G}_{d,f}(t)+2w_{\mathrm{f}}\rho_{4,1}\mathbb{G}_{f,2}(t))))\sigma_{\mathrm{d}}^2$$
$$+\mathbf{a}_{\mathrm{d}}\sigma_{\mathrm{f}}q_{f,1}(\sigma_{\mathrm{f}}(\gamma\sigma_{\mathrm{f}}q_{f,1}((\gamma-1)(\gamma+q_{d,1}\mathbb{G}_{d,1}(t)-1)\rho_{3,1}^2$$
$$-(\gamma-1)\rho_{2,1}\rho_{3,2}(\gamma+2q_{d,1}\mathbb{G}_{d,1}(t)-1)\rho_{3,1}+q_{d,1}(\gamma\rho_{2,1}^2$$
$$+(\gamma-1)\rho_{3,2}^2-\gamma)\mathbb{G}_{d,1}(t))\mathbb{G}_{d,f}(t)w_{\mathrm{d}}^2-(w_{\mathrm{f}}\sigma_{\mathrm{f}}q_{d,1}\rho_{2,1}(\rho_{3,1}$$
$$-\rho_{2,1}\rho_{3,2})\rho_{4,2}\mathbb{G}_{d,f}(t)(\gamma-1)^2+\gamma q_{f,1}(-2w_{\mathrm{f}}\sigma_{\mathrm{f}}\rho_{3,1}\rho_{4,1}\mathbb{G}_{f,2}(t)(\gamma-1)^2$$
$$+\rho_{2,1}(w_{\mathrm{f}}\sigma_{\mathrm{f}}q_{d,1}\rho_{3,2}\rho_{4,1}(\mathbb{G}_{d,f}(t)\mathbb{G}_{f,1}(t)+2\mathbb{G}_{d,1}(t)\mathbb{G}_{f,2}(t))$$
$$+\rho_{3,1}((\gamma-1)(2w_{\mathrm{f}}\sigma_{\mathrm{f}}\rho_{4,2}\mathbb{G}_{f,2}(t)+1)+q_{d,1}(w_{\mathrm{f}}\sigma_{\mathrm{f}}\rho_{4,2}\mathbb{G}_{d,f}(t)\mathbb{G}_{f,1}(t)$$
$$+\mathbb{G}_{d,1}(t)(2w_{\mathrm{f}}\sigma_{\mathrm{f}}\rho_{4,2}\mathbb{G}_{f,2}(t)+1))))(\gamma-1)$$
$$-\gamma w_{\mathrm{f}}\sigma_{\mathrm{f}}q_{d,1}\rho_{2,1}^2\rho_{4,3}(\mathbb{G}_{d,f}(t)\mathbb{G}_{f,1}(t)+2\mathbb{G}_{d,1}(t)\mathbb{G}_{f,2}(t))$$
$$+q_{d,1}(-w_{\mathrm{f}}\sigma_{\mathrm{f}}((\gamma-1)\rho_{3,1}\rho_{4,1}-\gamma\rho_{4,3})(\mathbb{G}_{d,f}(t)\mathbb{G}_{f,1}(t)$$
$$+2\mathbb{G}_{d,1}(t)\mathbb{G}_{f,2}(t))-(\gamma-1)\rho_{3,2}(w_{\mathrm{f}}\sigma_{\mathrm{f}}\rho_{4,2}\mathbb{G}_{d,f}(t)\mathbb{G}_{f,1}(t)$$
$$+\mathbb{G}_{d,1}(t)(2w_{\mathrm{f}}\sigma_{\mathrm{f}}\rho_{4,2}\mathbb{G}_{f,2}(t)+1)))))w_{\mathrm{d}}+q_{d,1}(\sigma_{\mathrm{f}}q_{f,1}(\rho_{2,1}^2-1)(q_{d,0}\mathbb{G}_{d,f}(t)$$
$$+q_{f,1}\mathbb{G}_{f,1}(t)+2q_{f,0}\mathbb{G}_{f,2}(t)+\mathbb{G}_{f,1}'(t))\gamma^2+(\gamma-1)w_{\mathrm{f}}((\gamma-1)\rho_{4,2}\rho_{2,1}^2$$
$$-\gamma q_{f,1}\rho_{4,1}\mathbb{G}_{f,1}(t)\rho_{2,1}+\gamma q_{f,1}\rho_{4,2}\mathbb{G}_{f,1}(t))+2w_{\mathrm{f}}^2\sigma_{\mathrm{f}}((q_{f,1}\mathbb{G}_{f,1}(t)\gamma^2$$
$$+(\gamma-1)^2\rho_{4,2}^2)\rho_{2,1}^2-(\gamma-1)\rho_{4,1}\rho_{4,2}(2q_{f,1}\mathbb{G}_{f,1}(t)\gamma+\gamma-1)\rho_{2,1}$$
$$+\gamma q_{f,1}((\gamma-1)\rho_{4,1}^2+(\gamma-1)\rho_{4,2}^2-\gamma)\mathbb{G}_{f,1}(t))\mathbb{G}_{f,2}(t)))$$
$$-(\gamma-1)\mathbf{a}_{\mathrm{f}}q_{d,1}q_{f,1}(-(w_{\mathrm{d}}\sigma_{\mathrm{f}}\rho_{3,2}\mathbb{G}_{d,f}(t)+2w_{\mathrm{f}}\sigma_{\mathrm{f}}\rho_{4,2}\mathbb{G}_{f,2}(t)+1)\rho_{2,1}^2$$

$$+(\gamma+1)\sigma_{\rm f}(w_{\rm d}\rho_{3,1}\mathbb{G}_{d,f}(t)+2w_{\rm f}\rho_{4,1}\mathbb{G}_{f,2}(t))\rho_{2,1}-\gamma(w_{\rm d}\sigma_{\rm f}\rho_{3,2}\mathbb{G}_{d,f}(t)$$
$$+2w_{\rm f}\sigma_{\rm f}\rho_{4,2}\mathbb{G}_{f,2}(t)+1)))\sigma_{\rm d}-(\gamma-1)\gamma\mathbf{a}_{\rm d}^2\sigma_{\rm f}^2q_{d,1}q_{f,1}^2(\rho_{2,1}(w_{\rm d}\sigma_{\rm f}\rho_{3,2}\mathbb{G}_{d,f}(t)$$
$$+2w_{\rm f}\sigma_{\rm f}\rho_{4,2}\mathbb{G}_{f,2}(t)+1)-\sigma_{\rm f}(w_{\rm d}\rho_{3,1}\mathbb{G}_{d,f}(t)+2w_{\rm f}\rho_{4,1}\mathbb{G}_{f,2}(t))))/$$
$$(\gamma\sigma_{\rm d}\sigma_{\rm f}q_{f,1}(\rho_{2,1}^2-1)(\mathbf{a}_{\rm d}q_{d,1}+(\gamma-1)w_{\rm d}\sigma_{\rm d}\rho_{3,1}))=0, \tag{7.5.19}$$

$$\frac{1}{\gamma\sigma_{\rm f}(\rho_{2,1}^2-1)}(w_{\rm d}^2(\gamma\rho_{2,1}^2-2(\gamma-1)\rho_{3,1}\rho_{3,2}\rho_{2,1}+(\gamma-1)\rho_{3,1}^2+\gamma\rho_{3,2}^2-\rho_{3,2}^2$$
$$-\gamma)\mathbb{G}_{d,f}(t)^2\sigma_{\rm f}^2+2(2w_{\rm f}^2(\gamma\rho_{2,1}^2-2(\gamma-1)\rho_{4,1}\rho_{4,2}\rho_{2,1}+(\gamma-1)\rho_{4,1}^2$$
$$+\gamma\rho_{4,2}^2-\rho_{4,2}^2-\gamma)\mathbb{G}_{f,2}(t)^2+2\gamma q_{f,1}(\rho_{2,1}^2-1)\mathbb{G}_{f,2}(t)$$
$$+\gamma(\rho_{2,1}^2-1)\mathbb{G}'_{f,2}(t))\sigma_{\rm f}^2-4(\gamma-1)w_{\rm f}(\rho_{2,1}\rho_{4,1}-\rho_{4,2})\mathbb{G}_{f,2}(t)\sigma_{\rm f}$$
$$+2w_{\rm d}\mathbb{G}_{d,f}(t)(2\gamma w_{\rm f}\sigma_{\rm f}\rho_{4,3}\mathbb{G}_{f,2}(t)\rho_{2,1}^2-(\gamma-1)(2w_{\rm f}\sigma_{\rm f}\rho_{3,2}\rho_{4,1}\mathbb{G}_{f,2}(t)$$
$$+\rho_{3,1}(2w_{\rm f}\sigma_{\rm f}\rho_{4,2}\mathbb{G}_{f,2}(t)+1))\rho_{2,1}+2w_{\rm f}\sigma_{\rm f}((\gamma-1)\rho_{3,1}\rho_{4,1}-\gamma\rho_{4,3})\mathbb{G}_{f,2}(t)$$
$$+(\gamma-1)\rho_{3,2}(2w_{\rm f}\sigma_{\rm f}\rho_{4,2}\mathbb{G}_{f,2}(t)+1))\sigma_{\rm f}+\gamma-1)=0, \tag{7.5.20}$$

$$\frac{2\rho_{3,1}^2\mathbb{G}_{d,1}(t)\mathbb{G}_{d,2}(t)w_{\rm d}^2}{\gamma}+2\left(-\rho_{3,1}^2+\frac{(\rho_{3,2}-\rho_{2,1}\rho_{3,1})^2}{\rho_{2,1}^2-1}+1\right)$$
$$\mathbb{G}_{d,1}(t)\mathbb{G}_{d,2}(t)w_{\rm d}^2+((\gamma-1)(\rho_{2,1}\rho_{3,1}-\rho_{3,2})\mathbb{G}_{d,1}(t)w_{\rm d})/(\gamma\sigma_{\rm f}(\rho_{2,1}^2-1))$$
$$+\frac{w_{\rm f}\rho_{3,1}\rho_{4,1}\mathbb{G}_{d,1}(t)\mathbb{G}_{d,f}(t)w_{\rm d}}{\gamma}+\frac{1}{\rho_{2,1}^2-1}w_{\rm f}(\rho_{4,3}\rho_{2,1}^2-\rho_{2,1}\rho_{3,2}\rho_{4,1}$$
$$+\rho_{3,2}\rho_{4,2}+\rho_{3,1}(\rho_{4,1}-\rho_{2,1}\rho_{4,2})-\rho_{4,3})\mathbb{G}_{d,1}(t)\mathbb{G}_{d,f}(t)w_{\rm d}$$
$$+\frac{2w_{\rm f}\rho_{3,1}\rho_{4,1}\mathbb{G}_{d,2}(t)\mathbb{G}_{f,1}(t)w_{\rm d}}{\gamma}+\frac{1}{\rho_{2,1}^2-1}2w_{\rm f}(\rho_{4,3}\rho_{2,1}^2-\rho_{2,1}\rho_{3,2}\rho_{4,1}$$
$$+\rho_{3,2}\rho_{4,2}+\rho_{3,1}(\rho_{4,1}-\rho_{2,1}\rho_{4,2})-\rho_{4,3})\mathbb{G}_{d,2}(t)\mathbb{G}_{f,1}(t)w_{\rm d}+q_{d,1}\mathbb{G}_{d,1}(t)$$
$$+2q_{d,0}\mathbb{G}_{d,2}(t)+(2(\gamma-1)w_{\rm d}\rho_{3,1}(-((\gamma-1)^2w_{\rm f}\rho_{4,2}(w_{\rm f}q_{d,1}\rho_{4,2}$$
$$-w_{\rm d}q_{f,1}\rho_{2,1}\rho_{3,1})\sigma_{\rm d}^2+(\gamma-1)\mathbf{a}_{\rm d}q_{f,1}(\gamma w_{\rm d}q_{f,1}\rho_{3,1}$$
$$-w_{\rm f}q_{d,1}\rho_{2,1}\rho_{4,2})\sigma_{\rm d}+\gamma\mathbf{a}_{\rm d}^2q_{d,1}q_{f,1}^2)\sigma_{\rm f}^2+\mathbf{a}_{\rm f}\sigma_{\rm d}q_{f,1}(\mathbf{a}_{\rm d}q_{d,1}q_{f,1}\rho_{2,1}$$
$$+(\gamma-1)\sigma_{\rm d}(w_{\rm d}q_{f,1}\rho_{2,1}\rho_{3,1}-2w_{\rm f}q_{d,1}\rho_{4,2}))\sigma_{\rm f}-\mathbf{a}_{\rm f}^2\sigma_{\rm d}^2q_{d,1}q_{f,1}^2)\mathbb{G}_{d,2}(t))/$$
$$(\gamma^2\sigma_{\rm d}\sigma_{\rm f}^2q_{f,1}^2(\mathbf{a}_{\rm d}q_{d,1}+(\gamma-1)w_{\rm d}\sigma_{\rm d}\rho_{3,1}))+q_{f,0}\mathbb{G}_{d,f}(t)$$
$$+((\gamma-1)w_{\rm f}\rho_{2,1}(\rho_{2,1}\rho_{4,1}-\rho_{4,2})(((\gamma-1)^2w_{\rm f}\rho_{4,2}(w_{\rm f}q_{d,1}\rho_{4,2}$$
$$-w_{\rm d}q_{f,1}\rho_{2,1}\rho_{3,1})\sigma_{\rm d}^2+(\gamma-1)\mathbf{a}_{\rm d}q_{f,1}(\gamma w_{\rm d}q_{f,1}\rho_{3,1}-w_{\rm f}q_{d,1}\rho_{2,1}\rho_{4,2})\sigma_{\rm d}$$
$$+\gamma\mathbf{a}_{\rm d}^2q_{d,1}q_{f,1}^2)\sigma_{\rm f}^2-\mathbf{a}_{\rm f}\sigma_{\rm d}q_{f,1}(\mathbf{a}_{\rm d}q_{d,1}q_{f,1}\rho_{2,1}+(\gamma-1)\sigma_{\rm d}(w_{\rm d}q_{f,1}\rho_{2,1}\rho_{3,1}$$

$$-2w_{\mathrm{f}}q_{d,1}\rho_{4,2}))\sigma_{\mathrm{f}}+\mathbf{a}_{\mathrm{f}}^2\sigma_{\mathrm{d}}^2q_{d,1}q_{f,1}^2)\mathbb{G}_{d,f}(t))/(\gamma^2\sigma_{\mathrm{d}}\sigma_{\mathrm{f}}^2q_{f,1}^2(\rho_{2,1}^2-1)(\mathbf{a}_{\mathrm{d}}q_{d,1}$$
$$+(\gamma-1)w_{\mathrm{d}}\sigma_{\mathrm{d}}\rho_{3,1}))+((\gamma-1)w_{\mathrm{f}}(\rho_{2,1}\rho_{4,1}-\rho_{4,2})\mathbb{G}_{f,1}(t))/(\gamma\sigma_{\mathrm{f}}(\rho_{2,1}^2-1))$$
$$+\frac{w_{\mathrm{f}}^2\rho_{4,1}^2\mathbb{G}_{d,f}(t)\mathbb{G}_{f,1}(t)}{\gamma}+(w_{\mathrm{f}}^2(\rho_{4,3}\rho_{2,1}^2-\rho_{2,1}\rho_{3,2}\rho_{4,1}+\rho_{3,2}\rho_{4,2}+\rho_{3,1}(\rho_{4,1}$$
$$-\rho_{2,1}\rho_{4,2})-\rho_{4,3})^2\mathbb{G}_{d,f}(t)\mathbb{G}_{f,1}(t))/((\rho_{2,1}^2-1)(\rho_{2,1}^2-2\rho_{3,1}\rho_{3,2}\rho_{2,1}$$
$$+\rho_{3,1}^2+\rho_{3,2}^2-1))+1/(\rho_{2,1}^2-2\rho_{3,1}\rho_{3,2}\rho_{2,1}+\rho_{3,1}^2+\rho_{3,2}^2-1)w_{\mathrm{f}}^2$$
$$(-(\rho_{4,3}^2-1)\rho_{2,1}^2-2(\rho_{4,1}(\rho_{4,2}-\rho_{3,2}\rho_{4,3})+\rho_{3,1}(\rho_{3,2}-\rho_{4,2}\rho_{4,3}))\rho_{2,1}+\rho_{3,2}^2$$
$$-\rho_{3,2}^2\rho_{4,1}^2+\rho_{4,1}^2+\rho_{4,2}^2+\rho_{4,3}^2-\rho_{3,1}^2(\rho_{4,2}^2-1)$$
$$+2\rho_{3,1}\rho_{4,1}(\rho_{3,2}\rho_{4,2}-\rho_{4,3})-2\rho_{3,2}\rho_{4,2}\rho_{4,3}-1)\mathbb{G}_{d,f}(t)\mathbb{G}_{f,1}(t)+\mathbb{G}'_{d,1}(t)$$
$$+1=\frac{1}{\gamma^2}(\gamma^3+(2(\gamma-1)w_{\mathrm{d}}\mathbf{a}_{\mathrm{f}}(\rho_{2,1}\rho_{3,1}-\rho_{3,2})\mathbb{G}_{d,2}(t)\gamma)/(\sigma_{\mathrm{f}}(\rho_{2,1}^2-1))$$
$$+(2w_{\mathrm{d}}^2(\rho_{3,2}-\rho_{2,1}\rho_{3,1})^2\mathbb{G}_{d,1}(t)\mathbb{G}_{d,2}(t)\gamma)/(\rho_{2,1}^2-1)$$
$$+((\gamma-1)w_{\mathrm{f}}\mathbf{a}_{\mathrm{f}}(\rho_{2,1}\rho_{4,1}-\rho_{4,2})\mathbb{G}_{d,f}(t)\gamma)/(\sigma_{\mathrm{f}}(\rho_{2,1}^2-1))+(w_{\mathrm{d}}w_{\mathrm{f}}(\rho_{2,1}\rho_{3,1}$$
$$-\rho_{3,2})(\rho_{2,1}\rho_{4,1}-\rho_{4,2})\mathbb{G}_{d,1}(t)\mathbb{G}_{d,f}(t)\gamma)/(\rho_{2,1}^2-1)+(2w_{\mathrm{d}}w_{\mathrm{f}}$$
$$(\rho_{2,1}\rho_{3,1}-\rho_{3,2})(\rho_{2,1}\rho_{4,1}-\rho_{4,2})\mathbb{G}_{d,2}(t)\mathbb{G}_{f,1}(t)\gamma)/(\rho_{2,1}^2-1)$$
$$+(w_{\mathrm{f}}^2(\rho_{4,2}-\rho_{2,1}\rho_{4,1})^2\mathbb{G}_{d,f}(t)\mathbb{G}_{f,1}(t)\gamma)/(\rho_{2,1}^2-1)+\frac{(\gamma-1)\mathbf{a}_{\mathrm{f}}\gamma}{\sigma_{\mathrm{f}}^2(\rho_{2,1}^2-1)}$$
$$+((\gamma-1)\rho_{2,1}(-((\gamma-1)^2w_{\mathrm{f}}\rho_{4,2}(w_{\mathrm{f}}q_{d,1}\rho_{4,2}-w_{\mathrm{d}}q_{f,1}\rho_{2,1}\rho_{3,1})\sigma_{\mathrm{d}}^2$$
$$+(\gamma-1)\mathbf{a}_{\mathrm{d}}q_{f,1}(\gamma w_{\mathrm{d}}q_{f,1}\rho_{3,1}-w_{\mathrm{f}}q_{d,1}\rho_{2,1}\rho_{4,2})\sigma_{\mathrm{d}}+\gamma\mathbf{a}_{\mathrm{d}}^2q_{d,1}q_{f,1}^2)\sigma_{\mathrm{f}}^2$$
$$+\mathbf{a}_{\mathrm{f}}\sigma_{\mathrm{d}}q_{f,1}(\mathbf{a}_{\mathrm{d}}q_{d,1}q_{f,1}\rho_{2,1}+(\gamma-1)\sigma_{\mathrm{d}}(w_{\mathrm{d}}q_{f,1}\rho_{2,1}\rho_{3,1}-2w_{\mathrm{f}}q_{d,1}\rho_{4,2}))\sigma_{\mathrm{f}}$$
$$-\mathbf{a}_{\mathrm{f}}^2\sigma_{\mathrm{d}}^2q_{d,1}q_{f,1}^2))/(\sigma_{\mathrm{d}}\sigma_{\mathrm{f}}^3q_{f,1}^2(\rho_{2,1}^2-1)(\mathbf{a}_{\mathrm{d}}q_{d,1}+(\gamma-1)w_{\mathrm{d}}\sigma_{\mathrm{d}}\rho_{3,1}))$$
$$+(2(\gamma-1)w_{\mathrm{d}}\rho_{2,1}(\rho_{2,1}\rho_{3,1}-\rho_{3,2})(-((\gamma-1)^2w_{\mathrm{f}}$$

$$\rho_{4,2}(w_{\mathrm{f}}q_{d,1}\rho_{4,2}-w_{\mathrm{d}}q_{f,1}\rho_{2,1}\rho_{3,1})\sigma_{\mathrm{d}}^2+(\gamma-1)\mathbf{a}_{\mathrm{d}}q_{f,1}(\gamma w_{\mathrm{d}}q_{f,1}\rho_{3,1}$$
$$-w_{\mathrm{f}}q_{d,1}\rho_{2,1}\rho_{4,2})\sigma_{\mathrm{d}}+\gamma\mathbf{a}_{\mathrm{d}}^2q_{d,1}q_{f,1}^2)\sigma_{\mathrm{f}}^2+\mathbf{a}_{\mathrm{f}}\sigma_{\mathrm{d}}q_{f,1}(\mathbf{a}_{\mathrm{d}}q_{d,1}q_{f,1}\rho_{2,1}$$
$$+(\gamma-1)\sigma_{\mathrm{d}}(w_{\mathrm{d}}q_{f,1}\rho_{2,1}\rho_{3,1}-2w_{\mathrm{f}}q_{d,1}\rho_{4,2}))\sigma_{\mathrm{f}}-\mathbf{a}_{\mathrm{f}}^2\sigma_{\mathrm{d}}^2q_{d,1}q_{f,1}^2)\mathbb{G}_{d,2}(t))/$$
$$(\sigma_{\mathrm{d}}\sigma_{\mathrm{f}}^2q_{f,1}^2(\rho_{2,1}^2-1)(\mathbf{a}_{\mathrm{d}}q_{d,1}+(\gamma-1)w_{\mathrm{d}}\sigma_{\mathrm{d}}\rho_{3,1}))$$
$$+((\gamma-1)w_{\mathrm{f}}\rho_{4,1}(((\gamma-1)^2w_{\mathrm{f}}\rho_{4,2}(w_{\mathrm{f}}q_{d,1}\rho_{4,2}-w_{\mathrm{d}}q_{f,1}\rho_{2,1}\rho_{3,1})\sigma_{\mathrm{d}}^2$$
$$+(\gamma-1)\mathbf{a}_{\mathrm{d}}q_{f,1}(\gamma w_{\mathrm{d}}q_{f,1}\rho_{3,1}-w_{\mathrm{f}}q_{d,1}\rho_{2,1}\rho_{4,2})\sigma_{\mathrm{d}}+\gamma\mathbf{a}_{\mathrm{d}}^2q_{d,1}q_{f,1}^2)\sigma_{\mathrm{f}}^2$$

$$-\mathbf{a}_{\mathrm{f}}\sigma_{\mathrm{d}}q_{f,1}(\mathbf{a}_{\mathrm{d}}q_{d,1}q_{f,1}\rho_{2,1}+(\gamma-1)\sigma_{\mathrm{d}}(w_{\mathrm{d}}q_{f,1}\rho_{2,1}\rho_{3,1}-2w_{\mathrm{f}}q_{d,1}\rho_{4,2}))\sigma_{\mathrm{f}}$$
$$+\mathbf{a}_{\mathrm{f}}^2\sigma_{\mathrm{d}}^2q_{d,1}q_{f,1}^2)\mathbb{G}_{d,f}(t))/(\sigma_{\mathrm{d}}\sigma_{\mathrm{f}}^2q_{f,1}^2(\mathbf{a}_{\mathrm{d}}q_{d,1}+(\gamma-1)w_{\mathrm{d}}\sigma_{\mathrm{d}}\rho_{3,1}))), \tag{7.5.21}$$

$$\frac{1}{\gamma\sigma_{\mathrm{f}}(\rho_{2,1}^2-1)}(2w_{\mathrm{d}}^2(\gamma\rho_{2,1}^2-2(\gamma-1)\rho_{3,1}\rho_{3,2}\rho_{2,1}+(\gamma-1)\rho_{3,1}^2+\gamma\rho_{3,2}^2$$
$$-\rho_{3,2}^2-\gamma)\mathbb{G}_{d,2}(t)\mathbb{G}_{d,f}(t)\sigma_{\mathrm{f}}^2+(2\gamma\rho_{2,1}^2\mathbb{G}_{d,f}(t)\mathbb{G}_{f,2}(t)w_{\mathrm{f}}^2+2\gamma\rho_{4,1}^2\mathbb{G}_{d,f}(t)$$
$$\mathbb{G}_{f,2}(t)w_{\mathrm{f}}^2-2\rho_{4,1}^2\mathbb{G}_{d,f}(t)\mathbb{G}_{f,2}(t)w_{\mathrm{f}}^2+2\gamma\rho_{4,2}^2\mathbb{G}_{d,f}(t)\mathbb{G}_{f,2}(t)w_{\mathrm{f}}^2$$
$$-2\rho_{4,2}^2\mathbb{G}_{d,f}(t)\mathbb{G}_{f,2}(t)w_{\mathrm{f}}^2-2\gamma\mathbb{G}_{d,f}(t)\mathbb{G}_{f,2}(t)w_{\mathrm{f}}^2$$
$$-4\gamma\rho_{2,1}\rho_{4,1}\rho_{4,2}\mathbb{G}_{d,f}(t)\mathbb{G}_{f,2}(t)w_{\mathrm{f}}^2+4\rho_{2,1}\rho_{4,1}\rho_{4,2}\mathbb{G}_{d,f}(t)\mathbb{G}_{f,2}(t)w_{\mathrm{f}}^2$$
$$+\gamma q_{d,1}(\rho_{2,1}^2-1)\mathbb{G}_{d,f}(t)+\gamma q_{f,1}(\rho_{2,1}^2-1)\mathbb{G}_{d,f}(t)+\gamma\rho_{2,1}^2\mathbb{G}'_{d,f}(t)$$
$$-\gamma\mathbb{G}'_{d,f}(t))\sigma_{\mathrm{f}}^2-(\gamma-1)w_{\mathrm{f}}(\rho_{2,1}\rho_{4,1}-\rho_{4,2})(\mathbb{G}_{d,f}(t)-2\mathbb{G}_{f,2}(t))\sigma_{\mathrm{f}}$$
$$+w_{\mathrm{d}}(\gamma w_{\mathrm{f}}\sigma_{\mathrm{f}}\rho_{4,3}(\mathbb{G}_{d,f}(t)^2+4\mathbb{G}_{d,2}(t)\mathbb{G}_{f,2}(t))\rho_{2,1}^2$$
$$-(\gamma-1)(w_{\mathrm{f}}\sigma_{\mathrm{f}}\rho_{3,2}\rho_{4,1}(\mathbb{G}_{d,f}(t)^2+4\mathbb{G}_{d,2}(t)\mathbb{G}_{f,2}(t))$$
$$+\rho_{3,1}(\mathbb{G}_{d,f}(t)(w_{\mathrm{f}}\sigma_{\mathrm{f}}\rho_{4,2}\mathbb{G}_{d,f}(t)-1)+\mathbb{G}_{d,2}(t)(4w_{\mathrm{f}}\sigma_{\mathrm{f}}\rho_{4,2}\mathbb{G}_{f,2}(t)+2)))\rho_{2,1}$$
$$+w_{\mathrm{f}}\sigma_{\mathrm{f}}((\gamma-1)\rho_{3,1}\rho_{4,1}-\gamma\rho_{4,3})(\mathbb{G}_{d,f}(t)^2+4\mathbb{G}_{d,2}(t)\mathbb{G}_{f,2}(t))$$
$$+(\gamma-1)\rho_{3,2}(\mathbb{G}_{d,f}(t)(w_{\mathrm{f}}\sigma_{\mathrm{f}}\rho_{4,2}\mathbb{G}_{d,f}(t)-1)$$
$$+\mathbb{G}_{d,2}(t)(4w_{\mathrm{f}}\sigma_{\mathrm{f}}\rho_{4,2}\mathbb{G}_{f,2}(t)+2)))\sigma_{\mathrm{f}}-\gamma+1)=0, \tag{7.5.22}$$

$$\frac{1}{\gamma\sigma_{\mathrm{f}}(\rho_{2,1}^2-1)}((\gamma w_{\mathrm{f}}^2\mathbb{G}_{d,f}(t)^2\rho_{2,1}^2+2\gamma\mathbb{G}'_{d,2}(t)\rho_{2,1}^2-2\gamma w_{\mathrm{f}}^2\rho_{4,1}\rho_{4,2}\mathbb{G}_{d,f}(t)^2\rho_{2,1}$$
$$+2w_{\mathrm{f}}^2\rho_{4,1}\rho_{4,2}\mathbb{G}_{d,f}(t)^2\rho_{2,1}+4w_{\mathrm{d}}^2(\gamma\rho_{2,1}^2-2(\gamma-1)\rho_{3,1}\rho_{3,2}\rho_{2,1}+(\gamma-1)\rho_{3,1}^2$$
$$+\gamma\rho_{3,2}^2-\rho_{3,2}^2-\gamma)\mathbb{G}_{d,2}(t)^2-\gamma w_{\mathrm{f}}^2\mathbb{G}_{d,f}(t)^2+\gamma w_{\mathrm{f}}^2\rho_{4,1}^2\mathbb{G}_{d,f}(t)^2$$
$$-w_{\mathrm{f}}^2\rho_{4,1}^2\mathbb{G}_{d,f}(t)^2+\gamma w_{\mathrm{f}}^2\rho_{4,2}^2\mathbb{G}_{d,f}(t)^2-w_{\mathrm{f}}^2\rho_{4,2}^2\mathbb{G}_{d,f}(t)^2$$
$$+4\gamma q_{d,1}(\rho_{2,1}^2-1)\mathbb{G}_{d,2}(t)+4w_{\mathrm{d}}w_{\mathrm{f}}(\gamma\rho_{4,3}\rho_{2,1}^2-(\gamma-1)\rho_{3,2}\rho_{4,1}\rho_{2,1}$$
$$+\gamma\rho_{3,2}\rho_{4,2}-\rho_{3,2}\rho_{4,2}+(\gamma-1)\rho_{3,1}(\rho_{4,1}-\rho_{2,1}\rho_{4,2})$$
$$-\gamma\rho_{4,3})\mathbb{G}_{d,2}(t)\mathbb{G}_{d,f}(t)-2\gamma\mathbb{G}'_{d,2}(t))\sigma_{\mathrm{f}}^2+2(\gamma-1)(2w_{\mathrm{d}}(\rho_{2,1}\rho_{3,1}$$
$$-\rho_{3,2})\mathbb{G}_{d,2}(t)+w_{\mathrm{f}}(\rho_{2,1}\rho_{4,1}-\rho_{4,2})\mathbb{G}_{d,f}(t))\sigma_{\mathrm{f}}+\gamma-1)=0, \tag{7.5.23}$$

for $t<T$, together with the terminal condition $\mathbb{G}_{d,2}(T)=\mathbb{G}_{d,1}(T)=\mathbb{G}_{d,f}(T)=\mathbb{G}_{f,1}(T)=\mathbb{G}_{f,2}(T)=0$, which can be solved numerically quite easily. We use

(7.5.18) in the pricing equation (4.9.7), and look for the solution in the form (7.5.27), obtaining three ODEs characterizing $\mathbb{V}_0(t), \mathbb{V}_d(t)$, and $\mathbb{V}_f(t)$:

$$\begin{aligned}
&\frac{\frac{\mathbf{a}_f}{\sigma_f} - \gamma\sigma_f + \frac{(\gamma-1)w_f\rho_{4,2}}{q_{f,1}}}{\gamma} + \frac{1}{\rho_{2,1}^2-1}\mathbb{V}_f(t)(w_f(\rho_{4,2}-\rho_{2,1}\rho_{4,1})\\
&\quad+\sigma_f(q_{f,1}(\rho_{2,1}^2-1)+w_f(2w_f(\rho_{2,1}^2-2\rho_{4,1}\rho_{4,2}\rho_{2,1}+\rho_{4,1}^2+\rho_{4,2}^2-1)\mathbb{G}_{f,2}(t)\\
&\quad+w_d(\rho_{3,2}(\rho_{4,2}-\rho_{2,1}\rho_{4,1})+\rho_{3,1}(\rho_{4,1}-\rho_{2,1}\rho_{4,2})+(\rho_{2,1}^2-1)\rho_{4,3})\mathbb{G}_{d,f,1}(t))))\\
&\quad+\frac{1}{\rho_{2,1}^2-1}w_d\mathbb{V}_d(t)(\sigma_f(2w_f\rho_{4,3}\mathbb{G}_{f,2}(t)+w_d\mathbb{G}_{d,f,1}(t))\rho_{2,1}^2\\
&\quad-(2w_f\sigma_f\rho_{3,2}\rho_{4,1}\mathbb{G}_{f,2}(t)+\rho_{3,1}(2\sigma_f(w_f\rho_{4,2}\mathbb{G}_{f,2}(t)+w_d\rho_{3,2}\mathbb{G}_{d,f,1}(t))\\
&\quad+1))\rho_{2,1}+\rho_{3,2}+\sigma_f(2w_f(\rho_{3,1}\rho_{4,1}+\rho_{3,2}\rho_{4,2}-\rho_{4,3})\mathbb{G}_{f,2}(t)\\
&\quad+w_d(\rho_{3,1}^2+\rho_{3,2}^2-1)\mathbb{G}_{d,f,1}(t)))+\sigma_f\mathbb{V}_f'(t)=0,
\end{aligned} \tag{7.5.24}$$

$$\begin{aligned}
&\frac{-\frac{\mathbf{a}_f}{\sigma_f} + \gamma\sigma_f - \frac{(\gamma-1)w_f\rho_{4,2}}{q_{f,1}}}{\gamma} + \frac{1}{\rho_{2,1}^2-1}\mathbb{V}_d(t)(w_d(\rho_{2,1}\rho_{3,1}-\rho_{3,2})\\
&\quad+\sigma_f(q_{d,1}(\rho_{2,1}^2-1)+w_d(2w_d(\rho_{2,1}^2-2\rho_{3,1}\rho_{3,2}\rho_{2,1}+\rho_{3,1}^2+\rho_{3,2}^2-1)\mathbb{G}_{d,2}(t)\\
&\quad+w_f(\rho_{3,2}(\rho_{4,2}-\rho_{2,1}\rho_{4,1})+\rho_{3,1}(\rho_{4,1}-\rho_{2,1}\rho_{4,2})+(\rho_{2,1}^2-1)\rho_{4,3})\mathbb{G}_{d,f,1}(t))))\\
&\quad+\frac{1}{\rho_{2,1}^2-1}w_f\mathbb{V}_f(t)(\sigma_f(2w_d\rho_{4,3}\mathbb{G}_{d,2}(t)+w_f\mathbb{G}_{d,f,1}(t))\rho_{2,1}^2\\
&\quad+(\rho_{4,1}(1-2\sigma_f(w_d\rho_{3,2}\mathbb{G}_{d,2}(t)+w_f\rho_{4,2}\mathbb{G}_{d,f,1}(t)))\\
&\quad-2w_d\sigma_f\rho_{3,1}\rho_{4,2}\mathbb{G}_{d,2}(t))\rho_{2,1}-\rho_{4,2}+\sigma_f(2w_d(\rho_{3,1}\rho_{4,1}+\rho_{3,2}\rho_{4,2}\\
&\quad-\rho_{4,3})\mathbb{G}_{d,2}(t)+w_f(\rho_{4,1}^2+\rho_{4,2}^2-1)\mathbb{G}_{d,f,1}(t)))+\sigma_f\mathbb{V}_d'(t)=0,
\end{aligned} \tag{7.5.25}$$

$$\begin{aligned}
&\frac{1}{2}w_d^2\mathbb{V}_d(t)^2+w_dw_f\rho_{4,3}\mathbb{V}_f(t)\mathbb{V}_d(t)+\frac{1}{2}w_f^2\mathbb{V}_f(t)^2+(\mathbb{V}_d(t)(w_d((\gamma-1)\\
&\quad(\gamma q_{d,0}q_{d,1}(\rho_{2,1}^2-1)\rho_{3,1}q_{f,1}^2+w_d^2\rho_{3,1}((\gamma-1)(\rho_{2,1}^2-1)\rho_{3,1}^2\\
&\quad+\gamma q_{d,1}(\rho_{2,1}^2-2\rho_{3,1}\rho_{3,2}\rho_{2,1}+\rho_{3,1}^2+\rho_{3,2}^2-1)\mathbb{G}_{d,1}(t))q_{f,1}^2\\
&\quad+w_dw_fq_{d,1}\rho_{3,1}(\gamma q_{f,1}(\rho_{3,2}(\rho_{4,2}-\rho_{2,1}\rho_{4,1})+\rho_{3,1}(\rho_{4,1}-\rho_{2,1}\rho_{4,2})\\
&\quad+(\rho_{2,1}^2-1)\rho_{4,3})\mathbb{G}_{f,1}(t)-(\gamma-1)(\rho_{2,1}\rho_{3,1}-\rho_{3,2})\rho_{4,2})q_{f,1}\\
&\quad+(\gamma-1)w_f^2q_{d,1}^2(\rho_{3,1}-\rho_{2,1}\rho_{3,2})\rho_{4,2}^2)\sigma_f^2-(\gamma-1)\mathbf{a}_fq_{d,1}q_{f,1}((\gamma+1)\\
&\quad w_dq_{f,1}\rho_{3,1}(\rho_{2,1}\rho_{3,1}-\rho_{3,2})-2w_fq_{d,1}(\rho_{3,1}-\rho_{2,1}\rho_{3,2})\rho_{4,2})\sigma_f\\
&\quad+\mathbf{a}_f^2q_{d,1}^2q_{f,1}^2(\rho_{3,1}-\rho_{2,1}\rho_{3,2}))\sigma_d^2+\mathbf{a}_d\sigma_fq_{d,1}q_{f,1}(\sigma_f(\gamma q_{d,0}q_{d,1}q_{f,1}(\rho_{2,1}^2-1)
\end{aligned}$$

$$
\begin{aligned}
&+w_{\mathrm{d}}(w_{\mathrm{d}}q_{f,1}((\gamma-1)\rho_{3,1}((2\rho_{2,1}^2+\gamma-2)\rho_{3,1}-\gamma\rho_{2,1}\rho_{3,2})\\
&+\gamma q_{d,1}(\rho_{2,1}^2-2\rho_{3,1}\rho_{3,2}\rho_{2,1}+\rho_{3,1}^2+\rho_{3,2}^2-1)\mathbb{G}_{d,1}(t))\\
&+w_{\mathrm{f}}q_{d,1}(\gamma q_{f,1}(\rho_{3,2}(\rho_{4,2}-\rho_{2,1}\rho_{4,1})+\rho_{3,1}(\rho_{4,1}-\rho_{2,1}\rho_{4,2})\\
&+(\rho_{2,1}^2-1)\rho_{4,3})\mathbb{G}_{f,1}(t)-(\gamma-1)(\rho_{2,1}\rho_{3,1}-\rho_{3,2})\rho_{4,2})))\\
&-(\gamma+1)w_{\mathrm{d}}\mathbf{a}_{\mathrm{f}}q_{d,1}q_{f,1}(\rho_{2,1}\rho_{3,1}-\rho_{3,2}))\sigma_{\mathrm{d}}+w_{\mathrm{d}}\mathbf{a}_{\mathrm{d}}^2\sigma_{\mathrm{f}}^2q_{d,1}^2q_{f,1}^2\\
&((\rho_{2,1}^2+\gamma-1)\rho_{3,1}-\gamma\rho_{2,1}\rho_{3,2})))/(\gamma\sigma_{\mathrm{d}}\sigma_{\mathrm{f}}^2q_{d,1}q_{f,1}^2(\rho_{2,1}^2-1)(\mathbf{a}_{\mathrm{d}}q_{d,1}\\
&+(\gamma-1)w_{\mathrm{d}}\sigma_{\mathrm{d}}\rho_{3,1}))+(\mathbb{V}_f(t)(((\gamma-1)w_{\mathrm{d}}^2w_{\mathrm{f}}\sigma_{\mathrm{f}}^2\rho_{3,1}((\gamma-1)(\rho_{2,1}^2-1)\rho_{3,1}\rho_{4,1}\\
&+\gamma q_{d,1}(\rho_{3,2}(\rho_{4,2}-\rho_{2,1}\rho_{4,1})+\rho_{3,1}(\rho_{4,1}-\rho_{2,1}\rho_{4,2})\\
&+(\rho_{2,1}^2-1)\rho_{4,3})\mathbb{G}_{d,1}(t))q_{f,1}^2+(\gamma-1)w_{\mathrm{d}}\sigma_{\mathrm{f}}q_{d,1}\rho_{3,1}(\sigma_{\mathrm{f}}((\gamma q_{f,1}(\rho_{2,1}^2\\
&-2\rho_{4,1}\rho_{4,2}\rho_{2,1}+\rho_{4,1}^2+\rho_{4,2}^2-1)\mathbb{G}_{f,1}(t)-(\gamma-1)(\rho_{2,1}\rho_{4,1}-\rho_{4,2})\rho_{4,2})w_{\mathrm{f}}^2\\
&+\gamma q_{f,0}q_{f,1}(\rho_{2,1}^2-1))-(\gamma+1)w_{\mathrm{f}}\mathbf{a}_{\mathrm{f}}q_{f,1}(\rho_{2,1}\rho_{4,1}-\rho_{4,2}))q_{f,1}\\
&+w_{\mathrm{f}}q_{d,1}^2(\mathbf{a}_{\mathrm{f}}q_{f,1}+(\gamma-1)w_{\mathrm{f}}\sigma_{\mathrm{f}}\rho_{4,2})^2(\rho_{4,1}-\rho_{2,1}\rho_{4,2}))\sigma_{\mathrm{d}}^2\\
&+\mathbf{a}_{\mathrm{d}}\sigma_{\mathrm{f}}q_{d,1}q_{f,1}(\sigma_{\mathrm{f}}((\gamma-1)w_{\mathrm{d}}w_{\mathrm{f}}q_{f,1}\rho_{3,1}((2\rho_{2,1}^2+\gamma-2)\rho_{4,1}-\gamma\rho_{2,1}\rho_{4,2})\\
&+q_{d,1}(\gamma q_{f,0}q_{f,1}(\rho_{2,1}^2-1)+w_{\mathrm{f}}(\gamma w_{\mathrm{d}}q_{f,1}(\rho_{3,2}(\rho_{4,2}-\rho_{2,1}\rho_{4,1})\\
&+\rho_{3,1}(\rho_{4,1}-\rho_{2,1}\rho_{4,2})+(\rho_{2,1}^2-1)\rho_{4,3})\mathbb{G}_{d,1}(t)\\
&+w_{\mathrm{f}}(\gamma q_{f,1}(\rho_{2,1}^2-2\rho_{4,1}\rho_{4,2}\rho_{2,1}+\rho_{4,1}^2+\rho_{4,2}^2-1)\mathbb{G}_{f,1}(t)-(\gamma-1)(\rho_{2,1}\rho_{4,1}\\
&-\rho_{4,2})\rho_{4,2}))))-(\gamma+1)w_{\mathrm{f}}\mathbf{a}_{\mathrm{f}}q_{d,1}q_{f,1}(\rho_{2,1}\rho_{4,1}-\rho_{4,2}))\sigma_{\mathrm{d}}\\
&+w_{\mathrm{f}}\mathbf{a}_{\mathrm{d}}^2\sigma_{\mathrm{f}}^2q_{d,1}^2q_{f,1}^2((\rho_{2,1}^2+\gamma-1)\rho_{4,1}-\gamma\rho_{2,1}\rho_{4,2})))/\\
&(\gamma\sigma_{\mathrm{d}}\sigma_{\mathrm{f}}^2q_{d,1}q_{f,1}^2(\rho_{2,1}^2-1)(\mathbf{a}_{\mathrm{d}}q_{d,1}+(\gamma-1)w_{\mathrm{d}}\sigma_{\mathrm{d}}\rho_{3,1}))+\mathbb{V}_0'(t)=0, \quad (7.5.26)
\end{aligned}
$$

for $t < T$, together with the terminal condition $\mathbb{V}_{\mathrm{d}}(T)=\mathbb{V}_f(T)=\mathbb{V}_0(T)=0$, which can be solved numerically quite easily. We summarize by stating the following proposition.

Proposition 7.5.2. *Under the stochastic interest rate model, assuming $\beta_{\mathrm{d}} = \beta_{\mathrm{f}} = 0$, and assuming that the hedging is done via trading in the foreign currency and foreign risky assets ($S = \{Y_{\mathrm{d,f}}, S_2\}$), an FX futures contract expiring at T has CRRA ($\kappa = 0$)-neutral (or equivalently, ($\kappa = 0$)-indifference) price at time $t < T$ equal to*

$$V_\gamma(t,S_1,S_2,r_{\mathrm{d}},r_{\mathrm{f}},Y_{\mathrm{d,f}}) = V_\gamma(t,r_{\mathrm{d}},r_{\mathrm{f}},Y_{\mathrm{d,f}}) = Y_{\mathrm{d,f}}e^{\mathbb{V}_0(t)+r_{\mathrm{d}}\mathbb{V}_{\mathrm{d}}(t)+r_{\mathrm{f}}\mathbb{V}_f(t)}, \quad (7.5.27)$$

with $\{\mathbb{G}_{d,2}(t), \mathbb{G}_{d,1}(t), \mathbb{G}_{d,f}(t), \mathbb{G}_{f,1}(t), \mathbb{G}_{f,2}(t), \mathbb{V}_0(t), \mathbb{V}_{\mathrm{d}}(t), \mathbb{V}_f(t)\}$ *the solution of the ODE system (7.5.19)–(7.5.26). Furthermore, the (most conservative) hedging formula for the position* $\kappa = \{\kappa_1\}$ *(see (5.2.3))reads*

$$\begin{aligned}
\Pi^{\mathbf{h}}(t, r_{\mathrm{d}}, r_{\mathrm{f}}, Y_{\mathrm{d,f}}) = \Big\{ & -(\mathrm{e}^{\mathbb{V}_0(t)+r_{\mathrm{d}}\mathbb{V}_{\mathrm{d}}(t)+r_{\mathrm{f}}\mathbb{V}_f(t)}\kappa_1 Y_{\mathrm{d,f}}(\mathbf{a}_{\mathrm{d}} q_{d,1}(\rho_{2,1}^2 - 1) \\
& +\sigma_{\mathrm{d}}(w_{\mathrm{d}}(\gamma q_{d,1}\rho_{2,1}\rho_{3,2}\mathbb{V}_{\mathrm{d}}(t) + \rho_{3,1}((\gamma-1)\rho_{2,1}^2 - \gamma \\
& -\gamma q_{d,1}\mathbb{V}_{\mathrm{d}}(t) + 1)) + \gamma w_{\mathrm{f}} q_{d,1}(\rho_{2,1}\rho_{4,2} - \rho_{4,1})\mathbb{V}_f(t))))/ \\
& ((\rho_{2,1}^2 - 1)(\mathbf{a}_{\mathrm{d}} q_{d,1} + (\gamma-1) w_{\mathrm{d}}\sigma_{\mathrm{d}}\rho_{3,1})), \\
& \frac{1}{\gamma\sigma_{\mathrm{f}}^2 q_{f,1}(\rho_{2,1}^2 - 1)} \mathrm{e}^{\mathbb{V}_0(t)+r_{\mathrm{d}}\mathbb{V}_{\mathrm{d}}(t)+r_{\mathrm{f}}\mathbb{V}_f(t)}\kappa_1 Y_{\mathrm{d,f}}(\mathbf{a}_{\mathrm{f}} q_{f,1}(\rho_{2,1}^2 - 1) \\
& +\sigma_{\mathrm{f}}(\gamma w_{\mathrm{d}} q_{f,1}(\rho_{3,2} - \rho_{2,1}\rho_{3,1})\mathbb{V}_{\mathrm{d}}(t) + w_{\mathrm{f}}(\rho_{4,2}((\gamma-1)\rho_{2,1}^2 \\
& -\gamma + \gamma q_{f,1}\mathbb{V}_f(t) + 1) - \gamma q_{f,1}\rho_{2,1}\rho_{4,1}\mathbb{V}_f(t))))\Big\}, \qquad (7.5.28)
\end{aligned}$$

where the first component represents the hedging position in the foreign currency, while the second component represents the corresponding position in the foreign risky assets.

Remark 7.5.1. Combining the above methods with Fourier transform methodology, one can solve the problem of pricing of some other kinds of FX derivatives, including options (see [43]).

Appendix A
Model Building: Correlations

A.1 Model Building

For the completeness of the presentation, and to facilitate applicability and empirical verification of the results presented in this book, we also present some useful long correlations formulas, derived and used symbolically throughout this book.

Numbers $\rho_{i,j}$ denote (mutual, imperfect) correlations, so that we shall assume that $-1 < \rho_{i,j} < 1$, or equivalently,

$$1 - \rho_{i,j}^2 > 0. \tag{A.1.1}$$

Being *mutual* correlations, they have to fulfill additional conditions, such as, for example,

$$1 - \rho_{3,1}^2 - \frac{(\rho_{3,2} - \rho_{2,1}\rho_{3,1})^2}{1 - \rho_{2,1}^2} > 0, \tag{A.1.2}$$

i.e.,

$$1 + 2\rho_{2,1}\rho_{3,1}\rho_{3,2} - \left(\rho_{2,1}^2 + \rho_{3,1}^2 + \rho_{3,2}^2\right) > 0 \tag{A.1.3}$$

(indeed, if two correlations, say $\rho_{2,1}$ and $\rho_{3,2}$ are "strong," i.e., close to ± 1, then $\rho_{3,1}$ has to be strong as well) etc., which will be evident from the formulas bellow.

Consider also corresponding correlation matrices P_n. For example, in dimension $n = 3$,

$$P_3 := \begin{pmatrix} 1 & \rho_{2,1} & \rho_{3,1} \\ \rho_{2,1} & 1 & \rho_{3,2} \\ \rho_{3,1} & \rho_{3,2} & 1 \end{pmatrix}. \tag{A.1.4}$$

Define

$$\mathbb{S}_1 := \left(\mathbf{s}_{1,1} \right) = \left(1 \right), \tag{A.1.5}$$

S. Stojanovic, *Neutral and Indifference Portfolio Pricing, Hedging and Investing: With Applications in Equity and FX*, DOI 10.1007/978-0-387-71418-9_8,

a 1×1 matrix, such that

$$\mathbb{S}_1.\mathbb{S}_1{}^T = \begin{pmatrix} 1 \end{pmatrix} = P_1, \tag{A.1.6}$$

the 1×1 correlation matrix. Then define

$$\mathbb{S}_2 := \begin{pmatrix} 1 & 0 \\ \rho_{2,1} & \mathbf{s}_{2,2} \end{pmatrix}, \tag{A.1.7}$$

i.e., choose $\mathbf{s}_{2,2}$ such that

$$\mathbb{S}_2.\mathbb{S}_2{}^T = \begin{pmatrix} 1 & \rho_{2,1} \\ \rho_{2,1} & \mathbf{s}_{2,2}^2 + \rho_{2,1}^2 \end{pmatrix} = P_2 = \begin{pmatrix} 1 & \rho_{2,1} \\ \rho_{2,1} & 1 \end{pmatrix}. \tag{A.1.8}$$

This yields quadratic equation

$$\mathbf{s}_{2,2}^2 + \rho_{2,1}^2 = 1, \tag{A.1.9}$$

and the usual choice of a solution for $\mathbf{s}_{2,2}$ is

$$\mathbf{s}_{2,2} = \sqrt{1 - \rho_{2,1}^2}, \tag{A.1.10}$$

so that

$$\mathbb{S}_2 = \begin{pmatrix} 1 & 0 \\ \rho_{2,1} & \sqrt{1 - \rho_{2,1}^2} \end{pmatrix}. \tag{A.1.11}$$

We notice that a matrix square root evaluates as

$$\sqrt{P_2} = \begin{pmatrix} \frac{1}{2}\sqrt{1-\rho_{2,1}} + \frac{1}{2}\sqrt{\rho_{2,1}+1} & \frac{1}{2}\sqrt{\rho_{2,1}+1} - \frac{1}{2}\sqrt{1-\rho_{2,1}} \\ \frac{1}{2}\sqrt{\rho_{2,1}+1} - \frac{1}{2}\sqrt{1-\rho_{2,1}} & \frac{1}{2}\sqrt{1-\rho_{2,1}} + \frac{1}{2}\sqrt{\rho_{2,1}+1} \end{pmatrix}, \tag{A.1.12}$$

which, being symmetric, satisfies the requirement

$$\sqrt{P_2} \cdot \left(\sqrt{P_2}\right)^T = P_2, \tag{A.1.13}$$

but it is much more complicated then $\mathbb{S}_2$, and similarly in higher dimensions. Now define

$$\mathbb{S}_3 = \begin{pmatrix} 1 & 0 & 0 \\ \rho_{2,1} & \sqrt{1 - \rho_{2,1}^2} & 0 \\ \rho_{3,1} & \mathbf{s}_{3,2} & \mathbf{s}_{3,3} \end{pmatrix}, \tag{A.1.14}$$

and choose $\mathbf{s}_{3,2}$ and $\mathbf{s}_{3,3}$ such that

$$\mathbb{S}_3.\mathbb{S}_3{}^T = \begin{pmatrix} 1 & \rho_{2,1} & \rho_{3,1} \\ \rho_{2,1} & 1 & \sqrt{1-\rho_{2,1}^2}\mathbf{s}_{3,2}+\rho_{2,1}\rho_{3,1} \\ \rho_{3,1} & \sqrt{1-\rho_{2,1}^2}\mathbf{s}_{3,2}+\rho_{2,1}\rho_{3,1} & \mathbf{s}_{3,2}^2+\mathbf{s}_{3,3}^2+\rho_{3,1}^2 \end{pmatrix}$$

$$= \begin{pmatrix} 1 & \rho_{2,1} & \rho_{3,1} \\ \rho_{2,1} & 1 & \rho_{3,2} \\ \rho_{3,1} & \rho_{3,2} & 1 \end{pmatrix}. \tag{A.1.15}$$

This yields a linear equation for $\mathbf{s}_{3,2}$,

$$\sqrt{1-\rho_{2,1}^2}\mathbf{s}_{3,2}+\rho_{2,1}\rho_{3,1} = \rho_{3,2}, \tag{A.1.16}$$

which can be solved first, and a quadratic equation for $\mathbf{s}_{3,3}$,

$$\mathbf{s}_{3,2}^2+\mathbf{s}_{3,3}^2+\rho_{3,1}^2 = 1, \tag{A.1.17}$$

so that

$$\mathbb{S}_3 = \begin{pmatrix} 1 & 0 & 0 \\ \rho_{2,1} & \sqrt{1-\rho_{2,1}^2} & 0 \\ \rho_{3,1} & \dfrac{\rho_{3,2}-\rho_{2,1}\rho_{3,1}}{\sqrt{1-\rho_{2,1}^2}} & \sqrt{1-\rho_{3,1}^2+\dfrac{(\rho_{3,2}-\rho_{2,1}\rho_{3,1})^2}{\rho_{2,1}^2-1}} \end{pmatrix}. \tag{A.1.18}$$

This procedure can be continued by induction,

$$\mathbb{S}_{n+1} := \begin{pmatrix} \mathbb{S}_n & \begin{pmatrix} 0 \\ \vdots \\ 0 \end{pmatrix} \\ \begin{pmatrix} \rho_{n+1,1} & \cdot\cdot & \mathbf{s}_{n+1,n} \end{pmatrix} & \begin{pmatrix} \mathbf{s}_{n+1,n+1} \end{pmatrix} \end{pmatrix}, \tag{A.1.19}$$

so that $\mathbb{S}_{n+1} \cdot \mathbb{S}_{n+1}{}^T$ is the $(n+1)\times(n+1)$ correlation matrix with given correlations $\rho_{i,j}, i > j$.

Also, multiplying the matrix $\mathbb{S}_n$ by a vector of standard deviations, we obtain a diffusion matrix. For example, in the 2-dimensional log-normal price model, set

$$\{S_1\mathbf{p}_1, S_2\mathbf{p}_2\}\mathbb{S}_2 = \begin{pmatrix} S_1\mathbf{p}_1 & 0 \\ S_2\mathbf{p}_2\rho_{2,1} & S_2\mathbf{p}_2\sqrt{1-\rho_{2,1}^2} \end{pmatrix} \tag{A.1.20}$$

and then

$$\begin{pmatrix} S_1\mathbf{p}_1 & 0 \\ S_2\mathbf{p}_2\rho_{2,1} & S_2\mathbf{p}_2\sqrt{1-\rho_{2,1}^2} \end{pmatrix} \cdot \begin{pmatrix} S_1\mathbf{p}_1 & 0 \\ S_2\mathbf{p}_2\rho_{2,1} & S_2\mathbf{p}_2\sqrt{1-\rho_{2,1}^2} \end{pmatrix}^T$$
$$= \begin{pmatrix} S_1^2\mathbf{p}_1^2 & S_1S_2\mathbf{p}_1\mathbf{p}_2\rho_{2,1} \\ S_1S_2\mathbf{p}_1\mathbf{p}_2\rho_{2,1} & S_2^2\mathbf{p}_2^2 \end{pmatrix},$$

which will be the price covariance matrix, with $\mathbf{p}_1$ and $\mathbf{p}_2$ the price volatilities.

Once the matrix $\mathbb{S}_n$ is known, it is easy to build models with n sources of randomness (the number of factors can be larger). For example, the 7×5 matrix c in (6.3.5) is generated as follows (using *Mathematica* syntax):

$$c = \text{Insert}\left[\text{Insert}\left[\{\mathbf{p}_0S_0, w_{1,\mathbf{i},0}+w_{1,\mathbf{i},1}i_1, w_{1,\mathbf{q}_0,0}+w_{1,\mathbf{q}_0,1}q_1, w_{2,\mathbf{i},0}+w_{2,\mathbf{i},1}\right.\right.$$
$$\left.\left. i_2, w_{2,\mathbf{q}_0,0}+w_{2,\mathbf{q}_0,1}q_2\}\,\mathbb{S}_5, \{0,0,0,0,0\},2\right], \{0,0,0,0,0\},5\right]. \quad \text{(A.1.21)}$$

A.2 Some Long Formulas

For completeness of presentation, and to facilitate immediate applicability of this book for the user, we list some long formulas for $\mathbf{s}_{j,i}$, for $1\le i\le j\le 5$ ($\mathbf{s}_{j,i}=0$, for $i>j$):

$$\mathbf{s}_{1,1} = 1, \quad \text{(A.2.22)}$$
$$\mathbf{s}_{2,1} = \rho_{2,1}, \quad \text{(A.2.23)}$$
$$\mathbf{s}_{2,2} = \sqrt{1-\rho_{2,1}^2}, \quad \text{(A.2.24)}$$
$$\mathbf{s}_{3,1} = \rho_{3,1}, \quad \text{(A.2.25)}$$
$$\mathbf{s}_{3,2} = \frac{-\rho_{2,1}\rho_{3,1}+\rho_{3,2}}{\sqrt{1-\rho_{2,1}^2}}, \quad \text{(A.2.26)}$$
$$\mathbf{s}_{3,3} = \sqrt{\left(\frac{1}{1-\rho_{2,1}^2}\left(1+2\rho_{2,1}\rho_{3,1}\rho_{3,2}-\left(\rho_{2,1}^2+\rho_{3,1}^2+\rho_{3,2}^2\right)\right)\right)}, \quad \text{(A.2.27)}$$
$$\mathbf{s}_{4,1} = \rho_{4,1}, \quad \text{(A.2.28)}$$
$$\mathbf{s}_{4,2} = \frac{-\rho_{2,1}\rho_{4,1}+\rho_{4,2}}{\sqrt{1-\rho_{2,1}^2}}, \quad \text{(A.2.29)}$$
$$\mathbf{s}_{4,3} = (-\rho_{2,1}\rho_{3,2}\rho_{4,1}+\rho_{3,2}\rho_{4,2}+\rho_{3,1}(\rho_{4,1}-\rho_{2,1}\rho_{4,2})-\rho_{4,3}+\rho_{2,1}^2\rho_{4,3})/$$

$$((-1+\rho_{2,1}^2)\surd((1+2\rho_{2,1}\rho_{3,1}\rho_{3,2}-(\rho_{2,1}^2+\rho_{3,1}^2+\rho_{3,2}^2))/(1-\rho_{2,1}^2))), \quad \text{(A.2.30)}$$

$$\begin{aligned}
\mathbf{s}_{4,4} = {} & \surd((-1+\rho_{3,2}^2+\rho_{4,1}^2-\rho_{3,2}^2\rho_{4,1}^2+\rho_{4,2}^2-\rho_{3,1}^2 \\
& (-1+\rho_{4,2}^2)+2\rho_{3,1}\rho_{4,1}(\rho_{3,2}\rho_{4,2}-\rho_{4,3})-2\rho_{3,2}\rho_{4,2}\rho_{4,3}+\rho_{4,3}^2-\rho_{2,1}^2 \\
& (-1+\rho_{4,3}^2)-2\rho_{2,1}(\rho_{4,1}(\rho_{4,2}-\rho_{3,2}\rho_{4,3}) \\
& +\rho_{3,1}(\rho_{3,2}-\rho_{4,2}\rho_{4,3})))/(-1+\rho_{2,1}^2+\rho_{3,1}^2-2\rho_{2,1}\rho_{3,1}\rho_{3,2}+\rho_{3,2}^2)), \quad \text{(A.2.31)}
\end{aligned}$$

$$\mathbf{s}_{5,1} = \rho_{5,1}, \quad \text{(A.2.32)}$$

$$\mathbf{s}_{5,2} = \frac{-\rho_{2,1}\rho_{5,1}+\rho_{5,2}}{\sqrt{1-\rho_{2,1}^2}}, \quad \text{(A.2.33)}$$

$$\begin{aligned}
\mathbf{s}_{5,3} = {} & (-\rho_{2,1}\rho_{3,2}\rho_{5,1}+\rho_{3,2}\rho_{5,2}+\rho_{3,1}(\rho_{5,1}-\rho_{2,1}\rho_{5,2})-\rho_{5,3}+\rho_{2,1}^2\rho_{5,3})/ \\
& ((-1+\rho_{2,1}^2)\surd((-1+\rho_{2,1}^2+\rho_{3,1}^2-2\rho_{2,1}\rho_{3,1}\rho_{3,2}+\rho_{3,2}^2)/(-1+\rho_{2,1}^2))) \quad \text{(A.2.34)}
\end{aligned}$$

$$\begin{aligned}
\mathbf{s}_{5,4} = {} & \Bigg(\Bigg(\rho_{4,1}^2+\frac{(-\rho_{2,1}\rho_{4,1}+\rho_{4,2})^2}{1-\rho_{2,1}^2}+(-\rho_{2,1}\rho_{3,2}\rho_{4,1}+\rho_{3,2}\rho_{4,2} \\
& +\rho_{3,1}(\rho_{4,1}-\rho_{2,1}\rho_{4,2})-\rho_{4,3}+\rho_{2,1}^2\rho_{4,3})^2/ \\
& ((-1+\rho_{2,1}^2)(-1+\rho_{2,1}^2+\rho_{3,1}^2-2\rho_{2,1}\rho_{3,1}\rho_{3,2}+\rho_{3,2}^2)) \\
& +(-1+\rho_{3,2}^2+\rho_{4,1}^2-\rho_{3,2}^2\rho_{4,1}^2+\rho_{4,2}^2-\rho_{3,1}^2(-1+\rho_{4,2}^2) \\
& +2\rho_{3,1}\rho_{4,1}(\rho_{3,2}\rho_{4,2}-\rho_{4,3})-2\rho_{3,2}\rho_{4,2}\rho_{4,3}+\rho_{4,3}^2-\rho_{2,1}^2(-1+\rho_{4,3}^2) \\
& -2\rho_{2,1}(\rho_{4,1}(\rho_{4,2}-\rho_{3,2}\rho_{4,3})+\rho_{3,1}(\rho_{3,2}-\rho_{4,2}\rho_{4,3})))/(-1+\rho_{2,1}^2 \\
& +\rho_{3,1}^2-2\rho_{2,1}\rho_{3,1}\rho_{3,2}+\rho\,_{3,2}^{2})\Bigg)\Bigg(-(\rho_{4,1}\rho_{5,1})/\Bigg(\rho_{4,1}^2+\frac{(-\rho_{2,1}\rho_{4,1}+\rho_{4,2})^2}{1-\rho_{2,1}^2} \\
& +(-\rho_{2,1}\rho_{3,2}\rho_{4,1}+\rho_{3,2}\rho_{4,2}+\rho_{3,1}(\rho_{4,1}-\rho_{2,1}\rho_{4,2})-\rho_{4,3}+\rho_{2,1}^2\rho_{4,3})^2/ \\
& ((-1+\rho_{2,1}^2)(-1+\rho_{2,1}^2+\rho_{3,1}^2-2\rho_{2,1}\rho_{3,1}\rho_{3,2}+\rho_{3,2}^2))+(-1+\rho_{3,2}^2 \\
& +\rho_{4,1}^2-\rho_{3,2}^2\rho_{4,1}^2+\rho_{4,2}^2-\rho_{3,1}^2(-1+\rho_{4,2}^2)+2\rho_{3,1}\rho_{4,1} \\
& (\rho_{3,2}\rho_{4,2}-\rho_{4,3})-2\rho_{3,2}\rho_{4,2}\rho_{4,3}+\rho_{4,3}^2-\rho_{2,1}^2(-1+\rho_{4,3}^2) \\
& -2\rho_{2,1}(\rho_{4,1}(\rho_{4,2}-\rho_{3,2}\rho_{4,3})+\rho_{3,1}(\rho_{3,2}-\rho_{4,2}\rho_{4,3})))/ \\
& (-1+\rho_{2,1}^2+\rho_{3,1}^2-2\rho_{2,1}\rho_{3,1}\rho_{3,2}+\rho_{3,2}^2\Bigg)\Bigg) \\
& -((-\rho_{2,1}\rho_{4,1}+\rho_{4,2})(-\rho_{2,1}\rho_{5,1}+\rho_{5,2}))/
\end{aligned}$$

$$\Bigg((1-\rho_{2,1}^2)\Bigg(\rho_{4,1}^2+\frac{(-\rho_{2,1}\rho_{4,1}+\rho_{4,2})^2}{1-\rho_{2,1}^2}+(-\rho_{2,1}\rho_{3,2}$$

$$\rho_{4,1}+\rho_{3,2}\rho_{4,2}+\rho_{3,1}(\rho_{4,1}-\rho_{2,1}\rho_{4,2})-\rho_{4,3}+\rho_{2,1}^2\rho_{4,3})^2/$$

$$((-1+\rho_{2,1}^2)(-1+\rho_{2,1}^2+\rho_{3,1}^2-2\rho_{2,1}\rho_{3,1}\rho_{3,2}+\rho_{3,2}^2))$$

$$+(-1+\rho_{3,2}^2+\rho_{4,1}^2-\rho_{3,2}^2\rho_{4,1}^2+\rho_{4,2}^2-\rho_{3,1}^2(-1+\rho_{4,2}^2)$$

$$+2\rho_{3,1}\rho_{4,1}(\rho_{3,2}\rho_{4,2}-\rho_{4,3})-2\rho_{3,2}\rho_{4,2}\rho_{4,3}+\rho_{4,3}^2$$

$$-\rho_{2,1}^2(-1+\rho_{4,3}^2)-2\rho_{2,1}(\rho_{4,1}(\rho_{4,2}-\rho_{3,2}\rho_{4,3})+\rho_{3,1}(\rho_{3,2}-\rho_{4,2}\rho_{4,3})))/$$

$$(-1+\rho_{2,1}^2+\rho_{3,1}^2-2\rho_{2,1}\rho_{3,1}\rho_{3,2}+\rho_{3,2}^2)\Bigg)\Bigg)-((-\rho_{2,1}\rho_{3,2}\rho_{4,1}+\rho_{3,2}$$

$$\rho_{4,2}+\rho_{3,1}(\rho_{4,1}-\rho_{2,1}\rho_{4,2})-\rho_{4,3}+\rho_{2,1}^2\rho_{4,3})(-\rho_{2,1}\rho_{3,2}\rho_{5,1}+\rho_{3,2}\rho_{5,2}$$

$$+\rho_{3,1}(\rho_{5,1}-\rho_{2,1}\rho_{5,2})-\rho_{5,3}+\rho_{2,1}^2\rho_{5,3}))/$$

$$\Bigg((-1+\rho_{2,1}^2)(-1+\rho_{2,1}^2+\rho_{3,1}^2-2\rho_{2,1}\rho_{3,1}\rho_{3,2}+\rho_{3,2}^2)$$

$$\Bigg(\rho_{4,1}^2+\frac{(-\rho_{2,1}\rho_{4,1}+\rho_{4,2})^2}{1-\rho_{2,1}^2}+(-\rho_{2,1}\rho_{3,2}\rho_{4,1}+\rho_{3,2}\rho_{4,2}+\rho_{3,1}$$

$$(\rho_{4,1}-\rho_{2,1}\rho_{4,2})-\rho_{4,3}+\rho_{2,1}^2\rho_{4,3})^2/((-1+\rho_{2,1}^2)$$

$$(-1+\rho_{2,1}^2+\rho_{3,1}^2-2\rho_{2,1}\rho_{3,1}\rho_{3,2}+\rho_{3,2}^2))$$

$$+(-1+\rho_{3,2}^2+\rho_{4,1}^2-\rho_{3,2}^2\rho_{4,1}^2+\rho_{4,2}^2-\rho_{3,1}^2(-1+\rho_{4,2}^2)$$

$$+2\rho_{3,1}\rho_{4,1}(\rho_{3,2}\rho_{4,2}-\rho_{4,3})-2\rho_{3,2}\rho_{4,2}\rho_{4,3}+\rho_{4,3}^2$$

$$-\rho_{2,1}^2(-1+\rho_{4,3}^2)-2\rho_{2,1}(\rho_{4,1}(\rho_{4,2}-\rho_{3,2}\rho_{4,3})$$

$$+\rho_{3,1}(\rho_{3,2}-\rho_{4,2}\rho_{4,3})))/(-1+\rho_{2,1}^2+\rho_{3,1}^2$$

$$-2\rho_{2,1}\rho_{3,1}\rho_{3,2}+\rho_{3,2}^2)\Bigg)\Bigg)+\rho_{5,4}\Bigg)\Bigg)/(\sqrt{}((-1+\rho_{3,2}^2$$

$$+\rho_{4,1}^2-\rho_{3,2}^2\rho_{4,1}^2+\rho_{4,2}^2-\rho_{3,1}^2(-1+\rho_{4,2}^2)+2\rho_{3,1}\rho_{4,1}(\rho_{3,2}\rho_{4,2}$$

$$-\rho_{4,3})-2\rho_{3,2}\rho_{4,2}\rho_{4,3}+\rho_{4,3}^2-\rho_{2,1}^2(-1+\rho_{4,3}^2)$$

$$-2\rho_{2,1}(\rho_{4,1}(\rho_{4,2}-\rho_{3,2}\rho_{4,3})+\rho_{3,1}(\rho_{3,2}-\rho_{4,2}\rho_{4,3})))/$$

$$(-1+\rho_{2,1}^2+\rho_{3,1}^2-2\rho_{2,1}\rho_{3,1}\rho_{3,2}+\rho_{3,2}^2))),$$

$$
\begin{aligned}
\mathbf{s}_{5,5} = \sqrt{}\Bigg(& 1-\rho_{5,1}^2-\frac{(-\rho_{2,1}\rho_{5,1}+\rho_{5,2})^2}{1-\rho_{2,1}^2}-(-\rho_{2,1}\rho_{3,2}\rho_{5,1}+\rho_{3,2}\rho_{5,2} \\
& +\rho_{3,1}(\rho_{5,1}-\rho_{2,1}\rho_{5,2})-\rho_{5,3}+\rho_{2,1}^2\rho_{5,3})^2/((-1+\rho_{2,1}^2)(-1+\rho_{2,1}^2 \\
& +\rho_{3,1}^2-2\rho_{2,1}\rho_{3,1}\rho_{3,2}+\rho_{3,2}^2))-\Bigg((-1+\rho_{2,1}^2+\rho_{3,1}^2-2\rho_{2,1}\rho_{3,1}\rho_{3,2}+\rho_{3,2}^2) \\
& \Bigg(\rho_{4,1}^2+\frac{(-\rho_{2,1}\rho_{4,1}+\rho_{4,2})^2}{1-\rho_{2,1}^2}+(-\rho_{2,1}\rho_{3,2}\rho_{4,1}+\rho_{3,2}\rho_{4,2} \\
& +\rho_{3,1}(\rho_{4,1}-\rho_{2,1}\rho_{4,2})-\rho_{4,3}+\rho_{2,1}^2\rho_{4,3})^2/ \\
& ((-1+\rho_{2,1}^2)(-1+\rho_{2,1}^2+\rho_{3,1}^2-2\rho_{2,1}\rho_{3,1}\rho_{3,2}+\rho_{3,2}^2)) \\
& +(-1+\rho_{3,2}^2+\rho_{4,1}^2-\rho_{3,2}^2\rho_{4,1}^2+\rho_{4,2}^2 \\
& -\rho_{3,1}^2(-1+\rho_{4,2}^2)+2\rho_{3,1}\rho_{4,1}(\rho_{3,2}\rho_{4,2}-\rho_{4,3}) \\
& -2\rho_{3,2}\rho_{4,2}\rho_{4,3}+\rho_{4,3}^2-\rho_{2,1}^2(-1+\rho_{4,3}^2) \\
& -2\rho_{2,1}(\rho_{4,1}(\rho_{4,2}-\rho_{3,2}\rho_{4,3})+\rho_{3,1}(\rho_{3,2}-\rho_{4,2}\rho_{4,3})))/ \\
& (-1+\rho_{2,1}^2+\rho_{3,1}^2-2\rho_{2,1}\rho_{3,1}\rho_{3,2}+\rho_{3,2}^2)\Bigg)^2 \\
& \Bigg(-(\rho_{4,1}\rho_{5,1})\Bigg/\Bigg(\rho_{4,1}^2+\frac{(-\rho_{2,1}\rho_{4,1}+\rho_{4,2})^2}{1-\rho_{2,1}^2}+(-\rho_{2,1}\rho_{3,2}\rho_{4,1}+\rho_{3,2}\rho_{4,2} \\
& +\rho_{3,1}(\rho_{4,1}-\rho_{2,1}\rho_{4,2})-\rho_{4,3}+\rho_{2,1}^2\rho_{4,3})^2/((-1+\rho_{2,1}^2)(-1+\rho_{2,1}^2 \\
& +\rho_{3,1}^2-2\rho_{2,1}\rho_{3,1}\rho_{3,2}+\rho_{3,2}^2))+(-1+\rho_{3,2}^2+\rho_{4,1}^2-\rho_{3,2}^2\rho_{4,1}^2+\rho_{4,2}^2 \\
& -\rho_{3,1}^2(-1+\rho_{4,2}^2)+2\rho_{3,1}\rho_{4,1}(\rho_{3,2}\rho_{4,2}-\rho_{4,3})-2\rho_{3,2}\rho_{4,2}\rho_{4,3} \\
& +\rho_{4,3}^2-\rho_{2,1}^2(-1+\rho_{4,3}^2)-2\rho_{2,1}(\rho_{4,1}(\rho_{4,2}-\rho_{3,2}\rho_{4,3}) \\
& +\rho_{3,1}(\rho_{3,2}-\rho_{4,2}\rho_{4,3})))/(-1+\rho_{2,1}^2+\rho_{3,1}^2-2\,\rho_{2,1}\rho_{3,1}\rho_{3,2}+\rho_{3,2}^2)\Bigg) \\
& -((-\rho_{2,1}\rho_{4,1}+\rho_{4,2})(-\rho_{2,1}\rho_{5,1}+\rho_{5,2}))/ \\
& \Bigg((1-\rho_{2,1}^2)\Bigg(\rho_{4,1}^2+\frac{(-\rho_{2,1}\rho_{4,1}+\rho_{4,2})^2}{1-\rho_{2,1}^2}+(-\rho_{2,1}\rho_{3,2}\rho_{4,1}+\rho_{3,2} \\
& \rho_{4,2}+\rho_{3,1}(\rho_{4,1}-\rho_{2,1}\rho_{4,2})-\rho_{4,3}+\rho_{2,1}^2\rho_{4,3})^2/((-1+\rho_{2,1}^2)(-1+\rho_{2,1}^2 \\
& +\rho_{3,1}^2-2\rho_{2,1}\rho_{3,1}\rho_{3,2}+\rho_{3,2}^2))
\end{aligned}
$$

$$
\begin{aligned}
&+(-1+\rho_{3,2}^2+\rho_{4,1}^2-\rho_{3,2}^2\rho_{4,1}^2+\rho_{4,2}^2-\rho_{3,1}^2(-1+\rho_{4,2}^2)\\
&+2\rho_{3,1}\rho_{4,1}(\rho_{3,2}\rho_{4,2}-\rho_{4,3})-2\rho_{3,2}\rho_{4,2}\rho_{4,3}+\rho_{4,3}^2-\rho_{2,1}^2(-1+\rho_{4,3}^2)\\
&-2\rho_{2,1}(\rho_{4,1}(\rho_{4,2}-\rho_{3,2}\rho_{4,3})+\rho_{3,1}(\rho_{3,2}-\rho_{4,2}\rho_{4,3})))/(-1+\rho_{2,1}^2+\rho_{3,1}^2\\
&\left.\left.-2\rho_{2,1}\rho_{3,1}\rho_{3,2}+\rho_{3,2}^2)\right)\right)-((-\rho_{2,1}\rho_{3,2}\rho_{4,1}+\rho_{3,2}\rho_{4,2}\\
&+\rho_{3,1}(\rho_{4,1}-\rho_{2,1}\rho_{4,2})-\rho_{4,3}+\rho_{2,1}^2\rho_{4,3})(-\rho_{2,1}\rho_{3,2}\rho_{5,1}+\rho_{3,2}\rho_{5,2}\\
&+\rho_{3,1}(\rho_{5,1}-\rho_{2,1}\rho_{5,2})-\rho_{5,3}+\rho_{2,1}^2\rho_{5,3}))/\\
&\left((-1+\rho_{2,1}^2)(-1+\rho_{2,1}^2+\rho_{3,1}^2-2\rho_{2,1}\rho_{3,1}\rho_{3,2}+\rho_{3,2}^2)\right.\\
&\left(\rho_{4,1}^2+\frac{(-\rho_{2,1}\rho_{4,1}+\rho_{4,2})^2}{1-\rho_{2,1}^2}+(-\rho_{2,1}\rho_{3,2}\rho_{4,1}+\rho_{3,2}\rho_{4,2}\right.\\
&+\rho_{3,1}(\rho_{4,1}-\rho_{2,1}\rho_{4,2})-\rho_{4,3}+\rho_{2,1}^2\rho_{4,3})^2/((-1+\rho_{2,1}^2)(-1+\rho_{2,1}^2+\rho_{3,1}^2\\
&-2\rho_{2,1}\rho_{3,1}\rho_{3,2}+\rho_{3,2}^2))+(-1+\rho_{3,2}^2+\rho_{4,1}^2-\rho_{3,2}^2\rho_{4,1}^2+\rho_{4,2}^2\\
&-\rho_{3,1}^2(-1+\rho_{4,2}^2)+2\rho_{3,1}\rho_{4,1}(\rho_{3,2}\rho_{4,2}-\rho_{4,3})-2\rho_{3,2}\rho_{4,2}\rho_{4,3}+\rho_{4,3}^2\\
&-\rho_{2,1}^2(-1+\rho_{4,3}^2)-2\rho_{2,1}(\rho_{4,1}(\rho_{4,2}-\rho_{3,2}\rho_{4,3})+\rho_{3,1}(\rho_{3,2}-\rho_{4,2}\rho_{4,3})))/\\
&\left.\left.\left.\left.(-1+\rho_{2,1}^2+\rho_{3,1}^2-2\ \rho_{2,1}\rho_{3,1}\rho_{3,2}+\rho_{3,2}^2)\right)\right)+\rho_{5,4}\right)^2\right)\Big/\\
&(-1+\rho_{3,2}^2+\rho_{4,1}^2-\rho_{3,2}^2\rho_{4,1}^2+\rho_{4,2}^2-\rho_{3,1}^2(-1+\rho_{4,2}^2)+2\rho_{3,1}\rho_{4,1}\\
&(\rho_{3,2}\rho_{4,2}-\rho_{4,3})-2\rho_{3,2}\rho_{4,2}\rho_{4,3}+\rho_{4,3}^2-\rho_{2,1}^2(-1+\rho_{4,3}^2)\\
&\left.-\ 2\rho_{2,1}(\rho_{4,1}(\rho_{4,2}-\rho_{3,2}\rho_{4,3})+\rho_{3,1}(\rho_{3,2}-\rho_{4,2}\rho_{4,3})))\right).
\end{aligned}
$$

Remark A.2.1. As is evident from the above, the formulas for $\mathbb{S}_n$ become quite involved as n increases. For example, while the formula for the matrix $\mathbb{S}_5$ shown above takes 24 KB to express, the formulas for matrices $\mathbb{S}_{10}$ and $\mathbb{S}_{15}$ have sizes of about 21.8 MB and 14.8 GB, respectively, or if $G(n)$ denotes the size of the formula for the matrix $\mathbb{S}_n$, then $G(n) \approx 0.0550382\mathrm{e}^{1.29383n}$KB. This can be taken as an indication that the methodology presented (or, more ambitiously, technology advocated) in this book, as the number of model factors increases, quickly enters the domain of the most powerful (super)computers.

References

1. D. Becherer, Utility-indifference hedging and valuation via reaction-diffusion systems, *Proc. R. Soc. Lond.* A 460 (2004), 27–51.
2. F. Black and M. Scholes, The pricing of options and corporate liabilities, *J. Political Econ.* 81 (1973), 637 –659.
3. F. Black, The dividend puzzle. *Journal of Portfolio Management*, 2 (1976), 5–8.
4. F. Black, The pricing of commodity contracts, *Journal of Financial Economics*, 3 (1976) 167–179.
5. D. Brigo, F. Mercurio, *Interest Rate Models: Theory and Practice*, Springer, Berlin (2001).
6. A. Černý, *Mathematical Techniques in Finance: Tools for Incomplete Markets*, Princeton University Press, Princeton (2004).
7. J.C. Cox, J.E. Ingersoll Jr., S.A. Ross, The relation between forward prices and futures prices, *Journal of Financial Economics*, 9(4) (1981) 321–346.
8. Y. Cui and S.D. Stojanovic, Equity valuation under stock dilution and buyback, submitted.
9. L.C. Evans, *Partial Differential Equations*, Graduate Studies in Mathematics Vol. 19, AMS (1998).
10. A. Friedman, *Stochastic Differential Equations*, Vols. 1 & 2, Academic Press, New York (1975).
11. M.B. Garman and S.W. Kohlhagen, Foreign currency option values, *J. International Money and Finance* 2 (1983) 231–237.
12. M. Grasselli, A stability result for the HARA class with stochastic interest rates, *Insurance: Mathematics and Economics* 33 (2003) 611–627.
13. S. Heston, A closed-form solution for options with stochastic volatility, *Review of Financial Studies*, 6 (1993) 327–343.
14. L. Jiang, *Mathematical Modeling and Methods of Option Pricing*, World Scientific Publishing, Singapore (2005).
15. J. Kallsen, Utility-based derivative pricing in incomplete markets, *Mathematical Finance—Bachelier Congress 2000*, Geman, H., Madan, D., Pliska, S. R., Vorst, T. (eds.), Springer, Berlin (2002).
16. Z. Kang, S.D. Stojanovic, Interest rate risk premium and equity valuation, *Journal of Systems Science and Complexity*, 23 (2010) 484–498.
17. Z. Kang, S.D. Stojanovic, General diffusive indifference pricing under CARA utility, preprint.
18. I. Karatzas, S.G. Kou, On the pricing of contingent claims under constraints, *Annals of Applied Probability*, 6 (1996), 321–369.
19. N. V. Krylov, *Controlled Diffusion Processes*, Springer-Verlag, New York (1980).
20. A. Lipton, *Mathematical Methods for Foreign Exchange*, World Scientific, Singapore (2001).
21. C. Liu, S.D. Stojanovic, Pricing futures for an underlying model with three factors, preprint.

S. Stojanovic, *Neutral and Indifference Portfolio Pricing, Hedging and Investing: With Applications in Equity and FX*, DOI 10.1007/978-0-387-71418-9,

22. J. Liu, Portfolio Selection in Stochastic Environments, *Review of Financial Studies*, 20(1) (2007), 1–39.
23. F. Longstaff, E.S. Schwartz, Interest rate volatility and the term structure: a two-factor general equilibrium model. *Journal of Finance* 47(1992) 1259–1282.
24. F. Menoncin, Investment strategies for HARA utility function: a general algebraic approximated solution, working paper.
25. R.C. Merton, Analytical Optimal Control Theory as Applied to Stochastic and Non-Stochastic Economics, Ph.D. thesis, Massachusetts Institute of Technology (1970).
26. R.C. Merton, Optimum consumption and portfolio rules in a continuous-time model, *Journal of Economic Theory* 3 (1971) 373–413.
27. R.C. Merton, Theory of rational option pricing, *Bell Journal of Economics and Management Science* 4(1) (1973) 141–183.
28. R.C. Merton, *Continuous-time finance*, Wiley-Blackwell (1990).
29. M.H. Miller, F. Modigliani, Dividend Policy, Growth, and the Valuation of Shares. *Journal of Business*, 34 (1961), 411–433.
30. M. Musiela, T. Zariphopoulou, An example of indifference prices under exponential preferences, *Finance and Stochastics* 8(2) (2004) 229–239.
31. H. Pham, N. Touzi, Equilibrium state prices in a stochastic volatility model, *Mathematical Finance*, 6 (1996), 215–236.
32. R. Rebonato, *Interest-Rate Option Models*, second edition, John Wiley & Sons, New York (1998).
33. R. Sircar, T. Zariphopoulou, Bounds and asymptotic approximations for utility prices when volatility is random, *SIAM Journal on Control and Optimization* 43(4) (2005) 1328–1353.
34. S.D. Stojanovic, *Computational Financial Mathematics Using Mathematica*, Birkhäuser, Boston (2003).
35. S.D. Stojanovic, Optimal momentum hedging via hypoelliptic reduced Monge–Ampère PDEs, *SIAM J. Control Optimization* 43 (2004) 1151–1173.
36. S.D. Stojanovic, Risk premium and fair option prices under stochastic volatility: the HARA solution, *C. R. Acad. Sci. Paris Ser. I* 340 (2005) 551–556.
37. S.D. Stojanovic, *Stochastic Volatility & Risk Premium,* Lecture Notes, GARP, New York (2005).
38. S.D. Stojanovic, Optimal portfolio series formula under dynamic appreciation rate uncertainty, *Journal for Computational Finance*, 8 (2) (2005).
39. S.D. Stojanovic, Higher dimensional fair option pricing and hedging under HARA and CARA utilities (preprint August 2005; revised June 28, 2006); SSRN: http://ssrn.com/abstract=912763.
40. S.D. Stojanovic, Pricing and hedging of multi type contracts under multidimensional risks in incomplete markets modeled by general Itô SDE systems, *Asia Pacific Financial Markets*, 13 (2006) 345–372.
41. S.D. Stojanovic, The dividend puzzle unpuzzled (preprint January 29, 2006); available at SSRN: http://ssrn.com/abstract=879514.
42. S.D. Stojanovic, Risk premium, pricing and hedging for variance swaps, *Volatility as an Asset Class*, edited by I. Nelken, Risk Books, London (2007) 259–285.
43. S.D. Stojanovic, *Advanced Financial Engineering for Interest Rates, Equity, and FX*, Lecture Notes, GARP, New York (2007).
44. S.D. Stojanovic, Any-utility neutral and indifference pricing and hedging, accepted by Risk and Decision Analysis.
45. S.D. Stojanovic, Z. Kang, General diffusive neutral pricing under non-zero portfolio position, preprint.
46. O. Vasicek, An equilibrium characterization of the term structure, *Journal of Financial Economics* 5(2) (1977) 177–188.

Index

S. Stojanovic, *Neutral and Indifference Portfolio Pricing, Hedging and Investing: With Applications in Equity and FX*, DOI 10.1007/978-0-387-71418-9,